CURRENT BASIC AND PATHOLOGICAL APPROACHES TO THE FUNCTION OF MUSCLE CELLS AND TISSUES – FROM MOLECULES TO HUMANS

Edited by **Haruo Sugi**

Current Basic and Pathological Approaches to the Function of Muscle Cells and Tissues – From Molecules to Humans
http://dx.doi.org/10.5772/3003
Edited by Haruo Sugi

Contributors

Haruo Sugi, Hiroki Minoda, Takuya Miyakawa, Suguru Tanokura, Shigeru Chaen, Takakazu Kobayashi, Larissa Lipskaia, Isabelle Limon, Regis Bobe, Roger Hajjar, Ricardo Espinosa-Tanguma, Paola Algara-Suárez, Rebeca Mejía-Elizondo, Víctor Saavedra-Alanís, Paul Fransen, Cor E. Van Hove, Johanna van Langen, Hidde Bult, Maoxian Deng, Lixia Deng, Yarong Xue, Marcelo J. Alfonzo, Fabiola Placeres-Uray, Walid Hassan-Soto, Adolfo Borges, Ramona González de Alfonzo, Itala Lippo de Becemberg, Saima Salim, Sharique A. Ali, Shintaro Nakano, Toshihiro Muramatsu, Shigeyuki Nishimura, Takaaki Senbonmatsu, Valérie Metzinger-Le Meuth, Eléonore M'Baya-Moutoula, Fatiha Taibi, Ziad Massy, Laurent Metzinger, Ho-Chang Kuo, Wei-Chiao Chang, Hafidh I. Al-Sadi, J.M. Ramírez, J.J. Salazar, R. de Hoz, B. Rojas, B.I. Gallego, A.I. Ramírez, A. Triviño, Carla Máximo Prado, Edna Aparecida Leick, Fernanda Degobbi Tenório Quirino dos Santos Lopes, Milton A. Martins, Iolanda de Fátima Lopes Calvo Tibério, Sho Shinohara, Satoko Shinohara, Takanori Kihara, Jun Miyake, Angel Vodenicharov, Canan G. Nebigil, Shiro Mizuno, Hirohisa Toga and Takeshi Ishizaki

Published by InTech

Janeza Trdine 9, 51000 Rijeka, Croatia

Publishing Process Manager Sandra Bakic
Technical Editor InTech Prepress, Novi Sad
Cover InTech Design team

Additional hard copies can be obtained from orders@intechopen.com

Current Basic and Pathological Approaches to the Function of Muscle Cells and Tissues – From Molecules to Humans, Edited by Haruo Sugi
p. cm.
ISBN 978-953-51-0679-1

For EU product safety concerns:
IN TECH d.o.o., Prolaz Marije Krucifikse Kozulić 3, 51000 Rijeka, Croatia, info@intechopen.com or visit our website at intechopen.com.

Meet the editor

Dr Haruo Sugi was born in 1933 in Tokyo, and graduated from Postgraduate School of Life Sciences at the University of Tokyo with Ph.D. degree in 1962, and appointed to be Instructor in Physiology at the University of Tokyo. From 1965 to 1967, Sugi stayed in USA as a research associate at Columbia University and as a visiting scientist in NIH. In 1973, he became Professor and Chairman in the Department of Physiology, Teikyo University. From 1994 to 2006, Dr Sugi was chairman of Muscle Commission in the International Union of Physiological Sciences. In 2004, he retired from his position in Teikyo University, and is now Emeritus Professor of Teikyo University.

Contents

Preface

This volume consists of 17 short review articles, originally submitted to the Editor under the theme of "Muscle Cell". Muscles are classified into three types, skeletal, cardiac and smooth muscles, according to their structure and function. In vertebrate animals including humans, skeletal muscle produces body movement, cardiac muscle is responsible for the heart function as a pump, and smooth muscle is distributed among various visceral organs and blood vessels to keep the animals alive. In all kinds of muscle, mechanical activity results from relative sliding of actin and myosin filaments coupled with ATP hydrolysis, though the mechanism of the myofilament sliding still remains to be a matter for debate and speculation. On the other hand, the mechanical activity of muscle is controlled by changes in the intracellular concentration of free Ca^{2+} ions. In skeletal muscle, contraction is initiated by the release of Ca^{2+} ions from the intracellular membranous structure, sarcoplasmic reticulum, while in cardiac muscle contraction is mainly coupled with influx of Ca^{2+} ions from the extracellular space. In smooth muscles, the origin of Ca^{2+} ions activating contraction, i.e. activator Ca^{2+}, is variable and is not yet fully understood, reflecting the complex structure of smooth muscle tissues. Fifty years ago, smooth muscles were sometimes called "headache muscle" because of extreme technical difficulties in studying their function. As the readers will become aware, considerable progress has now been achieved on the research field of smooth muscle cells and tissues, and smooth muscles are no longer "headache muscle".

For the sake of convenience for general readers, the book is divided into three parts according to the subjects of articles. Part I includes articles dealing with basic aspects of function of skeletal and smooth muscle cells, and also melanocytes which have many properties common to those of smooth muscles. Part II contains articles dealing with pathological aspects of cardiac and smooth muscle cell functions, while Part III consists of articles concerning factors influencing structure and function of cardiac and smooth muscle cells and tissues.

The Editor believes that these articles are extremely stimulating and informative for the readers who are interested not only in the basic mechanisms of muscle cell function, but also in the pathological and clinical aspects of muscle cells and tissues.

Dr. Haruo Sugi
Emeritus Professor, Teikyo University,
Japan

Section 1

Contractile and Regulatory Mechanisms of Contraction in Skeletal, Cardiac and Smooth Muscle Cells

Chapter 1

The Gas Environmental Chamber as a Powerful Tool to Study Structural Changes of Living Muscle Thick Filaments Coupled with ATP Hydrolysis

Haruo Sugi, Hiroki Minoda, Takuya Miyakawa,
Suguru Tanokura, Shigeru Chaen and Takakazu Kobayashi

Additional information is available at the end of the chapter

http://dx.doi.org/10.5772/39199

1. Introduction

The gas environmental chamber (or the hydration chamber) has been developed to observe chemical reactions in water solutions under high magnifications with an electron microscope (for an extensive review, see Buttler & Hale, 1981). The gas environmental chamber (EC) has been widely used for *in situ* observation of inorganic substances in the field of materials science. Fig.1 shows two different types of the EC. One is film-sealed EC, which is insulated from high vacuum of electron microscope with sealing film at is upper and lower windows to pass electron beam (Fig.1A). Water vapor (water gas) is constantly circulated through the EC to keep the specimen in hydrated state. The other is aperture-limited EC, which has apertures to pass electron beam without any sealing film. Water gas is constantly injected into the EC, and sucked out of the EC to keep the specimen in hydrated state (Fig.1B).

In the research field of medical and biological sciences, it was a dream of investigators to observe living microorganisms moving under an electron microscope with high magnifications. In order to realize this dream, a number of attempts have hitherto been made to observe living microorganisms by means of the EC attached to an electron microscope. Such attempts have been, however, found to be unsuccessful because the function of living microorganisms are readily impaired by electron beam irradiation. On the other hand, the function of biological macromolecules, such as proteins and lipids, are expected to be much more resistant against electron beam irradiation. The experiments to be described in this chapter were started to ascertain whether the EC was useful in studying dynamic structural changes of biological macromolecules related to their function. After many considerations, we decided to study molecular mechanism of muscle contraction using the EC, which was designed and constructed to be suitable for physiological experiments to investigate dynamic structural changes of hydrated muscle myosin filaments coupled with ATP hydrolysis.

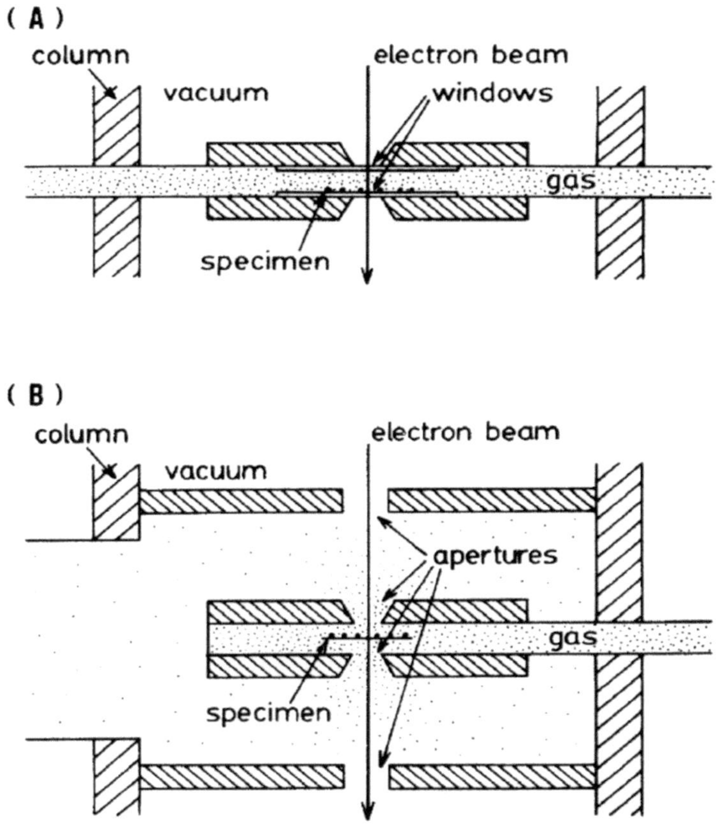

Figure 1. Two types of the EC. (A) Film-sealed EC. (B) Aperture-limited EC. (Fukushima, 1988)

As explained in detail in the following sections, the greatest mystery concerning the mechanism of muscle contraction is how the myosin heads extending from myosin filaments convert chemical energy derived from ATP hydrolysis into mechanical work producing force and motion in muscle. Despite extensive studies, the movement of the myosin heads still remains as a matter of debate and speculation. The reason for the present situation in the field of muscle research arises from the fact that the myosin head movement has been determined only indirectly. The most straightforward way to record the myosin head movement is to observe the myosin head movement in hydrated myosin filaments, which retain their physiological function. In the early 1980's, we had an opportunity to meet Professor Fukami in Nihon University, who succeeded in preparing the carbon sealing film for the film-sealed EC at that time and was looking for coworkers to study physiological function of biological tissues.

We started to work with Fukami's group using the EC, manufactured by the Japan Electron Optics Laboratory (JEOL, Ltd, Co., together with the carbon sealing film developed in Fukami's laboratory. After the period of trials and errors, encompassed over ten years, we succeeded in recording the ATP-induced myosin head movement in hydrated myosin filaments with a number of unexpected findings, which are described in this chapter.

2. The gas environmental chamber (EC)

Fig.2 is a schematic diagram of the film-sealed gas environmental chamber (EC). The EC consists of a metal compartment (diameter, 3.5mm; depth, 0.8mm) with upper and lower window frames (copper grids) to pass electron beam. Each window frame has nine apertures, each having a diameter of 0.1mm. The specimen is placed on the surface of lower sealing film, and covered by a thin layer of experimental solution by constantly circulating water vapor through the EC. To obtain clear specimen images, the internal pressure of the EC is made 60—80 Torr. The flow rate of water vapor is adjusted to 0.1—0.2l/min, so that thin layer of experimental solution covering the specimen is in equilibrium with the vapor pressure in the EC (Fukushima et al.,1985; Fukami et al.,1991). The EC was attached to a 200kV transmission electron microscope (JEM 2000EX, JEOL). (Sugi et al.,1997).

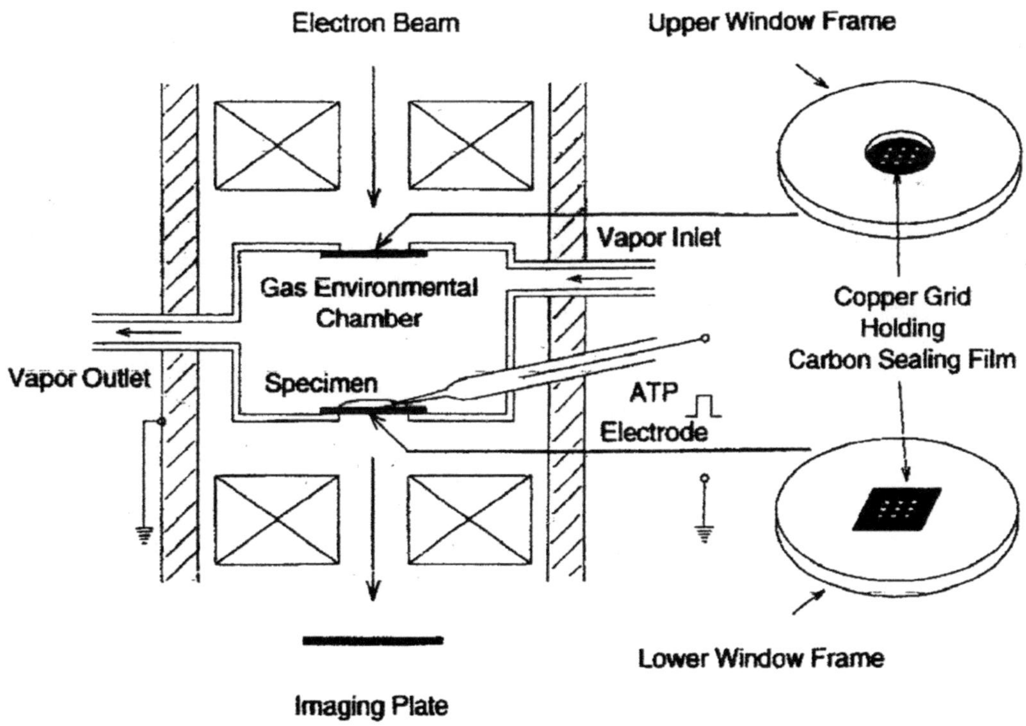

Figure 2. Diagram of the film-sealed EC. The upper and lower windows (copper grids with nine apertures) are covered with carbon sealing films held on copper grids. The EC contains an ATP-containing electrode to apply ATP to the specimen iontophoretically. The image of the specimen is recorded with the imaging plate (IP) (Sugi et al. , 1997).

3. Carbon sealing film

The most important element of the film-sealed EC is the carbon sealing film developed in Fukami's laboratory. In principle, both spatial resolution and contrast of electron micrographs taken by the EC increases with decreasing thickness of the sealing film. Preliminary experiments made in Fukami's laboratory indicated that, to obtain a spatial resolution < 1 nm, thickness of the sealing film should be 15—20nm. Meanwhile, resistivity of a sealing film against pressure difference decreases sharply with increasing its area; the thickness of a sealing film covering a circular aperture of 50μm diameter should be ~100nm to bear a practical pressure difference.

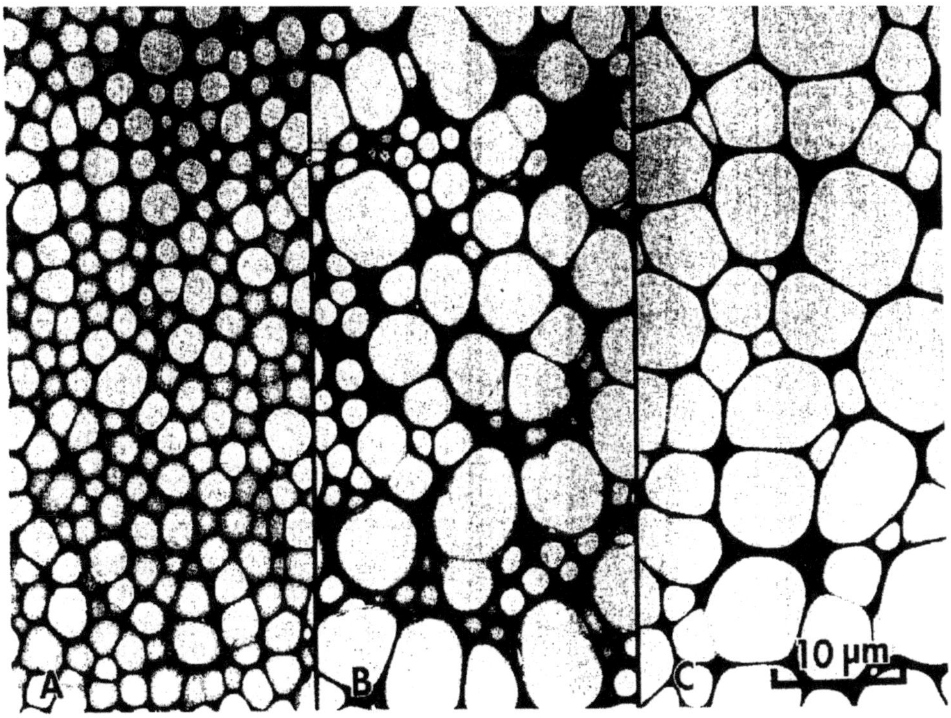

Figure 3. Photomicrographs of plastic microgrides with holes of small diameters (A), with holes of nonuniform diameters (B), and with holes of fairly uniform diameters (5—8nm)(C). (Fikushima, 1988).

As it is practically difficult to a hole < 50μm into metal wall of the EC, Fukami & Adachi (1965) plastic microgrids made from high-molecular organic compound (cellulose acetobutylate). Examples of microgrids are shown in Fig. 3. Microgrids with small (A) or nonuniform holes (B) were unsuitable, while microgrids with fairly uniform holes of 5—8nm diameters (C) were suitable for electron microscopic observation of the specimen.

Fig. 4 illustrates steps to prepare carbon sealing film by covering the microgrid with a thin layer of carbon film (thickness, ~20nm). First, plastic microgrids prepared on a glass slide is put onto water surface (a), where the microgrids (having trapezoidal cross-section) are floating with longer side dounwards (b). The position of the microgrids are inverted by means of triacetylcellurose (TAC) membrane, and again put oto water surface (c,d). The inverted microgrids are then placed on a mica surface, and exposed to evaporated carbon gas so that the grids are coated with thin carbon layer (e,f). The carbon sealing film prepared on a mica surface are cut into rectangular pieces of appropriate size, and put onto water surface (g,h,i). Finally, pieces of the carbon insulating film is placed onto the copper grid, in such a way that each piece of the insulating film covers nine apertures of copper grid (k).

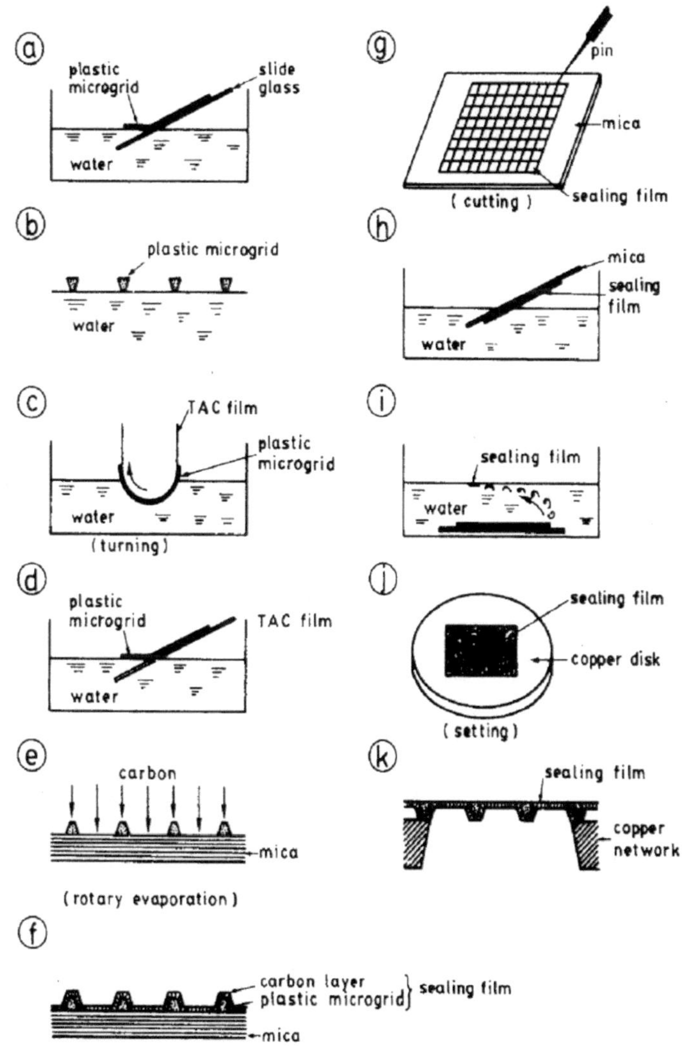

Figure 4. Diagram showing steps to prepare carbon insulating film supported by copper microgrids (Fukushima,1988). For explanation, see text...

The carbon insulating film prepared by the above method well resisted against pressure difference up to 1 atm (Fukushima, 1981).

4. Determination of the critical electron dose to impair function of contractile proteins

Although biological specimens mounted in the EC can be kept in living, hydrated state ,their function is gradually impaired by electron beam irradiation, thus giving a serious limitation in the use of the EC for physiological experiments. Therefore, the critical incident electron dose to impair physiological function of contractile proteins in muscle was determined in by Suda et al. (1992). They observed muscle myofibrils, consisting of hexagonal array of actin and myosin filaments, in the EC (magnification, 2500X), and activated them with ATP.

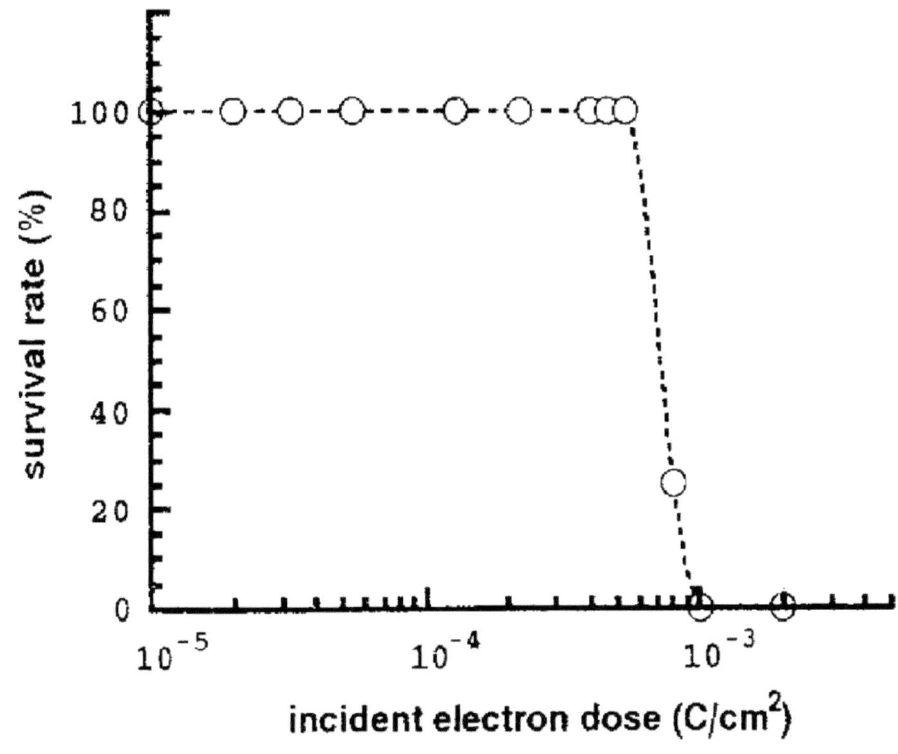

Figure 5. Relation between the total incident electron dose and the survival rate of muscle myofibrils, expressed as percentage of myofibrils contracted in response to ATP in the microscopic field (Suda et al.,1992). Note that contraction of myofibrils in response to ATP disappears when the electron dose exceeds 5×10^{-4}C/cm^2.

The results are summarized in Fig.5. When the total incident electron dose was < 5×10^{-4}C/cm^2, all the myofibrils in the electron microscopic field contracted in response to ATP. If, however, the total incident electron dose was further increased, the ATP-induced myofibril contraction disappeared in a nearly all-or-none manner, though the myofibrils showed no appreciable changes in appearance.

The critical electron dose to impair physiological function of contractile proteins was confirmed by us with respect to both the ATP-induced myosin head movement and the ATPase activity of hydrated myosin filaments mounted in the EC. Based on these results, electron microscopic observation and recording of the specimen was made with a total incident electron dose < 10^{-4}C/cm^2, being well below the critical dose to impair function of contractile proteins. In order to fulfill this condition, the specimen in the EC had to be observed with extremely weak electron beam intensities (at the fluorescent screen) < 5 x 10^{-13}A/cm^2. Therefore, observation and focusing of the specimen required enormous skill and patience. The electron beam intensity through the specimen under a magnification of 10,000x was 5 x 10^{-13}x $(10{,}000)^2$ = 5 x 10^{-5}A/cm^2. Immediately after the focusing of the specimen, electron beam was stopped until the time of recording.

5. Background of experiments with the EC

Before describing our experimental results, it seems necessary to give a brief overview of the experimental work to investigate mechanism of muscle contraction. In the middle1950s, H.E. Huxley & Hanson (1954) made a monumental discovery that a skeletal muscle consists of hexagonal lattice of actin and myosin filaments, and that muscle contraction results from relative sliding between actin and myosin filaments (Fig. 6).

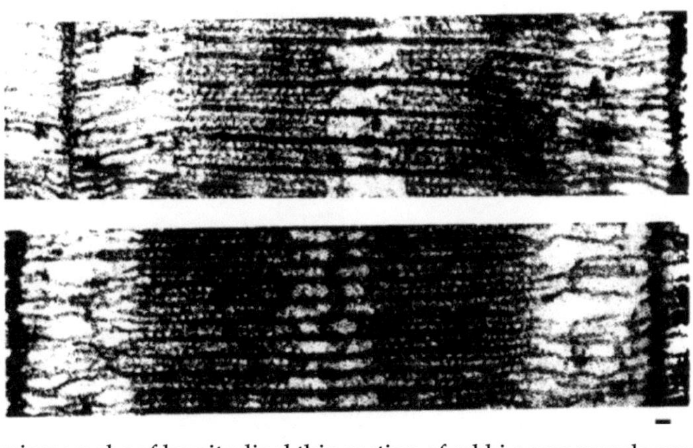

Figure 6. Electron micrographs of longitudinal thin section of rabbi psoas muscle myofibrils (H.E. Huxley, 1957).

Considerable progress has been made with respect to the structure and function of actin and myosin filaments after the discovery of sliding filament mechanism in muscle contraction. As shown in Fig.7A, a myosin molecule is divided into two parts; (1) a long rod called light meromyosin (LMM) and (2) the rest of myosin molecule consisting of a short rod (S2) and two heads (S1) is called heavy meromyosin (HMM). In myosin filaments (or thick filaments), LMM aggregates to form filament backbone, which is polarized in opposite directions on either side of the central part.

While the S1 heads extend laterally from the filament backbone with an axial interval of 14.3nm (Fig.7B). The central part of myosin filament is called the bare region (or bare zone), where the projection of myosin head is absent.

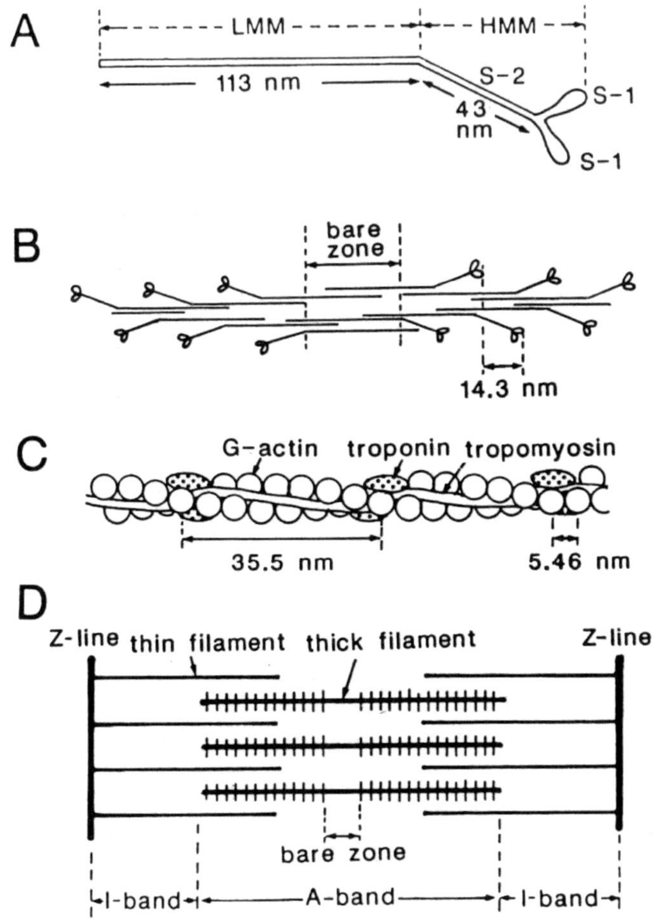

Figure 7. Ultrastructure of myosin (thick) and actin (thin) filaments and their arrangement within a sarcomere. (A) Diagram of a myosin molecule. (B) Arrangement of myosin molecules to form a myosin filament. (C) Arrangement of actin monomers (G-actin) in an actin filament. (D) Longitudinal arrangement of actin and myosin filaments within a sarcomere. Note that the half sarcomere is the structural and functional unit of muscle (Sugi, 1992).

On the other hand, actin filaments consist primarily of two helical strands of globular actin monomers (G-actin) , which are wound around each other with a pitch of 35.5nm. The axial separation of actin monomers in actin filaments is 5.46nm (Fig.7C). In vertebrate skeletal muscle, actin filaments contain tropomyosin and troponin.

As shown in Fig.7D, actin filaments extend from the Z-line to penetrate in between myosin filaments, which are located centrally in each sarcomere. Within a sarcomere, the region containing only actin filaments is called the I-band, whereas the region containing myosin filaments and part of actin filaments is called the A-band. It has been confirmed by a number of experimental methods (H.E. Huxley & Hanson,1954; Page & Huxley,1963; Wray & Holmes,1981) that the filament lengths remain constant irrespective of whether a muscle shortens or being stretched. Therefore, the central problem in understanding the molecular mechanism of muscle contraction is: what makes actin and myosin filaments slide past each other? Since both actin binding site and ATPase activity are localized in the S1 heads of myosin molecule, it is generally believed that the S1 heads, extending from myosin filament backbone towards actin filaments, play a key role in converting chemical energy of ATP hydrolysis into mechanical work producing force and motion in muscle.

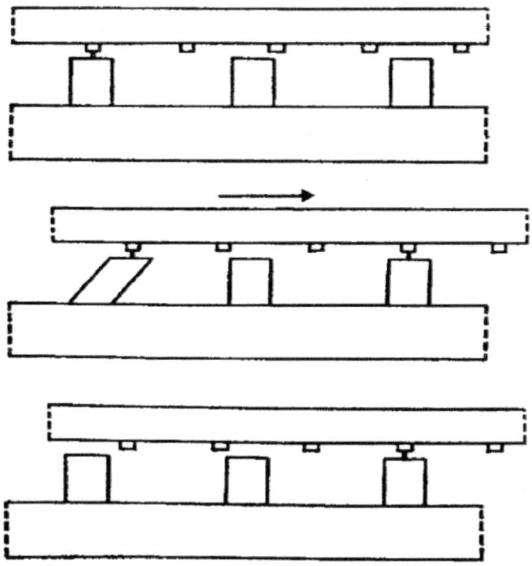

Figure 8. Diagrams showing hypothetical attachment-detachment cycle between the myosin S1 head extending from myosin filament and the sites on actin filament. The myosin head first attaches to actin filament (top diagram), changes its configuration to move actin filament to the right (middle diagram), and then detach from actin filament (bottom diagram). Axial spacing of the myosin heads on myosin filament differs from that of the sites on actin filament, so that the attachment-detachment cycle takes place asynchronously (H.E. Huxley,1969).

Fig.8 illustrates hypothetical attachment-detachment cycle between the S1 heads and the corresponding sites on actin filaments. Extensive studies have been made to prove conformational changes (or movement) of the myosin heads coupled with ATP during muscle contraction. Although experimental methods used include muscle mechanics, time-resolved X-ray diffraction, chemical probes attached to myosin heads, electron microscopy of quick frozen muscle fibers, and nucleotide-dependent changes of myosin head crystals, no clear conclusion has been obtained (Cooke,1986; Hibbard & Trentham,1986, Geeves & Holmes, 1999, A.F. Huxley,1998).

Thus, the myosin head movement coupled with ATP hydrolysis in muscle still remains to be a matter for debate and speculation. The difficulties in this research field seem to arise from the fact that numerous myosin heads undergo conformational changes asynchronously, so that experimental data are statistical to obscure behavior of individual myosin heads. Since the most straightforward way to study conformational changes in individual myosin heads electron microscopically, we attempted to record ATP-induced movement of individual myosin head in using the EC, enabling us to keep myofilaments in hydrated, living state. As described later, the EC has been proved to be extremely powerful tool in visualizing the behavior of individual myosin heads under the electron microscope with high magnifications.

6. Experimental methods

In order to achieve the purpose to record movements of myosin head in hydrated myosin filaments, the following problems in experimental technique had to be solved: (1) how to record images of the specimen with extremely weak electron beam intensities, (2) how to position-mark myosin heads without specimen staining used for conventional electron microscopy; and (3) how to apply ATP to the specimen without changing its position in the electron microscopic field. We solved these problems in the following ways.

6.1. Recording of specimen image

Based on the critical electron dose to impair function of contractile proteins (Fig.5), experiments were performed under electron microscopic magnification of 10,000x, and the specimen images were recorded on an imaging plate (IP) system (PIX system, JEOL). The IP is 10.2 x 7.7cm in size, and has a sensitivity ~60times that of X-ray film. The exposure time was 0.18s with an electron beam intensity of 1—2 x 10^{-12}A/cm^2. The number of pixels in the IP is ~12,000,000 to give a special resolution mdose, recording of the specimen image can only be repeated at most 4times. The IP system was developed by Fuji Photofilm Co., and is now used worldwide not noly for transmission electron microscope, but for other purposes like time-resolved X-ray diffraction.

6.2. Preparation of synthetic bipolar myosin filaments and position marking of myosin heads

We decided to use synthetic thick filaments, consisting of myosin-myosin rode mixture, prepared from rabbi psoas muscle. Myosin was prepared by the method of Perry (1955), while myosin rod was obtained by chymotryptic digestion of myosin by the method of Margossian & Lowey (1982). Myosin and myosin rod were mixed at a molar ratio of 1:1, and were slowly polymerized by dialysis against a solution of low ionic strength (KCl concentration, 120mM) to bipolar myosin filaments (1.5—3μm in length, and 50—200nm in diameter at the center) suitable for our experiments. As shown in Fig. 9, the synthetic filaments are spindle-shaped, and their polarity is reversed across their central region, as judged from the direction of extension of rod part of HMM (myosin S2) from the filaments. Though the myosin S1 heads are lost from the filaments, probably due to fixation and staining procedures, this indicates that the synthetic filaments are bipolar in structure, being similar to native myosin filaments in muscle.

To position-mark individual myosin heads in the hydrated myosin filaments without staining procedures, colloidal gold particles (diameter, 20nm; coated with protein A; EY laboratories) were attached to the myosin heads, using a site directed antibody (IgG) to the junctional peptide between 50- and 20-kDa segments of myosin heavy chain (Sutoh et al.,1989). The antibody attaches to only one of the two myosin heads near its distal end facing actin filaments. Technical details to position-mark individual myosin heads have been described elsewhere (Sugi et al., 1997). It was essential to position-mark myosin heads sparsely, so that each gold particle was reasonably separated from neighboring particles.

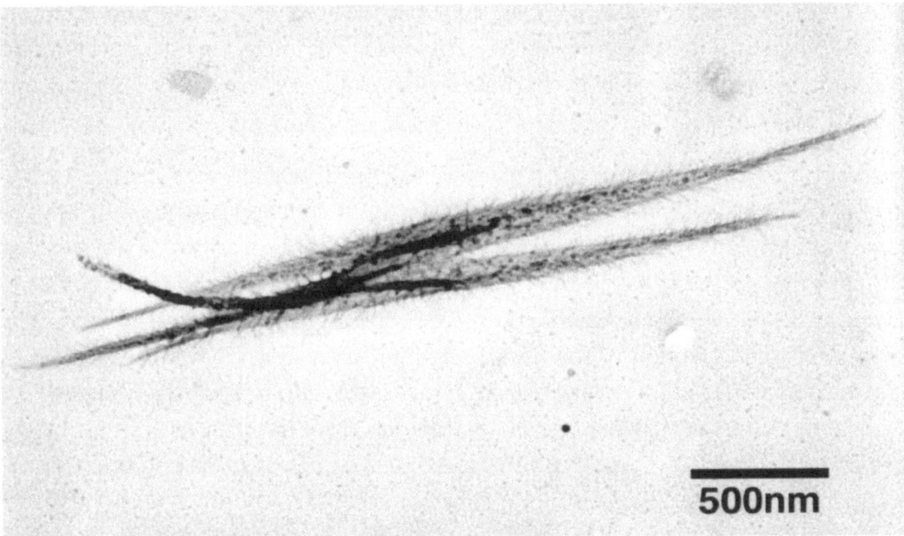

Figure 9. Conventional electron micrograph of synthetic bipolar myosin filaments. Note that the direction of extension of rod part of HMM (myosin subfragment 2) from the filaments is reversed across their central region

6.3. Application of ATP to the specimen

To apply ATP to the specimen without causing its displacement, we used conventional glass capillary microelectrodes containing 100mM ATP (see Fig.2). By passing current pulses through the electrode, negatively charged ATP ions are moved out of the electrode. The iontophoretically released ATP ions from the electrode reach to the specimen by diffusion in the experimental solution covering the specimen. Normally, a rectangular current pulse (intensity, 10nA; duration, 1s) from an electronic stimulator was applied to the electrode through a current clamp circuit (Oiwa et al.,1993). Total amount of ATP released from the microelectrode was estimated to be ~10—14mol (Oiwa et al.,1991). The time required for the released ATP to reach the specimen by diffusion was estimated to be <30s by video recording

ATP-induced shortening of myofibrils in the EC under a light microscope. Hexokinase (50units/ml) and D-glucose (2mM) were added to the experimental solution to eliminate contamination of ATP (Oiwa et al.,1991). In some experiments, ADP was also applied to the specimen with similar method.

6.4. Data analysis

Under an electron microscopic magnification of 10,000x, the pixel size on the IP is 2.5 x 2.5nm. In our experimental condition, the number of electrons reaching each pixel is estimated to be at most 7—8. Each IP record of the specimen was divided into a number of subframes, and each subframe was observed on the monitor screen of electron microscope. Due to electron statistics, the shape of gold particle images was variable. Particles with nearly circular shape were selected to be used for analysis, after an appropriate binning procedure, i.e. the procedure to determine each particle configuration consisting of particles with electron counts above a certain level. Particle shapes were not markedly altered by the level of binning.

Then, the center of mass position of each selected gold particle was determined with an image processor (Nexus Qube System, Nexsus) in the early experiments, and with an ordinary personal computer in the late experiments. The center of mass position was obtained as the coordinates (two significant figures) within a single pixel where the center of mass position was located, and the coordinates, representing the position of the particle, were also taken to represent the position of the myosin head. The position of the myosin head, determined by the above method, was compared between the two IP records. The absolute coordinates common to the two IP records were obtained from the position of natural markers, i.e. bright spots on the carbon sealing film. When the center of mass position was different between the two IP records, the distance (D) between the two center of mass positions (with the coordinates X1 and Y_1 and X2 and Y2, respectively) was calculated as $D = \sqrt{(X1—X2)^2 +(Y1-Y2)^2}$, and this value was taken as the amplitude of myosin head movement.

7. Experimental results and their interpretation

Prior to the experiments to be described in the following sections,we first made experiments with the EC using myosin-paramyosin hybrid filaments, in which rabbit skeletal muscle myosin was bound around the surface of long and thick paramyosin filaments obtained from molluscan somatic smooth muscle, because this hybrid filaments were very easy to handle experimentally. Although we established our experimental methods already described in the preceding sections during the course of experiments, and succeeded in recording the ATP-induced myosin head movement (Sugi et al.,1997), we do not mention the results obtained on this hybrid filaments because (1) the space available for this chapter is limited, and (2) the results obtained from the unusual material may not attract attention of general readers.

7.1. Stability of myosin head position in the absence of ATP

Fig.10 shows examples of spindle-shaped bipolar myosin filaments with a number of gold particles bound to individual myosin heads. The particle image consisted of 20—50 dark pixels with a wide range of gradation, reflecting electron statistics. We first examined whether the particle position, representing the myosin head position, was stable or changed with time in the absence of ATP, by comparing the center of mass position of the same particle between the two IP records of the same filament, taken at an interval of 5—10min, and then the two IP records were superimposed to detect differences in particle position.

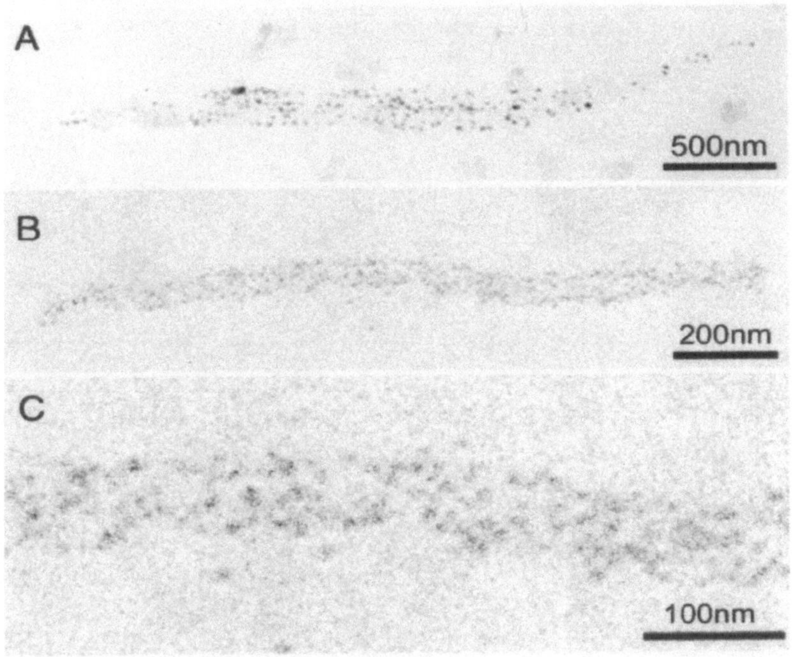

Figure 10. (a and b) Examples of IP records of single bipolar myosin filaments with a number of gold particles attached to individual myosin heads. (c) Enlarged view of myosin filament shown in (a) (Sugi et al.,2008).

An example of superimposed tracings of the two IP records is presented in Fig. 11a, in which open and filled circles of 20nm diameter are drawn around the center of mass position of particles in the first and the second records, respectively. It was found that filled circles in the second record are almost completely covered by open circles in the first record. This indicates that (1) the filament stick firmly to the surface of carbon sealing film, and that (2) the position of individual myosin heads on the filament remain almost unchanged with time. Fig.11b is a histogram showing distribution of the distance between the center of mass positions of particles in the first and the second records. Among 120 particles on three different pairs of IP records, 93 particles exhibited no significant changes in position ($D < 2.5$nm), while the rest 27 particles showed only small position changes (2.5nm $< D >$ 5nm).

The stability in position of both the filament and the myosin heads in the absence of ATP provided an extremely favorable condition for recording the myosin head movement in response to applied ATP. Although individual myosin heads are believed to continue thermal fluctuation, their mean position, time-averaged over the exposure time of IP recording (0.18s), remains almost unchanged with time. Since the same stability of myosin heads has also been observed in the hybrid filaments (Sugi et al.,1997), the stability in time-averaged myosin head mean position seems to be common to myosin heads extending from myosin filament in all kinds of muscle, and is consistent with the contraction model of A.F. Huxley, in which each myosin head fluctuates around a definite equilibrium position (A.F. Huxley, 1957).

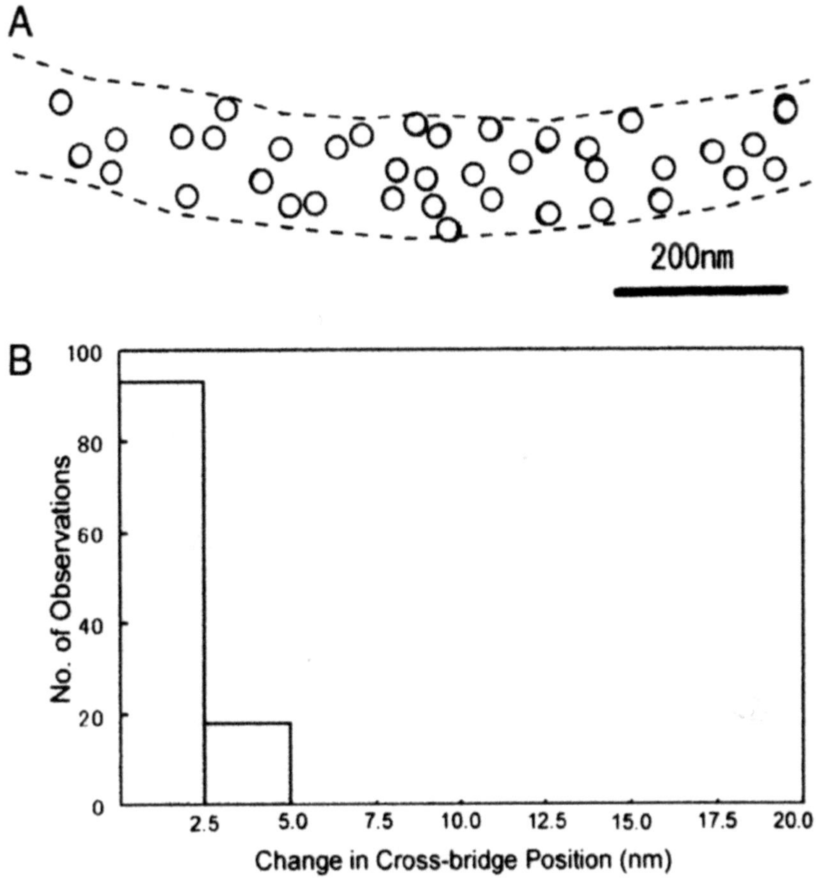

Figure 11. Stability of time-averaged myosin head position in the absence of ATP. (a) Comparison of the myosin head position between the two IP records of the same filament. Open and filled circles (diameter, 20nm) are drawn around the center of mass position of each particle in the first and the second IP records, respectively. In this and subsequent figures, broken lines indicate contour of the filament. Note that filled circles are barely visible because of almost complete overlap of open circles over filled circles. (b) Histogram showing distribution of distance between the center of mass positions of particles in the first and the second IP records (Sugi et al.,2008). Note also that, in Figs. 11 and 12, the term, cross-bridge, is used instead of the term, myosin heads.

7.2. ATP-induced myosin head movement

On the basis of the stability of time-averaged myosin head mean position with time, we explored myosin head movement in response to iontophoretically applied ATP, by comparing two IP records of the same filament, one taken 3—4min before while the other taken 40—60s after ATP application. Since it was not easy to focus part of myosin filament including the bare region (see Fig.7B) within the critical electron dose to impair function of myosin molecules, we first examined ATP-induced myosin head movement at one side of the bare region.

After ATP application, the position of individual myosin heads on the filament was found to move in one direction nearly parallel to the filament long axis, as shown in Fig. 12a (Sugi et al.,2008). Fig. 12b is a histogram showing distribution of the amplitude of ATP-induced myosin head movement, constructed from 1,285 measurements on 8 different pairs of IP records obtained from 8 different myosin filaments. The histogram exhibited a peak at 5—7nm, and the average amplitude of myosin head movement was 6.5±3.7nm (mean±SD, (n=1,210).

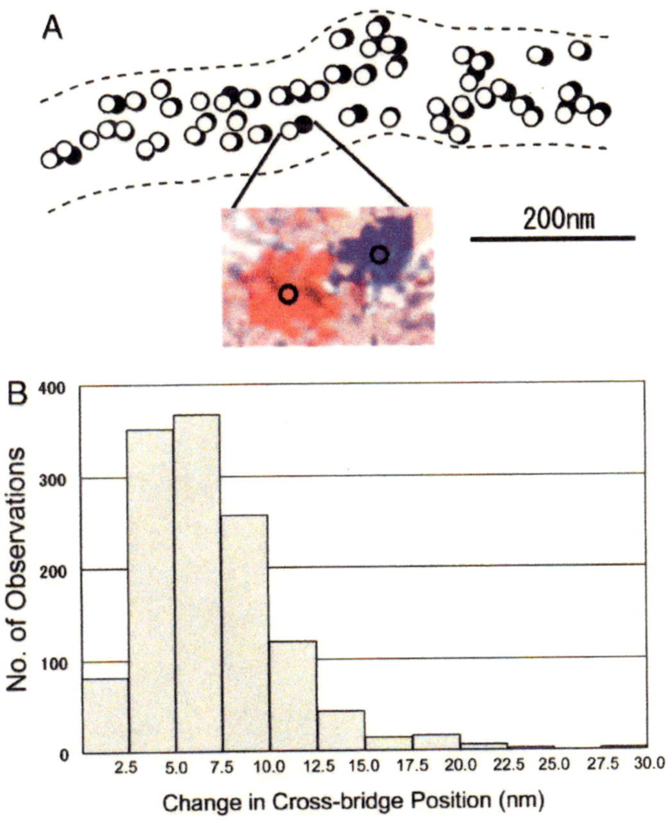

Figure 12. ATP-induced myosin head movement. (a) Comparison of the myosin head position between the two IP records. Open and filled circles (diameter, 20nm) are drawn around the center of mass positions of the same particles before and after ATP application, respectively.

(Inset) an example of superimposed IP records showing the change in position of the same particle, before (red) and after (blue) ATP application. (b) Histogram showing distribution of the amplitude of ATP-induced myosin head movement, determined from changes in the center of mass position of each particle (Sugi et al.,2008).

In our experimental condition, gold particles located on both upper and lower side of the filaments were equally in focus in the microscopic field. The myosin heads on the filament upper side may move freely in response to ATP, while the movement of myosin heads on the lower side of the filament may be largely or completely inhibited due to firm attachment of the filament to the carbon sealing film. If this explanation is correct, the mean amplitude of ATP-induced movement of myosin heads that can move freely would be > 7.5nm. As has been the case in the previous study (Sugi, 1997), the ATP-induced myosin head movement was eliminated by treatment with N-ethylmaleimide, indicating that the myosin head movement is associated with its reaction with ATP.

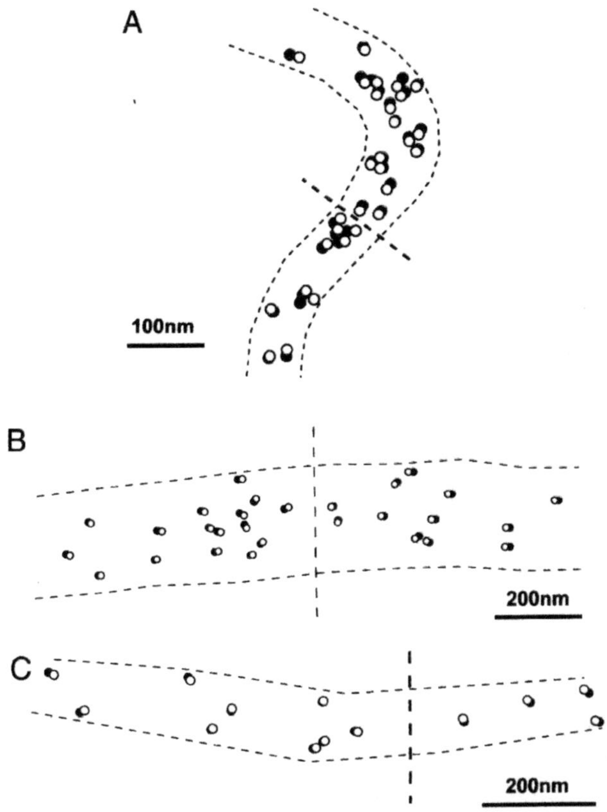

Figure 13. Examples of IP records showing the ATP-induced myosin head movement at both sides of the myosin filament bare region, across which the myosin head polarity is reversed. Open and filled circles (diameter, 20nm) are drawn around the center of mass positions of the same particles before and after ATP application, respectively. Note that the myosin heads move away from the bare region, indicated by vertical broken lines (Sugi et al., 2008).

7.3. Direct demonstration of myosin head recovery stroke

After enormous painstaking efforts, we finally succeeded in recording the ATP-induced myosin head movement at both sides of the myosin filament bare region, across which the myosin head polarity was reversed (see Figs. 7 and 9). It was found that, on application of ATP, myosin heads moved away from the bare region. Typical examples of IP records showing the reversal in the direction of myosin head movement are presented in Fig. 13.

Fig. 14 is a diagram illustrating generally accepted view on the attachment-detachment cycle between the myosin head (M) extending from myosin filament and actin monomer (A) in actin filament, based on biochemical studies on the kinetics of actomyosin ATPase reaction in water solution (Lymn & Taylor, 1971). M in the form of complex, M • ADP • Pi, attaches to A (A), and exerts a power stroke, associated with release of Pi and ADP (from A to B). After the end of power stroke, M remains attached to A, taking its post-power stroke configuration (B). Upon binding with ATP, M detaches from A, and exerts a recovery stroke, associated with reaction, M • ATP → M • ADP • Pi (from C to D). Then M • ADP • Pi again attaches to A (from D to A) and the cycle is repeated.

Though our experimental system does not contain actin filaments, it seems likely that myosin heads before ATP application may take configurations analogous to those at the end of power stroke (B in Fig. 14), and in response to applied ATP, they bind with ATP to form complex M • ADP • Pi, which is known to have average lifetime > 10s due to its slow Pi release (Lymn & Taylor,1971). Therefore, majority of myosin heads in the IP record, taken after ATP application, may be in the state of M • ADP • Pi, suggesting that the ATP-induced myosin head movement, recorded in our EC experiments, is coupled with reaction, M + ATP → M • ADP • Pi, and therefore may correspond to the recovery stroke (C to D, in the diagram of Fig. 14.

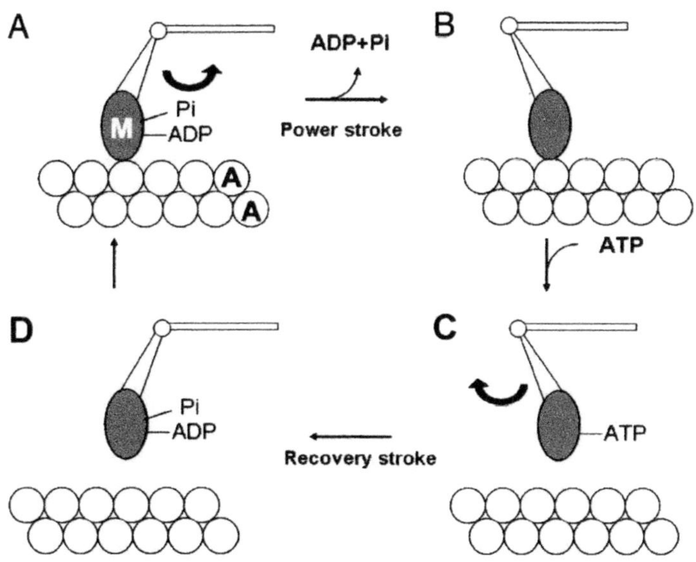

Figure 14. Diagram of the attachment-detachment cycle between myosin head (M) extending from myosin filament and actin monomer (A) in actin filament, based on biochemical studies on actomyosin ATPase reactions. For further explanations, see text. (Sugi et al.,2008).

In order that myosin heads in muscle repeat attachment-detachment cycles with actin filaments, the recovery stroke should be the same in amplitude as, but opposite in direction to, the power stroke, in which myosin heads should move towards the bare region of myosin filament. As a matter of fact, myosin heads that had moved away from the filament bare region, were found to return to their initial position after exhaustion of applied ATP with hexokinase and D-glucose serving as ATP scavenger.

Fig. 15 illustrates 9 examples of superimposed IP records, each record showing sequential changes in location of the pixels (2.5 x 2.5nm), in which the center of mass position of the corresponding gold particles is included. Red, blue and yellow pixels in each record indicate the center of mass positions of the same particle before ATP application, during ATP application, and after complete exhaustion of applied ATP, respectively. It can be seen that myosin heads returned exactly to their initial position in records a, b and i, and close to their initial position in records c to h. The return of myosin heads to their initial position may be associated with reaction, M • ADP • Pi → M + Pi + ADP, i.e. detachment of Pi and ADP from M. In the presence of actin filaments, this reaction corresponds to the myosin head power stroke (A to B in Fig.14).

To summarize, our findings on the ATP-induced myosin head movement in hydrated, living myosin filaments constitute the first direct demonstration of the myosin head recovery stroke. On the other hand, the return of myosin head to their initial position after exhaustion of applied ATP is not regarded to correspond to myosin head power stroke at present, as our experimental system does not contain actin filaments. Nevertheless, our results may be taken to indicate that, even in the absence of actin filaments, individual myosin head can exhibit cyclic movement coupled with ATP hydrolysis. In other words, individual myosin heads can perform cyclic movement analogous to that shown diagrammatically in Fig.14 without being guided by actin filaments. Recently, we have succeeded in recording the myosin head power stroke in the presence of actin filaments, and are obtaining extremely interesting preliminary results, further proving that the EC is a powerful tool in making breakthroughs in the field of molecular mechanism of muscle contraction.

8. Electron microscopic evidence for lever arm mechanism of myosin head movement

At the end of this chapter, we will describe our recent piece of work with EC concerning the myosin head lever arm mechanism. Fig. 16 is a diagram showing molecular structure of the myosin head, consisting of catalytic domain CAD) containing actin binding and ATPase sites, and lever arm domain (LD), connected to myosin filament backbone via myosin sub-fragment 2 (S2). The two domains are connected by small, flexible converter domain (CD). Mainly based on crystallographic studies on nucleotide-dependent structural changes in myosin head crystals, which are detached from myosin filaments (Geeves & Holmes,1999), it has been suggested that the myosin head power stroke is produced by active rotation of LD around CD, while CAD remains rigid. It is not clear, however, whether the myosin head power stroke is actually produced by the above lever arm mechanism in muscle, in which the myosin heads are not detached from, but are firmly connected to, myosin filament backbone.

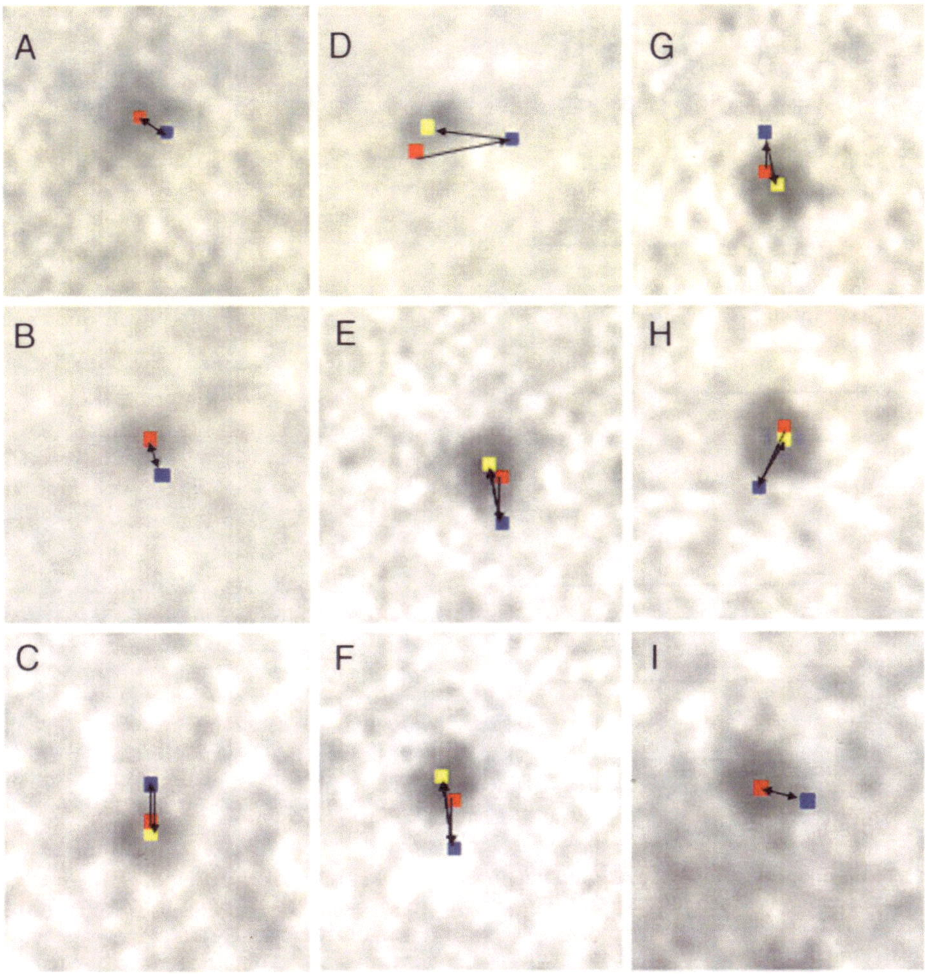

Figure 15. Examples showing sequential changes in position of 9 different pixels (each 2.5 x 2.5nm) where the center of mass positions of corresponding 9 particles are located. In each frame, pixel positions before ATP application (red), during ATP application (blue), and after exhaustion of ATP (yellow) are indicated. Note that myosin heads return towards their initial position after exhaustion of applied ATP (Sugi et al.,2008).

To give answer to this question, we prepared three different monoclonal antibodies (IgG) directed to three different regions within a single myosin head. Antibody 1 is identical with that used in our previous experiments already described in this chapter, and attaches to junctional peptides between 50k and 20k segments of myosin heavy chain. Antibody 2 attaches around reactive lysine residue (Lys 83) in CD. Antibody 3 attaches to two peptides (Met 58—Ala 70 and Leu 106−Phe 120) in myosin regulatory light chain in LD. The ATP-induced movement at three different parts within individual myosin heads was recorded using myosin filaments with myosin heads position-marked with antibodies 1, 2 or 3 and 3' by the method previously described.

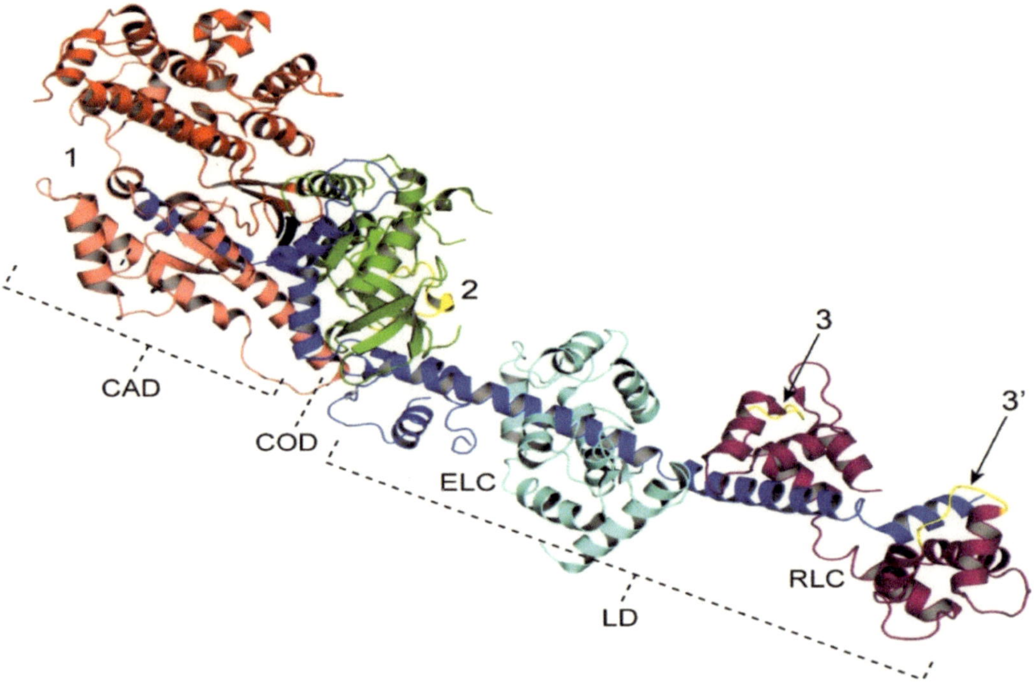

Figure 16. Myosin head structure showing approximate regions of attachment of antibody 1, 2 and 3, indicated by numbers 1, 2 and 3, respectively. The catalytic domain (CAD) comprises 25k (green), 50k (red) and part of 20k (dark blue) fragments of myosin heavy chain, while lever arm domain (LD) comprises the rest of 20k fragment and essential (ELC, light blue) and regulatory (RLC, magenta) light chains. CAD and LD are connected via converter domain (CD). Location of peptides around Lys 83 and that of two peptides (Met 58—Ala 70 and Leu 106—Phe 120) in LD are colored yellow. Regions of attachment of antibodies 1,2 and 3 are indicated by numbers 1, 2 and 3 and 3', respectively (Minoda et al.,2011).

Fig. 17 illustrated the results obtained as well as their interpretation. As can be seen in the three histograms. Fig. 17A, B and C are histograms of amplitude distribution of ATP-induced movement of myosin heads, position-marked with antibody 1, antibody 2 and antibody 3, respectively. The mean amplitude of ATP-induced movement was 6.14±0.09 (mean±s.e.m., n =1,692) at the distal part of CAD (A), and 6.14±0.22 (n = 1,112) at the CAD-CD boundary (B), indicating no significant difference between the two extreme regions of CAD. On the other hand, the average amplitude of ATP-induced movement at the regulatory light chain in LD was 3.55±0.11nm (n = 981), being significantly smaller than the corresponding values in CAD (t-test, $P < 0.01$).

If it is assumed that the cyclic conformational changes of myosin heads in the absence of actin filaments (Fig.17D) are in principle similar to the conformational changes of myosin heads in the presence of actin filaments in muscle (Fig.17E), the results shown in Fig.17 A—C can be accounted for by the lever arm mechanism in the following way.

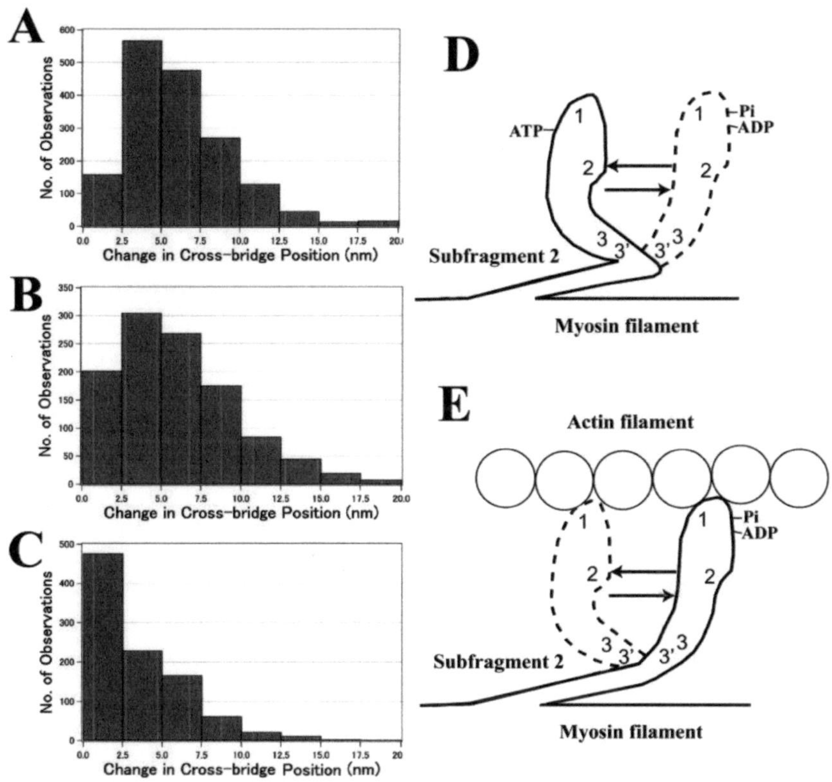

Figure 17. (A—C) Histograms showing amplitude distribution of ATP-induced myosin head movement, position-marked with antibody 1 (A(, antibody 2 (B), and antibody 3,3, respectively. (D,E) Diagrams illustrating myosin head lever arm mechanism in the absence (D) and in the presence (E) of actin filament. Attachment regions of 'of antibodies 1, 2 and 3 are indicated by numbers 1, 2 and 3,3', respectively (Minoda et al.,2011).

In the absence of actin filaments, the myosin head is initially thought to be in the post-power stroke configuration (solid line in D), and on binding with applied ATP it changes its conformation to reach the post-power stroke configuration with bound ATP hydrolysis products(Pi and ADP) (broken line in D). During this recovery stroke, the myosin head lever arm domain rotates not only around the converter domain, but also around the boundary between the lever arm domain and myosin S2, connecting the myosin head to myosin filament backbone. As a result, the amplitude of ATP-induced movement is definitely larger at both the distal and the proximal end of myosin head catalytic domain (indicated by numbers 1 and 2) than at the regulatory light chain in myosin lever arm domain (3 and 3').

During the myosin head power stroke taking place in muscle, the myosin head is initially in the pre-power stroke configuration with bound Pi and ADP, and attaches to actin filament (solid line in E). Then it undergoes power stroke releasing Pi and ADP, to take the post-power stroke configuration (broken line in E). To summarize, the measurement of ATP-induced movement at three different parts within individual myosin is not only consistent with the myosin head lever arm mechanism to produce force and motion in muscle, but may also constitute the first success in recording local structural changes taking place within a single macromolecule.

9. Conclusion

The experiments described in this chapter have proved that the EC is an extremely powerful tool in elucidating fundamental mysteries remaining in the research field on molecular mechanism of contraction. The greatest advantage of the use of EC for investigating muscle contraction is that it enables us to record movement of individual myosin heads coupled with ATP hydrolysis in hydrated myosin filaments, which retain their physiological function in an electron microscope.

In contrast, all other experimental methods hitherto used by a number of investigators, including time-resolved X-ray diffraction and chemical probe experiments (Cooke,1986; Hibbard & Trentham,1986), to study myosin head movement can only obtain averaged values since these methods inevitably sample numerous number of myosin heads acting asynchronously. Crystallographic and electron microscopic studies on myosin S1 crystal and acto-S1 complex (Geeves & Holmes,1999) are also concerned only with static structures and the results obtained are also statistical in nature. We believe that our work using the EC has made a breakthrough to open new horizon in this research field. As a matter of fact, we have already succeeded to study the myosin head power stroke in hydrated myosin filaments in the presence of actin filaments. A preliminary report of this work has appeared (Minoda et al.,2011).

Finally, we emphasize that the EC can be used not only for muscle research, but also for a number of other research fields to study function of biomolecules. We heartily hope that the EC will be used widely by life scientists to elucidate various mysteries in their respective research field. The EC system (JEOL,Ltd) is commercially available, and can be attached to any 100 or 200kV transmission microscope. Those who are interested in the carbon insulating film may consult JEOL or H.S. (sugi@kyf.biglobe.ne.jp) about its preparation.

Author details

Haruo Sugi*
Department of Physiology, School of Medicine, Teikyo University, Japan

Hiroki Minoda
Department of Applied Physics, Tokyo University of Agriculture and Technology, Japan

Takuya Miyakawa and Suguru Tanokura
Graduate School of Agriculture and Science, University of Tokyo, Japan

---

* Corresponding Author

Shigeru Chaen
Department of Human and Engineered Environmental Studies, Nihon University, Japan

Takakazu Kobayashi
Department of Electronic Engineering, Shibaura Institute of Technology, Japan

Acknowledgement

We would like to express our hearty thanks to President Kazuo Ito, President Terukazu Eto and President Yoshiyasu Harada of JEOL, Ltd. for providing generous support for our research work.

10. References

Buttler, E.P. & Hale, K.F.(eds) (1981). Dynamic Experiments in the Electron Microscope. In: *Practical Method in Electron Microscopy* Vol.9, North Holland, Amsterdam

Cooke, R. (1986). The mechanism of muscle contraction. CRC critical Reviwes in Biochemistry 21: 53—118

Fukami, A. & Adachi, K. (1965). A new method of preparation of a self-perforated microplastic grid and its applications. Journal of Electron Microscopy (Tokyo) 14: 112—118

Geeves, M.A. & Holmes, K.C. (1999). Structural mechanism of muscle contraction. Annual Review of Biochemistry 68: 687—728

Fukushima, K. (1988). Application of the gas environmental chamber for electron microscopy. Ph.D Thesis (Nagoya University, Nagoya) (in Japanese)

Hibbard, M.G. & Trentham, D.R. (1986). Relationships between chemical and mechanical events during muscular contraction. Annual Review of Biochemistry 15: 119—161

Huxley, A.F. (1957). Muscle structure and theories of contraction. Progress in Biophysics and Biophysical Chemistry 7: 255—318

Huxley, A.F. (1998). Support for the lever arm. Nature 396: 317—318

Huxley, H.E. (1969). The mechanism of muscular contraction. Science 164: 1356—1366

Huxley, H.E. (1957). The double array of filaments in cross-striated muscle. Journal of Biophysical and Biochemical Cytology 3: 631—648

Huxley, H.E. & Hanson, J. (1954). Changes in the cross-striations of muscle during contraction and stretch and their structural interpretation. Nature 173: 973—976

Lymn, R.W. & Taylor, E.W. (1971). Mechanism of adenosine triphosphate hydrolysis by actomyosin. Biochemistry 10: 4617—4624

Margossian S.S. & Lowey, S. (1982). Hybridization and reconstruction of thick filament structure. Methods in Enzymology 85: 20—55

Minoda, H., Okabe, T., Inayoshi, Y., Miyakawa, T., Miyauchi, Y., Tanokura, S., Katayama, E., Wakabayashi, T., Akimoto, T. & Sugi, H. (2011). Electron microscopic evidence for the myosin head lever arm mechanism in hydrated myosin filaments using the gas environmental chamber. Biochemical and Biophysical Research Communications. 405: 651—656

Oiw, K., Chaen, S. & Sugi, H. (1991). Measurement of work done by ATP-induced sliding between rabbit muscle myosin and algal cell actin cables in vitro. Journal of Physiology (London) 437: 751—763

Oiwa, K., Kawakami, T. & Sugi, H. (1993). Unitary distance of actin-myosin sliding studied using an in vitro force-movement assay system combined with ATP iontophoresis. Journal of Biochemistry (Tokyo) 114: 28—32

Page, S.G. & Huxley, H.E. (1963). Filament lengths in striated muscle. Journal of Cell Biology 19: 369—390

Perry, S.V. (1955). Myosin adenosine triphosphatase. Methods in Enzymology 2: 582—588

Suda, H., Ishikawa, A. & Fukami, A. (1992). Evaluation of the critical electron dose on the contractile activity of hydrated muscle fibers in the film-sealed environmental cell. Journal of Electron Microscopy 41: 223—229

Sugi, H. (1992). Molecular mechanism of actin-myosin interactionin muscle contraction. In: *Muscle contraction and Cell Motility*, Sugi,H (ed), Advances in Comparative & Environmental Physiology Vol.12, Springer, Berlin

Sugi, H., Akimoto, T., Sutoh, K., Chaen, S., Oishi, N. & Suzuki, S. (1997). Dynamic electron microscopy of ATP-induced myosin head movement in living muscle thick filaments. Proceedings of the National Academy of Sciences of the USA. 94: 4378—4382

Sutoh, K., Tokunaga, M. & Wakabayashi, T. (1989). Electron microscopic mapping of myosin head with site-directed antibodies. Journal of Molecular Biology 206: 357—363

Calcium Cycling in Synthetic and Contractile Phasic or Tonic Vascular Smooth Muscle Cells

Larissa Lipskaia, Isabelle Limon, Regis Bobe and Roger Hajjar

Additional information is available at the end of the chapter

http://dx.doi.org/10.5772/48240

1. Introduction

Calcium ions (Ca^{2+}) are present in low concentrations in the cytosol (~100 nM) and in high concentrations (in mM range) in both the extracellular medium and intracellular stores (mainly sarco/endo/plasmic reticulum, SR). This differential allows the calcium ion to be a ubiquitous 2nd messenger that carries information essential for cellular functions as diverse as contraction, metabolism, apoptosis, proliferation and/or hypertrophic growth. The mechanisms responsible for generating a Ca^{2+} signal greatly differ from one cell type to another. In the different types of vascular smooth muscle cells (VSMC), enormous variations do exist with regard to the mechanisms responsible for generating Ca^{2+} signal. In each VSMC phenotype (synthetic/proliferating[1] and contractile[2] [1], tonic or phasic), the Ca^{2+} signaling system is adapted to its particular function and is due to the specific patterns of expression and regulation of Ca^{2+} handling molecules (**Figure 1**). For instance, in contractile VSMCs, the initiation of contractile events is driven by membrane depolarization; and the principal entry-point for extracellular Ca^{2+} is the voltage-operated L-type calcium channel (LTCC). In contrast, in synthetic/proliferating VSMCs, the principal way-in for extracellular Ca^{2+} is the store-operated calcium (SOC) channel. Whatever the cell type, the calcium signal consists of limited elevations of cytosolic free calcium ions in time and space. The calcium pump, sarco/endoplasmic reticulum Ca^{2+} ATPase (SERCA), has a critical role in determining the frequency of SR Ca^{2+} release by controlling the velocity of Ca^{2+} upload into the sarcoplasmic reticulum (SR) and the Ca^{2+} sensitivity of SR calcium channels, Ryanodin Receptor, RyR and Inositol tri-Phosphate

[1]Synthetic VSMCs have a fibroblast appearance, proliferate readily, and synthesize increased levels of various extracellular matrix components, particularly fibronectin, collagen types I and III, and tropoelastin [1].
[2]Contractile VSMCs have a muscle-like or spindle-shaped appearance and well-developed contractile apparatus resulting from the expression and intracellular accumulation of thick and thin muscle filaments [1].

Receptor, IP_3R. Therefore, it is a major player in determining the spacio-temporal patterns of intracellular calcium signaling. This chapter focuses on the changes in Ca^{2+} signaling associated with different VSMC phenotypes. We will discuss the physiological implications of altered expressions of Ca^{2+} channels and pumps (referred to as Ca^{2+} handling proteins) and how they contribute to VSMC dysfunction in vascular disease.

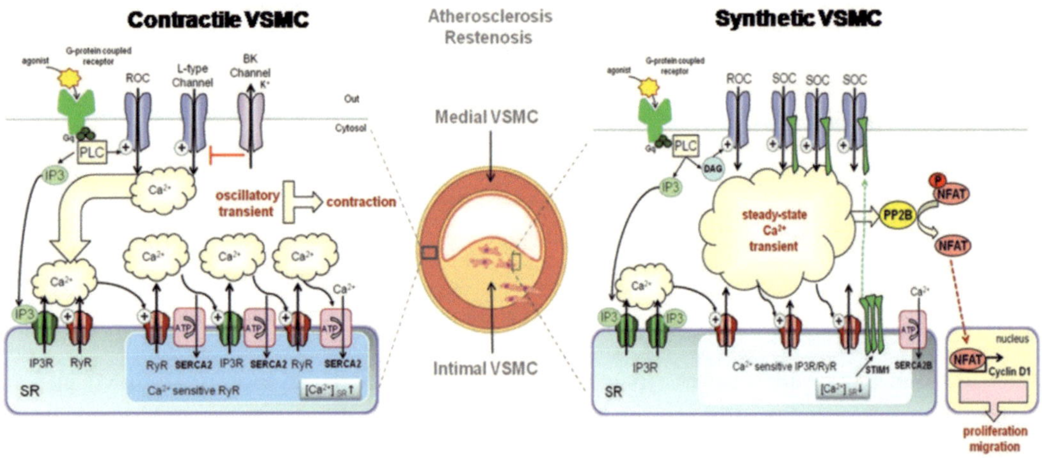

Figure 1. Schematic representation of Calcium Cycling in Contractile and Proliferating VSMCs. Left panel: schematic representation of calcium cycling in quiescent /contractile VSMCs. Contractile response is initiated by extracellular Ca^{2+} influx due to activation of Receptor Operated Ca^{2+} channels (through phosphoinositol-coupled receptor) or to activation of L-Type Calcium channels (through an increase in luminal pressure). Small increase of cytosolic due IP_3 binding to IP_3R (puff) or RyR activation by LTCC or ROC-dependent Ca^{2+} influx leads to large SR Ca^{2+} release due to the activation of IP_3R or RyR clusters ("Ca^{2+}-induced Ca^{2+}release" phenomenon). Cytosolic Ca^{2+} is rapidly reduced by SR calcium pumps (both SERCA2a and SERCA2b are expressed in quiescent VSMCs), maintaining high concentration of cytosolic Ca^{2+} and setting the sensitivity of RyR or IP_3R for the next spike. Contraction of VSMCs occurs during oscillatory Ca^{2+} transient. Middle panel: schematic representation of atherosclerotic vessel wall. Contractile VSMC are located in the media layer, synthetic VSMC are located in sub-endothelial intima. Right panel: schematic representation of calcium cycling in quiescent /contractile VSMCs. Agonist binding to phosphoinositol-coupled receptor leads to the activation of IP_3R resulting in large increase in cytosolic Ca^{2+}. Calcium is weakly reduced by SR calcium pumps (only SERCA2b, having low turnover and low affinity to Ca^{2+} is expressed). Store depletion leads to translocation of SR Ca^{2+} sensor STIM1 towards PM, resulting in extracellular Ca^{2+} influx though opening of Store Operated Channel (CRAC). Resulted steady state Ca^{2+} transient is critical for activation of proliferation-related transcription factors 'NFAT). Abbreviations: PLC - phospholipase C; PM - plasma membrane; PP2B - Ca^{2+}/calmodulin-activated protein phosphatase 2B (calcineurin); ROC- receptor activated channel; IP_3 - inositol-1,4,5-trisphosphate, IP_3R - inositol-1,4,5-trisphosphate receptor; RyR - ryanodine receptor; NFAT - nuclear factor of activated T-lymphocytes; VSMC - vascular smooth muscle cells; SERCA - sarco(endo)plasmic reticulum Ca^{2+} ATPase; SR - sarcoplasmic reticulum.

2. General aspects of calcium cycling and signaling in vascular smooth muscle cells

Besides maintaining vascular tone in mature vessels, VSMCs also preserve blood vessel integrity [2]. In other words, VSMCs are also instrumental for vascular remodeling and repair associated with VSMCs proliferation and migration. Interestingly, Ca^{2+} plays a central role in both physiological processes. In VSMCs, calcium signaling involves a cross-regulation of Ca^{2+} influx, sarcolemmal membrane signaling molecules and Ca^{2+} release and uptake from the sarco/endo/plasmic reticulum and mitochondria, which plays a central role in both vascular tone and integrity.

2.1. Calcium handling by the plasma membrane's calcium channels and pumps

Membrane depolarization is believed to be a key process for the activation of calcium events in mature VSMCs. Thus, much attention has been given to uncovering the various mechanisms responsible for triggering this depolarization. Increased intra-vascular pressure of resistance arteries stimulates gradual membrane depolarization in VSMCs, increasing the probability of opening L-type high voltage-gated Ca^{2+} channels (Cav1.2) (LTCC) [3, 4]. Alternatively, the calcium-dependent contractile response can be induced through the activation of specific membrane receptors coupled to phospholipase C (PLC) isoforms[3]. The various isoforms of transient receptor potential (TRP) ion channel family, particularly TRPC3, TRPC6 and TRPC7 possibly activated directly by diacyl glycerol (DAG), can also contribute to initial plasma membrane Ca^{2+} influx and subsequent membrane depolarization [5-8]. Non-selective receptor-activated canonical TRPC6 channel, that conduct large sodium (Na^{2+}) currents was also suggested to contribute to membrane depolarization and subsequent L-type channel activation [9, 10]. Membrane depolarization can spread to neighboring cells by current flow through gap junctions providing a synchronization mechanism for VSMC membrane depolarization within the vessel wall [11, 12].

Among voltage-insensitive calcium influx pathways, the store-operated Ca^{2+} channels (SOC), maintain a long-term cellular Ca^{2+} signal. They are activated upon a decrease of internal store Ca^{2+} concentration resulting from a Ca^{2+} release via the opening of SR Ca^{2+} release channels. SOC was first hypothesized in 1986 [13], a paradigm that was confirmed by the identification of its two essential regulatory components, the SR/ER located Ca^{2+} sensor STIM1 (stromal interaction molecule) and the Ca^{2+} channels Orai1 [14-17]. Upon decrease of $[Ca^{2+}]$ in the reticulum (<500μM), Ca^{2+} dissociates from STIM1; then STIM1 molecules oligomerize and translocate to specialized cortical reticulum compartments adjacent to the plasma membrane [18, 19]. There, the STIM1 cytosolic activating domains bind to and cluster the Orai proteins into an opened archaic Ca^{2+} channel known as $\underline{C}a^{2+}$-$\underline{r}$elease $\underline{a}$ctivated $\underline{C}a^{2+}$ $\underline{c}$hannel (CRAC) [4]. Furthermore, transient receptor potential ion

---

[3]All isoforms of PLC, catalyze the hydrolysis of phosphatidylinositol4,5-biphosphate (PIP_2) to produce the intracellular messengers IP_3 increase and diacylglycerol (DAG); both of which promote cytosolic Ca^{2+} rise through activation of plasma membrane or sarcoplasmic reticulum calcium channels.

[4] The CRAC is responsible for the "2h cytosolic Ca^{2+} increase" required to induce VSMCs proliferation [57].

channel (TRPC) family members have also been demonstrated to participate in SOC channels functioning via interactions with STIM1 and Orai proteins [20-22].

The calcium signal is terminated by membrane hyper-polarization and cytosolic Ca^{2+} removal. First, calcium sparks resulting from the opening of sub-plasmalemmal clusters of RyR activate large-conductance Ca^{2+} sensitive K^+ (BK) channels. Then, the resulting spontaneous transient outward currents (STOC) hyperpolarize the membrane and decrease the open probability of L-type Ca^{2+} channels [23]. Cytosolic calcium is extruded at the level of plasma membrane by plasma membrane Ca^{2+} ATPase (PMCA) and the Na^+/Ca^{2+} exchanger (NCX) [24, 25]. The principal amount of cytosolic Ca^{2+} (> 70%) is re-uploaded to the internal store.

2.2. Calcium handling by the sarco/endoplasmic reticulum's calcium channels and pumps

The initial entry of Ca^{2+} through plasma membrane channels triggers large Ca^{2+} release from the internal store via the process of Ca^{2+}-induced Ca^{2+}-release (CICR). The mechanism responsible for initiating Ca^{2+} release depends on Ca^{2+} sensitive SR calcium channels, the ryanodin receptor (RyR)[5] or the IP_3 receptor (IP_3R). Indeed, IP_3R and RyR are highly sensitive to cytosolic Ca^{2+} concentrations and when cytosolic Ca^{2+} concentration ranges from nM to μM, they open up. On the contrary, a higher cytosolic Ca^{2+} concentration (from μM to mM) closes them [26]. In other words, cytosolic Ca^{2+} increase first exerts a positive feedback and facilitates SR channels opening whereas a further increase has an opposite effect and actually inhibits the SR channels opening [27-29]. Importantly enough to be mentioned, RyR phosphorylation by the second messenger cyclic ADP ribose (cADPR) and protein kinase A (PKA) enhances Ca^{2+} sensitivity, the phosphorylation induced by the protein kinase C (PKC) decreases RyR sensitivity to Ca^{2+} [29, 30]. The initial release occurs in the vicinity of the plasma membrane. It spreads into the cell through the regenerative release of Ca^{2+} by the RyR and /or the IP_3R in the form of an intracellular Ca^{2+} wave travelling down the length of the cell [31-33]. When $[Ca^{2+}]i$ is integrated over an entire cell with time, these Ca^{2+} waves appear as rhythmical oscillations [34].

Sarco/Endoplasmic Ca^{2+}ATPases (SERCA), the only calcium transporters expressed within sarco/endoplasmic reticulum (SR), serve to actively return calcium into this organelle. In mammals, three SERCA genes ATP2A1, ATP2A2 and ATP2A3 coding for SERCA1, SERCA2 and SERCA3 isoforms respectively have been identified [35]. Each gene gives rise to a different SERCA isoform through alternative splicing (**Figure 2**); they all have discrete tissue distributions and unique regulatory properties, providing a potential focal point within the cell for the integration of diverse stimuli to adjust and fine-tune calcium homeostasis in the SR/ER [36]. In VSMCs, SERCA2a and the ubiquitous SERCA2b isoforms are expressed; besides vascular smooth muscle, SERCA2a is preferentially expressed in cardiac and skeletal muscles. SERCA2b differs from SERCA2a by an extension of 46 amino acids that forms an

---

[5] RyR are structurally and functionally analogous to IP_3R, although they are approximately twice as large and have twice the conductance of IP_3R [27]; RyR channels are sensitive to store loading and IP_3R channels are sensitized by the agonist-dependent formation of IP_3.

additional trans-membrane domain setting the SERCA2b C terminus in the SR lumen [37]. Functional characterization of SERCA2 isoforms performed in transfected HEK-298 cells clearly indicated that the SERCA2a isoform displays a lower affinity for Ca^{2+} ($K_{0.5}$ = 0.985 μM) but has a higher turnover rate (ATP hydrolysis 70 s^{-1}) compared to SERCA2b ($K_{0.5}$ = 0.508 μM; 35 s^{-1}) [38]. Diversity of SERCA isoforms in the same cell suggests that each of them could be responsible for controlling unique cell functions.

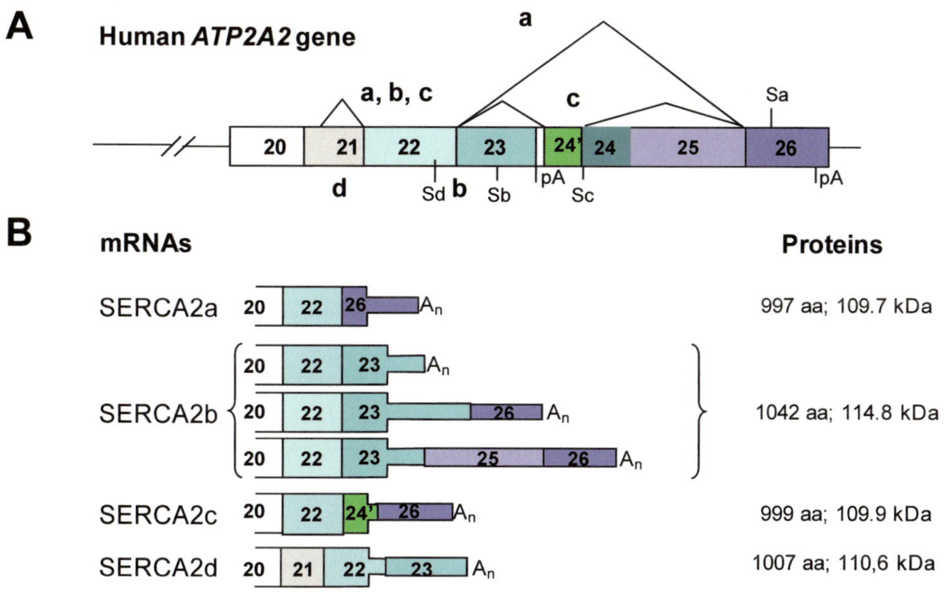

Figure 2. Alternative splicing of the human ATP2A2 genes, protein sizes of the SERCA2 isoforms. A. Representation of the 3'-end of the human ATP2A2 gene. Broken lines with letters represent alternative splicings. The position of stop codons for the corresponding isoforms and polyadenylation site are indicated by Sa-Sd & pA, respectively. B. Representation of the 3'-end of SERCA2a, SERCA2b, SERCA2c and SERCA2d mRNAs expressed in human cells and the sizes of the corresponding proteins. Wide boxes represent translated segments; the less wide boxes represent untranslated segments.

SERCA2's activity depends on its interaction with phospholamban and is inhibitory in its de-phosphorylated form. PKA phosphorylation of phospholamban results in its dissociation from SERCA2, thus activating the Ca^{2+} pumps [39]. Cyclic ADP-ribose was also reported to stimulate SERCA pump activity [40].

As previously mentioned, SR Ca^{2+} content controls the sensitivity of SR Ca^{2+} channels, RyR and IP_3R, as well as functioning of SOC-mediated Ca^{2+} entry, thereby determining the type of intracellular calcium transient. Since SOCs opening depends on Ca^{2+} content of the store, one may suggest that SERCA participates to its regulation. Consistent with this, SOCs open up when the leak of Ca^{2+} from intracellular stores is not compensated with SERCA activity; SERCA inhibitors such as thapsigargin which prevent Ca^{2+} uptake are commonly used to chemically induce SOC currents; several works have established that SERCA can cluster with STIM1 and Orai1 in various cellular types [41, 42].

2.3. Mechanisms of cytosolic Ca^{2+} oscillations in VSMC

Ca^{2+} oscillations are one of the ways that VSMCs respond to agonists [43-47]. These Ca^{2+} oscillations are maintained during receptor occupancy and are driven by an endogenous pacemaker mechanism, called the cellular Ca^{2+} oscillator [12, 33, 34]. Ca^{2+} oscillators were classified into two main types, the membrane oscillators and the cytosolic oscillators [48].

Membrane oscillators are those which generate oscillations at the cell membrane by successive membrane depolarization. In most small resistance arteries, inhibitors of plasma membrane voltage-dependent channels reduce or even abolish the membrane potential oscillations which precede rhythmical contractions. This suggests that rhythmic extracellular Ca^{2+} influx can be required for calcium oscillatory transient [34]. Besides, membrane oscillators greatly depend on Ca^{2+} entry in order to provide enough Ca^{2+} to charge up the intracellular stores for each oscillatory cycle. Some of the entry mechanisms characterized include the Na^{+}/Ca^{2+} exchanger (NCX) operating in its reverse mode [32, 44, 49], the ATP-sensitive P2X receptor responsible for generating junctional Ca^{2+} transients (jCaTs) [50] or the L-type Cav1.2 channels operating as clusters in a "high open probability mode" to produce persistent Ca^{2+} sparklets [51]. In the case of the latter, entry of Ca^{2+} through these L-type channel clusters does not directly activate RyR to produce sparks, but provides the necessary amount of Ca^{2+} to charge up the ER and sensitize the RyR [52]. Another way of internal store reloading is achieved by the various isoforms of the transient receptor potential (TRP) ion channel family, activated directly by DAG [5, 6].

Cytosolic oscillators do not depend on the cell membrane to generate oscillations. Instead, they arise from intracellular store membrane instability. The pacemaker mechanism of cytosolic Ca^{2+} oscillator is based on the velocity of luminal Ca^{2+} loading and luminal Ca^{2+} content [53, 54]. The mechanism responsible for initiating Ca^{2+} release depends either on RyRs or IP_3R activation. As soon as stores are sufficiently charged with Ca^{2+}, the SR Ca^{2+} channels become sensitive to cytosolic Ca^{2+} and can participate to the process of Ca^{2+}-induced Ca^{2+}-release, which is responsible for orchestrating the regenerative release of Ca^{2+} from the SR/ER. Importantly, extracellular Ca^{2+} influx is not required for cytosolic oscillator function. Indeed, the Ca^{2+} oscillations can be observed in the absence of extracellular Ca^{2+} [43, 45, 55, 56].

Finally, it should be mentioned that SERCA proteins play a major role in the establishment of oscillatory Ca^{2+} transient required for phasic contractile response[6], independently from the type of oscillator. Indeed, the frequency of Ca^{2+} oscillations depends solely on the velocity with which Ca^{2+} is re-loaded to the SR. This allows the sensitization of SR release channels, which determine the timing of the next Ca^{2+} spike [54, 57]. Thus, as long as the IP_3 membrane receptor/Ca^{2+} channel is activated, the next spike is initiated as soon as the sensitivity of IP_3Rs (RyRs) is restored, resulting in oscillatory mode of calcium cycling.

[6]Phasic contractions are apparent as rapid peaks, whereas tonic contractions cause gradual changes in force that can be maintained for prolonged periods.

3. Calcium cycling and signalling in vascular smooth muscle cells

3.1. Vascular smooth muscle cell phenotype diversity

VSMCs maintain a considerable plasticity throughout life, exhibiting a diverse range of phenotypes in response to local environmental changes [58, 59]. Because studies of smooth muscle phenotype have principally focused on mechanisms which control VSMC proliferation and differentiation [58], VSMC phenotypes are classified into two main categories: the synthetic/proliferating/migratory/inflammatory and the contractile/quiescent/differentiated phenotype. In mature vessels, most VSMCs exhibit a quiescent/contractile phenotype and control the vascular tone [58, 59]. Transition of contractile VSMCs towards a proliferating/migratory/inflammatory phenotype is one of the initial mechanisms leading to pathological vascular remodeling[7] [58, 60]. Culturing VSMCs *in vitro* mimics this progression, as primary cultures rapidly lose differentiation markers and exhibit a synthetic phenotype [60].

Another possibility for classifying VSMCs is to categorize them based on their contractile properties which determine whether the smooth muscle is considered phasic or tonic. Phasic vascular smooth muscle refers to blood vessels displaying rhythmic contractile activity whereas tonic vascular smooth muscle refers to blood vessels displaying continuous contractions [61]. Phasic contraction is the characteristic of small resistance arteries (SRA, 20-50 μm diameters) which predominantly regulates vascular functions such as pressure and flow. *In vivo* small arteries exhibit a mixture of tonic and phasic contractions (RW: [34, 45, 62, 63]) and/or conducted vasomotor response termed vasomotion[8] [64].

The contractile capacities of various VSMC phenotypes are determined by expression of different isoforms of contractile proteins. For instance, phasic contractions displayed by small arteries of the heart and lung, muscular femoral artery, small mesenteric arteries and renal afferent arteriole result from the VSMC expression of the fast isoform of smooth muscle myosin heavy chain (MHC) which determines the velocity of shortening during vasomotion (rev. [61]). Of note, lent isoform of smooth muscle myosin, so called "non muscular myosin, (NM-B)" is expressed in all types of VSMCs including synthetic/proliferating VSMC; it plays a significant role in force maintenance during tonic contraction [58, 61].

Contractile response is initiated by the rise in cytosolic $[Ca^{2+}]i$ leading to the activation of Ca^{2+}/calmodulin-dependent myosin light chain kinase (MLCK) [65]. However the mode of contraction, phasic or tonic, is determined by the type of cytosolic Ca^{2+} cycling. Different types of calcium cycling were observed in synthetic and contractile tonic or phasic VSMCs, in accordance with differential expression of calcium handling proteins [59, 60, 66].

---

[7]Vascular remodeling was initially defined as the process of arterial enlargement to accommodate the plaque and maintain constant flow despite increases in atherosclerotic lesion mass. Experimental and clinical observations indicate that blood flow properties influence remodeling after angioplasty, hypertension, and flow diversion as well as atherosclerotic plaque progression [60].

[8] Vasomotion or pulsative flow is suggested to enhance blood flow or tissue oxygenation (rev.[61])

3.2. Calcium cycling in synthetic/proliferating VSMC

The hallmark of synthetic status of VSMCs is a lack of functional proteins entity associated with the contractile response [60]; we refer to voltage activated L-type calcium channels (LTCC), SR calcium release channel RyR and "fast" isoform of SR calcium pump SERCA2a (**Figure 3**). In line with this, large conductance K^+ channels (BK_{Ca}), which are involved in negative feedback regulation of LTCC activity through plasma membrane hyperpolarization, are also down-regulated in synthetic VSMC [67, 68]. On the other hand, the expression of the molecular entities modulating the plasma membrane Ca^{2+}-release activated Ca^{2+} channel (CRAC) functioning [57, 69] are highly up-regulated; we refer to the proteins forming the CRAC complex or regulating the I_{CRAC} (such as ORAI1-3 and STIM1) and to the IP_3R [70, 71]. Besides, the expression of TRPCs family members, particularly TRPC1 and TRPC6, dramatically increases in synthetic cells leading to the increase of whole cell Ca^{2+} current [72, 73].

In synthetic VSMC, agonist binding to PLC-coupled membrane receptors activates IP_3R, resulting in a drastic increase of cytosolic Ca^{2+} which is weakly pumped by the "slow" calcium pump SERCA2b (the only isoform of SERCA expressed in synthetic VSMCs). The depleted store triggers the translocation of STIM1 towards the plasma membrane, which, through the opening of CRAC, induces an extracellular Ca^{2+} influx. This translates into a long lasting increase of cytosolic calcium critical for the activation of Ca^{2+}-sensitive transcription factor NFAT (nuclear factor of activated T lymphocytes), required for proliferation and migration of VSMCs [56, 74]. Since these cells express contractile proteins, such as NM-B [58, 61], one may suggest that long lasting increase of cytosolic calcium can also produces tonic contraction.

Importantly enough to be mentioned, the restoration of SERCA2a expression by gene transfer in synthetic VSMCs blocks their proliferation and migration via inhibition of transcription factor NFAT [56, 75]. Molecular mechanisms of this effect are related to the prevention of functional association between STIM1 and Orai1 (CRAC protein entity) which lead to the suppression of store-operated calcium influx [56]. It is worth mentioning that SOC influx following agonist stimulation is not observed in contractile VSMCs, naturally expressing SERCA2a (Bobe & Lipskaia, unpublished data), highlighting again the importance of the SERCA isoform(s) expressed in VSMCs.

3.3. Calcium cycling in contractile tonic and phasic VSMC

In mature vessels, VSMCs mainly exhibit a tonic or phasic contractile phenotype. In contractile VSMCs extracellular calcium influx predominantly takes place through the voltage-dependent L-type calcium channel, LTCC[9] (**Figure 3**). Extracellular Ca^{2+} influx causes a small increase of cytosolic Ca^{2+} generated by the opening of IP_3R clusters, called puff and/or RyR2 clusters, called spark [28, 57]. These local rises of cytosolic Ca^{2+} generate a larger SR Ca^{2+} release through the Ca^{2+}-induced Ca^{2+} release phenomenon. Elevation of free cytosolic calcium triggers VSMC contraction. The mode of intracellular calcium transient determines the type of contraction,

[9] In contractile VSMCs, NFAT can be activated by sustained Ca^{2+} influx (persistent Ca^{2+} sparklets) mediated by clusters of L-type Ca^{2+} channels operating in a high open probability mode [76, 77].

tonic or phasic. Steady state increase in cytosolic Ca^{2+} triggers tonic contraction; oscillatory type of Ca^{2+} transient triggers phasic contraction. [34, 76]. It is worth mentioning that accumulating evidence indicate that SR Ca^{2+}ATPase functioning/location within the cell (which greatly influences the velocity of calcium upload) determines the mode of Ca^{2+} transient in VSMCs. Consistent with this, i) "phasic" VSMCs display a greater number of peripherally located SR than "tonic" VSMCs; indeed "tonic" VSMCs exhibit centrally located SR; (rev in [61, 77]); ii) drugs which interfere with the IP_3 pathway or intracellular stores abolish spontaneous vasomotion [11, 78]; iii) blocking SERCA strongly inhibits the Ca^{2+} oscillations, demonstrating that they are induced by SR Ca^{2+} release; this latter argument is further supported by the fact that oscillations are present even in the absence of extracellular Ca^{2+} [43, 45, 55, 56].

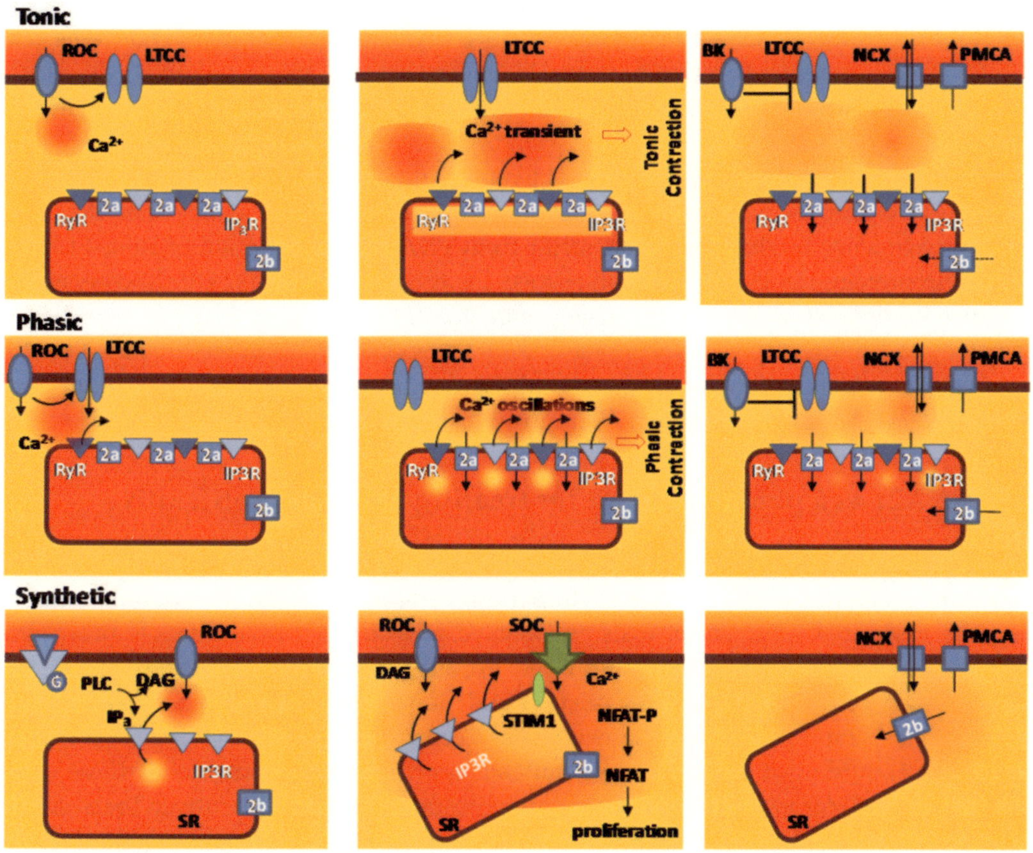

Figure 3. Dynamic schematic representation of calcium cycling in contractile phasic or tonic and synthetic VSMCs. Left panel: initiation of calcium event. Middle panel: resulting calcium transient and related physiological function. Right panel: termination of calcium event. The color intensity reflects Ca^{2+} concentrations. Abbreviations: BK - potassium channel; DAG - diacylglycerol; IP3 - inositol-1,4,5-trisphosphate; IP3R - inositol-1,4,5-trisphosphate receptor; LTCC - voltage-gated L-Type Calcium channel; NCX - Na^{+}/Ca^{2+} exchanger; NFAT - nuclear factor of activated T-lymphocytes; PLC - phospholipase C; PMCA – plasma membrane Ca^{2+} ATPase; ROC - receptor activated channel; RyR - ryanodine receptor; SERCA2a and SERCA2b - sarco(endo)plasmic reticulum Ca^{2+} ATPase type 2a and 2b; SOC - store operated Ca^{2+} channel; SR - sarcoplasmic reticulum, STIM1 - stromal interaction molecule 1.

The "fast" calcium pump SERCA2a, specifically expressed in contractile VSMCs can be responsible for the establishment of the "cytosolic oscillator". Several arguments are in agreement with this proposal: i) SERCA2a has a higher catalytic turnover when compared to SERCA2b due to a higher rate of de-phosphorylation and a lower affinity for Ca^{2+}; ii) SERCA2a is absent in synthetic VSMCs, which only exhibit tonic contraction, iii) transferring the SERCA2a gene to synthetic cultured VSMCs modifies the agonist-induced calcium transient from steady-state to oscillatory mode [56]. Therefore, one might suggest that the physiological role of SERCA2a in VSMCs consists of controlling the "cytosolic oscillator", thereby determining phasic vs tonic type of smooth muscle contraction.

Despite the fact that agonist-induced Ca^{2+} oscillations are a characteristic feature of the activation mechanisms of VSMCs [43-47], oscillatory type of Ca^{2+} transient is poorly associated with phasic contractile response. In some vessels, asynchronous oscillations of individual VSMC maintain a particular vascular tonus. However, in small resistance vessels the oscillations of groups of cells are synchronized through gap-junctions resulting in the pulsative contractile response [34, 45, 79, 80]. This oscillatory activity can be regulated by variations of neurotransmitters following sympathetic activation and can affect contractile tone through the increase of frequency, thereby increasing blood flow or tissue oxygenation [43, 45, 46, 81]. This frequency modulation could result from PKA phosphorylation of RyR, PLB and contractile proteins, as it has been established for cardiomyocytes [82-84]. Micro-vascular dysfunction, defined as the intrinsic changes in VSMCs contractility (such as reduction of frequency and shortening velocity of phasic contractions), observed in the context of cardiovascular [85, 86], may be related to reduced PKA phosphorylation of Ca^{2+} handling and contractile proteins, as observed in failing cardiomyocytes [82].

4. SERCA2a as a potential target for treating vascular proliferative diseases

Abundant proliferation of VSMCs is an important component of the chronic inflammatory response associated to atherosclerosis and related vascular occlusive diseases (intra-stent restenosis, transplant vasculopathy, and vessel bypass graft failure). Great efforts have been made to prevent/reduce trans-differentiation and proliferation of synthetic VSMCs. Antiproliferative therapies including the use of pharmacological agents and gene therapy approaches are, until now, considered as a suitable approach in the treatment of these disorders [87]. Indeed, coronary stenting is the only procedure that has been proven to reduce the incidence of late restenosis after percutaneous transluminal coronary angioplasty ([88]). Nevertheless, post-interventional intra-stent restenosis, characterized by the re-narrowing of the arteries caused by VSMC proliferation, occurs in 10 to 20 % of patients. These disorders remain the major limitation of revascularization by percutaneous transluminal angioplasty and artery bypass surgery. The use of drug-eluting stents (stent eluting anti-proliferative drug) significantly reduces restenosis but impairs the re-endothelialization process and subsequently often induces late thrombosis [89, 90]. In human, trans-differentiation of contractile VSMCs towards a synthetic/proliferating inflammatory/migratory phenotype after percutaneous transluminal angioplasty appears to be a fundamental process of vascular proliferative disease [91]. In contrast, phenotypic re-differentiation of neo-intimal VSMCs after bare metal stent implantation was reported to be associated with a decline in platelet activation and inflammatory cell infiltration, and the regeneration of the endothelial cell layer [92]. Thus, defining novel molecular target(s) of DESs, that can simultaneously prevents VSMC proliferation and adverse vascular remodeling while facilitating re-endothelialization, is crucial. SERCA2a gene transfer prevents neo-intimal proliferation and intimal thickening in the rat carotid injury model by normalizing calcium cycling and inhibiting NFAT activity [75]. Furthermore, SERCA2a gene transfer prevents VSMC trans-differentiation in injured segments while allowing re-endothelialization [75]. Thus, SERCA2a can be considered as a potential and powerful target for treating vascular proliferative disease.

5. Concluding remarks

Over the last decade, great progress has been made in the understanding of the various intracellular molecular mechanisms in VSMCs which control calcium cycling and excitation/contraction or excitation/transcription coupling. VSMCs employ a great variety of Ca^{2+} signaling systems that are adapted to control their different contractile functions. Alterations in the expressions of Ca^{2+} handling molecules are closely associated with VSMC phenotype modulation. Furthermore, these changes in expression are inter-connected and each acquired or lost Ca^{2+} signaling molecule represents a component of signaling module functioning as a single unit.

In non-excitable synthetic VSMCs, calcium cycling results from the protein module ROC/IP_3R/STIM1/ORAI1 which controls SOC influx. Agonist stimulation of synthetic VSMCs translates into a sustained increase in cytosolic Ca^{2+}. This increase is required for the activation of NFAT downstream cellular signaling pathways inducing proliferation, migration and possibly an inflammatory response. Calcium cycling in excitable contractile VSMCs is governed by the protein module composed of ROC/LTCC/RyR2/SERCA2a and controls the contractile response. The location of particular ion channels within the smooth muscle cell with regards to internal stores, other membrane ion channels, gap junctions as well as the expression of fast isoforms of contractile proteins have a significant impact on the resulting phasic or tonic contractile response. Future studies unraveling the correlation between the dynamic changes in Ca^{2+} signaling protein expression and specific subcellular localization are needed to delineate the mechanisms by which Ca^{2+} signaling molecules produce a phasic or tonic contractile response.

Author details

Larissa Lipskaia
Mount Sinai School of Medicine, Department of Cardiology, New York, NY, USA

Isabelle Limon
Univ Paris 6, UR4 stress inflammation and aging, Paris, France

Regis Bobe
INSERM U770, CHU Bicêtre, Le Kremlin-Bicêtre, France

Roger Hajjar
Mount Sinai School of Medicine, Department of Cardiology, New York, NY, USA

Acknowledgments

LL is supported by AHA SDG 0930116N (USA), IL is supported by Pierre and Marie Curie University (Univ Paris 6, France) and ANR grant 11BSV103401; RB is supported by Association Française Contre les Myopathies (AFM, France); RJH is supported by NIH grants (USA) HL088434, HL080498, HL100396 and NIH/NHLBI Contract HHSN268201000045C. We thank Zela Keuylian for editing.

Abbreviations

BK – large-conductance Ca^{2+} sensitive K^{+} channel

cADPR – cyclic Adenosine Diphosphate Ribose

CICR - Ca^{2+}- Induced Ca^{2+} Release

CRAC - Ca^{2+}- Release Activated Ca^{2+} Channels

DAG - Diacyl Glycerol

IP_3R - sarco/endoplasmic reticulum Ca^{2+} channel Inositol tri-Phosphate Receptor

LTCC - voltage-dependent L-type Ca^{2+} channels

NCX – Na^{+}/Ca^{2+} exchanger

PKA – Protein Kinase A (activated by cAMP, cyclic adenosine monophosphate)

PLC – Phospholipase C

PMCA – Plasmic Membrane Ca^{2+} ATPase

RyR - sarco/endoplasmic reticulum Ca^{2+} channel Ryanodin Receptor

SOC - Store-Operated Ca^{2+} Channels

SERCA - Sarco/Endoplasmic Reticulum Ca^{2+} ATPase

SRA - Small Resistance Arteries

SR/ER – Sarco/Endoplasmic Reticulum

STIM1 – Stromal Interaction Molecule 1, SR Ca^{2+} sensor

TRPC - Transient Receptor Potential ion Channel

VSMCs - Vascular Smooth Muscle Cells

6. References

[1] Pauly RR, Bilato C, Cheng L, Monticone R, Crow MT. Vascular smooth muscle cell cultures. Methods Cell Biol. 1997;52:133-54.
[2] Yoshida T, Owens GK. Molecular determinants of vascular smooth muscle cell diversity. Circulation research. 2005 Feb 18;96(3):280-91.
[3] Harder DR, Gilbert R, Lombard JH. Vascular muscle cell depolarization and activation in renal arteries on elevation of transmural pressure. The American journal of physiology. 1987 Oct;253(4 Pt 2):F778-81.
[4] Fleischmann BK, Murray RK, Kotlikoff MI. Voltage window for sustained elevation of cytosolic calcium in smooth muscle cells. Proceedings of the National Academy of Sciences of the United States of America. 1994 Dec 6;91(25):11914-8.
[5] Saleh SN, Albert AP, Peppiatt-Wildman CM, Large WA. Diverse properties of store-operated TRPC channels activated by protein kinase C in vascular myocytes. The Journal of physiology. 2008 May 15;586(10):2463-76.
[6] Peppiatt-Wildman CM, Albert AP, Saleh SN, Large WA. Endothelin-1 activates a Ca2+-permeable cation channel with TRPC3 and TRPC7 properties in rabbit coronary artery myocytes. The Journal of physiology. 2007 May 1;580(Pt.3):755-64.
[7] Hofmann T, Obukhov AG, Schaefer M, Harteneck C, Gudermann T, Schultz G. Direct activation of human TRPC6 and TRPC3 channels by diacylglycerol. Nature. 1999 Jan 21;397(6716):259-63.
[8] Okada T, Inoue R, Yamazaki K, Maeda A, Kurosaki T, Yamakuni T, et al. Molecular and functional characterization of a novel mouse transient receptor potential protein homologue TRP7. Ca(2+)-permeable cation channel that is constitutively activated and enhanced by stimulation of G protein-coupled receptor. The Journal of biological chemistry. 1999 Sep 24;274(39):27359-70.
[9] Soboloff J, Spassova M, Xu W, He LP, Cuesta N, Gill DL. Role of endogenous TRPC6 channels in Ca2+ signal generation in A7r5 smooth muscle cells. The Journal of biological chemistry. 2005 Dec 2;280(48):39786-94.
[10] Welsh DG, Morielli AD, Nelson MT, Brayden JE. Transient receptor potential channels regulate myogenic tone of resistance arteries. Circulation research. 2002 Feb 22;90(3):248-50.
[11] Haddock RE, Hill CE. Differential activation of ion channels by inositol 1,4,5-trisphosphate (IP3)- and ryanodine-sensitive calcium stores in rat basilar artery vasomotion. The Journal of physiology. 2002 Dec 1;545(Pt 2):615-27.
[12] Imtiaz MS, Zhao J, Hosaka K, von der Weid PY, Crowe M, van Helden DF. Pacemaking through Ca2+ stores interacting as coupled oscillators via membrane depolarization. Biophysical journal. 2007 Jun 1;92(11):3843-61.
[13] Putney JW. A model for receptor-regulated calcium entry. Cell Calcium. 1986;7:1-12.
[14] Liou J, Kim ML, Heo WD, Jones JT, Myers JW, Ferrell JE, Jr., et al. STIM is a Ca2+ sensor essential for Ca2+-store-depletion-triggered Ca2+ influx. Curr Biol. 2005 Jul 12;15(13):1235-41.

[15] Roos J, DiGregorio PJ, Yeromin AV, Ohlsen K, Lioudyno M, Zhang S, et al. STIM1, an essential and conserved component of store-operated Ca2+ channel function. J Cell Biol. 2005 May 9;169(3):435-45.

[16] Peinelt C, Vig M, Koomoa DL, Beck A, Nadler MJ, Koblan-Huberson M, et al. Amplification of CRAC current by STIM1 and CRACM1 (Orai1). Nat Cell Biol. 2006 Jul;8(7):771-3.

[17] Soboloff J, Spassova MA, Tang XD, Hewavitharana T, Xu W, Gill DL. Orai1 and STIM reconstitute store-operated calcium channel function. J Biol Chem. 2006 Jul 28;281(30):20661-5.

[18] Wu MM, Buchanan J, Luik RM, Lewis RS. Ca2+ store depletion causes STIM1 to accumulate in ER regions closely associated with the plasma membrane. The Journal of cell biology. 2006 Sep 11;174(6):803-13.

[19] Shen WW, Demaurex N. Morphological and functional aspects of STIM1-dependent assembly and disassembly of store-operated calcium entry complexes. Biochemical Society transactions. 2012 Feb 1;40(1):112-8.

[20] Yuan JP, Lee KP, Hong JH, Muallem S. The closing and opening of TRPC channels by Homer1 and STIM1. Acta Physiol (Oxf). 2012 Feb;204(2):238-47.

[21] Ong HL, Cheng KT, Liu X, Bandyopadhyay BC, Paria BC, Soboloff J, et al. Dynamic assembly of TRPC1-STIM1-Orai1 ternary complex is involved in store-operated calcium influx. Evidence for similarities in store-operated and calcium release-activated calcium channel components. J Biol Chem. 2007 Mar 23;282(12):9105-16.

[22] Liao Y, Erxleben C, Yildirim E, Abramowitz J, Armstrong DL, Birnbaumer L. Orai proteins interact with TRPC channels and confer responsiveness to store depletion. Proceedings of the National Academy of Sciences of the United States of America. 2007 Mar 13;104(11):4682-7.

[23] Nelson MT, Cheng H, Rubart M, Santana LF, Bonev AD, Knot HJ, et al. Relaxation of arterial smooth muscle by calcium sparks. Science. 1995;270:633-7.

[24] Abramowitz J, Aydemir-Koksoy A, Helgason T, Jemelka S, Odebunmi T, Seidel CL, et al. Expression of plasma membrane calcium ATPases in phenotypically distinct canine vascular smooth muscle cells. Journal of molecular and cellular cardiology. 2000 May;32(5):777-89.

[25] Strehler EE, Caride AJ, Filoteo AG, Xiong Y, Penniston JT, Enyedi A. Plasma membrane Ca2+ ATPases as dynamic regulators of cellular calcium handling. Annals of the New York Academy of Sciences. 2007 Mar;1099:226-36.

[26] Bootman MD, Collins TJ, Peppiatt CM, Prothero LS, MacKenzie L, De Smet P, et al. Calcium signalling--an overview. Semin Cell Dev Biol. 2001;12(1):3-10.

[27] Bootman MD, Lipp P, Berridge MJ. The organisation and functions of local Ca^{2+} signals. J Cell Sci. 2001;114(Pt 12):2213-22.

[28] Berridge MJ, Bootman MD, Roderick HL. Calcium signalling: dynamics, homeostasis and remodelling. Nat Rev Mol Cell Biol. 2003 Jul;4(7):517-29.

[29] Lipskaia L, Lompre AM. Alteration in temporal kinetics of Ca2+ signaling and control of growth and proliferation. Biol Cell. 2004 Feb;96(1):55-68.

[30] Jaggar JH, Porter VA, Lederer WJ, Nelson MT. Calcium sparks in smooth muscle. Am J Physiol Cell Physiol. 2000 Feb;278(2):C235-56.

[31] Brain KL, Cuprian AM, Williams DJ, Cunnane TC. The sources and sequestration of Ca(2+) contributing to neuroeffector Ca(2+) transients in the mouse vas deferens. The Journal of physiology. 2003 Dec 1;553(Pt 2):627-35.
[32] Dai JM, Kuo KH, Leo JM, van Breemen C, Lee CH. Mechanism of ACh-induced asynchronous calcium waves and tonic contraction in porcine tracheal muscle bundle. Am J Physiol Lung Cell Mol Physiol. 2006 Mar;290(3):L459-69.
[33] Sanderson MJ, Delmotte P, Bai Y, Perez-Zogbhi JF. Regulation of airway smooth muscle cell contractility by Ca2+ signaling and sensitivity. Proc Am Thorac Soc. 2008 Jan 1;5(1):23-31.
[34] Haddock RE, Hill CE. Rhythmicity in arterial smooth muscle. The Journal of physiology. 2005 Aug 1;566(Pt 3):645-56.
[35] Burk SE, Lytton J, MacLennan DH, Shull GE. cDNA cloning, functional expression, and mRNA tissue distribution of a third organellar Ca^{2+} pump. J Biol Chem. 1989;264(31):18561-8.
[36] Bobe R, Bredoux R, Corvazier E, Lacabaratz-Porret C, Martin V, Kovacs T, et al. How many Ca^{2+}ATPase isoforms are expressed in a cell type? A growing family of membrane proteins illustrated by studies in platelets. Platelets. 2005 May-Jun;16(3-4):133-50.
[37] Campbell AM, Kessler PD, Fambrough DM. The alternative carboxyl termini of avian cardiac and brain sarcoplasmic reticulum/endoplasmic reticulum Ca(2+)-ATPases are on opposite sides of the membrane. J Biol Chem. 1992;267(13):9321-5.
[38] Dally S, Bredoux R, Corvazier E, Andersen JP, Clausen JD, Dode L, et al. Ca^{2+}ATPases in non-failing and failing heart: evidence for a novel cardiac sarco/endoplasmic reticulum Ca^{2+}ATPase 2 isoform (SERCA2c). Biochem J. 2006 Apr 15;395(2):249-58.
[39] MacLennan DH, Asahi M, Tupling AR. The regulation of SERCA-type pumps by phospholamban and sarcolipin. Annals of the New York Academy of Sciences. 2003 Apr;986:472-80.
[40] Bradley KN, Currie S, MacMillan D, Muir TC, McCarron JG. Cyclic ADP-ribose increases Ca2+ removal in smooth muscle. Journal of cell science. 2003 Nov 1;116(Pt 21):4291-306.
[41] Sampieri A, Zepeda A, Asanov A, Vaca L. Visualizing the store-operated channel complex assembly in real time: Identification of SERCA2 as a new member. Cell Calcium. 2009 Mar 25.
[42] Manjarres IM, Rodriguez-Garcia A, Alonso MT, Garcia-Sancho J. The sarco/endoplasmic reticulum Ca(2+) ATPase (SERCA) is the third element in capacitative calcium entry. Cell Calcium Mar 25.
[43] Iino M, Kasai H, Yamazawa T. Visualization of neural control in intracellular Ca2+ concentration in single vascular smooth muscle cells in situ. cl. 1994;13:5026-31.
[44] Lee CH, Poburko D, Sahota P, Sandhu J, Ruehlmann DO, van Breemen C. The mechanism of phenylephrine-mediated [Ca(2+)](i) oscillations underlying tonic contraction in the rabbit inferior vena cava. J Physiol. 2001 Aug 1;534(Pt 3):641-50.
[45] Peng H, Matchkov V, Ivarsen A, Aalkjaer C, Nilsson H. Hypothesis for the initiation of vasomotion. Circulation research. 2001 Apr 27;88(8):810-5.

[46] Perez JF, Sanderson MJ. The contraction of smooth muscle cells of intrapulmonary arterioles is determined by the frequency of Ca2+ oscillations induced by 5-HT and KCl. J Gen Physiol. 2005 Jun;125(6):555-67.
[47] Shaw FZ, Liao YF. Relation between activities of the cortex and vibrissae muscles during high-voltage rhythmic spike discharges in rats. J Neurophysiol. 2005 May;93(5):2435-48.
[48] Berridge MJ, Rapp PE. A comparative survey of the function, mechanism and control of cellular oscillators. J Exp Biol. 1979;81:217-79.
[49] Rebolledo A, Speroni F, Raingo J, Salemme SV, Tanzi F, Munin V, et al. The Na+/Ca2+ exchanger is active and working in the reverse mode in human umbilical artery smooth muscle cells. Biochemical and biophysical research communications. 2006 Jan 20;339(3):840-5.
[50] Lamont C, Wier WG. Evoked and spontaneous purinergic junctional Ca2+ transients (jCaTs) in rat small arteries. Circulation research. 2002 Sep 20;91(6):454-6.
[51] Amberg GC, Navedo MF, Nieves-Cintron M, Molkentin JD, Santana LF. Calcium sparklets regulate local and global calcium in murine arterial smooth muscle. The Journal of physiology. 2007 Feb 15;579(Pt 1):187-201.
[52] Essin K, Welling A, Hofmann F, Luft FC, Gollasch M, Moosmang S. Indirect coupling between Cav1.2 channels and ryanodine receptors to generate Ca2+ sparks in murine arterial smooth muscle cells. The Journal of physiology. 2007 Oct 1;584(Pt 1):205-19.
[53] Berridge MJ. Calcium microdomains: organization and function. Cell Calcium. 2006 Nov-Dec;40(5-6):405-12.
[54] Berridge MJ. Smooth muscle cell calcium activation mechanisms. The Journal of physiology. 2008 Nov 1;586(Pt 21):5047-61.
[55] Ruehlmann DO, Lee CH, Poburko D, van Breemen C. Asynchronous Ca(2+) waves in intact venous smooth muscle. Circulation research. 2000 Mar 3;86(4):E72-9.
[56] Bobe R, Hadri L, Lopez JJ, Sassi Y, Atassi F, Karakikes I, et al. SERCA2a controls the mode of agonist-induced intracellular Ca^{2+} signal, transcription factor NFAT and proliferation in human vascular smooth muscle cells. Journal of molecular and cellular cardiology. 2011 Apr;50(4):621-33.
[57] Berridge MJ. Inositol trisphosphate and calcium signalling mechanisms. Biochimica et biophysica acta. 2009 Jun;1793(6):933-40.
[58] Owens GK, Kumar MS, Wamhoff BR. Molecular regulation of vascular smooth muscle cell differentiation in development and disease. Physiological reviews. 2004 Jul;84(3):767-801.
[59] Wamhoff BR, Bowles DK, Owens GK. Excitation-transcription coupling in arterial smooth muscle. Circulation research. 2006 Apr 14;98(7):868-78.
[60] House SJ, Potier M, Bisaillon J, Singer HA, Trebak M. The non-excitable smooth muscle: calcium signaling and phenotypic switching during vascular disease. Pflugers Archiv : European journal of physiology. 2008 Aug;456(5):769-85.
[61] Fisher SA. Vascular smooth muscle phenotypic diversity and function. Physiol Genomics. 2010 Nov 15;42A(3):169-87.
[62] Nilsson H, Aalkjaer C. Vasomotion: mechanisms and physiological importance. Mol Interv. 2003 Mar;3(2):79-89, 51.

[63] Parthimos D, Haddock RE, Hill CE, Griffith TM. Dynamics of a three-variable nonlinear model of vasomotion: comparison of theory and experiment. Biophysical journal. 2007 Sep 1;93(5):1534-56.
[64] Figueroa XF, Isakson BE, Duling BR. Connexins: gaps in our knowledge of vascular function. Physiology (Bethesda). 2004 Oct;19:277-84.
[65] Wray S, Burdyga T, Noble K. Calcium signalling in smooth muscle. Cell Calcium. 2005 Sep-Oct;38(3-4):397-407.
[66] Thorneloe KS, Nelson MT. Ion channels in smooth muscle: regulators of intracellular calcium and contractility. Can J Physiol Pharmacol. 2005 Mar;83(3):215-42.
[67] Neylon CB, Lang RJ, Fu Y, Bobik A, Reinhart PH. Molecular cloning and characterization of the intermediate-conductance Ca(2+)-activated K(+) channel in vascular smooth muscle: relationship between K(Ca) channel diversity and smooth muscle cell function. Circulation research. 1999 Oct 29;85(9):e33-43.
[68] Kohler R, Wulff H, Eichler I, Kneifel M, Neumann D, Knorr A, et al. Blockade of the intermediate-conductance calcium-activated potassium channel as a new therapeutic strategy for restenosis. Circulation. 2003 Sep 2;108(9):1119-25.
[69] Lewis RS. The molecular choreography of a store-operated calcium channel. Nature. 2007 Mar 15;446(7133):284-7.
[70] Berra-Romani R, Mazzocco-Spezzia A, Pulina MV, Golovina VA. Ca2+ handling is altered when arterial myocytes progress from a contractile to a proliferative phenotype in culture. Am J Physiol Cell Physiol. 2008 Sep;295(3):C779-90.
[71] Potier M, Gonzalez JC, Motiani RK, Abdullaev IF, Bisaillon JM, Singer HA, et al. Evidence for STIM1- and Orai1-dependent store-operated calcium influx through ICRAC in vascular smooth muscle cells: role in proliferation and migration. The FASEB journal : official publication of the Federation of American Societies for Experimental Biology. 2009 Aug;23(8):2425-37.
[72] Bergdahl A, Gomez MF, Wihlborg AK, Erlinge D, Eyjolfson A, Xu SZ, et al. Plasticity of TRPC expression in arterial smooth muscle: correlation with store-operated Ca2+ entry. American journal of physiology Cell physiology. 2005 Apr;288(4):C872-80.
[73] Kumar B, Dreja K, Shah SS, Cheong A, Xu SZ, Sukumar P, et al. Upregulated TRPC1 channel in vascular injury in vivo and its role in human neointimal hyperplasia. Circulation research. 2006 Mar 3;98(4):557-63.
[74] Dolmetsch RE, Lewis RS, Goodnow CC, Healy JI. Differential activation of transcription factors induced by Ca^{2+} response amplitude and duration. Nature. 1997;386(6627):855-8.
[75] Lipskaia L, del Monte F, Capiod T, Yacoubi S, Hadri L, Hours M, et al. Sarco/endoplasmic reticulum Ca2+-ATPase gene transfer reduces vascular smooth muscle cell proliferation and neointima formation in the rat. Circ Res. 2005 Sep 2;97(5):488-95.
[76] Aalkjaer C, Nilsson H. Vasomotion: cellular background for the oscillator and for the synchronization of smooth muscle cells. Br J Pharmacol. 2005 Mar;144(5):605-16.
[77] Wray S, Burdyga T. Sarcoplasmic reticulum function in smooth muscle. Physiological reviews. 2010 Jan;90(1):113-78.
[78] Bartlett IS, Crane GJ, Neild TO, Segal SS. Electrophysiological basis of arteriolar vasomotion in vivo. Journal of vascular research. 2000 Nov-Dec;37(6):568-75.

[79] Mauban JR, Lamont C, Balke CW, Wier WG. Adrenergic stimulation of rat resistance arteries affects Ca(2+) sparks, Ca(2+) waves, and Ca(2+) oscillations. American journal of physiology Heart and circulatory physiology. 2001 May;280(5):H2399-405.

[80] Lamboley M, Schuster A, Beny JL, Meister JJ. Recruitment of smooth muscle cells and arterial vasomotion. American journal of physiology Heart and circulatory physiology. 2003 Aug;285(2):H562-9.

[81] Kuo KH, Dai J, Seow CY, Lee CH, van Breemen C. Relationship between asynchronous Ca2+ waves and force development in intact smooth muscle bundles of the porcine trachea. Am J Physiol Lung Cell Mol Physiol. 2003 Dec;285(6):L1345-53.

[82] Lipskaia L, Chemaly ER, Hadri L, Lompre AM, Hajjar RJ. Sarcoplasmic reticulum Ca(2+) ATPase as a therapeutic target for heart failure. Expert Opin Biol Ther. 2010 Jan;10(1):29-41.

[83] Choudhury N, Khromov AS, Somlyo AP, Somlyo AV. Telokin mediates Ca2+-desensitization through activation of myosin phosphatase in phasic and tonic smooth muscle. J Muscle Res Cell Motil. 2004;25(8):657-65.

[84] Schubert R, Lehmann G, Serebryakov VN, Mewes H, Hopp HH. cAMP-dependent protein kinase is in an active state in rat small arteries possessing a myogenic tone. The American journal of physiology. 1999 Sep;277(3 Pt 2):H1145-55.

[85] Zhang H, Fisher SA. Conditioning effect of blood flow on resistance artery smooth muscle myosin phosphatase. Circulation research. 2007 Mar 16;100(5):730-7.

[86] Payne MC, Zhang HY, Shirasawa Y, Koga Y, Ikebe M, Benoit JN, et al. Dynamic changes in expression of myosin phosphatase in a model of portal hypertension. American journal of physiology Heart and circulatory physiology. 2004 May;286(5):H1801-10.

[87] Andres V, Castro C. Antiproliferative strategies for the treatment of vascular proliferative disease. Curr Vasc Pharmacol. 2003 Mar;1(1):85-98.

[88] Foley DP, Melkert R, Umans VA, de Jaegere PP, Strikwerda S, de Feyter PJ, et al. Differences in restenosis propensity of devices for transluminal coronary intervention. A quantitative angiographic comparison of balloon angioplasty, directional atherectomy, stent implantation and excimer laser angioplasty. CARPORT, MERCATOR, MARCATOR, PARK, and BENESTENT Trial Groups. European heart journal. 1995 Oct;16(10):1331-46.

[89] Luscher TF, Steffel J, Eberli FR, Joner M, Nakazawa G, Tanner FC, et al. Drug-eluting stent and coronary thrombosis: biological mechanisms and clinical implications. Circulation. 2007 Feb 27;115(8):1051-8.

[90] Fukuta Y, Joshizumi M, Kitagawa T, Hori T, Katoh I, Houchi H, et al. Effect of angiotensin II on Ca $^{2+}$ efflux from freshly isolated adult rat cardiomyocytes. Possible involvement of Na+/ Ca $^{2+}$ exchanger. Biochem Pharmacol. 1998;55:481-7.

[91] Ueda M, Becker AE, Naruko T, Kojima A. Smooth muscle cell de-differentiation is a fundamental change preceding wound healing after percutaneous transluminal coronary angioplasty in humans. Coron Artery Dis. 1995 Jan;6(1):71-81.

[92] Nakagawa M, Naruko T, Ikura Y, Komatsu R, Iwasa Y, Kitabayashi C, et al. A decline in platelet activation and inflammatory cell infiltration is associated with the phenotypic redifferentiation of neointimal smooth muscle cells after bare-metal stent implantation in acute coronary syndrome. J Atheroscler Thromb. 2010 Jul 30;17(7):675-87.

Chapter 3

The Role of Sodium-Calcium Exchanger in the Calcium Homeostasis of Airway Smooth Muscle

Ricardo Espinosa-Tanguma, Paola Algara-Suárez,
Rebeca Mejía-Elizondo and Víctor Saavedra-Alanís

Additional information is available at the end of the chapter

http://dx.doi.org/10.5772/47758

1. Introduction

ASM is a widespread component of the respiratory system. The lung parenchyma, like the airways, is a contractile tissue that responds to agonists like histamine and its muscular behavior highly impacts respiratory physiology. Asthma, for example, is a common disorder characterized by an excessive narrowing of the airways and inflammation in response to certain stimulants. Although the relative contribution of each element in this pathology is not precisely known, it is clear that smooth muscle relaxants alleviate acute asthmatic episodes. All of these points to ASM as an important target for study and therapy related to asthma.

The contractility of ASM highly depends on intracellular Ca^{2+} concentration and sensitization to Ca^{2+}, which in turn depend on several transport and signaling mechanisms. Ca^{2+} homeostasis can be understood as a balance between Ca^{2+} entry and exit pathways governed by a dynamic web of physical and chemical signals. Ca^{2+} entry pathways in ASM include: voltage activated Ca^{2+} channels, non-selective cationic channels, IP_3 activated Ca^{2+} channel and Ryanodine receptor-channel. On the other hand, Ca^{2+}- ATPase pumps located in the plasma membrane and the Sarcoplasmic Reticulum account for the Ca^{2+} exit pathways. A very peculiar transporter, the Na^+-Ca^{2+} exchanger (NCX), which is the main subject of this chapter, accounts for both Ca^{2+} exit and entry pathways due to its dual mode of operation. The many features and regulation of the NCX have been described mainly for the cardiac isoform which was the first to be cloned and characterized. Nevertheless, new research has been directed to other isoforms found in several tissues, including ASM, since a physiological role in contractility is now evident.

In this chapter, several aspects of ASM and the NCX will be addressed including: its role in Ca^{2+} homeostasis, contraction and proliferation; history of research related to the NCX; molecular and functional characteristics; and clinical implications.

2. The role of Calcium in airway smooth muscle

As with other smooth muscles, cytosolic Ca^{2+} concentration underlies the most important features of ASM: contractility, proliferation and phenotype. Various complex mechanisms regulate cytosolic Ca^{2+} concentration and are strongly influenced by neurotransmitters, cytokines and physical forces, to name a few. Since the ASM cell interacts closely with tissues such as nervous terminals, epithelium and lymphocytes that secrete all such substances, it is not surprising that pathologies such as asthma and COPD are strongly linked to alterations in ASM Ca^{2+} homeostasis. In this section, the role of Calcium in contraction, phenotype acquisition and proliferation will be reviewed.

2.1. ASM Contraction

ASM contraction induced by agonist stimulation results mainly from two phenomena: elevation in cytosolic Ca^{2+} concentration and sensitization of the contractile machinery to Ca^{2+} [1]. This tissue is constricted directly by agonists such as histamine, cysteinylleukotrienes, thromboxanes and acetylcholine released by mast cells or airway nerves [2]. The primary signaling mechanism coupled to most contractile receptors is the activation of phospholipase Cγ (PLCγ) via a pertussis toxin-insensitive $G_{q/11}$-protein [3]. Activation of PLC leads to hydrolysis of phosphatidylinositol-bis-phosphate to inositol-1, 4, 5-trisphospate (IP_3) and diacylglycerol (DAG) to, respectively, cause Ca^{2+} release from the sarcoplasmic reticulum (SR) and activate protein kinase C (PKC). It is known that prolonged stimulation by histamine causes intracellular Ca^{2+} increase in ASM strips which correlates with a sustained contraction [4]. As stated above, this is the sum of both sensitization and activation mechanisms involved in the contraction of smooth muscle.

Two key events in smooth muscle contraction are the phosphorylation and dephosphorylation of the regulatory light chains of myosin II (rMLC). These reactions are partly catalyzed by the Ca^{2+} and calmodulin-activated myosin light-chain kinase (MLCK) and the type 1 myosin phosphatase (MLCP), respectively. The balance of the activity of these enzymes results in the extent of contraction or relaxation of smooth muscle. After agonist stimulation, intracellular Ca^{2+} binds to calmodulin and changes its conformation, enabling it to activate MLCK. MLCK then phosphorylates rMLC, predominantly at Ser-19, allowing the myosin ATPase to be activated by actin. This leads to crossbridge formation between myosin and actin, and generates muscle contraction. The coupling between force and rMLC phosphorylation is quite variable and non-linear, however dephosphorylation of rMLC generally produces relaxation.

In addition to their effects of Ca^{2+} concentration changes, contractile agonists increase Ca^{2+} sensitivity of contraction. There are two ways to modulate such Ca^{2+} sensitivity: 1) altering the balance between the activities of MLCK and MLCP at a constant Ca^{2+} concentration, and 2) by rMLC phosphorylation-independent mechanisms as in the case of calponin, caldesmon and heat shock proteins.

MLCP is an heterotrimeric enzyme that contains a regulatory subunit referred to as the myosin phosphatase target subunit (MYPT) which helps to form the active heterotrimer as well as to increase the substrate specificity of MLCP toward myosin. This subunit is known to be phosphorylated by Rho Kinase becoming inhibited and thus favoring contraction. On the other hand, CPI-17 is an endogenous inhibitory protein of MLCP expressed in smooth muscle tissues which is itself regulated by RhoK and PKC phosphorylation [5-7].

On the other hand there are other mechanisms independent of phosphorylation of rMLC and pertain only to the thin filament actin. Calponin and caldesmon interact with F-actin and myosin and inhibit actomyosin ATPase activity. Both are regulated by PKC and ERK-activities [8, 9]. Altogether, the specific state of sensitization and intracellular Ca^{2+} concentration during agonist stimulation result in force development or relaxation of smooth muscle.

2.2. ASM phenotype and proliferation

ASM remodeling is an important aspect of many atopic respiratory diseases. Among the structural changes that the airways undergo include epithelial fibrosis, increase in ASM mass, mucous gland hyperplasia and edema. Wall thickening due to ASM hyperplasia and hypertrophy are a common hallmark of asthma and constitute the mayor obstruction for air flow during a crisis. It is believed that these changes occur as a response to chronic airway inflammation and mechanical stretch in which ASM cells take an active role by migrating to the epithelium and secreting various adhesion molecules and cytokines. Airway remodeling is proposed to begin with ASM cell phenotype change from a contractile to a synthetic and migrating type. It is still not very clear how Ca^{2+} homeostasis is associated with the many features of ASM remodeling. It has been reported that the activity of Cav1.2 channels and the SERCA pump may underlie this process by an up-regulation of intracellular Ca^{2+}. Downstream signaling mechanisms that lead to phenotype change include: Ca^{2+}/calmodulin-dependent protein kinase IV (CaMK.IV), peroxisome proliferator-activated receptor γ coactivator-1α (PGC-α), nuclear respiratory factor-1 (NRF-1) and mitochondrial transcription factor A (mtTFA). Among these, mtTFA up-regulates mitochondrial DNA replication and biogenesis probably leading to ASM cell proliferation. Tissues from asthmatic patients tend to show increased intracellular Ca^{2+} levels, which may render ASM cells hyper-proliferative as well as hyperreactive to contractile stimuli [10, 11].

3. Calcium transport mechanisms

Ca^{2+} is a fundamental second messenger in the mechanisms of remodeling and contraction of smooth muscle cells. Under basal conditions, the intracellular Ca^{2+} concentration in smooth muscle cells ranges from 100 to 200 nM [6]. Upon activation by agonists such as acetylcholine or histamine, there is a biphasic intracellular Ca^{2+} response in ASM cells consisting of an initial Ca^{2+} rise followed by a fast decline to a steady-state level that remains above basal concentration until agonist is washed out [12]. This biphasic profile reflects the Ca^{2+} release from the SR as well as Ca^{2+} influx from the extracellular space. On the other hand, there are mechanisms that remove intracellular Ca^{2+} such as the Sarco/Endoplasmic Reticulum Ca^{2+}-ATPase(SERCA), the plasmalemal Ca^{2+} ATPase (PMCA) and the mitochondria which become evident once agonist is removed and Ca^{2+} entry is abolished.

3.1. Ca^{2+} release from the SR

In airway smooth muscle, Ca^{2+} release from SR depends on IP_3 and RyR receptor-channels [13, 14]. These channels belong to two different families and share significant homology, especially in the sequences that are proposed to form the channels′ pore. These Ca^{2+} release channels are large oligomeric structures formed by association of either four IP_3R proteins (300 KDa each) or four RyR proteins (565 KDa each) [15-17].

The IP_3R channel requires binding of IP_3 for Ca^{2+} release with each monomer of the channel binding one molecule of IP_3 in a non-cooperative fashion with a K_D around 50 nM. An endogenous ligand for the RyR channels has remained elusive, and so far it has been proposed that in smooth muscle tissue, Ca^{2+} released by IP_3R activates RyR channels; an event referred to as Ca^{2+} -Induced Ca^{2+} Release (CICR). Ca^{2+} release channels are regulated by various factors. Cytoplasmic Ca^{2+} shows a biphasic effect on the IP_3 -induced Ca^{2+} release with a maximum rate at 300 nM. From that concentration on, the channel is inhibited by Ca^{2+}. This inhibitory effect is also shared by the RyR channel, although it is not physiologically relevant, since millimolar concentrations of Ca^{2+} (1 – 10 mM) are required to inhibit the channel. Therefore, the Ca^{2+} release from the RyR channel shows only Ca^{2+}-dependent activation in the physiological range of Ca^{2+} concentration (1 – 10 μM). Both channels are activated by free ATP, around 10 μM for the IP_3R and 300 μM for the RyR channel.

An important modulator of the RyR channel is the plant alkaloid ryanodine which binds to each monomer with high affinity (Kd $<$ 50 nM). Low doses of ryanodine (around 10 nM) are reported to increase the frequency of single RyR channel opening. Intermediate ryanodine doses (around 1 μM) are reported to induce very long – duration open events and simultaneously reduce ion conductance through the pore. High doses of ryanodine (around 100 μM) are reported to lock the channel in a closed configuration.

3.2. Ca^{2+} uptake into the SR

The SR is continually re-filled with Ca^{2+} with the aid of the SERCA pumps located in its membrane. The SERCA are 110 kDa proteins that belong to the P-type ion pumps family. Their activity is a cycle of chemical reactions that lead to conformational changes and Ca^{2+} transport powered by phosphorylation. The activity of SERCA is largely regulated by phospholamban which in its phosphorylated state increases Ca^{2+} uptake. Even in the absence of an agonist, SR Ca^{2+} uptake appears to be critical for ASM basal tone maintenance. This has been explored in experiments where the SERCA pumps are inhibited by Cyclopiazonic acid (CPA) or Thapsigargin (TG) in resting smooth muscle. Under such conditions, ASM spontaneously contracts following an increase in intracellular Ca^{2+}, which can be explained as a leak from the SR which cannot be handled by the impaired SERCA function. SERCA function is also evident when its inhibition is followed by incubation of ASM in Ca^{2+}- fee solutions and agonist stimulation. Upon Ca^{2+} re-addition in these circumstances, a transitory contraction can be observed which is characteristic of a depleted SR and reflects the store-operated Ca^{2+} entry (SOCC) which is discussed below.

3.3. Ca^{2+} influx from the extracellular space

Besides the SR, the other source for Ca^{2+} in ASM is the extracellular media, but in this respect the involved mechanisms are unclear. Three main sources of Ca^{2+} have been proposed to be active in ASM: voltage activated Cav1.2 channels, transient receptor potential channels (TRP) and the NCX in reverse mode. The signaling mechanisms that activate each of these channels and transporters have not been completely understood, given that some of them have been just recently described in this tissue. Another interesting feature is the functional interaction among these systems as well as interactions with other channels such as the Ca^{2+}-activated chloride channels (Cl_{Ca}) and the Ca^{2+} activated potassium channels (BK_{Ca}). As will be discussed later, these channels regulate membrane potential and thus alter the activity of Cav1.2 channels, TRPC channels and the NCX. On the other hand, the ionic concentration reached within the vicinity of these channels upon stimulation, has also been observed to alter their function regardless of whole membrane potential.

3.4. ASM resting membrane potential

It has been established that ASM membrane potential relies mainly on the activity of Cl_{Ca}, BK_{Ca} and TRPC channels. Cl_{Ca} and BK_{Ca} channels are activated by intracellular Ca^{2+} showing slightly different sensitivities between the ranges 100 to 900 nM [18, 19], and influence the membrane potential in opposite ways. Cl_{Ca} channels allow Cl^- to exit the cell, therefore depolarizing the membrane, while BK channels allow K^+ to exit causing membrane hyperpolarization. It has been reported that Ca^{2+} released by the SR during histamine stimulation causes activation of Cl_{Ca} channels and membrane depolarization [20]. TRPC have also been proposed to impact membrane potential at rest and after agonist stimulation causing depolarization. Histamine evokes an inward Na^+ current in equine tracheal myocytes together with an outward Cl^- current. In that work, it was suggested that NSCC of the TRPC family were responsible for the cationic current observed [21].

3.5. Cav1.2 channels

In many tissues, smooth muscle dihydropyridine-sensitive channels activated by membrane depolarization comprise an important source for external Ca^{2+} [22]. Interestingly, blockade of these channels has not served as a therapeutic tool in asthma, and thus their physiological relevance remains unclear. These channels are composed of pore-forming (α subunit) and accessory subunits that regulate expression, gating and channel kinetics. The α subunit carries the Ca^{2+} current and provides the voltage- and DHP-sensitivity of these channels [23]. Research has shown that voltage-related Ca^{2+} currents in ASM reflect Cav1.2 channel activity [12] which is the main isoform expressed in this tissue [24, 25]. Electrophysiological studies have found the threshold potential for these currents to be around -40mV and the peak activation between +10 and +20 mV. Although no precise value for resting ASM membrane potential has been described, several studies have reported values ranging between -60 and -30 mV [26, 27]. Thus, depolarization must occur before Cav1.2 channels could participate in agonist ASM induced contraction.

3.6. Non-selective cationic channels of the TRP family

It has been reported that agonists like histamine and carbachol provoke a small and inward cationic current through non-selective cationic channels (NSCC) in tracheal smooth muscle of different species [28, 29] .There is evidence that points to the transient receptor potential channels (TRP) as candidates for this conductance. TRP channels were first described in *Drosophila melanogaster* and then, homologues for these channels in at least 20 mammalian species were found to the point that almost all mammalian TRP channels are now known. Unlike most ion channels, TRP channels are identified by their homology rather than by ligand function or selectivity, because their functions are diverse and mostly unknown. The canonical (TRPC) subfamily of these channels comprises seven isoforms: TRPC1-7, and which have been detected in guinea pig and human ASM [30, 31]. It is generally accepted that TRPC channels are activated downstream by agonist-stimulated PIP_2 hydrolysis, but still their exact mode of activation and operation is unclear. Both store-dependent and independent mechanisms of activation have been proposed, in cases, even for the same channel in different preparations. All mammalian TRPC channels can be activated by GPCRs including muscarinic type 1 receptors (TRPC1, TRPC4, TRPC5 heteromers or TRPC4 and TRPC5 homomers); histaminergic type 1 receptors (TRPC3, TRPC6) and purinergic receptors (TRPC7) [32].

It is important to note at this point that Ca^{2+} release from the SR results in lowering of the Ca^{2+} content in intracellular stores to a certain degree. This lowering in turn activates a signaling mechanism that allows Ca^{2+} entry from the extracellular space. This mechanism was originally called [33] store-operated Ca^{2+} entry (SOCE). Since then, this phenomenon has received much attention, but still the complete mechanism and molecular identity of SOC channels remain unclear [34]. Experiments performed in our laboratory on guinea pig epithelial-free tracheal rings suggest that SOCs are non-selective cation channels that mainly permit Na^{+} entry causing depolarization (unpublished results). We proposed that such depolarization and increased levels in Na^{+} induce the NCX to allow Ca^{2+} influx which in turn activates Cl_{Ca}, and opening of the $Ca_v1.2$ channels [35]. The molecular identity of SOC channels points to the protein ORAI1 as well as to TRPC channels [36, 37]. ORAI1 is a four-transmembrane spanning protein that forms a pore with high selectivity for Ca^{2+}. More recent advances have been made regarding the signaling mechanisms that induce SOC current. The stromal interacting molecule (STIM) 1 has been found to sense Ca^{2+} concentration within the SR. STIM1 contains an EF-hand Ca^{2+} binding domain on the N-terminal ER luminal portion. When Ca^{2+} diminishes in the SR, STIM1 suffers a change in its distribution on the SR membrane and forms discrete clusters called puncta that interact with the plasma membrane. It is now clear that STIM1 couples to Orai1 to refill SR in some cell types. Nevertheless, interaction between STIM1 and other channels such as TRPC has also been observed and could account for SR refilling [38, 39].

3.7. The Na^{+}/Ca^{2+} exchanger in ASM

NCX is a membrane associated protein that catalyzes electrogenic exchange of 3 Na^{+} ions and 1 Ca^{2+} ion across the plasma membrane in a high capacity, and low Ca^{2+} affinity fashion.

This transporter can operate in either the Ca^{2+} -efflux or Ca^{2+} -influx mode depending on the electrochemical gradients of the substrate ions. Its physiological relevance became apparent as the role of extracellular Na^{+} in regulating contraction of smooth muscle was studied. Experiments performed in our laboratory pointed to the NCX as a crucial transporter involved in ASM contraction. In our hands, Na^{+} substitution by N-Methyl D-glucamine (NMDG) or inhibition of Na^{+}/K^{+} pump with Ouabain produced an increase in intracellular Ca^{2} in cultivated ASM cells [40]. Also, Na^{+} substitution with NMDG or inhibition of Na^{+}/K^{+} pump with Ouabain increase muscle tension [41], and histamine-pre-contracted guinea-pig tracheal rings show decreased relaxation rate when washed in a Na^{+} -free solution [42]. The role of NCX in ASM is still unclear, although much evidence has been mounting towards its importance. In further sections of this chapter, various aspects of this transporter will be reviewed in detail.

3.8. The PMCA and Ca^{2+} extrusion

Similar to SERCA function during SR Ca^{2+} uptake, the PMCA is constantly extruding Ca^{2+} outside the cell. The PMCA is a membrane protein that also belongs to the P-type pump family. It operates with high Ca^{2+} affinity and low transport capacity with a Kd ranging from 10-30 nM at rest to 0.2-0.5 μM at its optimal activation. It is thus considered to be the fine tuner of cytosolic Ca^{2+} concentration [43]. The PMCA is inhibited by Lanthanum ions and Vanadate, as many other transport systems, and for a long time there was no specific inhibitor available. Recently, some peptides such as caloxin have been synthesized which bind to the extracellular domains of the pump significantly reducing its activity.

4. Brief history of the NCX

The existence of the NCX exchanger was proposed in 1963 [44] as a result of studies in cardiac muscle contraction in low Na^{+} concentrated solutions. However the proposal was the result of reports from many other investigators who had previously documented the important role that Ca^{2+} and Na^{+} played in cardiac contraction. For example, it was reported [45, 46] that cardiac muscle contraction depended on extracellular Ca^{2+}. On the other hand it was [47] described that the force of cardiac contraction increased in the presence of low concentrated sodium solutions and years later other group [48] reported that the increase in the cardiac contraction force was associated with the quotient between the extracellular Ca^{2+} concentration and the extracellular Na^{+} concentration. Also, it was reported [44] that the decrease in the extracellular Na^{+} concentration is related to an increase in the Ca^{2+} content in cardiac muscle cells. The aforementioned papers and others which will not be mentioned here due to lack of space, lead to the conclusion [44] which says that Na^{+} and Ca^{2+} ions had to be transported by a transporter. During the same decade several papers were published, which allowed to suggest the presence of a Na^{+}-Ca^{2+} transport mechanism in other muscular tissues: skeletal muscle [49, 50], vascular smooth muscle [51] and intestinal muscle [52]. It was not until the end of the sixties when two groups of researchers working separately proposed the existence of a contratransport system coupled for Na^{+} and Ca^{2+} [53-58]. Similarly the existence of the Na^{+}-Ca^{2+} exchanger in the smooth airway muscle was suggested by [59, 60].

The strongest evidence of the existence of the Na^{+}-Ca^{2+} exchanger was given by the partial purification of the protein [61] and their posterior molecular cloning [62] from cardiac muscle. Other exchangers have been completely or partially cloned in other tissues such as photoreceptors [63], airway smooth muscle [64, 65], brain [66], kidney [67], etc.

We should point out that thanks to the discovery of the Na^{+}-Ca^{2+} exchanger it was possible to give a rational explanation of the inotropic effect of cardiac glycosides and, also, lead the group of M.P. Blaustein and J. M. Hamlyn to the discovery of the endogenous ouabain [68] a compound that is indistinguishable from plant ouabain, a Na^{+}/K^{+} ATPase specific blocker.

5. Molecular aspects of the NCX

The mammalian Na^{+}-Ca^{2+} exchanger belongs to a family of at least 3 genes NCX1, NCX2 and NCX3 which share a high degree of homology at the DNA and protein sequence level. NCX1 is the best characterized and is expressed in most tissues but mainly in heart brain and kidney, whereas NCX2 is expressed mainly in brain and NCX3 in brain and skeletal muscle [62, 69, 70].

An additional member of the mammalian NCX gene family recently identified is the mitochondrial Na^{+}- Ca^{2+} exchanger (NCLX). This gene is expressed in various tissues, such as pancreas, skeletal muscle and stomach smooth muscle and encodes a 70 KDa protein distinct from the other members of the family. This molecule is localized to the inner mitochondrial membrane and mediates the mitochondrial Ca^{2+} efflux in exchange for Na^{+} or Li^{+}, contributing apparently to intracellular Ca^{2+} homeostasis [71].

NCX 1 is the best characterized at the molecular level and its gene consists of a coding region of 12 exons that encode a protein of 938 amino-acids, and a large upstream regulatory region of more than 2000 bp that contains binding sites for several transcription factors such as GATA 4 SRF, NF-Y, CREB, C/EBP and AP1, among others [72, 73]. The H1 promoter regulates the expression in the heart, K1 in the kidney and Br1 in the brain. The use of each of these promoters in a tissue specific manner produces transcripts with different length of exon 1 and might enable the response to different stimuli. Exon 1 is part of the 5′-untranslated regions (5′-UTR) and each of these alternate exons is spliced to the common coding exon 2. Although this process does not change the coding sequence, it changes the length of the 5′-UTR which might be important for a posttranscriptional regulation.

In cardiac hypertrophy NCX expression is induced by α-adrenergic stimulation mediated in part by p38 MAPK activation and this is dependent on the presence of the proximal CArG promoter. Moreover it seems possible that the activation of the ERK kinase induced by hypertrophic stimuli plays a role in the transcriptional up-regulation of cardiac NCX. Within this region there are at least 3 alternate promoters that confer tissue specificity for NCX expression [73-75].

The major isoform of NCX1 encodes a protein of 120 KDa and NCX2 and NC3 encode proteins of approximately 100 kDa, respectively. The actual topological model suggests five transmembrane helices followed by a large intracellular loop of about 550 amino-acids, flanked by 2 α-repeats, and then the last four transmembrane segments. Spanning the large intracellular loop of NCX1 there is an alternatively spliced region that encompasses 6 exons (A, B, C, D, E, and F), which are expressed in a relatively tissue specific manner. Exons A and B are mutually exclusive and the others are combined with either of these two to produce at least 17 spliced isoforms of the exchanger [69-71]. Exon A appears in excitable tissues (heart, brain, skeletal muscle) and exon B mainly in no-excitable tissues (76-78). The longest spliced isoform is expressed in heart with exon A, C, D, E, F (NCX1.1) and the shortest in brain with the B, D (NCX1.3) isoform. Adjacent to this region there are 2 Ca^{2+} binding sites, CBD1 and CBD2, and close to the latter there is a alternatively spliced variable region. The interaction between CBD2 and the variable region seems to influence the sensitivity of the NCX isoforms to the regulation by intracellular Na^{+} and Ca^{2+} (79).

Other domains in the intracellular loop include an XIP site (eXchanger Inhibitory Peptide), an α-catenin homology region and a putative binding site, within the alternatively spliced region, for the ganglioside GM1. This interaction seems to be specific for exon B expressing isoforms and allow the localization of NCX1 to the nuclear envelope, where might influence not only nuclear Ca^{2+} but endoplasmic reticulum lumen stores as well, through its vicinity with the nuclear envelope. This interaction has been observed in neuronal and non-neuronal cells, and it has been suggested that the presence of 4 arginine residues in exon B, instead of 1 in exon A could favor a major interaction of the negatively charged ganglioside with these NCX1 isoforms (80).

Apparently NCX may be phosphorylated by protein kinase A (PKA) and protein kinase C (PKC) but it is still unclear whether these posttranslational modifications confer physiological effects directly or indirectly through the interaction with other proteins. It has been reported evidence that the large intracellular loop forms a complex with the PKC and PKA kinases subunits, PP1 and 2 phosphatases, and the PKA-anchoring protein AKAP and although there are compelling evidence for in vitro NCX phosphorylation by PKA and PKC, debate about the functional significance of these findings, still remains. It seems that the intracellular loop is necessary for agonist stimulation of NCX activity, but not necessarily the direct phosphorylation (81-83).

Airway smooth muscle cells and tissue express mainly NCX1, and absence of expression of the NCX2 and NCX3 isoforms. The first molecular evidence of NCX1 expression in ASM was realized in human trachea smooth muscle tissue where the alternatively spliced isoform revealed through reverse transcription coupled polymerase chain reaction (RT-PCR) method was the alternatively spliced isoform NCX1.3 (64).

This same isoform was later found in guinea pig tracheal tissue, showing a high grade homology in the alternatively spliced region among both species, with only minor aminoacid conservative changes. This isoform is predominantly expressed in kidney and contains a 102 amino-acids B exon linked to an 8 aminoacids D exon (84, 65).

At the protein level, NCX1 expression was demonstrated in bovine tracheal smooth muscle, where apparently a 120 kDa and a 110 kDa proteins corresponding to NCX1.1 and NCX1.3 isoforms, respectively, were identified by Western blotting [85]. Functional and comparative studies of the major NCX1.1 and the NCX1.3 isoforms have shown aminoacid differences within these variable exons that influence the inhibitory sensitivity de NCX to intracellular Na^+ (77-79).

Recently, an advanced molecular approach based on protein expression knocking down at the messenger RNA (mRNA) level by interference with small RNA molecules (siRNA), has been successfully applied to human airway smooth muscles allowing a better correlation of expression level with function in this tissue. Interestingly, these studies show that histamine and cytokines, like TNFα and IL-13, are able to induce the expression of NCX. When these cells are transfected with siRNA specific for NCX, the protein levels of the exchanger are decreased, as well as the Ca^{2+} influx elicited by these stimuli [86]. Moreover, in one of these studies it has been shown that cytokine induction of NCX1 is at the transcription level, mediated apparently by a mitogen activated protein kinase (MAPK) and NFκB pathways [87].

6. Functional aspects of the NCX

In this section, the basic aspects of NCX function will be reviewed considering the molecular mechanics of ion transport, activation of ion transport, interaction between NCX and ionic channels and pharmacology of the NCX.

6.1. Mechanism of Na^+/Ca^{2+} transport

The mechanism of transport of NCX1 has been widely studied and reveals a consecutive mechanism in which only 1 substrate ion is translocated at a time. Interaction of NCX1 with Na^+ or Ca^{2+} is asymmetric since the apparent affinity for intracellular Ca^{2+} is several hundred times higher than that for extracellular Ca^{2+}, although affinities for Na^+ differ little. Besides being transported substrates, Ca^{2+} and Na^+ regulate the NCX1 activity. In both modes of operation, the NCX1 is activated only when regulatory intracellular Ca^{2+} binds to a high-affinity site showing $K_{1/2}$ values of 0.1 to 0.4 μM [88]. In contrast, intracellular Na^+ exerts an inhibitory process upon NCX that occurs when the transport sites in NCX1 are fully loaded with Na^+ from the cytoplasmic side. This inactivation process is influenced by a variety of factors: it is enhanced at low pH but attenuated by intracellular Ca^{2+}, millimolar ATP or PIP_2. The steady-state activity of the NCX1 also exhibits intracellular pH dependence. At pH 6, activity is almost null; whereas at pH 9, activity is maximal [89]. Also, NCX activity shows voltage dependence, attributed mostly to voltage dependence on behalf of the Na^+ translocation step, or Na^+ binding to the NCX, which is rate limiting in overall reaction [90].

A very interesting effect occurs when alkali metal ions such as Na^+, K^+ or Li^+ are present on the extracellular side: all of them increase NCX activity 2 to 3 times with low affinity. Apart from Na^+, these cations bind to sites which are different from the transport sites and are not transported by the NCX [91]. Intracellular metal cations also stimulate the NCX, but apparently they need to be present as well in the internal side of NCX to show such effect. Gadsby et al. [92] found a striking difference between the outward NCX current-voltage relationships obtained in isolated guinea-pig myocytes when extracellular Na^+ was completely replaced with

Li^+ as compared to replacing with $NMDG^+$. This increase in outward current observed when Li^+ replaced Na^+ suggested that the voltage sensitivity and the magnitude of Na^+-Ca^{2+} exchange depend on the nature of the extracellular monovalent cation present. Our group previously observed that Na^+ substitution with Li^+ in force experiments performed on guinea pig tracheal rings produced a small reproducible increase in tension (Figure 1. upper trace). Once histamine was added, a further sustained contraction was observed and the peak tension, measured from the previous basal level, showed no significant decrease as compared to control. We suggested that histamine stimulation produces Li^+ influx through TRPC, membrane depolarization and activation of Cav1.2 channels. This depolarization is apparently enough to completely explain contraction, as observed when verapamil was added causing almost complete relaxation [35]. It is worth noting that Li^+ is not transported through the NCX in either direction [93] and therefore, NCX function under this condition is expected to be null.

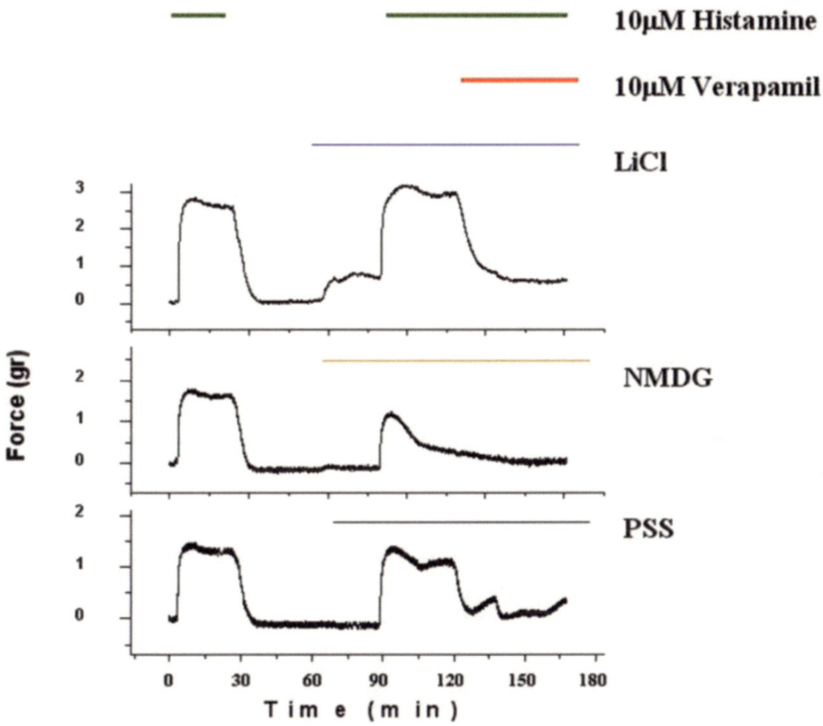

Figure 1. Representative traces of isometric force measurements of histamine-stimulated guinea-pig tracheal rings in Na^+-free with LiCl (upper trace), Na^+ - free with $NMDG^+$ (middle trace) and PSS (lower trace). PSS = Saline Solution. Force measurements were considered 5 min after histamine stimulation for comparison between contractions.

6.2. Functional relationship between NCX and ion channels.

As mentioned before, the NCX is a transporter whose activity and mode of operation can be finely modulated by the electrochemical gradient for Na^+ and Ca^{2+}. It has also been suggested that its localization and physical association with ion channels and cell organelles (as the SR) might determine its modulation [94]. Activation of NCX in the Ca^{2+} influx mode after agonist stimulation has been observed by different groups. Rosker et al. reported that in HEK 293 cells over-expressing TRPC3 channels, stimulation with carbachol was associated with an increase of intracellular Ca^{2+} concentration, which depended on extracellular Na^{+} since its substitution or the NCX inhibition with KB-R7943 reduced such effect. In the same cell line they also reported that NCX and TRPC3 are physically associated after cellular fractionation in low-density sucrose gradients and co-immunoprecipitation. The same group using glutathione S-transferase pull-down technique, which revealed that NCX interacts with the carboxy-terminal of TRPC3, confirmed these data. They also showed by co-immunoprecipitation experiments that NCX and TRPC3 are physically associated [95].

Later, this group tested a similar hypothesis on rat cardiomyocytes, and found that inhibition of the NCX in reverse mode by KB-R7943 also diminished the Ca^{2+} entry associated with agonist stimulation [96]. Other groups have also observed this functional association using models that better resemble in vivo conditions. It was reported that in rat aortic smooth muscle cells, the NCX inhibitor KB-R7943 as well as the TRPC inhibitor SKF-96365 abolished Ca^{2+} influx after ATP stimulation. They also observed a similar effect when cells were transfected with a dominant negative transcript for TRPC6 [97].

Hirota et al. [98] have also suggested a functional association between the NCX and TRPC activated by store depletion in dog ASM. In that work, several agonists were tested on tracheal rings together with NCX inhibitor KB-R7943 and sensitivity to this drug was observed. Also, contraction depended on extracellular Na^{+}, corresponding to our own observations. Dai et al. [99] observed in porcine tracheal smooth muscle bundle that Ca^{2+} waves typically obtained by acetylcholine stimulation were sensitive to TRPC blocker SKF-96365 as well as to KB-R7943. They also observed this sensitivity at the level of muscle contraction, where these drugs, together with nifedipine completely relaxed the muscle.

In our own experience, this relationship between the NCX and TRPC channels became evident in guinea pig ASM. We performed several force experiments on guinea pig tracheal rings and observed that histamine contraction depends on extracellular Na^{+} and is sensitive to the non-specific NCX blocker KBR-7943 as well as to TRPC non-specific blockers SKF-96365 and 2-APB. These findings have led us to propose that histamine causes Ca^{2+} entry mediated by the NCX operating in its reverse mode secondary to a cationic influx (primarily Na^{+}) through TRPC. We then proposed that histamine stimulates Na^{+} influx through TRPC, cell depolarization and the increase in subplasmalemmal Na^{+} concentration. These conditions might favor first the reverse mode of the NCX and later the activation of Cav1.2 channels which have been characterized in this tissue previously [24]. On the other hand, as Na^{+} is replaced by $NMDG^{+}$, neither depolarization nor Na^{+} influx through TRPC could be possible. Thus Ca^{2+} release from SR, Ca^{+2} entries through TRPC and Ca^{+2} entries through reverse mode NCX would provide for the small histamine-induced contraction observed [35]. Preincubation with KB-R7943 allowed us to explore the role of NCX during the beginning of histamine contraction as well as throughout tonic force development. In the presence of KB-R7943 a significant diminishment in maximal force developed as observed, whereas pre-incubation with 70 nM nifedipine had no effect. We thus suggest that NCX is active in its reverse mode at an early stage of contraction after Ca^{2+} release from the SR, while Cav1.2 channels participate somewhat later during stimulation. In addition, this contraction is also similar to the one observed when Na^+ is substituted for NMDG. Our interpretation for this is that when NCX has been inhibited, not only Ca^{2+} entry is blocked but also membrane depolarization does not reach the threshold for Cav1.2 channel activation.

Ca^{2+} imaging experiments performed on freshly isolated tracheal smooth muscle cells pointed to the same direction as the tension experiments. We first obtained isolated muscle cells able to contract and to show an increase in FURA-2 fluorescence ratio after histamine stimulation. Stimulation resulted in a peak in fluorescence ratio followed by a plateau of fluorescence just above the basal value (15% of the peak in Ca^{2+} rise), which persisted until the agonist was washed out. Addition of 100 μM KB-R7943 significantly decreased the change in peak fluorescence ratio in a second stimulation with histamine as well as during the sustained phase (Figure 2.). External Na^+ substitution by $NMDG^+$ showed a significant decrease in fluorescence ratio in the sustained phase suggesting that the NCX is operating in the Ca^{2+} influx mode and that the KB-R7943-insensitive component is due to Ca^{2+} release from the stores and perhaps Ca^{2+} entry through TRPC. We also observed that application of SKF-93635 and thus inhibition of TRPC significantly lowers the peak fluorescence ratio and completely abolishes fluorescence in the sustained phase of the curve. This is in agreement with results reported by Dai et al. [99] where SKF-96365 inhibits contraction and Ca^{2+} waves in porcine tracheal smooth muscle cells.

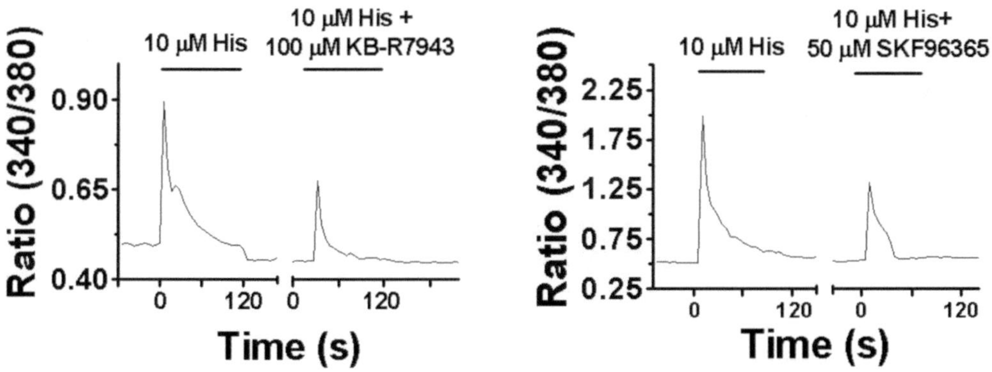

A B

Figure 2. Representative traces from fluorescence ratio changes observed during stimulation of isolated smooth muscle cells with 10 μM histamine. Cells were stimulated twice with histamine and given a 20 min recovery time between stimulations. Fluorescence ratio was measured as indicated by arrows. During the second stimulation histamine was added together with 100 μM KB-R7943 in PS, A) or 50 μM SKF-96365 in PS, B).

These results led us to propose the following model: Activation of histamine receptors triggers a signaling cascade leading to formation of IP_3 and DAG which causes Ca^{2+} release from SR generating initial contraction. Emptying of the SR by such Ca^{2+} release, activates TRPC channel opening leading to Na^+ influx. This Na^+ current in turn causes membrane depolarization as well as a local increase in $[Na^+]_i$ in the vicinity of NCX which promote its reverse mode of operation. Ca^{2+} entry mediated by the NCX may add to Ca^{2+} released from the SR and activate Ca^{2+}-activated Cl^- channels [20, 100]. This in turn should cause enough depolarization to activate a larger population of Cav1.2 channels and, together with sensitization events, give rise to a characteristic histamine contraction.

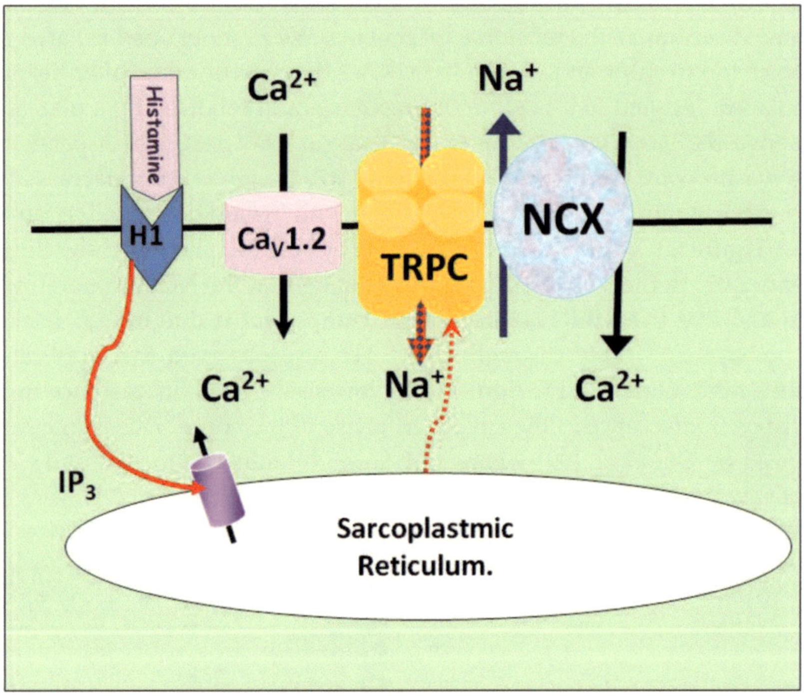

Figure 3. Proposed model that explains the functional interaction between NCX and TRPC during histamine stimulation. Histamine acts on its specific H1 receptor and initiates a signaling cascade leading to formation of IP_3. IP_3 produces Ca^{2+} release from SR and this in turn causes TRPC opening. The Na^+ current entering through these channels depolarizes the membrane and locally increases $[Na^+]_i$ in the vicinity of NCX. These conditions would then promote NCX operation in reverse mode as well as Cav1.2 channel activation. Ca^{2+} entry through NCX might activate Ca^{2+}-dependent Cl^- channels and cause even greater depolarization resulting in an activation of a greater Cav1.2 channel population.

Recently, the functional interaction between NCX and the SOCC channel activator STIM1 was observed in human bronchial smooth muscle cells [86]. In this work, electrophysiological recordings of isolated cells revealed an outwardly rectifying current characteristic of the NCX in reverse mode which was completely abolished by KB.R7943. Interestingly, the current was activated by histamine addition and inhibited completely by STIM1 knockdown. STIM is proposed as a sensor of SR emptying which interacts with membrane channels, it is possible that the TRPC channels are activated by STIM1 causing Na^+ influx and NCX activation as was previously proposed. This evidence shows again the tight relationship between the NCX and other channels which are activated in response to agonist stimulation or SR emptying.

6.3. Role of the NCX during ASM relaxation

The NCX working in reverse mode promotes ASM contraction according to the evidence just described. Nevertheless, its mode of operation once agonist is washed seems to be different. In contrast to its alleged role in heart as a fundamental Ca^{2+} extrusion system [101], we observed that inhibition of the NCX with KB-R7943 during relaxation of guinea pig tracheal rings does not alter the process at all [65]. On the other hand, Na+ substitution by NMDG during washing does retard the relaxation of the rings, indicating that the NCX is turned to the Ca2+ influx mode. This delay in the relaxation process in Na+-free washing was abolished by KB-R943, suggesting the participation of the NCX under these conditions. In accordance with our results, it has also been shown that the NCX plays at most a minor role as a Ca^{2+} extrusion system during canine ASM relaxation [102]. This is in agreement with our results, suggesting that the NCX found in ASM is active during contraction in the Ca^{2+} entry mode, but not during relaxation.

6.4. Pharmacology of the NCX

Specific inhibitors of the NCX are not yet available for research or therapeutic use. Many divalent and trivalent cations such as La^{+}, Ni^{2+} and Cd^{2+}, as well as amiloride derivatives or the substituted pyrrolidineethanamine have long been used, although their lack of specificity remain a great handicap for their use. The isothiourea derivative KB-R7943 has been used as a potent inhibitor of the NCX [103]. It is 3-fold less potent on NCX1 and NCX2 than on NCX3 and has a preferential effect on the Ca^{2+} influx mode of NCX1 [104]. This drug seems to act on specific residues of the NCX1: Val 820, Gln 826 and Gly 833 which lie in a reentrant membrane loop [105]. An important handicap for the use of KB-R7943 is the lack of specificity for NCX, since it has been reported to block ion channels [106], neuronal nicotinic acetylcholine receptor [107], N-methyl-D-aspartate receptor [108] and norepinephrine transporter at relatively low doses. Another more potent and specific inhibitor of NCX is SEA0400. This drug has been reported to be 30 times more powerful than KB-R7943 and to block predominantly NCX1 in CCL39 cells [109,110]. Analysis performed with NCX1 and NCX3 chimeras showed that multiple amino acids are involved in SEA0400 sensitivity encompassing residues 73-108 and 193-230. Regarding its specificity, SEA0400 at 1 μM does not affect Cav1.2 channels, Cav2.2 channels or Na^{+} channels. The affinity reported for NCX in cultured neurons, astrocytes and microglia has IC_{50} values from 5 to 33 nM [111]. An important drawback for the use of SEA0400 is that it is not yet commercially available, limiting its use to the general research public. Two other NCX blockers are SN-6 and YM-244769 and are under investigation. The blockers mentioned before have the characteristic that they are poorly active when the exchanger is working in the forward mode under normal conditions (low intracellular Na^{+}) but very active when the exchanger is working in the reverse mode under pathological conditions [112].

7. Clinical implications

As it has been mentioned in previous sections, the NCX plays a critical role in the regulation of intracellular Ca^{2+} concentration. The direction of Na^{+} in exchange for Ca^{2+} depends on the membrane potential and the Na^+ and Ca^{2+} transmembrane ionic gradient. Out of the two types of exchange, the reverse mode (Ca^{2+} influx mode) has received more attention due to the fact that its function determines extracellular Ca^{2+} influx into the cell. When this is given in normal conditions is used for either the contraction process or to refill the SR. However, there is experimental evidence that suggests that in certain pathological conditions such as essential hypertension [113], ischemia-reperfusion injury [113] and certain types of cardiac arrhythmia [114] the NCX transports a bigger amount of Ca^{2+} than necessary.

The usage of NCX blockers such as KB-R7943, SEA0400, and SN-6 [115] have helped to understand the physiological role that the exchanger plays in different tissues. Moreover, they are now seen as potential therapeutic drugs. Indeed, the experimental evidence accumulated over the past few years has allowed to establish that, at least in experimental models, the blockers of the reverse mode of the NCX are useful to diminish the high blood pressure, to abolish cardiac arrhythmias or to reduce the tissue damage after ischemia-reperfusion damage models.

However, as far as we know no reports regarding the use of the NCX blockers have been published in models of airway disease such as Asthma or Chronic Obstructive Pulmonary Disease. It is well known that Asthma is an inflammatory chronic disease characterized by reversible airflow obstruction and nonspecific airway hyperresponsiveness. In spite of the drugs for Asthma treatments available such systemic or local steroids, leukotrienes inhibitors and/or smooth muscle airway relaxants (B_2 adrenergic agonist) no absolute control of the disease is obtained. Thus, and because the prevalence of Asthma worldwide has increased in the past few years it challenges the discovery of new and better pharmacological treatments for it. Throughout the past years it has been suggested that the use of NCX blockers could be of some help in the therapeutic management of this disease but the experimental information is scarce. Our group reported [35] that the tonic phase of the contraction induced by histamine is partially blocked by KB-R7943 and its effect is not due to the Ca^{2+} voltage dependent channel blockage since these had been previously inhibited by Nifedipine. Other groups have reported similar results to ours [116-118]. Therefore, it is expected that in the next years different research groups will proceed to investigate if NCX blockers have any kind of therapeutic use in animal Asthma models.

Author details

Ricardo Espinosa-Tanguma*
Departamento de Fisiología y Biofísica, Facultad de Medicina, UASLP, San Luis Potosí, México

Paola Algara-Suárez
Departamento de Núcleo Básico, Facultad de Enfermería, UASLP, San Luis Potosí, México

Rebeca Mejía-Elizondo and Víctor Saavedra-Alanís
Departamento de Bioquímica, Facultad de Medicina, UASLP, San Luis Potosí, México

* Corresponding Author

Acknowledgement

This work was supported by CONACyT grant No. 62220 and CIHR grant MOP10019.

8. References

[1] Janssen LJ, Killian K. (2006) Airway smooth muscle as a target of asthma therapy: history and new directions. Respir. Res. 7:123.

[2] Barnes PJ. (1998) Pharmacology of airway smooth muscle. Am. J. Respir. Crit. Care Med. 158(5):S123-132.

[3] Hill SJ, Ganellin CR, Timmerman H, Schwartz JC, Shankley NP, Young JM, Schunack W, Levi R, Haas HL. (1997) International Union of Pharmacology. XIII. Classification of histamine receptors. Pharmacol. Rev. 49(3):253-278.

[4] Carbajal V, Vargas MH, Flores-Soto E, Martínez-Cordero E, Bazán-Perkins B, Montaño LM. (2005) LTD4 induces hyperresponsiveness to histamine in bovine airway smooth muscle: role of SR-ATPase Ca^{2+} pump and tyrosine kinase. Am. J. Physiol. Lung Cell. Mol. Physiol. 288: L84-L92.

[5] Pfitzer G. (2001) Invited review: regulation of myosin phosphorylation in smooth muscle. J. Appl. Physiol. 91(1):497-503.

[6] Somlyo AP, Somlyo AV. (2003) Ca^{2+} sensitivity of smooth muscle and nonmuscle myosin II: modulated by G proteins, kinases, and myosin phosphatase. Physiol. Rev. 83(4):1325-1358.

[7] Hirano K. (2007) Current topics in the regulatory mechanism underlying the Ca^{2+} sensitization of the contractile apparatus in vascular smooth muscle. J. Pharmacol. Sci. 104(2):109-11.

[8] Gusev NB. (2001) Some properties of caldesmon and calponin and the participation of these proteins in regulation of smooth muscle contraction and cytoskeleton formation. Biochemistry (Mosc). 66(10):1112-1121.

[9] Morgan KG, Gangopadhyay SS. (2001) Invited review: cross-bridge regulation by thin filament-associated proteins. J. Appl Physiol. 91(2):953-962.

[10] Girodet PO, Ozier A, Bara I, Tunon de Lara JM, Marthan R, Berger P. (2011) Airway remodeling in asthma: new mechanisms and potential for pharmacological intervention. Pharmacol. Ther. 130(3):325-37

[11] Halwani R, Al-Muhsen S, Hamid Q. (2010) Airway remodeling in asthma. Curr. Opin. Pharmacol. 10(3):236-45

[12] Murray RK and Kotlikoff MI. (1993) Receptor-activated calcium influx in human airway smooth muscle cells. J. Physiol. (London) 435:123-144.

[13] Bergner A, Sanderson MJ. (2002) Acetylcholine-induced calcium signaling and contraction of airway smooth muscle cells in lung slices. J. Gen. Physiol. 119:187-198.

[14] Kannan MS, Prakash YS, Brenner T, Mickelson JR, Sieck GC. (1997) Role of ryanodine receptor channels in Ca^{2+} oscillations of porcine tracheal smooth muscle. Am. J. Physiol. 272: L659-L664.

[15] Patel S, Joseph SK, Thomas AP. (1999) Molecular properties of inositol 1, 4, 5-trisphosphate receptors. Cell Calcium. 25(3):247-264.

[16] Yoshida Y, Imai S. (1997) Structure and function of inositol 1,4,5-trisphosphate receptor. Jpn. J. Pharmacol. 74:125-137.

[17] Fill M, Copello JA. (2002) Ryanodine receptor calcium release channels. Physiol. Rev.82:893-922.

[18] Eggermont J. (2004) Calcium-activated chloride channels: (un)known, (un)loved? Proc. Am. Thorac. Soc. 1(1):22-27.

[19] Latorre R, Brauchi S. (2006) Large conductance Ca^{2+}-activated K^+ (BK) channel: activation by Ca^{2+} and voltage. Biol. Res. 39(3):385-401.

[20] Janssen LJ, Sims SM. (1993) Histamine activates Cl- and K^+ currents in guinea-pig tracheal myocytes: convergence with muscarinic signalling pathway. J. Physiol. 465:661-677.

[21] Wang YX, Kotlikoff MI. (2000) Signalling pathway for histamine activation of non-selective cation channels in equine tracheal myocytes. J. Physiol. (London) 523:131-138.

[22] Karaki H, Ozaki H, Hori M, Mitsui-Saito M, Amano K, Harada K, Miyamoto S, Nakazawa H, Won KJ, Sato K. (1997) Calcium movements, distribution, and functions in smooth muscle. Pharmacol. Rev. 49:157-230.

[23] Catterall WA, Perez-Reyes E, Snutch TP, Striessnig J. (2005) International Union of Pharmacology. XLVIII. Nomenclature and structure-function relationships of voltage-gated calcium channels. Pharmacol. Rev. 57:411-25.

[24] Worley JF, Kotlikoff MI. (1990) Dihydropyridine-sensitive single calcium channels in airway smooth muscle cells. Am. J. Physiol. 259:L468-L480.

[25] Du W, McMahon TJ, Zhang ZS, Stiber JA, Meissner G, Eu JP. (2006) Excitation-contraction coupling in airway smooth muscle. J. Biol. Chem. 281:30143-30151.

[26] Cloutier M, Campbell S, Basora N, Proteau S, Payet MD, Rousseau E. (2003) 20-HETE inotropic effects involve the activation of a nonselective cationic current in airway smooth muscle. Am. J. Physiol. Lung Cell Mol. Physiol. 285:L560-568.

[27] Sausbier M, Zhou XB, Beier C, Sausbier U, Wolpers D, Maget S, Martin C, Dietrich A, Ressmeyer AR, Renz H, Schlossmann J, Hofmann F, Neuhuber W, Gudermann T, Uhlig S, Korth M, Ruth P. (2007) Reduced rather than enhanced cholinergic airway constriction in mice with ablation of the large conductance Ca^{2+}-activated K^+ channel. FASEB J. 21:812-822.

[28] Beech DJ, Muraki K, Flemming A. (2004) Non-selective cationic channels of smooth muscle and the mammalian homologues of Drosophila TRP. J. Physiol (London) 559:685-706.

[29] Ito S, Kume H, Honjo H, Katoh H, Kodama I, Yamaki K, Hayashi H. (2001) Possible involvement of Rho kinase in Ca^{2+} sensitization and mobilization by MCh in tracheal smooth muscle. Am. J. Physiol. Lung Cell. Mol. Physiol. 280:L1218-24.

[30] Corteling R, Li S, Giddings J, Westwick J, Poll C, Hall I. (2004) Expression of transient receptor potential C6 and related transient receptor potential family members in human airway smooth muscle and lung tissue. Am. J. Respir. Cell Mol. Biol. 30:145-154.

[31] Ong HL, Brereton HM, Harland ML, Barritt GJ. (2003) Evidence for the expression of transient receptor potential proteins in guinea pig airway smooth muscle cells. Respirology. 8:23-32.
[32] Clapham DE. (2003) TRP channels as cellular sensors. Nature 426:517-524.
[33] Putney JW Jr.(1986) A model for receptor-regulated calcium entry. Cell Calcium 7:1–12.
[34] Parekh AB, Putney JW Jr. (2005) Store-operated calcium channels. Physiol. Rev. 85(2):757-810.
[35] Algara-Suárez P, Romero-Méndez C, Chrones T, Sánchez-Armass S, Meza U, Sims SM, Espinosa-Tanguma R. (2007) Functional coupling between the Na+/Ca2+ exchanger and nonselective cation channels during histamine stimulation in guinea pig tracheal smooth muscle. Am. J. Physiol. Lung Cell Mol. Physiol. 293(1):L191-8.
[36] Ambudkar IS, Bandyopadhyay BC, Liu X, Lockwich TP, Paria B, Ong HL. (2006) Functional organization of TRPC-Ca^{2+} channels and regulation of calcium microdomains. Cell Calcium.40:495-504.
[37] Hewavitharana T, Deng X, Soboloff J, Gill DL. (2007) Role of STIM and Orai proteins in the store-operated calcium signaling pathway. Cell Calcium. 42:173-182.
[38] Wang Y, Deng X, Hewavitharana T, Soboloff J, Gill DL. (2008) Stim, ORAI and TRPC channels in the control of calcium entry signals in smooth muscle. Clin. Exp. Pharmacol. Physiol. 35(9):1127-33.
[39] Wang Y, Deng X, Gill DL. (2010) Calcium signaling by STIM and Orai: intimate coupling details revealed. Sci. Signal. 3(148):pe42.
[40] Espinosa-Tanguma R, Guevara-Lopez C, Ortega F, Ramírez-Zacarias JL, Hernández-Salinas AE, Mandeville P, Sánchez-Armass S. (2003) Changes in cytosolic calcium in tracheal smooth muscle cells in culture: effects of extracellular sodium and Ouabain. J. Biochem. Physiol. 59(1):25-33.
[41] Espinosa-Tanguma R, Valle-Aguilera JR, Zarazúa-Garcia O, Navarro-Huerta MP, Pecina C, Sánchez-Armass S. (2004) Mechanism of ouabain-induced contractions in guinea-pig tracheal rings. Clin. Exp. Pharmacol. Physiol. 31: 710-715.
[42] Espinosa-Tanguma R, Espericueta-Monsivais VM, Herrera-Mendoza P. (2004) Physiological Role of the Sodium-Calcium Exchanger in Tracheal Smooth Muscle of Guinea-Pig. J. Mus. Res. Cell Mot. 25:620-621.
[43] Brinni M, Carafoli E. (2009) Calcium Pumps in Health and Disease. Physiol. Rev. 89:1341-1378.
[44] Ringer S. (1883) A further contribution regarding the influence of the different constituents of the blood on the contraction of the heart. J. Physiol. (Lond) 4:29-42.
[45] Ringer S. (1885). Further observations regarding antagonism between calcium salts and sodium, potassium and ammonium salts. J. Physiol. (Lond) 18:425-429.
[46] Daly ID, Clark AJ. (1921) The action of ions upon the frog's heart. J. Physiol. (Lond) 54:367-383.
[47] Willibrandt W, Koller H. (1948) Die Calcium wirkung am Frosch herzen als Funktion des Ionengleichgewichts swischen Zellmembran and Umgebun. Helv. Physiol. Pharmacol. Acta. 6:208-221.

[48] Niedergerke, R. (1963) Movements of Ca in beating ventricle of the frog heart. J. Physiol. (Lond) 167:551-580.

[49] Schaechtelin, G. (1961) Der Einflub von Calcium und Natrium auf die Kontracktur des M. rectus abdominis. Pflugers Arch. 273:164-181.

[50] Cosmos EE, Harris EJ. (1961) In vitro studies of the gain and exchange of calcium in frog skeletal muscle. J. Gen. Physiol. 44:1121-1130.

[51] Briggs AH, Melvin S. (1961) Ion movement in isolated rabbit aortic strips. A. J. Physiol 201:365-368.

[52] Goodford PJ. (1967) The calcium content of the smooth muscle of the guinea-pig taenia coli. J. Physiol. (Lond) 192:145-157.

[53] Reuter H, Seitz N, (1968) The dependence of calcium efflux from cardiac muscle on temperature and external ion composition. J. Physiol. (Lond) 195: 451-470.

[54] Glitsch HG, Reuter H, Scholz H. (1970) The effect of the internal sodium concentration on calcium fluxes in isolated guinea-pig auricles. J. Physiol. (Lond) 209:25-43.

[55] Baker PF, Blaustein MP. (1968) Sodium dependent uptake of calcium by crab nerve. Biochim. Biophys. Acta 150:167-170.

[56] Baker PF, Balustein MP, Hodgkin AL, Steinhardt RA. (1969) The influence of calcium ions on sodium efflux in squid axons. J. Physiol. (Lond) 200:431-458, 1969.

[57] Blaustein MP, Hodgkin AL. (1969) The effect of cyanide on the efflux of calcium from squid axons. J. Physiol. (Lond) 200:497-527.

[58] Martin DL, De Luca HF. (1969) Influence of sodium on calcium transport by rat small intestine. Am. J. Physiol. 216:1351-1359.

[59] Kawanishi M, Baba K, Tomita T. (1984) Effects of Na removal and readmission on the mechanical response in the guinea-pig tracheal smooth muscle. Jpn. J. Physiol. 34:127-139.

[60] Chideckel EW, Frost JL, Mike P, Fedan JS. (1987) The effect of ouabain on tension in isolated respiratory tract smooth muscle of humans and other species. Br. J. Pharmacol. 92:609-614.

[61] Philipson KD, Longoni S, Ward R. (1988) Purification of the cardiac Na^+ - Ca^{2+} exchange protein. Biophys. Acta 945:298-306.

[62] Nicoll DA, Longoni S, Philipson KD (1990) Molecular cloning and functional expression of the cardiac sarcolemmal Na+-Ca2+ exchanger. Science 250:562–565;

[63] Reiländer H, Achilles A, Friedel U, Maul G, Lottspeich F, Cook NJ. (1992) Primary structure and functional expression of the Na/Ca,K-exchanger from bovine rod photorecepts. EMBO J. 11:1689-1695.

[64] Pitt A, Knox AJ. (1996) Molecular characterization of the human airway smooth muscle Na^+/Ca^{2+} exchanger, Am. J. Respir. Cell. Mol. Biol. 15:726-730.

[65] Algara-Suárez P, Mejía-Elizondo R, Sims SM, Saavedra-Alanis VM, Espinosa-Tanguma R. (2010) The 1.3 isoform of Na+-Ca 2+ exchanger expressed in guinea pig tracheal smooth muscle is less sensitive to KB-R7943. J. Physiol. Biochem. 66(2):117-25.

[66] Tsoi M, Rhee KH, Bungard D, Li XF, Lee SL, Auer RN, Lytton J. (1998) Molecular cloning of a novel potassium dependent sodium,-calcium exchange from rat brain. J Biol. Chem. 273:4155-4162.

[67] Reill RF, Shugrue CA. (1992) cDNA cloning of a renal Na^{+}-Ca^{2+} exchanger. Am. J. Physiol. 262(6):1105-1109.

[68] Hamlyn JM, Blaustein MP, Bova S, Ducharme DW, Harris DW, Mandel F. (1991) Identification and characterization of a ouabain-like compound from human plasma. Proc. Natl. Acad. Sci. USA. 88:6259-6263.

[69] Philipson KD, Nicoll DA. (2000) SODIUM-CALCIUM EXCHANGE: A Molecular Perspective. Annu. Rev. Physiol. 62:111–133.

[70] Lytton J. (2007) Na+/Ca2+ exchangers: three mammalian gene families control Ca2+ transport. Biochem. J. 406:365–382

[71] Palty R, Silverman WF, Hershfinkel M, Caporale T, Sensi SL, Parnis J, Nolte C, Fishman D, Shoshan-Barmatz V, Herrmann S. (2010) NCLX is an essential component of mitochondrial Na+/Ca2+ exchange. Proc. Natl. Acad. Sci. USA. 107:436–441.

[72] Barnes KV, Cheng G, Dawson MM, Menick DR. (1997) Cloning of cardiac, kidney, and brain promoters of the feline ncx1 gene. J. Biol. Chem. 272:11510–11517.

[73] Nicholas SB, Yang W, Lee SL, Zhu H, Philipson KD, Lytton J.(1998) Alternative promoters and cardiac muscle cell-specific expression of the Na+/Ca2+ exchanger gene. Am. J.Physiol. 274 (Heart Circ. Physiol. 43):H217–H233.

[74] Xu L, Renaud L, Müller JG, Baicu CF, Bonnema DD, Zhou H, Kappler CS, Kubalak SW, Zile MR, Conway SJ, Menick DR. (2006) Regulation of Ncx1 Expression. J. Biol. Chem. 281(45):34430–34440.

[75] Donald R, Menick DR, Renaud L, Buchholza A, Müllera JG, Zhouc H, Kapplera CS, Kubalak SW, Conway SJ, Xu L. (2007) Regulation of Ncx1 Gene Expression in the Normal and Hypertrophic Heart. Ann. N. Y. Acad. Sci. 1099:195–203

[76] Quednau BD, Nicoll DA, Philipson KD. (1997) Tissue specificity and alternative splicing of theNa+/Ca2+ exchanger isoforms NCXl, NCX2, and NCX3 in rat. Am. J. Physiol. 272: C1250-C126.

[77] Dunn J, Elias CL, Le HD, Omelchenko A, Hryshko LV, Lytton J. (2002) The Molecular Determinants of Ionic Regulatory Differences between Brain and Kidney Na+/Ca2+ Exchanger (NCX1) Isoforms. J. Biol. Chem. 277(37):33957-33962.

[78] Hurtado C, Prociuk M, Maddaford TG, Dibrov E, Mesaeli N, Hryshko LV, Pierce GN. (2006) Cells expressing unique Na+/Ca2+ exchange (NCX1) splice variants exhibit different susceptibilities to Ca2+ overload. Am. J. Physiol. Heart. Circ. Physiol. 290:H2155-H2162.

[79] Hilge M, Aelen J, Vuister GW. (2006) Ca2+ Regulation in the Na+/Ca2+ Exchanger Involves Two Markedly Different Ca2+ Sensors. Molecular Cell. 22:15-25.

[80] Ledeen R, Wu G. (2007) GM1 in the nuclear envelope regulates nuclear calcium through association with a nuclear sodium-calcium exchanger. J. Neurochemistry.103:126-134

[81] Schulze DH, Muqhal M, Lederer WJ, Ruknudin AM. (2003) Sodium/calcium exchanger (NCX1) macromolecular complex, J. Biol. Chem. 278:28849-28855

[82] Iwamoto T, Pan Y, Nakamura TY, Wakabayashi S, Shigekawa M. (1998) Protein kinase C-dependent regulation of Na+/Ca2+ exchanger isoforms NCX1 and NCX3 does not require their direct phosphorylation. Biochemistry 37:17230-17238.

[83] Zhanga YH, Hancox JC. (2009) Regulation of cardiac Na+-Ca2+ exchanger activity by protein kinase phosphorylation-still a paradox? Cell Calcium. 45: 1-10

[84] Mejía-Elizondo R, Espinosa-Tanguma R, Saavedra-Alanis VM. (2002) Molecular identification of the NCX isoform expressed in tracheal smooth muscle of guinea pig. Ann. N.Y. Acad.Sci. 976:73-76.

[85] Hirota S, Janssen LJ. (2007) Store-refilling involves both L-type calcium channels and reverse-mode sodium-calcium exchange in airway smooth muscle. Eur. Respir. J. 30: 269–278

[86] Liu B, Peel SE, Fox J, Hall IP (2010) Reverse mode Na+/Ca2+ exchange mediated by STIM1 contributes to Ca2+ influx in airway smooth muscle following agonist stimulation. Respir. Res. 11:168.

[87] Sathish V, Delmotte PF, Thompson MA, Pabelick CM, Sieck GC, Prakash YS. (2011) Sodium-Calcium Exchange in Intracellular Calcium Handling of Human Airway Smooth Muscle. PLoS ONE 6(8):e23662.

[88] Matsuoka S, Hilgemann DW. (1994) Inactivation of outward $Na(^{+})$-$Ca2^{+}$ exchange current in guinea-pig ventricular myocytes. J. Physiol. 476:443-458.

[89] Doering AE, Lederer WJ. (1993) The mechanism by which cytoplasmic protons inhibit the sodium-calcium exchanger in guinea-pig heart cells. J. Physiol. 466:481-499.

[90] Matsuoka S, Hilgemann DW. (1992) Steady-state and dynamic properties of cardiac sodium-calcium exchange. Ion and voltage dependencies of the transport cycle. J. Gen. Physiol. 100:963-1001.

[91] Iwamoto T, Shigekawa M. (1998) Differential inhibition of Na^{+}/Ca^{2+} exchanger isoforms by divalent cations and isothiourea derivative. Am. J. Physiol. 275:C423-430.

[92] Gadsby DC, Noda M, Shepherd RN, Nakao M. (1991) Influence of external monovalent cations on Na-Ca exchange current-voltage relationships in cardiac myocytes. Ann. N.Y. Acad. Sci. 639:140-146.

[93] Blaustein M, Lederer W. (1999) Na^{+}/Ca^{2+} Exchange: Its Physiological Implications. Physiol. Rev. 79:763-854.

[94] Eder P, Poteser M, Romanin C, Groschner K. (2005) $Na(^{+})$ entry and modulation of $Na(^{+})/Ca(^{2+})$ exchange as a key mechanism of TRPC signaling. Pflugers Arch. 451:99-104.

[95] Rosker C, Graziani A, Lukas M, Eder P, Zhu M, Romanin C, Groschner A. (2004) Ca^{2+} signaling by TRPC3 involves Na^{+} entry and local coupling to the Na^{+}/Ca^{2+} exchanger. J. Biol. Chem. 279: 13696-13704.

[96] Eder P, Probst D, Rosker C, Poteser M, Wolinski H, Kohlwein SD, Romanin C, Groschner K. (2007) Phospholipase C-dependent control of cardiac calcium homeostasis involves a TRPC3-NCX1 signaling complex. Cardiovasc. Res. 73(1):111-119.

[97] Lemos VS, Poburko D, Liao CH, Cole WC, van Breemen C. (2007) Na^{+} entry via TRPC6 causes Ca^{2+} entry via NCX reversal in ATP stimulated smooth muscle cells. Biochem. Biophys. Res. Commun. 352(1):130-134.

[98] Hirota S, Pertens E, Janssen L. (2007) The reverse-mode of the Na^{+}/Ca^{2+} exchanger provides a source of calcium for store-refilling following agonist-induced calcium mobilization. Am. J. Physiol. Lung Cell Mol. Physiol. 292:L438-447.

[99] Dai JM, Kuo KH, Leo JM, van Breemen C, Lee CH. (2006) Mechanism of ACh-induced asynchronous calcium waves and tonic contraction in porcine tracheal muscle bundle. Am. J. Physiol. Lung Cell Mol. Physiol. 290(3):L459-469.

[100] Hirota S, Trimble N, Pertens E, Janssen L. (2006) Intracellular Cl- fluxes play a novel role in Ca^{2+} handling in airway smooth muscle. Am. J. Physiol. Lung Cell Mol. Physiol. 290:L1146-L1153.

[101] Hilgemann DW. (2004) New insights into the molecular and cellular workings of the cardiac Na+/Ca2+ exchanger. Am. J. Physiol. Cell Physiol. 287:C1167-1172

[102] Janssen LJ, Walters DK, Wattie J. (1997) Regulation of $(Ca^{2+})_i$ in canine airway smooth muscle by Ca^{2+}-ATPase and Na^+/Ca^{2+} exchange mechanisms. Am. J. Physiol. 273:L322-L330

[103] Iwamoto T, Watano T, Shigekawa M. (1996) A novel isothiourea derivative selectively inhibits the reverse mode of Na^+/Ca^{2+} exchange in cells expressing NCX1, J. Biol. Chem. 271:22391-22397.

[104] Iwamoto T. (2004) Forefront of Na^+/Ca^{2+} exchanger studies: molecular pharmacology of Na^+/Ca^{2+} exchange inhibitors. J. Pharmacol. Sci. 96:27-32.

[105] Iwamoto T, Kita S, Uehara A, Inoue Y, Taniguchi Y, Imanaga I, Shigekawa M. (2001) Structural domains influencing sensitivity to isothiourea derivative inhibitor KB-R7943 in cardiac $Na(^+)/Ca(^{2+})$ exchanger. Mol. Pharmacol. 59:524-531.

[106] Birinyi P, Acsai K, Banyasz T, Toth A, Horvath B, Virag L, Szentandrassy N, Magyar J, Varro A, Fulop F, Nanasi PP. (2005) Effects of SEA0400 and KB-R7943 on Na^+/Ca^{2+} exchange current and L-type Ca^{2+} current in canine ventricular cardiomyocytes. Naunyn-Schmiedeberg's Arch. Pharmacol. 372:63-70.

[107] Pintado AJ, Herrero CJ, Garcia AG, Montiel C. (2000) The novel $Na^{(+)}/Ca^{(2+)}$ exchange inhibitor KB-R7943 also blocks native and expressed neuronal nicotinic receptors. Br. J. Pharmacol. 130:1893-1902.

[108] Sobolevsky AI, Khodorov BI. (1999) Blockade of NMDA channels in acutely isolated rat hippocampal neurons by the Na^+/Ca^{2+} exchange inhibitor KB-R7943. Neuropharmacology. 38 :1235-1242.

[109] Matsuda T, Arakawa N, Takuma K, Kishida Y, Kawasaki Y, Sakaue M, Takahashi K, Takahashi T, Suzuki T, Ota T, Hamano-Takahashi A, Onishi M, Tanaka Y, Kameo K, Baba A. (2001) SEA0400, a novel and selective inhibitor of the Na^+-Ca^{2+} exchanger, attenuates reperfusion injury in the in vitro and in vivo cerebral ischemic models. J. Pharmacol. Exp. Ther. 298:249-256.

[110] Iwamoto T, Kita S, Uehara A, Imanaga I, Matsuda T, Baba A, Katsuragi A. (2004) Molecular determinants of Na^+/Ca^{2+} exchange (NCX1) inhibition by SEA0400. J. Biol. Chem. 279:7544-7553.

[111] Nagano T, Osakada M, Ago Y, Koyama Y, Baba A, Maeda S, Takemura M, Matsuda T.(2005) SEA0400, a specific inhibitor of the Na^+-Ca^{2+} exchanger, attenuates sodium nitroprusside-induced apoptosis in cultured rat microglia. Br. J. Pharmacol. 144:669-679.

[112] Iwamoto T, Watanabe Y, Kita S, Blaustein MP. (2007) Na+/Ca2+ exchange inhibitors: a new class of calcium regulators. Cardiovas. Hematol. Disord. Drug Targets 7(3):188-198.

[113] Iwamoto T. (2007) Na+/Ca2+ exchange as a drug target-insights from molecular pharmacology and genetic engineering. Ann. N.Y. Acad. Sci. 1099:516-528.

[114] Pott C, Eckardt L, Goldhaber JI. (2011) Triple Threat: The Na+/Ca2+ exchanger in the pathophysiology of cardiac arrhythmia, ischemia and heart failure. Curr. Drug Targets 12(5):737-747.

[115] Iwamoto T. (2004) Forefront of Na+/Ca2+ exchanger studies: molecular pharmacology and Na+/Ca2+ exchange inhibitors. J. Pharmacol. Sci. 96:27-32.

[116] Dai JM, Kuo KH, Leo JM, van Breemen C.,Lee CH. (2006) Mechanism of Ach-induced asynchronous calcium waves and tonic contraction in porcine tracheal muscle bundle. Am. J. Physiol. Lung Cell Mol. Physiol. 290:L459-L469.

[117] Dai JM, Kuo KH, Leo JM, Paré PD, van Breemen C, Lee CH. (2007) Acetylcholine-incuded asynchronous calcium waves in intact human bronchial muscle bundles. Am. J. Physiol. Cell Mol. Biol. 36:600-608.

[118] Hirota SA, Pertens E, Janssen LJ. (2007) The reverse mode of the sodium-calcium exchanger provides a source of calcium for store-refilling following agonist-induced calcium mobilization. Am. J. Physiol. Lung Cell Mol. Physiol. 292:L438-L447.

Chapter 4

Contraction by Ca^{2+} Influx via the L-Type Ca^{2+} Channel Voltage Window in Mouse Aortic Segments is Modulated by Nitric Oxide

Paul Fransen, Cor E. Van Hove, Johanna van Langen and Hidde Bult

Additional information is available at the end of the chapter

http://dx.doi.org/10.5772/47771

1. Introduction

L-type Ca^{2+} channels play a dominant role in blood pressure regulation as suggested by the fall in mean arterial blood pressure in mice with selective ablation of the channel gene (CACN1C) in vascular smooth muscle cells (VSMCs) (Moosmang et al., 2003), the increased CACN1C gene expression in VSMCs in rodent models of hypertension (Pesic et al., 2004; Rhee et al., 2009), the anti-hypertensive effect of L-type Ca^{2+} channel blockers (CaBs) (Mancia et al., 2007) and the relationship between an autoantibody against vascular L-type Ca^{2+} channels and clinical characteristics of hypertensive patients (Zhou et al., 2008). Transcripts and protein expression of CACN1C are found widely in the cardiovascular system, where the ion channels serve the time- and voltage-dependent influx of Ca^{2+} ions to initiate muscle contraction (Akata, 2007; Bers, 2002). It is now recognized, however, that not only peripheral resistance but also arterial compliance is of great importance, especially in old age (systolic) hypertension (Belz, 1995; Mitchell, 1999; Westerhof et al., 2009). In the large conduit arteries, L-type Ca^{2+} channels are important determinants of their mechanical properties and compliance, which are such that blood pressure and flow are propagated between the heart and arterioles and that thereby pulsatile flow is transformed into steady flow due to the "windkessel" effect (Westerhof et al., 2009). For example, CaBs increase vascular compliance of large elastic vessels and may be of importance for the pathogenesis and prognosis of cardiovascular complications such as atherosclerosis, left ventricular hypertrophy and heart failure (Bellien et al., 2010; Belz, 1995; Essalihi et al., 2007; Safar et al., 1989; Slama et al., 1995; Vayssettes-Courchay et al., 2011). Further evidence for a role of L-type Ca^{2+} channels in atherosclerosis was obtained in carotid and femoral VSMCs, where the L-type Ca^{2+} channel gene expression differs between atherosclerotic versus non-atherosclerotic regions (Tiwari et al., 2006). Not only Ca^{2+} channels but also endothelial released nitric oxide

(NO) may be involved in these processes. Inhibition of endothelial NO synthase (eNOS) with N^{Ω}-nitro-L-arginine methyl ester (L-NAME) or N^{Ω}-nitro-L-arginine (L-NNA) causes hypertension and the decrease of basal endothelial NO release or availability may be at the basis of increased reactivity to vasoconstrictors in hypertension (Panza et al., 1993). Furthermore, in the beginning of plaque development in animal models and in patients with atherosclerotic symptoms or risk factors, eNOS activity and concomitant NO release is altered, especially in atherosclerosis-prone aortic segments (Fransen et al., 2008; Kauser et al., 2000; Vanhoutte et al., 2009). Recently, we showed that L-type Ca^{2+} influx and its inhibition with CaBs may also have consequences for the capacity of NO to relax constricted mouse aorta. The relaxing efficacy of NO in mouse aorta was dependent on the contractile agonist, and more specifically, decreased when the contraction was mainly elicited via L-type Ca^{2+} influx, but increased when Ca^{2+} influx was partially inhibited with CaBs (Van Hove et al., 2009). The above observations may suggest an interaction between basal NO and VSMC L-type Ca^{2+} channels. An inhibitory effect of NO on Ca^{2+} currents in vascular smooth muscle and cardiomyocytes has been described (Blatter & Wier, 1994; Fischmeister et al., 2005; Tsai & Kass, 2009), but has not been associated with L-type Ca^{2+} channel-mediated contractions in mouse aorta. This chapter will focus on the specific role of L-type Ca^{2+} channels in vasoconstriction and dilation and the interplay between NO and these L-type Ca^{2+} channels.

2. Interaction between endothelial and vascular smooth muscle cells in intact mouse aorta

It is generally assumed that the increase of intracellular Ca^{2+} in vascular endothelial cells results in release of NO via complex interactions with calmodulin, caveolin, endothelial NOS (eNOS) and tetrahydrobiopterin. NO released from the endothelial cells stimulates guanylate cyclase (sGC) to produce cGMP, which causes relaxation of the VSMCs via reduction of intracellular Ca^{2+} and the Ca^{2+}-sensitivity of the contractile elements. Reduction of intracellular Ca^{2+} can occur via different mechanisms: inhibition of IP_3-mediated Ca^{2+} release from intracellular Ca^{2+} stores; removal and sequestration of intracellular Ca^{2+} pump mechanisms and/or both direct and indirect inhibition of influx of extracellular Ca^{2+} through voltage-gated Ca^{2+} channels (Tsai & Kass., 2009). In baseline conditions, i.e. in the absence of receptor-stimulation with agonists such as acetylcholine (ACh), there is basal release of NO through constitutive activity of endothelial eNOS in the mouse aorta. In isolated arteries, the basal NO production can be assessed by comparing force development by the SMCs in response to a vasoconstrictor such as phenylephrine in the absence and presence of NOS inhibitors. On the other hand, receptor-stimulated eNOS activity can be determined based on its sensitivity for agonists like ACh to induce endothelium-dependent relaxation of pre-constricted segments. Remarkably, basal and stimulated eNOS activity are differentially regulated. In rat aorta, basal NO release is more sensitive to destruction by superoxide anion, and can be selectively inhibited by N^{Ω}-monomethyl-L-arginine (L-NMMA) and asymmetric N^{Ω}-dimethyl-L-arginine (ADMA) without ACh responses being affected (Al-Zobaidy et al., 2011; Frew et al., 1993; Mian & Martin, 1995).

Differences between basal and stimulated eNOS activity were also observed in apolipoprotein E-deficient (apoE$^{-/-}$) mice which develop atherosclerotic lesions in the thoracic aorta spontaneously at the age of 6 months, a process, which can be effectively accelerated by feeding these mice a high cholesterol diet. The mice exhibit preferential lesion formation in the aortic root and arch, and the proximal and distal part of the thoracic aorta (Crauwels et al., 2003; Nakashima et al., 1998; Reddick et al., 1994). This makes the apoE$^{-/-}$ mouse an interesting model to study differences between atherosclerosis-prone and atherosclerosis-resistant regions. We have previously shown that at the age of 4 months, hence, before development of any visible lesions in the apoE$^{-/-}$ mouse model, basal and agonist-stimulated NO release decreased, respectively increased in atherosclerosis-prone aortic segments of apoE$^{-/-}$ in comparison with wild-type (WT) mice (Fransen et al., 2008). Because no differences in eNOS expression were found between aorta of WT and apoE$^{-/-}$ mice, this indicated that the release or efficacy of NO differed between both mouse strains. In non-stimulated conditions the compromised basal NO release in the apoE$^{-/-}$ mouse might be related to lower intra-endothelial Ca^{2+} concentrations because it could be restored to "normal" wild-type levels by experimentally increasing intracellular Ca^{2+}. Agonist (ACh)-stimulated VSMC relaxation was temporally and dose-dependently related to the increase of endothelial Ca^{2+}. Although the agonist-stimulated increase of endothelial Ca^{2+} was similar in both strains, apoE$^{-/-}$ segments were significantly more sensitive than WT segments in their relaxation to ACh. Results of these studies might suggest an important role of basal and stimulated eNOS activity in the pathophysiology of atherosclerotic lesions.

In functional and molecular biological studies we have further demonstrated that before development of any visible lesions, not only endothelial cells but also the SMCs of the thoracic aorta of apoE$^{-/-}$ mice displayed altered intracellular Ca^{2+} homeostasis in comparison with WT VSMCs (Van Assche et al., 2007) and that within the apoE$^{-/-}$ strain the smooth muscle transcriptome is altered at atherosclerosis-prone versus atherosclerosis-resistant locations (Van Assche et al., 2011). Hence, not only endothelial, but also VSMC function is altered in atherosclerosis-prone versus -resistant segments. Whether the cross-talk between both cell types is affected in both directions during the process of atherosclerosis, is far less studied. At least, we have shown before that NO-dependent vasodilation is dependent on the agonist causing contraction and whenever VSMC function alters during development of atherosclerosis, this is also expected to affect the efficacy of endothelial cell stimulation.

3. Relaxation of VSMCs depends on the agonist causing contraction

Contractions of isolated arteries are often studied by initiating VSMC intracellular Ca^{2+} increase with two widely used and different stimuli. On the one hand, elevated external K^+ depolarises the VSMCs and elicits multiphasic Ca^{2+} signalling and force development through influx of extracellular Ca^{2+} via L-type Ca^{2+} channels. On the other hand, α_1-adrenoceptor stimulation with phenylephrine also causes multiphasic Ca^{2+} signalling and force development, but the signalling is different from the depolarization-induced signalling. The vasodilator effects of NO differ significantly for both contractile agents. Bigger contractions produced by increasing concentrations of phenylephrine were all equally sensitive to relaxation by ACh (figure 1) or the NO donor DEANO. In contrast, when the VSMCs were clamped to depolarized membrane potential (V_m) with high extracellular K^+ concentrations, relaxation by ACh or DEANO became attenuated as K^+-induced force augmented (figure 1) (Van Hove et al., 2009).

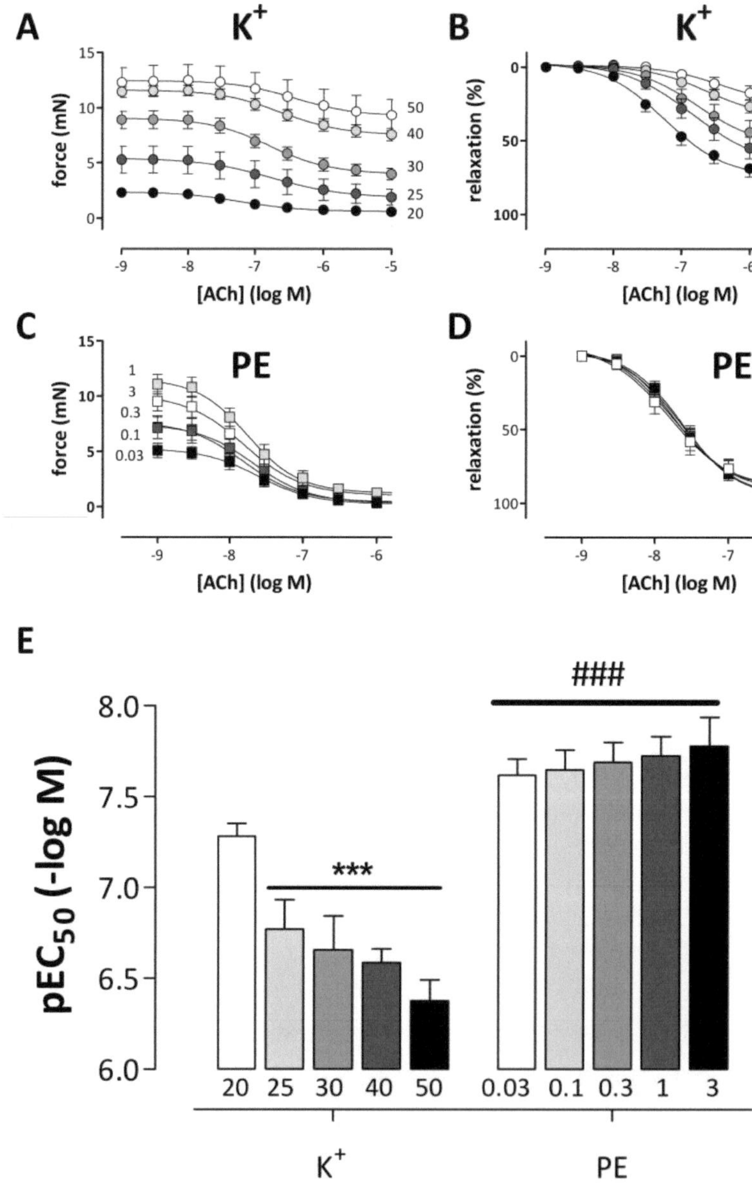

Figure 1. Relaxation induced by ACh-stimulation of mouse aortic segments at different levels of contraction induced by depolarization (K^+) or phenylephrine (PE). A, C: absolute values; B, D: normalized values. Relaxation curves were fitted with sigmoidal concentration-response equations with variable slope, which revealed maximal responses (E_{max}) and the negative logarithm of the concentration resulting in 50% of the maximal effect (pEC_{50}). Relaxation curves after contraction by increasing depolariza-

tion were shifted to the right as external K^+ concentration (and force level, A) increased. Not only the maximal amplitude produced by ACh declined significantly, but also the $logEC_{50}$ of ACh was significantly shifted to the right (E). In contrast, the normalized relaxation curves after contraction elicited by increasing phenylephrine concentrations were identical at each contraction level, and there was only a minor shift of the $logEC_{50}$ (E). ***: P< 0.001 versus 20 mM K^+ ###: P<0.001 versus K^+; n=5; mean±SEM. Modified after figure 2 in (Van Hove et al., 2009)

Simultaneous measurement of intracellular Ca^{2+} and force development in SMCs of de-endothelialised aortic segments revealed important differences between Ca^{2+} signals in depolarization (50 mM K^+)- or phenylephrine-constricted segments upon addition of DEANO, a donor of exogenous NO (figure 2). For phenylephrine-induced contractions and

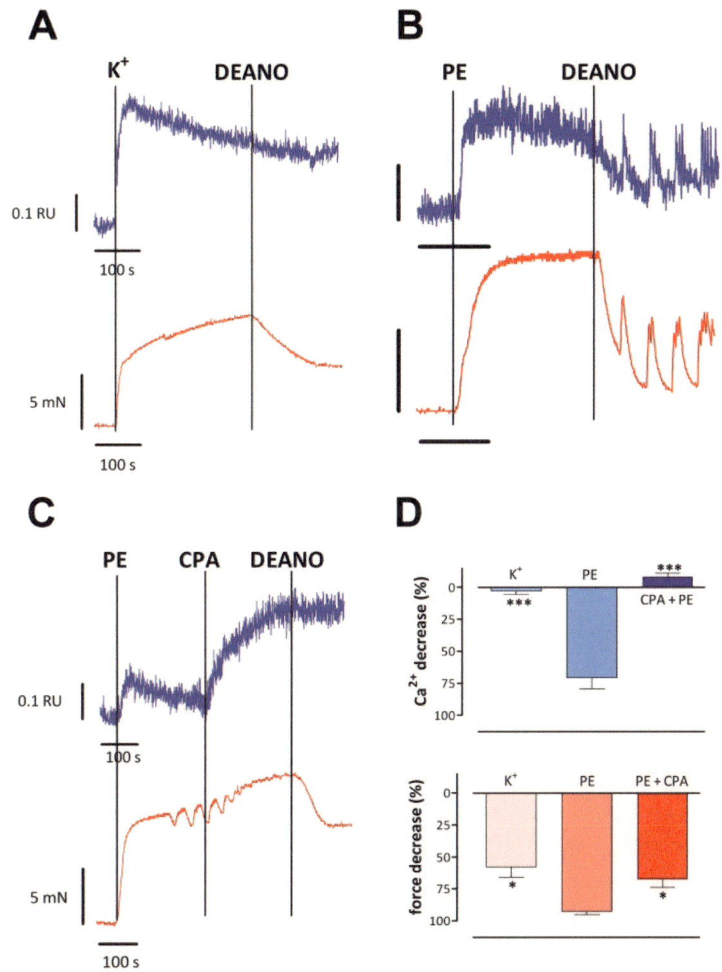

Figure 2. Mobilisation of intra-VSMC Ca^{2+} (Fura-2 technique) and concomitant isometric force before and after addition of 10 μM DEANO to 50 mM K^+ (A) and 3 μM phenylephrine (PE) in the absence (B) and presence of 10 μM cyclopiazonic acid (CPA, C). D shows the mean±SEM of Ca^{2+} and force decrease after addition of DEANO. Data in D: mean±SEM, n=4 endothelium-denuded segments; *, ***: P<0.05, 0.001 versus phenylephrine. Modified after figure 6 in (Van Hove et al., 2009).

VSMC Ca^{2+} signals, addition of NO resulted in an abrupt decrease of intracellular Ca^{2+}, which could be blocked by adding cyclopiazonic acid (CPA), an inhibitor of the sarcoplasmic reticulum (SR) Ca^{2+} pump (SERCA), but not by inhibition of sGC with 1H-[1,2,4]oxadiazolo[4,3-a]quinoxalin-1-one (ODQ). Results indicated a direct stimulation by NO of Ca^{2+} re-uptake to the SR by SERCA, which is in accordance with previous reports (Cohen et al., 1999; Cohen & Adachi, 2006; Van Hove et al., 2009). Contractions, but not Ca^{2+} signals of VSMCs, depolarized with 50 mM K^+, declined upon addition of exogenous NO. Nevertheless, as seen for contractions by depolarization with elevated K^+, also phenylephrine-induced contractions are associated with depolarization and opening of L-type Ca^{2+} channels (Akata, 2007; Plane et al., 1998; Quignard et al., 2000; Richards et al., 2001; Van Hove et al., 2009). If L-type Ca^{2+} influx is an important determinant of the vasodilator capacity of NO, why then is there an abrupt decrease of intracellular Ca^{2+} in the VSMCs after addition of exogenous NO to phenylephrine-elicited contractions and not to 50 mM K^+-evoked contractions? Reduction of Ca^{2+} influx in depolarized segments with CaBs increased NO's capacity to dilate the segments, indicating that not only the depolarized V_m of the VSMCs, but also the amount of Ca^{2+} influx determines the efficacy of NO to cause complete relaxation (Van Hove et al., 2009). Therefore, it was decided to investigate the phenylephrine- and K^+-induced isometric contractions in more detail.

4. Phenylephrine-induced contractions and NO

Addition of phenylephrine causes intracellular Ca^{2+} release from intracellular SR Ca^{2+} stores via activation of IP_3-receptors (Karaki et al., 1997). Indeed, in the absence of extracellular Ca^{2+} phenylephrine causes emptying of the SR Ca^{2+} stores and elicits a transient contraction (figure 3). This transient contraction in Ca^{2+}-free conditions is higher with the constitutive activity of eNOS because inhibition of basal NO release decreased the IP_3-mediated contraction by phenylephrine (figure 3). These results are compatible with previous observations that basal NO stimulates SERCA activity (Cohen et al., 1999; Cohen & Adachi., 2006; Van Hove et al., 2009). Thereby, the Ca^{2+} content of the SR stores of the VSMCs may be increased and may then lead to higher IP_3-mediated contractions. The transient contractions by phenylephrine in the absence of extracellular Ca^{2+} were dose-dependently inhibited by exogenous NO (DEANO, figures 4A, B). This suggests that although basal NO might stimulate Ca^{2+} uptake to the SR via stimulation of SERCA, it inhibits the IP_3-mediated release of Ca^{2+} or the concomitant transient contraction. When these experiments were repeated with addition of ACh to promote endothelial NO release, however, relaxation of phenylephrine-induced contractions by increasing concentrations of ACh in the absence of external Ca^{2+} was completely absent (figures 4C, D). These experiments demonstrate the absolute necessity of external Ca^{2+} to induce ACh-stimulated release of NO from the endothelial cells. Although ACh has been described to release Ca^{2+} from the SR (Fransen et al., 1998), this release does not stimulate eNOS to inhibit IP_3-mediated contractions in mouse aortic SMCs. Hence, release of endogenous NO is dependent upon influx of extracellular Ca^{2+} into the endothelial cells. It has been observed before that Ca^{2+} influx into the endothelial cells stimulated NO production more potently than Ca^{2+} released from internal stores, which evokes little NO production (Isshiki et al., 2002; Isshiki et al., 2004). This is also consistent with old observations that the resting level of cGMP falls after removal of external Ca^{2+} (White & Martin, 1989). In other studies, it has been shown that the TRPV4-mediated Ca^{2+} signal was required for eNOS activation by ACh (Adapala et al., 2011; Zhang et al., 2009), also illustrating the necessity of Ca^{2+} influx for endothelial NO release. Isshiki et al. (Isshiki et al., 2004) suggested that agonist-stimulated eNOS activity in the endothelial cells was sensitive to external Ca^{2+}-dependent acute changes in intracellular subcortical Ca^{2+} signals and that basal eNOS activity was maintained and regulated by subplasmalemmal Ca^{2+} equilibrated with extracellular Ca^{2+}.

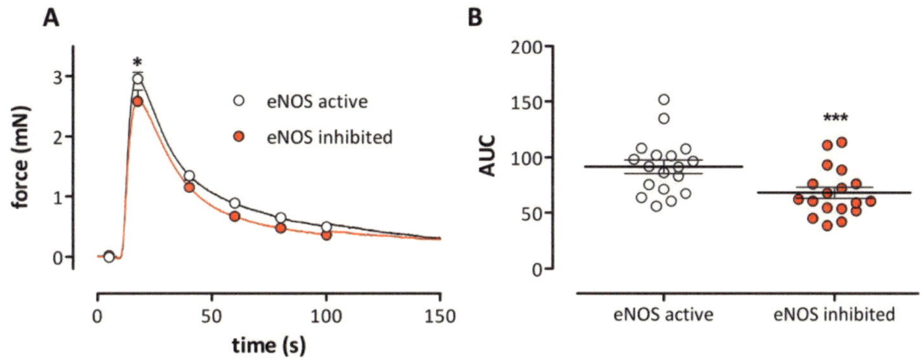

Figure 3. Effect of basal NO release on transient contractions by phenylephrine in the absence of external Ca^{2+}. A: IP_3-mediated contractions by 2 μM phenylephrine in the absence of extracellular Ca^{2+} (0 mM Ca^{2+} + 2 mM EGTA) before (black, open symbol) and after inhibition (red, closed symbol) basal eNOS activity with 300 μM LNAME/LNNA. In B the area under the curve (AUC) shows that the transient contraction is significantly larger with eNOS active than with eNOS inhibited. (A: mean contraction (n=18) with mean force value±SEM at certain time points).

Concomitant with the transient contraction, due to IP_3-mediated Ca^{2+} release, α_1-adrenoceptor stimulation with phenylephrine causes also tonic contractions, which are measureable only in the presence of external Ca^{2+} (figure 5A). These contractions are mediated by Ca^{2+} influx from the extracellular medium via Ca^{2+}-permeable channels (SOCE or store-operated Ca^{2+}-entry). The Ca^{2+} entry occurs via L-type Ca^{2+} channels, which can be inhibited with CaBs (figure 5B) and via non-selective cation channels, which can be inhibited with 50 μM 2-aminoethoxydiphenylborane (2-APB) (Bootman et al., 2002; Peppiatt et al., 2003) (figure 5C). Addition of verapamil (CaB) or 2-APB between the phasic and tonic contraction (figure 5B, C) results in reduced SOCE and SOCE-related contraction. The SOCE contraction in the presence of 2-APB can be completely inhibited with CaBs, indicating it is mediated by L-type Ca^{2+} influx only, whereas the SOCE contraction in the presence of verapamil can be completely inhibited with 2-APB, indicating it is mediated by non-selective cation Ca^{2+} influx only. Figure 6 compares relaxation by ACh of phenylephrine-preconstricted aortic segments in control conditions (contractions in the presence of external Ca^{2+}), upon re-addition of external Ca^{2+} (SOCE contractions) and following inhibition of the cation channel-mediated Ca^{2+} influx with 2-APB.

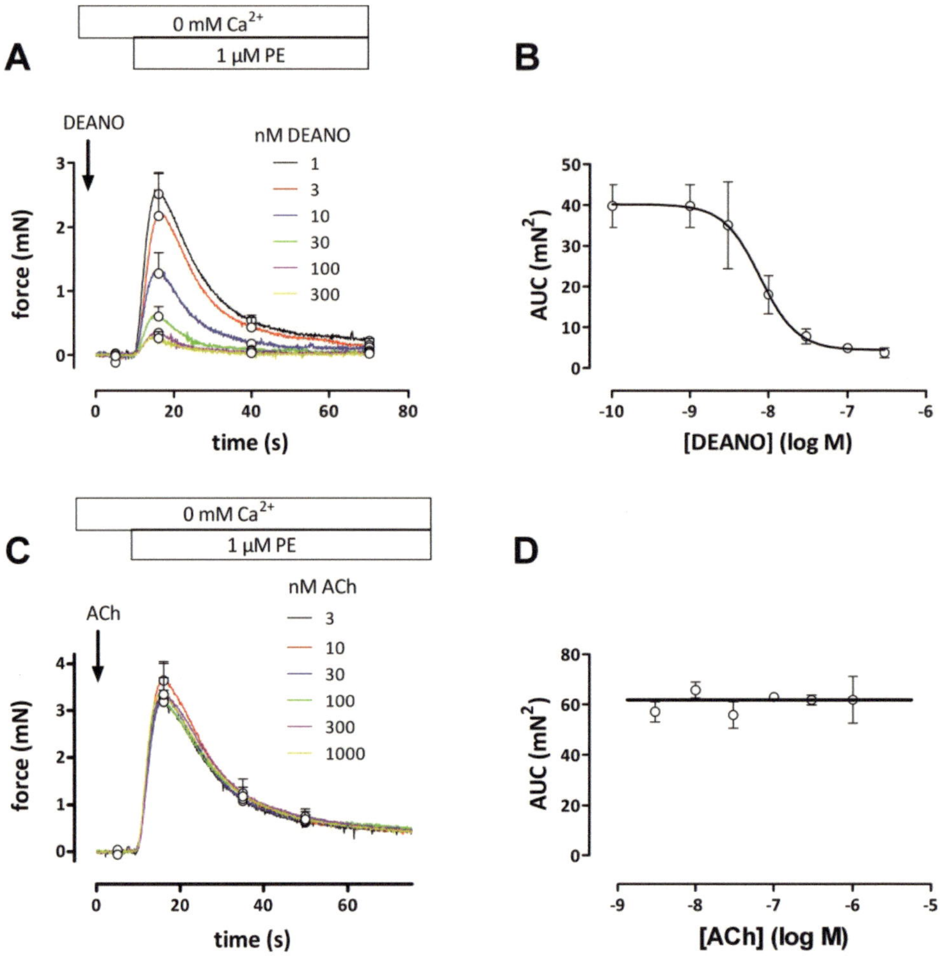

Figure 4. Exogenous NO, but not ACh inhibits IP_3-mediated contraction evoked by 1 μM phenylephrine in the absence of external Ca^{2+}. A: Force development by 1 μM phenylephrine (PE) after eNOS inhibition with 300 μM LNAME/LNNA. Force was measured 3 minutes after switching to 0 mM Ca^{2+}. Before the transient contraction increasing concentrations of DEANO (indicated in nM) were applied. B: Dose-effect relationship for the inhibition by DEANO of the area under the curve (AUC) for the different IP_3-mediated phenylephrine-elicited contractions. C: IP_3-mediated phenylephrine-elicited contractions in segments with eNOS active were not influenced by increasing concentrations of ACh (indicated in nM). D: Dose-effect relationship for the inhibition by ACh of the AUC for the IP3-mediated phenylephrine-elicited contractions. $LogEC_{50}$ for DEANO was -8.12±0.20 logM. Data: mean±SEM at certain time points; n= 4.

Relaxation by ACh of these tonic phenylephrine-induced contractions were not different for control contractions (normal Krebs-Ringer solution with external Ca^{2+}) or SOCE contractions (elicited by re-addition of external Ca^{2+} after emptying the stores with phenylephrine) (figure 6A). When Ca^{2+} influx via non-selective cation channels during the phenylephrine-induced SOCE contraction was inhibited with 2-APB, the tonic contraction induced by phenylephrine was significantly smaller and relaxation by ACh was reduced by about 50%

(figure 6B). Results indicate that the relaxing capacity of NO is small when contractions are mediated by Ca^{2+} influx via L-type Ca^{2+} channels as shown before for depolarization-mediated contractions. These results further confirmed the hypothesis that the difference in NO-mediated relaxation of segments depolarized with K^+ versus phenylephrine, as shown in figure 1, is due to the amount of Ca^{2+} influx via L-type Ca^{2+} channels and the membrane potential (V_m) of the VSMCs. Whereas relaxation of contractions elicited with high external K^+ was not complete, was not associated with re-uptake of Ca^{2+} to the SR or with a significant decrease of intracellular Ca^{2+}, relaxation of phenylephrine-induced contractions was accompanied by inhibition of IP_3-mediated contractions, by NO-mediated re-uptake of Ca^{2+} to the SR, by repolarising the VSMC V_m and by inhibition of Ca^{2+} influx during the tonic contraction (Van Hove et al., 2009). To investigate whether the effects of NO on relaxation of VSMCs were voltage-dependent, the effects of NO were studied at different external K^+ concentrations or V_m of the VSMCs.

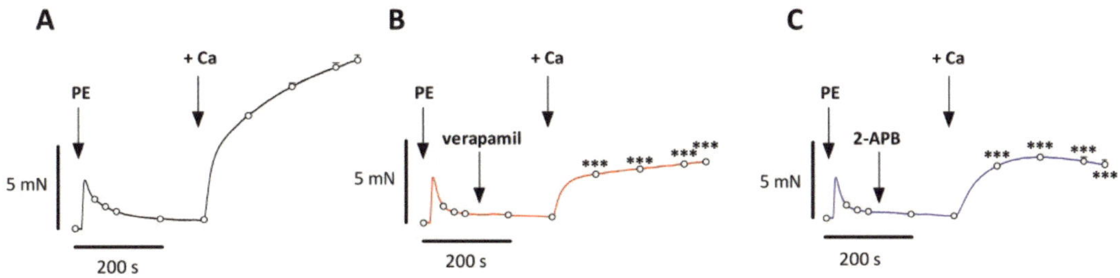

Figure 5. Phasic (transient) IP_3-mediated contraction by 2 μM (PE) in the absence of extracellular Ca^{2+}, followed by the tonic (SOCE) contraction upon re-addition of external Ca^{2+} (+3.5 mM Ca^{2+}) in control (A) and after addition of 10 μM verapamil (B) or 50 μM 2-APB (C) between the phasic and tonic contraction. Phasic force developed in B is due to non-selective cation Ca^{2+} influx and in C to L-type Ca^{2+} influx. The figure shows mean force traces with data±SEM at different time intervals; (n=4).

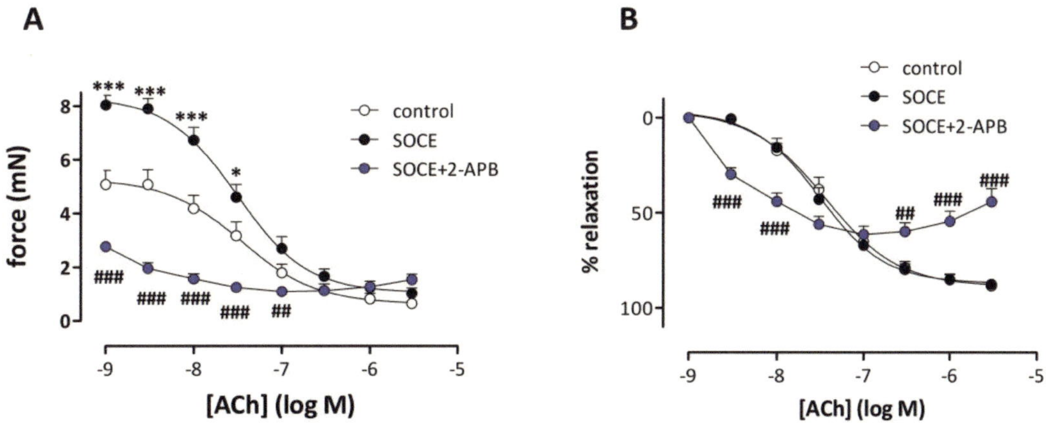

Figure 6. Absolute (A) and relative (B) ACh-induced relaxation of segments constricted with 2 μM phenylephrine in control conditions (white) and after eliciting SOCE by previously emptying the Ca^{2+} stores and re-adding external Ca^{2+} (black). Relaxation curves were fitted with sigmoidal concentration-response equations with variable slope. After inhibition of SOCE via non-selective cation channels with

50 μM 2-APB (blue), relaxation of phenylephrine-induced SOCE contraction by ACh, although very small, was completely different and compromised and could not be fitted with sigmoidal concentration-response equations (B). Data: mean±SEM; n=7; *, ***: P<0.05, 0.001 SOCE versus control; ##, ###: P<0.01, 0.001 2-APB versus SOCE.

5. Depolarization-induced contractions and NO

Aortic segments are unable to produce force upon depolarization with high K^+ in 0 mM Ca^{2+}. In the presence of extracellular Ca^{2+}, depolarization-induced contractions could be completely inhibited with CaBs, but were not affected by blocking the ryanodine receptor (15 μM ryanodine) or IP_3 receptor of the SR (50 μM 2-APB) (Peppiatt et al., 2003). These observations suggested that K^+-induced contractions of mouse aortic segments are solely supported by influx of Ca^{2+} via L-type Ca^{2+} channels. Figure 1 showed that ACh-induced relaxations of contractions produced by 20 mM extracellular K^+ (mild depolarization) were near complete and occurred with a sensitivity close to the ACh-sensitivity for relaxations of phenylephrine-induced contractions. However, for stronger depolarization, relaxation was severely compromised. The latter results suggest that relaxation of depolarization-induced contractions depended upon V_m of the VSMCs. V_m of the endothelial cells did not significantly contribute to the sensitivity to ACh because ACh-induced relaxations of phenylephrine-induced contractions at high extracellular K^+ in the presence of CaB were not attenuated (Van Hove et al., 2009). Why are relaxations of depolarized segments dependent on V_m of the VSMCs, i.e. complete for phenylephrine-induced contractions, which are also accompanied by depolarization of the segments, nearly complete for mild K^+-induced depolarization and severely attenuated for high K^+-induced depolarization? To solve this question we investigated the depolarization-induced contraction in more detail.

5.1. Depolarization-induced window contractions

Electrophysiological studies in isolated SMCs revealed the occurrence of a voltage range (window), in which L-type Ca^{2+} channels do not inactivate, leading to a "time-independent" influx of Ca^{2+} ions (Curtis & Scholfield, 2001; Fleischmann et al., 1994; Ganitkevich & Isenberg, 1990; Matsuda et al., 1990; Smirnov & Aaronson, 1992). This window influx of Ca^{2+} ions has been shown to lead to an increase of intracellular Ca^{2+} within the SMCs (Fleischmann et al., 1994), but has never been associated with the tonus of blood vessels. By clamping V_m of the endothelial and SMCs in aortic segments to depolarized potentials by elevating extracellular K^+, we were able to show a window contraction within the voltage range of overlap of activation and inactivation curves of the L-type Ca^{2+} channels (manuscript submitted, see also figure 8). In order to correlate the extracellular K^+ with V_m of the VSMCs of the aortic segments, V_m was measured with sharp glass intracellular microelectrodes (filled with 2 M KCl, tip resistances between 65 and 90 MΩ, HEKA EPC9 amplifier in the zero current-clamp mode). Deviation of measured V_m from the K^+ equilibrium potential (Nernstian V_K) was largest at physiological K^+ concentrations, indicating that in non-stimulated VSMCs other ions than K^+ contribute in setting the resting V_m (figure 7A). However, by adding levcromakalim, a drug that sets V_m to V_K by activating

ATP-dependent K^+ channels (Knot & Nelson, 1998; Weston et al., 2002), V_m of the VSMCs could be hyperpolarised to V_K. Indeed, when measured with intracellular microelectrodes, the resting V_m of VSMCs of aortic segments was -60.1±2.6 mV (n=8), which could be repolarized with levcromakalim to -84 mV.

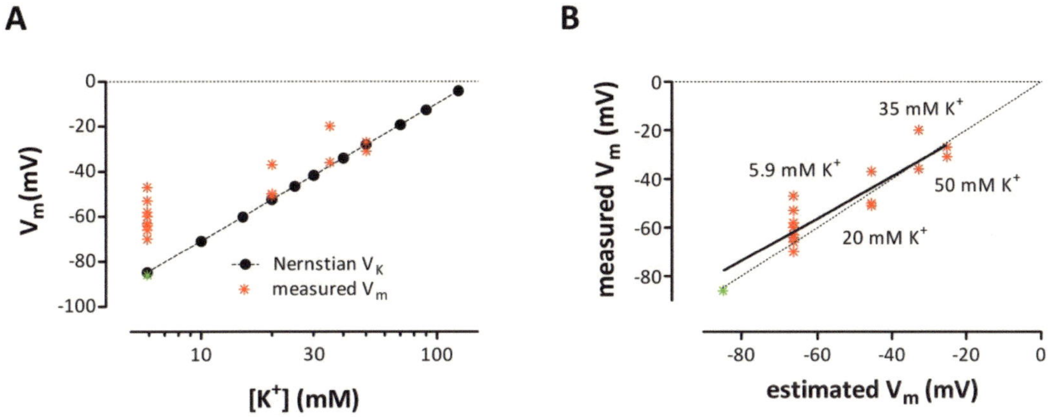

Figure 7. A: Relationship between extracellular K^+ and measured V_m (red asterisks) of mouse aortic VSMCs. Levcromakalim (green asterisk) caused hyperpolarization of V_m to V_K. The Nernstian relationship between V_K and extracellular K^+ is indicated by the black symbols and dotted line. B: Relationship between estimated and measured membrane potential (V_m) of mouse aortic VSMCs. Linear regression of the relationship between estimated and measured V_m (open circles, straight line) reveals slope of 0.90 (95% confidence interval 0.69-1.12), which was not significantly different from unity (dotted line) (R^2= 0.84). Full circle represents V_m in the presence of 200 nM levcromakalim at 5.9 mM K^+.

Taking into account the shift of the contraction curves at different K^+ in the presence of levcromakalim (see figure 8), V_m of the VSMCs evoked by increasing extracellular K^+ (V_{clamp}) could be estimated. V_{clamp} changed with K^+ concentration according to $V_{clamp} = 61*\log(([K^+]_o + 6)/[K^+]_i))$ with $[K^+]_o$ the extracellular K^+ concentration in mM, which is elevated by 6 mM as revealed by the shift induced by levcromakalim (figure 8), and $[K^+]_i$ the intracellular K^+ concentration (assumed to be 140 mM). Figure 7B shows that the relationship between the estimated and measured V_m was close to unity. This means that in the following graphs the K^+ axis could be replaced by a V_{clamp} axis. This is shown in figure 8, where the effects of levcromakalim on K^+-induced isometric contractions at different K^+ concentrations are considered. By adding levcromakalim, the aortic segments were all set to the same V_m, i.e. V_K.

Depolarization by elevated K^+ causes the isometric contraction to increase until maximal force of 100% was attained at 50 mM K^+. At higher K^+, force decreased again, leading to a bell-shape of the force-K^+ relationship, which agrees well with the bell-shaped voltage dependency of the L-type Ca^{2+} currents measured with voltage-clamp in single, isolated SMCs. The "steady-state" contractions observed at each K^+ concentration are window contractions, which are due to influx of Ca^{2+} via non-inactivating L-type Ca^{2+} channels. Differentiation of these contractions (%force/mM K^+ change) reveals a K^+-dependent contraction curve, which fits very well with the intracellular Ca^{2+} mobilized during the depolarizing voltage clamp steps in single SMCs (Fleischmann et al., 1994). It is clear that levcromakalim, which causes repolarization of V_m from -60 to -80 mV, shifts the window contraction to higher extracellular K^+, indicating that in the presence of levcromakalim higher extracellular K^+ is needed to attain the same isometric contraction as in control. When contractions at different external K^+ concentrations were plotted as a function of V_{clamp}, contraction curves in the absence and presence of levcromakalim were identical (figure 8C, D). Hence, levcromakalim shifted the K^+-dependence of the isometric contraction, but not the voltage-dependence, suggesting that levcromakalim does not affect L-type Ca^{2+} channel gating properties. Results further suggest that at resting V_m of the VSMCs (-50 to -60 mV), V_{clamp} situates within the window voltage range, which results in continuous baseline Ca^{2+} influx and concomitant tonus via open L-type Ca^{2+} channels.

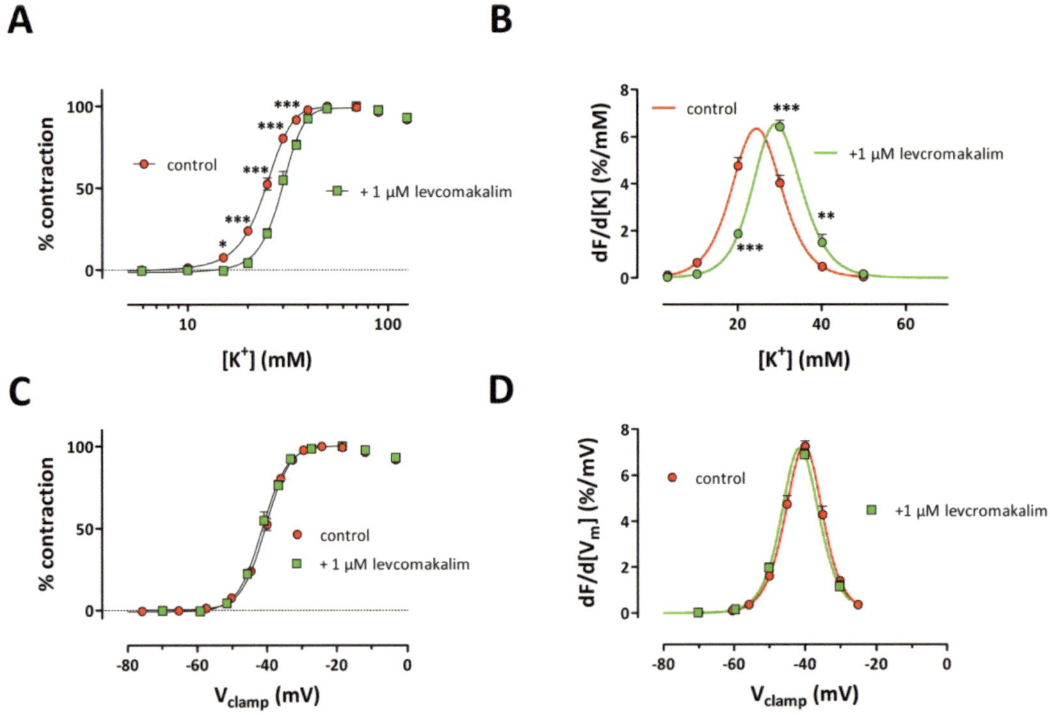

Figure 8. A and B: Effects of 1 μM levcromakalim on the K^+-dependence of the isometric contraction. Extracellular K^+ was increased from 2 mM to 124 mM in control (+300 μM LNAME/LNNA to inhibit eNOS) and in the presence of 1 μM levcromakalim to hyperpolarize the VSMCs to V_K. The fitted (sigmoidal concentration-response curves with variable slope) curves in A were differentiated in B to show K^+-dependence of the change of force development per mM change of K^+. Levcromakalim caused a significant rightward shift (± 6 mM) of the curves in A and B. C and D: Effects of 1 μM levcromakalim on the voltage-dependence of the isometric contraction. V_{clamp} was estimated as indicated in the text for control, i.e. absence of levcromakalim, and in the presence of levcromakalim, where V_{clamp} equals V_K. There was no shift of the voltage-dependence of the L-type Ca^{2+} channel-mediated contraction and the differentiated curves in D were equal. Data: mean±SEM; n=6; *, **, ***: $P<0.05$, 0.01, 0.001 levcromakalim versus control.

5.2. Depolarization-induced window contractions and NO

To investigate the effects of basal NO release on voltage-dependent L-type Ca^{2+} channel-mediated contractions, the contraction evoked by the increase of extracellular K^+ was measured before and after inhibition of eNOS with 300 μM LNAME/LNNA. NO has been described to activate voltage-gated K^+ channels, to hyperpolarize VSMCs and to decrease the intracellular Ca^{2+} concentration (Edwards et al., 2010; Quignard et al., 2000; Yuan et al., 1996). To compensate for the hyperpolarizing effects of NO or depolarizing effects of eNOS inhibition, 1 μM levcromakalim was added to set V_m of all the segments to V_K (figure 9). Inhibition of basal NO release shifted the contraction window to hyperpolarized potentials. A possible explanation is that S-nitrosylation of cysteine residues of the channel changes the voltage-dependence of the L-type Ca^{2+} channel-mediated contraction. Effects of NO and cGMP on L-type Ca^{2+} current have been described before. They might occur via an indirect mechanism through a NO-cGMP-PKG pathway and/or a direct mechanism mediated by S-nitrosylation (Almanza et al., 2007). The α1C-subunit of the L-type Ca^{2+} channel contains more than 10 cysteine residues that modulate channel gating and is constitutively S-nitrosylated in the mouse heart (Sun et al., 2006; Tamargo et al., 2010). According to this hypothesis, basal NO release should diminish Ca^{2+} influx via the L-type Ca^{2+} window at physiological V_m, leading to vasodilation. It will be interesting to test this hypothesis with intracellular VSMC Ca^{2+} measurements. Is intracellular Ca^{2+} decreasing upon addition of NO to segments pre-contracted with different K^+ (different V_{clamp})? Indirect evidence for the hypothesis is provided by the experiments of figure 1, where it was shown that the relaxing capacity of NO increased when external K^+ was decreased.

Figure 9. The contribution of basal NO release to window contractions. V_{clamp}-force curves for segments with active eNOS (blue, control segments) and segments with eNOS inhibited (red, 300 μM LNAME/LNNA combination) were constructed in the presence of 1 μM levcromakalim, which allowed to express force data as a function of V_{clamp} (= V_K). The fitted (sigmoidal concentration-response curves with variable slope) curves in A were differentiated in B to show the voltage-dependence of the change of force development per mV change of V_{clamp}. eNOS-inhibition caused a significant leftward shift of the eNOS active curves in A and B. Data: mean±SEM; n=6; **, ***: P<0.01, 0.001 eNOS inactive versus eNOS active.

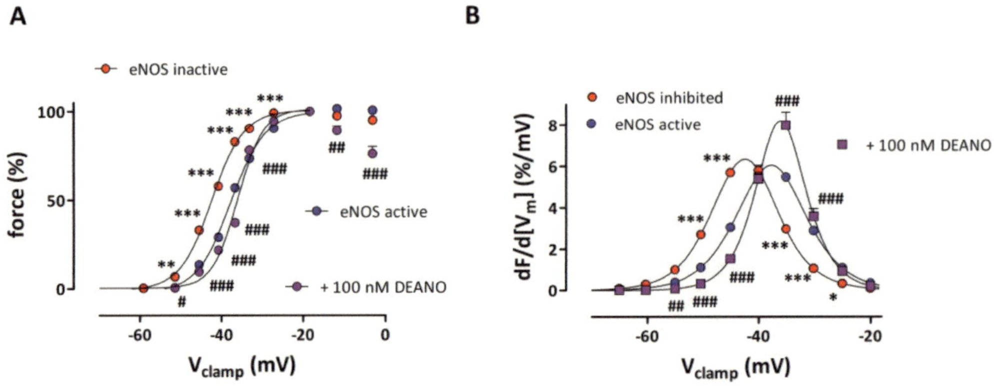

Figure 10. Effects of exogenous NO on contraction window in mouse aortic segments following repolarization of V_m to V_K with 1 μM levcromakalim. eNOS was active (blue circles) or inhibited (red circles) and in this situation exogenous NO could be applied (100 nM DEANO, purple circles). The fitted (sigmoidal concentration-response curves with variable slope) curves in A were differentiated in B to show the voltage-dependence of the change of force development per mV change of V_{clamp}. Data: mean±SEM; n=4; *, **, ***: P<0.05, 0.01, 0.001 eNOS inactive versus eNOS active; #, ##, ###: P<0.05, 0.01, 0.001 +100 nM DEANO versus eNOS inactive.

The shift of the contraction window by eNOS inhibition could be completely reversed by adding exogenous NO (100 nM DEANO) (figure 10). The addition of exogenous NO to segments, in which eNOS activity was inhibited, caused a rightward (depolarising) shift of the voltage-dependency of contraction. Also in this situation, effects of eNOS inhibition or exogenous NO addition on V_m were avoided by performing the experiments in the presence of 1 μM levcromakalim. Results confirmed that NO altered the voltage-dependency of the L-type Ca^{2+} channel-mediated contraction, probably by direct effects of NO on the channel´s gating properties. It should be noted that for strong depolarization above -20 mV (90 and 124 mM K^+), addition of exogenous NO caused significant relaxation of the pre-constricted segments. Hence, the window contraction in the presence of NO is narrower than in the absence of NO (figure 10B). Whether these results can also be explained by NO-dependent changes of the voltage dependence of activation of the L-type Ca^{2+} channels needs further investigation.

The above data may provide an explanation for the absence of a decrease of intracellular VSMC Ca^{2+} with relaxation upon addition of NO to segments depolarized with 50 mM K^+, whereas similar experiments in phenylephrine-constricted segments displayed relaxation with an abrupt decrease of intracellular Ca^{2+} upon addition of NO (see figure 2). Firstly, depolarization with elevated K^+ did not empty the SR Ca^{2+} stores, and, hence, did not stimulate SOCE via NO-sensitive non-selective cation channels. Secondly, at 50 mM K^+ V_m of the VSMCs was clamped at depolarized potentials and NO cannot elicit hyperpolarisation of V_m. Thirdly, at 50 mM K^+, the window Ca^{2+} influx and concomitant contraction were maximal and addition of NO will only cause small relaxations, probably via effects on Ca^{2+}-sensitivity and not on intracellular Ca^{2+} concentrations.

6. NO efficacy and VSMC L-type Ca^{2+} channels along the thoracic aorta

Increased stiffness of elastic arteries represents an early risk factor for cardiovascular diseases (O'Rourke & Mancia, 1999) and, therefore, the assessment of mechanical properties of the aorta is important to understand the mechanisms of cardiovascular disease. It has been shown that in the aorta of C57Bl6 mice, the circumferential modulus is greatest (most rigid) near the diaphragm, and that about 85% of volume compliance is in the thoracic compared with abdominal aorta (Guo & Kassab, 2003). Because as well NO as CaBs have been described to de-stiffen large arteries (Fitch et al., 2006; Safar et al., 1989; Safar et al., 2011), we wondered whether NO release (basal and stimulated) and L-type Ca^{2+} channel activity differed along the length of the thoracic aorta.

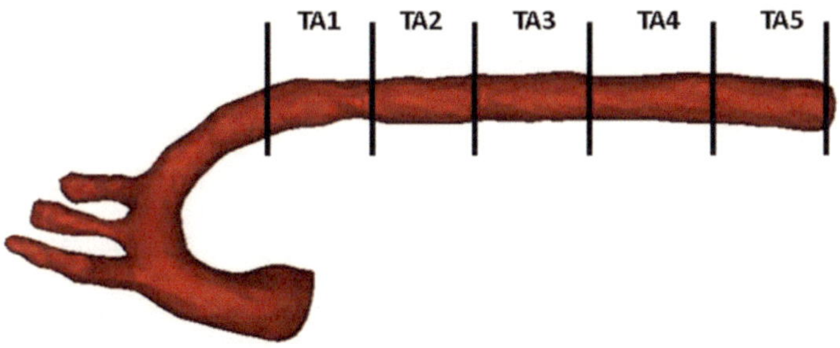

Figure 11. Dissection of the mouse thoracic aorta in five 2 mm wide segments. The mouse aorta was removed from the animal, stripped of adherent tissue and dissected systematically. Starting at the diaphragm, the ascending thoracic aorta was cut in segments of 2 mm width (TA5 up to TA1).

In this part of the chapter, we show some preliminary data on aortic segment differences with respect to NO and L-type Ca^{2+} channels. In order to test whether the release or efficacy of NO differs between different locations along the thoracic aorta of C57Bl6 mice, the aorta was divided into 5 segments (figure 11) and in each segment endogenous and exogenous NO efficacy was determined. Basal release of NO and its efficacy to counteract contraction was measured in each aortic segment by measuring isometric contractions to the α_1-adrenoceptor agonist phenylephrine before and after inhibition of eNOS with 300 μM LNAME/LNNA. The increase of force (Δ force) and the shift of $logEC_{50}$ (Δ logEC50) by eNOS inhibition are an index of the basal release of endogenous NO in each segment (figures 12A and B).

Although not significant, the effects of eNOS inhibition on phenylephrine-induced contractions were smaller in the atherosclerosis-resistant segments TA2, TA3 and TA4 as compared with the atherosclerosis-prone segments TA1 and TA5 (figure 12A). In accordance, inhibition of eNOS caused a significantly larger shift of the segment's sensitivity to phenylephrine in the atherosclerosis-prone segment TA1 compared with the other segments (figure 12B). Results may point to a higher basal eNOS activity or release of basal NO in the atherosclerosis-prone segments TA1 and TA5. Subsequently, endogenous release of NO was evoked by adding increasing concentrations of ACh to segments pre-contracted with 1 μM phenylephrine. Figures 12C and D show maximal relaxation evoked by and sensitivity ($logEC_{50}$)

for ACh in each segment. Although TA1 seemed to have the highest NO efficacy without endothelial stimulation (basal NO, figures A and B), this atherosclerosis-prone segment caused only relaxations of about 65% and displayed the lowest sensitivity for ACh-induced relaxations. This was due to endothelial dysfunction because the effects of exogenous NO (DEANO) were not significantly different between the different aortic segments (figures 12 E and F). Results of these experiments show that endothelial function (basal and stimulated release of NO) may differ between atherosclerosis-prone and –resistant segments along the thoracic aorta of wild-type mice, which may have important consequences for the tonus and compliance of the thoracic aorta.

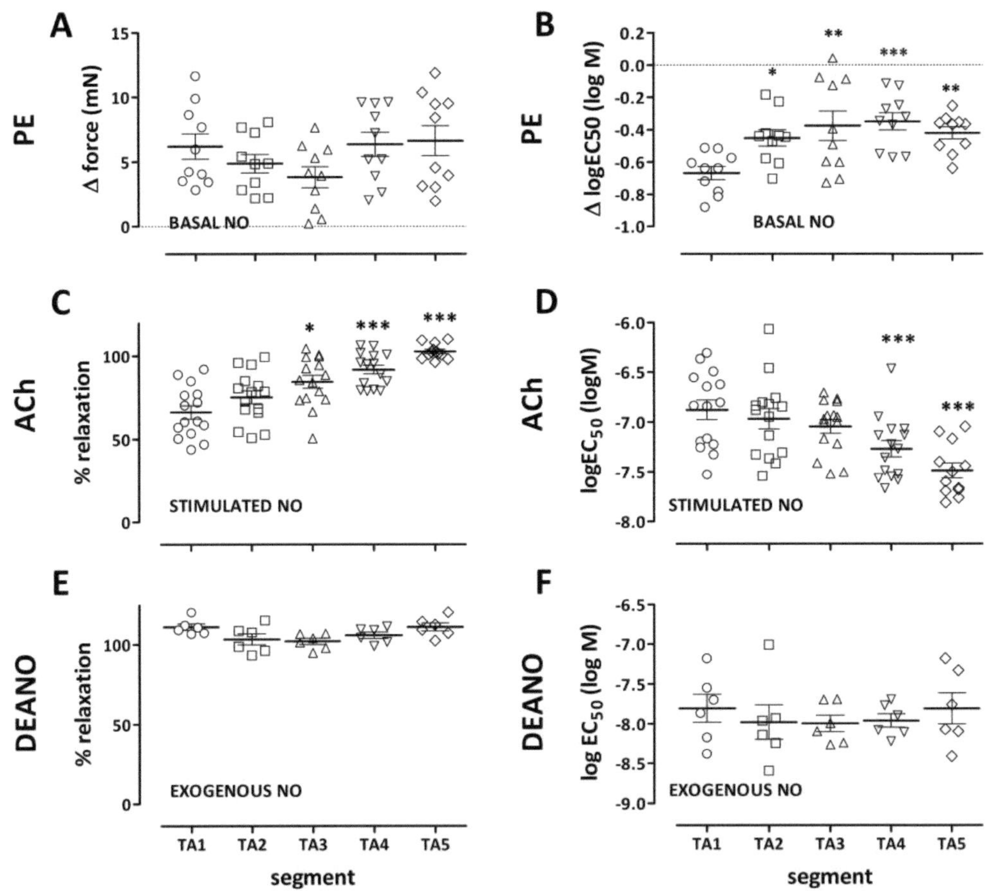

Figure 12. Basal and agonist-stimulated NO efficacy for different aortic segments. A and B show maximal increase of force development (Δforce) and the shift of the $logEC_{50}$ ($\Delta logEC_{50}$) for phenylephrine after inhibition of eNOS with the 300 μM LNAME/LNNA combination (measure of basal NO release). C and D show % relaxation and $logEC_{50}$ for ACh-induced relaxation of the contraction elicited with 1 μM phenylephrine (stimulated endogenous NO), hence in the absence of LNAME/LNNA. E and F show % relaxation and $logEC_{50}$ for DEANO-induced relaxation of the contraction elicited with 1 μM phenylephrine (exogenous NO) after inhibition of eNOS with 300 μM LNAME/LNNA. Data: mean ±SEM; n=6 or more; *, **, ***: P<0.05, 0.01, 0.001 versus TA1.

When the impact of ACh-stimulated eNOS activity on L-type Ca^{2+} channels was investigated further (see figure 6), but now in atherosclerosis-prone and -resistant segments, figures 13 A and C show that as well in the presence of external Ca^{2+} as after eliciting contraction by re-addition of Ca^{2+} to zero Ca^{2+} (SOCE contraction), relaxation in prone segments in comparison with resistant segments was attenuated. When SOCE via cation-selective channels was inhibited with 50 μM 2-APB, the SOCE contraction by phenylephrine was smaller in the resistant than in the prone segments (figure 13B, values at -9 logM ACh). Because this SOCE-contraction by phenylephrine is due to influx of Ca^{2+} via L-type Ca^{2+} channels only (see figure 5), this suggests that the L-type Ca^{2+} channel-mediated SOCE contraction in prone segments is larger than in resistant segments. This L-type Ca^{2+} channel-mediated contraction by phenylephrine in prone segments displayed attenuated relaxation to ACh (figure 13D), again pointing to the lower capacity of NO to relax contractions evoked by L-type Ca^{2+} influx.

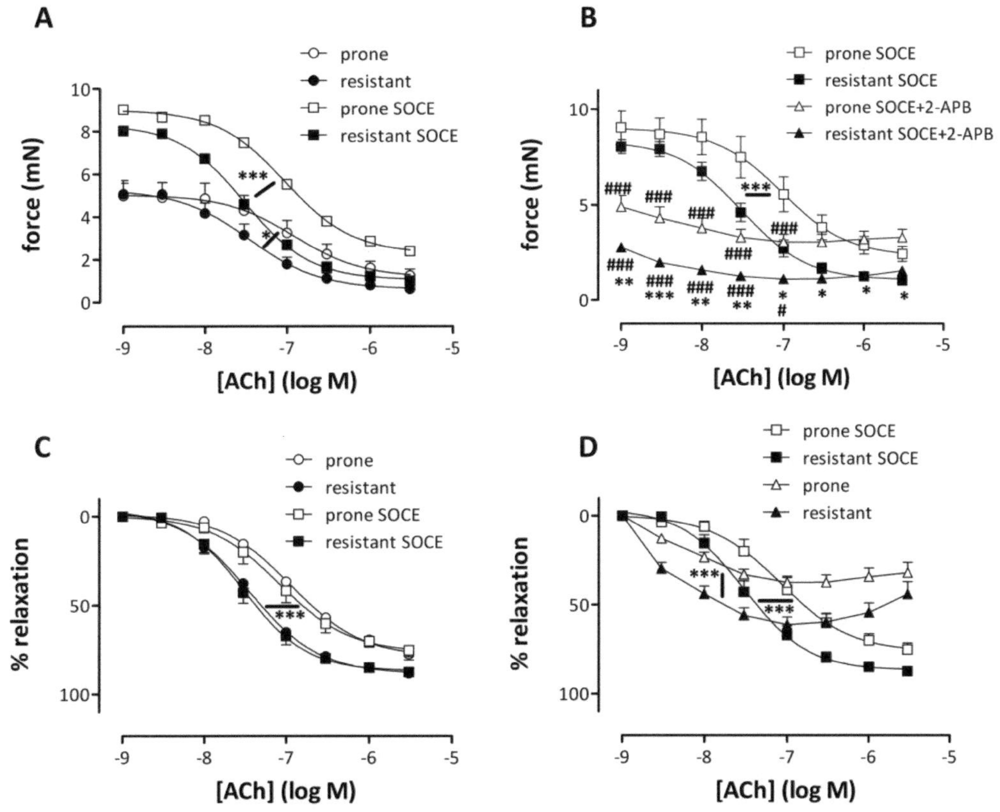

Figure 13. Absolute (mN, A, B) and relative (%, C, D) ACh-induced relaxation of atherosclerosis-prone (TA1,white symbols) and atherosclerosis-resistant (TA2/TA3, black symbols) segments constricted with 2 μM phenylephrine in the presence of extracellular Ca^{2+} (control conditions, circles) and after eliciting contraction by re-addition of external Ca^{2+} to zero Ca^{2+} (SOCE, squares). Relaxation curves were fitted with sigmoidal concentration-response equations with variable slope and significantly differed between prone and resistant segments. After inhibition of SOCE via non-selective cation channels with 50 μM 2-APB (triangles), the SOCE contraction, which is mediated through L-type Ca^{2+} influx only, was larger in prone than in resistant segments. Relaxation of these phenylephrine-induced SOCE contractions by

ACh were compromised more in prone than in resistant segments and could not be fitted with sigmoidal concentration-response equations (B). Data: mean±SEM; n=7; *, **, ***: P<0.05, 0.01, 0.001 prone versus resistant; #, ###: 2-APB versus SOCE.

7. Conclusions

The present chapter investigated the relaxing efficacy of NO for contractions induced by α_1-adrenoceptor stimulation with phenylephrine or by depolarization with elevated extracellular K^+ and confirmed previous observations that relaxation by NO is attenuated when the contraction was due to L-type Ca^{2+} influx (Van Hove et al., 2009). Contractions, which were initiated by Ca^{2+} release from the SR as with phenylephrine were very sensitive to endogenous or exogenous NO. Here, NO caused relaxation by inhibiting the IP_3-mediated contraction, by stimulating Ca^{2+} re-uptake to the SR via stimulation of SERCA, by inhibiting SOCE via cation channels and by reducing SOCE via L-type Ca^{2+} channels. Although phenylephrine has been described to cause depolarization of the VSMCs, V_m is not clamped at depolarized values and addition of NO is expected to cause hyperpolarization, thereby reducing Ca^{2+} influx via L-type Ca^{2+} channels. By clamping V_m to depolarized values with elevation of K^+, we were able to show contractions due to Ca^{2+} influx via L-type Ca^{2+} channels only and more specifically, to Ca^{2+} influx via non-inactivating Ca^{2+} channels (window contraction). This window contraction could be decreased by repolarizing V_m of VSMCs with levcromakalim (an opener of ATP-dependent K^+ channels) or by adding exogenous NO as long as the extracellular K^+ concentration was below 50 mM. Because NO also shifted the voltage-dependence of the window contraction in the presence of levcromakalim, it was hypothesized that NO, but not levcromakalim exerted a direct effect on the gating properties of the L-type Ca^{2+} channel. Although it has been described that NO affects L-type Ca^{2+} channels directly via S-nitrosylation (Almanza et al., 2007), an effect on the voltage-dependent contraction has not been directly demonstrated yet. Hence, it is hypothesized that in mouse aortic segments NO changed the voltage-dependence of the contraction mediated by L-type Ca^{2+} influx, probably via S-nitrosylation of the cysteine residues located in the gating part of the channel´s α1C-subunit.

An interplay between NO and L-type Ca^{2+} channels was also suggested by a number of other observations: a) When treated with the CaB, lacidipine, Western-type diet-evoked hypertension and atherosclerosis development in apoE-deficient mice was reduced, whereas endothelial function was preserved (Kyselovic et al., 2005), b) after partly inhibiting L-type Ca^{2+} influx induced by high K^+ with the CaB nifedipine, relaxation by exogenous NO was ameliorated (Van Hove et al., 2009) and c) $ApoE^{-/-}$ mice, which spontaneously develop atherosclerotic lesions at the age of 6 months when fed a normal diet, displayed altered Ca^{2+} homeostasis at the age of 4 months, hence, before development of plaques. In comparison with wild-type C57Bl6 mice, they showed higher baseline intracellular Ca^{2+} in the SMCs (Van Assche et al., 2007), lower baseline Ca^{2+} in endothelial cells and decreased basal but normal or even enhanced agonist-evoked NO release and efficacy (Fransen et al., 2008).

The present study might also provide an explanation for the unique features of the antihypertensive classes of CaBs with different chemical structure (phenylalkylamines such as verapamil, benzothiazepines such as diltiazem and dihydropyridines such as nifedipine). They reduce blood pressure more effectively in hypertensive than in normotensive subjects (Leonetti et al., 1982) and they inhibit L-type Ca^{2+} influx more effectively in vascular tissue than in heart (Godfraind et al., 1984; Godfraind, 2005; Striessnig et al., 1998). Several attempts have been made to explain these unique features. In hypertension, there is an increase of reactivity to vasoconstrictors, which is not only due to a higher number of L-type Ca^{2+} channels (Godfraind, 2005; Pesic et al., 2004; Pratt et al., 2002), but also to depolarization of the resting potential of hypertensive VSMCs (Morel & Godfraind, 1994; Pesic et al., 2004). Depolarization leads to an increased proportion L-type Ca^{2+} channels in the inactivated state, which according to the "modulated receptor theory" may have higher affinity to CaBs than channels in the resting state (Bean et al., 1986; Godfraind, 2005; Morel & Godfraind, 1987). Cardiac muscle cells are hyperpolarized with respect to VSMC and hence less susceptible to block by CaBs. Another specific feature of L-type Ca^{2+} channels is that their population is not homogeneous because of the occurrence of alternatively spliced isoforms (Koch et al., 1990). Among the 55 known human L-type Ca^{2+} channel exons 19 undergo alternative splicing and display differences in tissue distribution, physiology, pharmacology and disease-related up- and/or down-regulation (Liao et al., 2005; Tang et al., 2007; Tiwari et al., 2006). Some of these isoforms were dominant in aorta (> 50%) and less abundant in heart (<5%). Moreover, the VSM-specific splice variant of the L-type Ca^{2+} channel displayed hyperpolarised window current at voltages where there is overlap between activation and inactivation curves and enhanced inhibition by nifedipine in comparison with the predominant cardiac isoform (Liao et al., 2007). The selective affinity of CaBs for the different L-type Ca^{2+} channel subpopulations may further contribute to the different susceptibility of VSMC or cardiac cells to CaBs. Because the L-type Ca^{2+} channel window in VSMCs may be responsible for "time-independent" baseline Ca^{2+} influx at normal resting or slightly depolarized membrane potentials (Fleischmann et al., 1994; Poburko et al., 2004), depolarization of the resting potential by hypertension not only favours the inactivated state of the L-type Ca^{2+} channel, but is also expected to increase L-type Ca^{2+} influx via the L-type Ca^{2+} channel window and to evoke inhibition by CaBs (Fleischmann et al., 1994).

Preliminary results of this chapter, finally, indicate that NO efficacy and L-type Ca^{2+} channel distribution may differ along the length of the thoracic aorta. Basal and stimulated eNOS activation or NO release occurred differently along the thoracic aorta. Close to the aortic arch, segments released more basal NO but less stimulated NO than segments close to the diaphragm. Similarly, L-type Ca^{2+} channel distribution and related window contraction were higher in atherosclerosis-prone than in -resistant segments. These preliminary data need to be further explored, but may have important consequences for the compliance of the different aortic segments and their susceptibility to the development of atherosclerosis.

Author details

Paul Fransen
Laboratory of Physiopharmacology, University of Antwerp, Wilrijk, Belgium

Cor E. Van Hove, Johanna van Langen and Hidde Bult
Laboratory of Pharmacology, University of Antwerp, Wilrijk, Belgium

8. References

Adapala, R. K.; Talasila, P. K.; Bratz, I. N.; Zhang, D. X.; Suzuki, M.; Meszaros, J. G.; & Thodeti, C. K. (2011) PKCalpha mediates acetylcholine-induced activation of TRPV4-dependent calcium influx in endothelial cells. *Am.J.Physiol Heart Circ.Physiol*,Vol.301, No.3, pp. H757-H765

Akata, T. (2007) Cellular and molecular mechanisms regulating vascular tone. Part 1: basic mechanisms controlling cytosolic Ca^{2+} concentration and the Ca^{2+}-dependent regulation of vascular tone. *J.Anesth.*,Vol.21, No.2, pp. 220-231

Al-Zobaidy, M. J.; Craig, J.; Brown, K.; Pettifor, G.; & Martin, W. (2011) Stimulus-specific blockade of nitric oxide-mediated dilatation by asymmetric dimethylarginine (ADMA) and monomethylarginine (L-NMMA) in rat aorta and carotid artery. *Eur.J.Pharmacol.*,Vol.673, No.1-3, pp. 78-84

Almanza, A.; Navarrete, F.; Vega, R.; & Soto, E. (2007) Modulation of voltage-gated Ca^{2+} current in vestibular hair cells by nitric oxide. *J.Neurophysiol.*,Vol.97, No.2, pp. 1188-1195

Bean, B. P.; Sturek, M.; Puga, A.; & Hermsmeyer, K. (1986) Calcium channels in muscle cells isolated from rat mesenteric arteries: modulation by dihydropyridine drugs. *Circ.Res.*,Vol.59, No.2, pp. 229-235

Bellien, J.; Favre, J.; Iacob, M.; Gao, J.; Thuillez, C.; Richard, V.; & Joannides, R. (2010) Arterial stiffness is regulated by nitric oxide and endothelium-derived hyperpolarizing factor during changes in blood flow in humans. *Hypertension*,Vol.55, No.3, pp. 674-680

Belz, G. G. (1995) Elastic properties and Windkessel function of the human aorta. *Cardiovasc.Drugs Ther.*,Vol.9, No.1, pp. 73-83

Bers, D. M. (2002) Cardiac excitation-contraction coupling. *Nature*,Vol.415, No.6868, pp. 198-205

Blatter, L. A. & Wier, W. G. (1994) Nitric oxide decreases $[Ca^{2+}]_i$ in vascular smooth muscle by inhibition of the calcium current. *Cell Calcium*,Vol.15, No.2, pp. 122-131

Bootman, M. D.; Collins, T. J.; Mackenzie, L.; Roderick, H. L.; Berridge, M. J.; & Peppiatt, C. M. (2002) 2-aminoethoxydiphenyl borate (2-APB) is a reliable blocker of store-operated Ca2+ entry but an inconsistent inhibitor of InsP3-induced Ca2+ release. *FASEB J.*,Vol.16, No.10, pp. 1145-1150

Cohen, R. A. & Adachi, T. (2006) Nitric-oxide-induced vasodilatation: regulation by physiologic s-glutathiolation and pathologic oxidation of the sarcoplasmic endoplasmic reticulum calcium ATPase. *Trends Cardiovasc.Med.*,Vol.16, No.4, pp. 109-114

Cohen, R. A.; Weisbrod, R. M.; Gericke, M.; Yaghoubi, M.; Bierl, C.; & Bolotina, V. M. (1999) Mechanism of nitric oxide-induced vasodilatation: refilling of intracellular stores by sarcoplasmic reticulum Ca^{2+} ATPase and inhibition of store-operated Ca^{2+} influx. *Circ.Res.*,Vol.84, No.2, pp. 210-219

Crauwels, H. M.; Van Hove, C. E.; Holvoet, P.; Herman, A. G.; & Bult, H. (2003) Plaque-associated endothelial dysfunction in apolipoprotein E-deficient mice on a regular diet. Effect of human apolipoprotein AI. *Cardiovasc.Res.*,Vol.59, No.1, pp. 189-199

Curtis, T. M. & Scholfield, C. N. (2001) Nifedipine blocks Ca^{2+} store refilling through a pathway not involving L-type Ca^{2+} channels in rabbit arteriolar smooth muscle. *J.Physiol*,Vol.532, No.Pt 3, pp. 609-623

Edwards, G.; Feletou, M.; & Weston, A. H. (2010) Endothelium-derived hyperpolarising factors and associated pathways: a synopsis. *Pflugers Arch.*,Vol.459, No.6, pp. 863-879

Essalihi, R.; Zandvliet, M. L.; Moreau, S.; Gilbert, L. A.; Bouvet, C.; Lenoel, C.; Nekka, F.; McKee, M. D.; & Moreau, P. (2007) Distinct effects of amlodipine treatment on vascular elastocalcinosis and stiffness in a rat model of isolated systolic hypertension. *J.Hypertens.*,Vol.25, No.9, pp. 1879-1886

Fischmeister, R.; Castro, L.; Abi-Gerges, A.; Rochais, F.; & Vandecasteele, G. (2005) Species- and tissue-dependent effects of NO and cyclic GMP on cardiac ion channels. *Comp Biochem.Physiol A Mol.Integr.Physiol*,Vol.142, No.2, pp. 136-143

Fitch, R. M.; Rutledge, J. C.; Wang, Y. X.; Powers, A. F.; Tseng, J. L.; Clary, T.; & Rubanyi, G. M. (2006) Synergistic effect of angiotensin II and nitric oxide synthase inhibitor in increasing aortic stiffness in mice. *Am.J.Physiol Heart Circ.Physiol*,Vol.290, No.3, pp. H1190-H1198

Fleischmann, B. K.; Murray, R. K.; & Kotlikoff, M. I. (1994) Voltage window for sustained elevation of cytosolic calcium in smooth muscle cells. *Proc.Natl.Acad.Sci.U.S.A*,Vol.91, No.25, pp. 11914-11918

Fransen, P.; Katnik, C.; & Adams, D. J. (1998) ACh- and caffeine-induced Ca2+ mobilization and current activation in rabbit arterial endothelial cells. *Am.J.Physiol*,Vol.275, No.5 Pt 2, pp. H1748-H1758

Fransen, P.; Van Assche, T.; Guns, P. J.; Van Hove, C. E.; De Keulenaer, G. W.; Herman, A. G.; & Bult, H. (2008) Endothelial function in aorta segments of apolipoprotein E-deficient mice before development of atherosclerotic lesions. *Pflugers Arch.*,Vol.455, No.5, pp. 811-818

Frew, J. D.; Paisley, K.; & Martin, W. (1993) Selective inhibition of basal but not agonist-stimulated activity of nitric oxide in rat aorta by NG-monomethyl-L-arginine. *Br.J.Pharmacol.*,Vol.110, No.3, pp. 1003-1008

Ganitkevich, V. Y. & Isenberg, G. (1990) Contribution of two types of calcium channels to membrane conductance of single myocytes from guinea-pig coronary artery. *J.Physiol*,Vol.426, pp. 19-42

Godfraind, T. (2005) Antioxidant effects and the therapeutic mode of action of calcium channel blockers in hypertension and atherosclerosis. *Philos.Trans.R.Soc.Lond B Biol.Sci.*,Vol.360, No.1464, pp. 2259-2272

Godfraind, T.; Finet, M.; Lima, J. S.; & Miller, R. C. (1984) Contractile activity of human coronary arteries and human myocardium in vitro and their sensitivity to calcium entry blockade by nifedipine. *J.Pharmacol.Exp.Ther.*,Vol.230, No.2, pp. 514-518

Guo, X. & Kassab, G. S. (2003) Variation of mechanical properties along the length of the aorta in C57bl/6 mice. *Am.J.Physiol Heart Circ.Physiol*,Vol.285, No.6, pp. H2614-H2622

Isshiki, M.; Mutoh, A.; & Fujita, T. (2004) Subcortical Ca2+ waves sneaking under the plasma membrane in endothelial cells. *Circ.Res.*,Vol.95, No.3, pp. e11-e21

Isshiki, M.; Ying, Y. S.; Fujita, T.; & Anderson, R. G. (2002) A molecular sensor detects signal transduction from caveolae in living cells. *J.Biol.Chem.*,Vol.277, No.45, pp. 43389-43398

Karaki, H.; Ozaki, H.; Hori, M.; Mitsui-Saito, M.; Amano, K.; Harada, K.; Miyamoto, S.; Nakazawa, H.; Won, K. J.et al. (1997) Calcium movements, distribution, and functions in smooth muscle. *Pharmacol.Rev.*,Vol.49, No.2, pp. 157-230

Kauser, K.; da, C., V; Fitch, R.; Mallari, C.; & Rubanyi, G. M. (2000) Role of endogenous nitric oxide in progression of atherosclerosis in apolipoprotein E-deficient mice. *Am.J.Physiol Heart Circ.Physiol*,Vol.278, No.5, pp. H1679-H1685

Knot, H. J. & Nelson, M. T. (1998) Regulation of arterial diameter and wall [Ca^{2+}] in cerebral arteries of rat by membrane potential and intravascular pressure. *J.Physiol*,Vol.508 (Pt 1), pp. 199-209

Koch, W. J.; Ellinor, P. T.; & Schwartz, A. (1990) cDNA cloning of a dihydropyridine-sensitive calcium channel from rat aorta. Evidence for the existence of alternatively spliced forms. *J.Biol.Chem.*,Vol.265, No.29, pp. 17786-17791

Kyselovic, J.; Martinka, P.; Batova, Z.; Gazova, A.; & Godfraind, T. (2005) Calcium channel blocker inhibits Western-type diet-evoked atherosclerosis development in ApoE-deficient mice. *J.Pharmacol.Exp.Ther.*,Vol.315, No.1, pp. 320-328

Leonetti, G.; Cuspidi, C.; Sampieri, L.; Terzoli, L.; & Zanchetti, A. (1982) Comparison of cardiovascular, renal, and humoral effects of acute administration of two calcium channel blockers in normotensive and hypertensive subjects. *J.Cardiovasc.Pharmacol.*,Vol.4 Suppl 3, pp. S319-S324

Liao, P.; Yong, T. F.; Liang, M. C.; Yue, D. T.; & Soong, T. W. (2005) Splicing for alternative structures of $Ca_v1.2$ Ca^{2+} channels in cardiac and smooth muscles. *Cardiovasc.Res.*,Vol.68, No.2, pp. 197-203

Liao, P.; Yu, D.; Li, G.; Yong, T. F.; Soon, J. L.; Chua, Y. L.; & Soong, T. W. (2007) A smooth muscle $Ca_v1.2$ calcium channel splice variant underlies hyperpolarized window current and enhanced state-dependent inhibition by nifedipine. *J.Biol.Chem.*,Vol.282, No.48, pp. 35133-35142

Mancia, G.; De, B. G.; Dominiczak, A.; Cifkova, R.; Fagard, R.; Germano, G.; Grassi, G.; Heagerty, A. M.; Kjeldsen, S. E.et al. (2007) 2007 Guidelines for the Management of Arterial Hypertension: The Task Force for the Management of Arterial Hypertension of the European Society of Hypertension (ESH) and of the European Society of Cardiology (ESC). *J.Hypertens.*,Vol.25, No.6, pp. 1105-1187

Matsuda, J. J.; Volk, K. A.; & Shibata, E. F. (1990) Calcium currents in isolated rabbit coronary arterial smooth muscle myocytes. *J.Physiol*,Vol.427, pp. 657-680

Mian, K. B. & Martin, W. (1995) Differential sensitivity of basal and acetylcholine-stimulated activity of nitric oxide to destruction by superoxide anion in rat aorta. *Br.J.Pharmacol.*,Vol.115, No.6, pp. 993-1000

Mitchell, G. F. (1999) Pulse pressure, arterial compliance and cardiovascular morbidity and mortality. *Curr.Opin.Nephrol.Hypertens.*,Vol.8, No.3, pp. 335-342

Moosmang, S.; Schulla, V.; Welling, A.; Feil, R.; Feil, S.; Wegener, J. W.; Hofmann, F.; & Klugbauer, N. (2003) Dominant role of smooth muscle L-type calcium channel $Ca_v1.2$ for blood pressure regulation. *EMBO J.*,Vol.22, No.22, pp. 6027-6034

Morel, N. & Godfraind, T. (1987) Prolonged depolarization increases the pharmacological effect of dihydropyridines and their binding affinity for calcium channels of vascular smooth muscle. *J.Pharmacol.Exp.Ther.*,Vol.243, No.2, pp. 711-715

Morel, N. & Godfraind, T. (1994) Selective interaction of the calcium antagonist amlodipine with calcium channels in arteries of spontaneously hypertensive rats. *J.Cardiovasc.Pharmacol.*,Vol.24, No.4, pp. 524-533

Nakashima, Y.; Raines, E. W.; Plump, A. S.; Breslow, J. L.; & Ross, R. (1998) Upregulation of VCAM-1 and ICAM-1 at atherosclerosis-prone sites on the endothelium in the ApoE-deficient mouse. *Arterioscler.Thromb.Vasc.Biol.*,Vol.18, No.5, pp. 842-851

O'Rourke, M. F. & Mancia, G. (1999) Arterial stiffness. *J.Hypertens.*,Vol.17, No.1, pp. 1-4

Panza, J. A.; Casino, P. R.; Kilcoyne, C. M.; & Quyyumi, A. A. (1993) Role of endothelium-derived nitric oxide in the abnormal endothelium-dependent vascular relaxation of patients with essential hypertension. *Circulation*,Vol.87, No.5, pp. 1468-1474

Peppiatt, C. M.; Collins, T. J.; Mackenzie, L.; Conway, S. J.; Holmes, A. B.; Bootman, M. D.; Berridge, M. J.; Seo, J. T.; & Roderick, H. L. (2003) 2-Aminoethoxydiphenyl borate (2-APB) antagonises inositol 1,4,5-trisphosphate-induced calcium release, inhibits calcium pumps and has a use-dependent and slowly reversible action on store-operated calcium entry channels. *Cell Calcium*,Vol.34, No.1, pp. 97-108

Pesic, A.; Madden, J. A.; Pesic, M.; & Rusch, N. J. (2004) High blood pressure upregulates arterial L-type Ca^{2+} channels: is membrane depolarization the signal? *Circ.Res.*,Vol.94, No.10, pp. e97-104

Plane, F.; Wiley, K. E.; Jeremy, J. Y.; Cohen, R. A.; & Garland, C. J. (1998) Evidence that different mechanisms underlie smooth muscle relaxation to nitric oxide and nitric oxide donors in the rabbit isolated carotid artery. *Br.J.Pharmacol.*,Vol.123, No.7, pp. 1351-1358

Poburko, D.; Lhote, P.; Szado, T.; Behra, T.; Rahimian, R.; McManus, B.; van, B. C.; & Ruegg, U. T. (2004) Basal calcium entry in vascular smooth muscle. *Eur.J.Pharmacol.*,Vol.505, No.1-3, pp. 19-29

Pratt, P. F.; Bonnet, S.; Ludwig, L. M.; Bonnet, P.; & Rusch, N. J. (2002) Upregulation of L-type Ca^{2+} channels in mesenteric and skeletal arteries of SHR. *Hypertension*,Vol.40, No.2, pp. 214-219

Quignard, J.; Feletou, M.; Corriu, C.; Chataigneau, T.; Edwards, G.; Weston, A. H.; & Vanhoutte, P. M. (2000) 3-Morpholinosydnonimine (SIN-1) and K(+) channels in smooth muscle cells of the rabbit and guinea pig carotid arteries. *Eur.J.Pharmacol.*,Vol.399, No.1, pp. 9-16

Reddick, R. L.; Zhang, S. H.; & Maeda, N. (1994) Atherosclerosis in mice lacking apo E. Evaluation of lesional development and progression. *Arterioscler.Thromb.*,Vol.14, No.1, pp. 141-147

Rhee, S. W.; Stimers, J. R.; Wang, W.; & Pang, L. (2009) Vascular smooth muscle-specific knockdown of the noncardiac form of the L-type calcium channel by microRNA-based short hairpin RNA as a potential antihypertensive therapy. *J.Pharmacol.Exp.Ther.*,Vol.329, No.2, pp. 775-782

Richards, G. R.; Weston, A. H.; Burnham, M. P.; Feletou, M.; Vanhoutte, P. M.; & Edwards, G. (2001) Suppression of K(+)-induced hyperpolarization by phenylephrine in rat mesenteric artery: relevance to studies of endothelium-derived hyperpolarizing factor. *Br.J.Pharmacol.*,Vol.134, No.1, pp. 1-5

Safar, M. E.; Blacher, J.; & Jankowski, P. (2011) Arterial stiffness, pulse pressure, and cardiovascular disease-is it possible to break the vicious circle? *Atherosclerosis*,Vol.218, No.2, pp. 263-271

Safar, M. E.; Pannier, B.; Laurent, S.; & London, G. M. (1989) Calcium-entry blockers and arterial compliance in hypertension. *J.Cardiovasc.Pharmacol.*,Vol.14 Suppl 10, pp. S1-S6

Slama, M.; Safavian, A.; Tual, J. L.; Laurent, S.; & Safar, M. E. (1995) Effects of antihypertensive drugs on large artery compliance. *Neth.J.Med.*,Vol.47, No.4, pp. 162-168

Smirnov, S. V. & Aaronson, P. I. (1992) Ca^{2+} currents in single myocytes from human mesenteric arteries: evidence for a physiological role of L-type channels. *J.Physiol*,Vol.457, pp. 455-475

Striessnig, J.; Grabner, M.; Mitterdorfer, J.; Hering, S.; Sinnegger, M. J.; & Glossmann, H. (1998) Structural basis of drug binding to L Ca^{2+} channels. *Trends Pharmacol.Sci.*,Vol.19, No.3, pp. 108-115

Sun, J.; Picht, E.; Ginsburg, K. S.; Bers, D. M.; Steenbergen, C.; & Murphy, E. (2006) Hypercontractile female hearts exhibit increased S-nitrosylation of the L-type Ca^{2+} channel alpha1 subunit and reduced ischemia/reperfusion injury. *Circ.Res.*,Vol.98, No.3, pp. 403-411

Tamargo, J.; Caballero, R.; Gomez, R.; & Delpon, E. (2010) Cardiac electrophysiological effects of nitric oxide. *Cardiovasc.Res.*,Vol.87, No.4, pp. 593-600

Tang, Z. Z.; Hong, X.; Wang, J.; & Soong, T. W. (2007) Signature combinatorial splicing profiles of rat cardiac- and smooth-muscle $Ca_v1.2$ channels with distinct biophysical properties. *Cell Calcium*,Vol.41, No.5, pp. 417-428

Tiwari, S.; Zhang, Y.; Heller, J.; Abernethy, D. R.; & Soldatov, N. M. (2006) Atherosclerosis-related molecular alteration of the human $Ca_v1.2$ calcium channel alpha1C subunit. *Proc.Natl.Acad.Sci.U.S.A*,Vol.103, No.45, pp. 17024-17029

Tsai, E. J. & Kass, D. A. (2009) Cyclic GMP signaling in cardiovascular pathophysiology and therapeutics. *Pharmacol.Ther.*,Vol.122, No.3, pp. 216-238

Van Assche, T.; Fransen, P.; Guns, P. J.; Herman, A. G.; & Bult, H. (2007) Altered Ca^{2+} handling of smooth muscle cells in aorta of apolipoprotein E-deficient mice before development of atherosclerotic lesions. *Cell Calcium*,Vol.41, No.3, pp. 295-302

Van Assche, T.; Hendrickx, J.; Crauwels, H. M.; Guns, P. J.; Martinet, W.; Fransen, P.; Raes, M.; & Bult, H. (2011) Transcription profiles of aortic smooth muscle cells from atherosclerosis-prone and -resistant regions in young apolipoprotein E-deficient mice before plaque development. *J.Vasc.Res.*,Vol.48, No.1, pp. 31-42

Van Hove, C. E.; Van der Donckt, C.; Herman, A. G.; Bult, H.; & Fransen, P. (2009) Vasodilator efficacy of nitric oxide depends on mechanisms of intracellular calcium mobilization in mouse aortic smooth muscle cells. *Br.J.Pharmacol.*,Vol.158, No.3, pp. 920-930

Vanhoutte, P. M.; Shimokawa, H.; Tang, E. H.; & Feletou, M. (2009) Endothelial dysfunction and vascular disease. *Acta Physiol (Oxf)*,Vol.196, No.2, pp. 193-222

Vayssettes-Courchay, C.; Ragonnet, C.; Isabelle, M.; & Verbeuren, T. J. (2011) Aortic stiffness in vivo in hypertensive rat via echo-tracking: analysis of the pulsatile distension waveform. *Am.J.Physiol Heart Circ.Physiol*,Vol.301, No.2, pp. H382-H390

Westerhof, N.; Lankhaar, J. W.; & Westerhof, B. E. (2009) The arterial Windkessel. *Med.Biol.Eng Comput.*,Vol.47, No.2, pp. 131-141

Weston, A. H.; Richards, G. R.; Burnham, M. P.; Feletou, M.; Vanhoutte, P. M.; & Edwards, G. (2002) K^+-induced hyperpolarization in rat mesenteric artery: identification, localization and role of Na^+/K^+-ATPases. *Br.J.Pharmacol.*,Vol.136, No.6, pp. 918-926

White, D. G. & Martin, W. (1989) Differential control and calcium-dependence of production of endothelium-derived relaxing factor and prostacyclin by pig aortic endothelial cells. *Br.J.Pharmacol.*,Vol.97, No.3, pp. 683-690

Yuan, X. J.; Tod, M. L.; Rubin, L. J.; & Blaustein, M. P. (1996) NO hyperpolarizes pulmonary artery smooth muscle cells and decreases the intracellular Ca2+ concentration by activating voltage-gated K+ channels. *Proc.Natl.Acad.Sci.U.S.A*,Vol.93, No.19, pp. 10489-10494

Zhang, D. X.; Mendoza, S. A.; Bubolz, A. H.; Mizuno, A.; Ge, Z. D.; Li, R.; Warltier, D. C.; Suzuki, M.; & Gutterman, D. D. (2009) Transient receptor potential vanilloid type 4-deficient mice exhibit impaired endothelium-dependent relaxation induced by acetylcholine in vitro and in vivo. *Hypertension*,Vol.53, No.3, pp. 532-538

Zhou, Z. H.; Wang, J.; Xiao, H.; Chen, Z. J.; Wang, M.; Cheng, X.; & Liao, Y. H. (2008) A novel autoantibody in patients with primary hypertension: antibody against L-type Ca^{2+} channel. *Chin Med.J.(Engl.)*,Vol.121, No.16, pp. 1513-1517

MAP Kinase-Mediated and MLCK-Independent Phosphorylation of MLC20 in Smooth Muscle Cells

Maoxian Deng, Lixia Deng and Yarong Xue

Additional information is available at the end of the chapter

http://dx.doi.org/10.5772/47732

1. Introduction

Smooth muscle cells constitute the walls of various organs and tubes in the body, including the blood vessels, gastrointestinal tract, respiratory tract, bladder, and reproductive tracts. The primary function of smooth muscle contraction is to generate force, which is utilized to perform many physiological processes such as blood flow and blood pressure maintenance, gastrointestinal motility, bronchial diameter regulation, bladder evacuation, and fetus expulsion. Smooth muscle contraction is caused by the sliding of myosin and actin filaments over each other. Movement of the two types of filaments happens when the globular myosin heads protruding from myosin filaments attach and interact with actin filaments to form crossbridges. The myosin head first attaches to actin together with the products of ATP hydrolysis, performs a power stroke associated with release of hydrolysis products, and detaches from actin upon binding with a new ATP. The myosin interacts with the actin to convert chemical energy, in the form of ATP, to mechanical energy. The coordinated regulation of contraction is a key property of the smooth muscle. When the smooth muscle functions normally, it contributes to general health and wellness. Contractile abnormalities of the smooth muscle are considered to underlie many diseases and disorders, including hypertension, vasospasm, diabetes-associated microvascular abnormalities, bronchial asthma, preterm labor, urinary incontinence, megacolon, and irritable bowel syndrome. Not surprisingly, inadequate contraction and relaxation of smooth muscle may cause the dysfunction of these hollow organs, which is usually associated with morbidity and mortality. Hence, the precise regulation of smooth muscle contraction is much more important in smooth muscles than in striated muscle. The muscle cells can respond to physiological and pathological signals from the environment to adapt to the environmental demands. This adaptation is accomplished through signal transduction, which activates factors that signal pathways and ultimately lead to muscle contraction or relaxation. The regulation of smooth muscle contraction has two main mechanisms: neuromuscular and myogenic. In the neuromuscular mechanism, the smooth muscle receives principally neural innervation from the autonomic nervous system inside the same tissue, even though the central nervous system may be involved in the regulation of smooth muscle contraction. The myogenic mechanism, on the other hand, plays a more important role in the regulation of smooth muscle contraction. Under this mechanism, smooth muscle cell contraction is regulated principally by the mechanical (stretch) activation of the contractile proteins myosin and the actin in the intact body. A change in membrane potential, brought by the firing of action potentials or by the activation of stretch-dependent ion channels in the plasma membrane, can also trigger contraction. These smooth muscle cells also develop tonic and phasic contractions in response to changes in load or length. In addition, the contractile state of the smooth muscle is controlled by the hormones, autocrine/paracrine agents, and other local chemical signals. Regardless of the stimulus, the smooth muscle cells use crossbridge cycling between the actin and myosin to develop force and calcium ions (Ca^{2+}) that serve to initiate contraction in light of the current view. The deep mechanism for both myogenic and neuromuscular regulations is involved in cell signaling. Because current therapies for disorders and diseases of smooth muscles are costly and primarily palliative and do not target the cause of the disease, attention is being focused on the detailed understanding of the molecular basis of smooth muscle function and regulation and identification of abnormalities (dysfunctional proteins and signaling pathways) that lead to contractile pathologies and the development of strategies to reverse such abnormalities. The muscle contraction regulation mediated by signaling pathways is the key event. In recent years, a variety of signaling pathways has been implicated in the regulation of muscle contraction. A growing body of evidence shows a fine vista for the research of pathway-mediated regulation of the smooth muscle. One of the widely accepted notions is that the increase in Ca^{2+} concentration activates the Ca^{2+}-calmodulin (CaM)–myosin light chain kinase (MLCK) pathway and stimulates the 20-kDa myosin light chain (MLC20) phosphorylation. The MLC20 phosphorylation at Ser^{19} causes a conformational change that increases the angle in the neck domain of the myosin heavy chain which mobilizes the crossbridges and causes the actin thin filament to slide along the myosin thick filament. Through an unknown mechanism, this interaction between the myosin and actin activates the ATPase activity of the myosin head region and leads to the development of a contractile force. Therefore, in theory, the factors that cause the inhibition of the MLCK activity and expression, such as calcium deprivation and calmodulin inactivation, could abolish consequentially MLC20 phosphorylation and smooth muscle cell contraction. However, increasing evidence supports Ca^{2+}-independent contraction of smooth muscle (Deng et al., 2001; Harnett & Biancani, 2003; Ratz et al., 2009). Here, we will describe an MLCK-independent and mitogen-activated protein (MAP) kinase-mediated phosphorylation of MLC20.

2. Expression and function of MLC20 and MLCK in smooth muscle

The myosin II (also called conventional myosin), a hexameric protein complex, is composed of two identical 200-kD heavy chains and two sets of light chains: the 17-kD light chain (MLC17, also known as ELC) and the 20-kD regulatory light chain (MLC20). The exact function of the MLC17 is unclear but it may contribute to the structural stability of the myosin head along with the MLC20. The MLC20 is a small ring around the neck region of heavy chains of myosin and is also known as the regulatory light chain (also called RLC20, RLC2, and LC2). It is believed to participate actively in muscle contraction. The MLC20 is expressed not only in smooth muscle, but also in cardiac, skeletal, and nonmuscle cells. Only the MLC20 in the smooth muscle appears to play a unique role in increasing the actin-activated ATPase activity. The MLC20 is activated by phosphorylation at multiple serine and threonine residues. The phosphorylation of the MLC20 is thought to play a pivotal role in regulating muscle contraction.

Up to now, the MLC20 is the only known physiological substrate of the MLCK. Protein kinase C (PKC), CaM-kinase II, Rho-kinase, p21-activated kinase, and p34cdc2 kinase also phosphorylate three residues on the N-terminus of the MLC20, Ser-1, Ser-2, and Thr-9. In addition, the integrin-linked kinase and ZIP-kinase are reported to phosphorylate the MLC20 in the absence of Ca^{2+}, at Ser19 and Thr18, thus activating the myosin activity (Niiro & Ikebe, 2001). *In vivo* phosphorylation of the MLC20 isoforms is accomplished by raising free intracellular Ca^{2+} and subsequent activation of the Ca^{2+}-calmodulin-dependent MLCK. Ca^{2+}/calmodulin-dependent MLCK is considered the primary regulator of MLC20 phosphorylation among potential regulators of MLC20 phosphorylation. The dephosphorylation of phosphorylated MLC20 is catalyzed by the myosin light chain phosphatase (MLCP), which counters the MLCK that promotes contraction by phosphorylating MLC20. The degree of phosphorylation of the RLC depends on the ratio of the activities of the MLCK and MLCP.

The myosin light chain kinase (MLCK or MYLK) is a Ca^{2+}/CaM-activated kinase found in smooth, cardiac, and skeletal muscles as well as in many mammalian nonmuscle cells. It is a serine/threonine-specific protein kinase that phosphorylates the MLC20 of myosin II. There are three isoforms of the MLCK, i.e., smooth muscle (smMLCK, ~130-kDa), skeletal muscle (skMLCK, ~220-kDa), and cardiac muscle (cMLCK). The short MLCK (smMLCK, ~130kDa) is best known as the conventional smooth muscle MLCK. The smMLCK is encoded by the mylk1 gene, which expresses three transcripts in a cell-specific manner due to the alternate promoters, long MLCK (210- to 220-kDa), short MLCK (130-kDa), and the noncatalytic gene product, called telokin. The short smooth muscle MLCK (130-kDa) is ubiquitous in all adult tissues with the highest amounts in smooth muscle tissues. Since both the short and long MLCK are found in smooth muscles or cultured/embryonic smooth muscle tissues, they are called smooth muscle MLCK (smMLCK). The two smMLCKs have been extensively described, and their function is to regulate the activity of the nonmuscle and smooth muscle myosin II. Here we will deal only with the short smooth muscle MLCK (smMLCK).

The activation of myosin motors by the MLCK modulates a variety of contractile processes, including smooth muscle contraction, cell adhesion, migration, and proliferation. The dysregulation of these processes contribute to a number of diseases. It is widely accepted that the MLCK phosphorylates the MLC20 of smooth and nonmuscle myosin II in the presence of Ca^{2+} and calmodulin. The phosphorylation of the MLC20 facilitates myosin interaction with actin filaments. Once there is an influx of calcium cations into the muscle, either from the sarcoplasmic reticulum or, more important, from the extracellular space, the contraction of smooth muscle fibers may begin. First, the elevation of the Ca^{2+} concentration in smooth muscles causes the Ca^{2+} to bind to the calmodulin (CaM). Then, the complex of the Ca^{2+} and CaM (Ca^{2+}/CaM) activates the MLCK, which would phosphorylate the myosin light chain (MLC20) at serine residue 19 to generate phospho-MLC20. The phosphorylation of the MLC20 enables the myosin crossbridge to bind to the actin filament and allows contraction to begin (through the crossbridge cycle). The Ca^{2+}-CaM activates the MLCK by reversal of an auto-inhibited state. In contrast, reducing intracellular calcium concentration inactivates MLCK but does not stop smooth muscle contraction since the MLC20 has been physically modified through phosphorylation. This regulatory model of myosin phosphorylation is widely accepted as the intracellular path for the induction of smooth muscle contraction.

However, to a certain extent, this viewpoint about the Ca^{2+}-, MLCK- and MLC20-dependent regulatory mechanism is founded on a presumption. Several observations of smooth muscle contraction cannot be explained by the mode of phosphorylation (Kohama K & K., 1995). For example, when the uterine smooth muscle is subjected to prolonged incubation in a Ca^{2+}-free medium, the oxytocin induces the contraction of the muscle without any signs of MLC20 phosphorylation (Oishi et al., 1991). Another example is the MLC20 phosphorylated in rat embryo fibroblasts, which contains no detectable MLCK (Emmert et al., 2004). In addition, these are increasing evidence to support the calcium- and MLCK-independent mechanism of MLC20 phosphorylation (McFawn et al., 2003; El-Toukhy et al., 2006; Cho et al., 2011). According to the current mode, the expression of the MLCK in tissues/cells is necessarily consistent with the MLC20 phosphorylation. However, there has not been any report about the consistence between the MLCK expression and MLC20 phosphorylation. Herein, we first investigated the consistency between the MLCK expression and phosphor MLC20. In the following section we will describe a recent finding, the significant inconsistence between MLCK expression, and MLC20 phosphorylation in multiple smooth muscles in mice.

3. Inconsistence between MLCK expression and MLC20 phosphorylation

According to the current view, the MLCK expression must co-localize with the MLC20 phosphorylation in cells or tissues. However, by a series of experiments, we recently found that phosphorylation (Ser19) of the MLC20 is inconsistent with 130-kDa MLCK expression in mouse aorta, bladder, large/small intestines, stomach, and uterus (Deng et al., 2011) (Figure.1A and B).

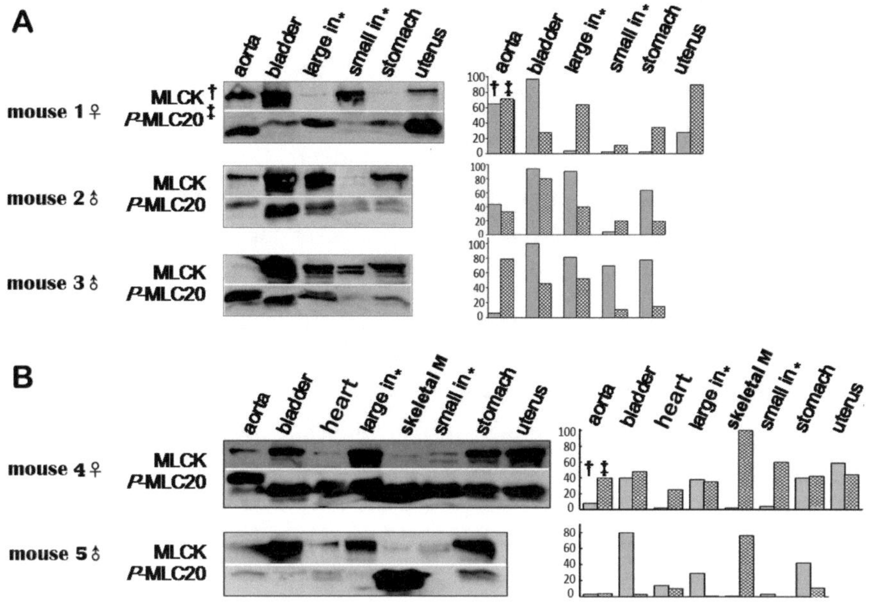

Figure 1. Inconsistence between MLC20 phosphorylation and MLCK expression in smooth muscle tissues. *in, intestine; †, MLCK expression; ‡, phosphor MLC20.

In the experiments, fresh smooth muscle tissues were homogenized on ice with pre-cooled small glass homogenizers within ice-cold lysis buffer (50 mM Tris, 300 mM NaCl, 3 mM EGTA, 0.1 mM sodium orthovanadate, 10% glycerol v/v, 1% NP-40 v/v, 0.3% SDS w/v, protease and phosphatase inhibitor cocktails from Sigma Co., Germany, pH 7.6). Twenty micrograms of protein were loaded onto 10% SDS-PAGE gels that were subjected to electrophoresis in running buffer (BioRad Co., Hercules, CA). The proteins were then transferred to the Immobilon-P Transfer Membrane (Millipore Co., MA), followed by blocking it with 5% fat-free milk in TBST buffer. The MLCK was detected with a specific antibody. After it was washed with 1×PBS for three times, the membranes were re-probed with antibody anti-phospho MLC20. The results show that the levels of MLCK expression and phosphorylated MLC20 in different smooth muscles from the same mouse are markedly different (Figure 1). For instance, the large intestine of mouse 1 contains a high level of phosphor MLC20 but the MLCK expression is low. In contrast, the small intestine has low phosphor MLC20 but the MLCK expression is high. Besides, the MLCK expression and phosphor MLC20 in the same type of smooth muscle tissue are inconsistent among different individuals. For example, the large intestine of mouse 1 contains a low expression of MLCK and high phosphor MLC20. Conversely, mice 2 and 5 express high levels of MLCK in their large intestines but the MLC20 phosphorylation is low. The conspicuous inconsistence between the MLC20 phosphorylation and MLCK expression was subsequently verified by Western blotting in independent experiments. These interesting results cannot be explained by the current regulatory mode. Obviously, they suggest that an alternative regulation system may play an active role in the MLC20 phosphorylation.

4. Unknown substance(s) other than MLCK phosphorylates MLC20

Even though some of the smooth muscle does not show phosphor MLC20 bands (Figure 1), those tissues still contain the MLC20 because we noticed in subsequent experiments that all these frozen lysates had similar levels of phosphor MLC20 after storage in -80°C and several cycles of freezing and thawing. The MLC20 in some tissues was not visualized by any specific antibody against the phosphor MLC20 only because it was not phosphorylated. The MLC20 phosphorylation in MLCK-free tissues suggests that unknown molecule(s) may phosphorylate the MLC20 in phosphor MLC20-containing tissues.

Thus, we proposed a hypothesis that something else other than the MLCK phosphorylates MLC20 in the smooth muscle tissues, which contain no detectable MLCK. Even though we did not know what kind of substance(s) the phosphorylated MLC20 is/are in the phosphor MLC20-containing and MLCK-free tissue, it may mobilize the phosphorylative process of the MLC20. To test our hypothesis, we designed an unconventional experiment, and added a small amount of tissue extract which was freshly prepared from the phosphor MLC20-containing tissues without any detectable MLCK (extract 2), to another extract that contained the unphosphorylated but had no phosphorylated MLC20 and MLCK (extract 1). In other words, we theorized that extract 2 is a catalyst and extract 1 is a substrate of extract 2. By rapid Western blotting assay (with one-hour incubation at room temperature), we chose the tissue extract containing the high phosphor MLC20 but no detectable MLCK as a catalyst (extract 2), and the one containing no phosphor MLC20 and detectable MLCK as the substrate (extract 1). Then, a small amount of exact 2 was added directly into extract 1, as shown in Figure 2A. After the one-hour 37°C incubation in water, as expected, the added extract 2 phosphorylated strongly MLC20 of the extract 1 (lines 2 and 3 in Figure 2A). This suggests that high phosphor MLC20 tissue content may have non-MLCK substance(s) that induce/s MLC20 phosphorylation.

To evaluate the further roles of the MLCK and the possible non-MLCK substance(s) in the MLC20 phosphorylation, we assayed the effects of the isolated MLCK on MLC20 phosphorylation (Figure 2B). Firstly, the MLCK was isolated from the fresh smooth muscle tissue by the anti-MLCK monoclonal antibody from Sigma Co., following conventional immunoprecipitation protocol. Secondly, the isolated MLCK was immediately added to the freshly prepared extract 1, which contains unphosphorylated MLC20 but no detectable MLCK (line 2 in Figure 2B). In the meantime, a small amount of extract 2 that contains a high level phosphor MLC20 but no detectable MLCK was added into the same 'substrate' extract 1 as a control (line 3). The mixtures were incubated in 37°C water for one hour, followed by Western blotting assay. Extracts 1, 2 and isolated MLCK were loaded to lines 1, 5 and 4, respectively, for assays of its own MLCK and phosphor MLC20. No protease and phosphatase inhibitor were added into the reaction system. After the membrane was washed with PBS for three times, the actin and phospho-MLC20 were simultaneously probed by mixture solution of the antibodies against actin and phospho-MLC20 (Membrane was not striped by striping buffer because the sizes of the three targeting proteins were much different). The results show that the immunoprecipitated MLCK failed to phosphorylate MLC20, while the extract without detectable MLCK phosphorylated MLC20 of extract 1 contains no detectable MLCK (line 3). Several independent experiments were conducted to verify the results. These experiments suggest that MLC20 can be phosphorylated *in vitro* and that non-MLCK substance(s) may trigger the MLC20 phosphorylation.

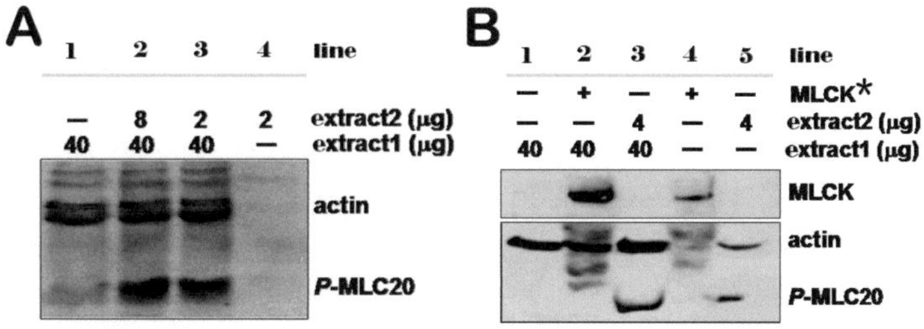

Figure 2. Unknown substance(s) in smooth muscle tissue but not immunoprecipitated MLCK induced MLC20 phosphorylation. (A) Small amount of extract 2 without MLCK induced MLC20 phosphorylation of MLC20 in extract 1 (lines 2 and 3). (B) Precipitated MLCK did not induce MLC20 phosphorylation (line 2). The extract 2 containing no detectable MLCK increased significantly the phosphorylation of MLC 20 in extract 1(line 3). MLCK*, MLCK freshly isolated by co-immunoprecipitation from smooth muscle tissue; *p*-MLC20, phosphor MLC20.

5. Inhibition of MLCK expression and activity do not affect MLC20 phosphorylation

According to the current regulatory model, intracellular calcium ion is the trigger of smooth muscle contraction. The contractile initiation, maintenance, and strength are dependent on the control of intracellular free Ca^{2+} level. The canonical excitation-contraction coupling pathway is triggered by neural, hormonal or myogenic stimulation to elicit the influx of extracellular or intracellular Ca^{2+} (from the sarcoplasmic reticulum) into the cytosol. The Ca^{2+} binds to calmodulin to form the Ca^{2+}/CaM complex. At an increased level of Ca^{2+}, rapid binding of the Ca^{2+} to CaM occurs and triggers MLCK activation. Once activated, the MLCK phosphorylates MLC20 leads to crossbridges cycling and generation of the contractile force. Therefore, several factors may affect the MLC20 phosphorylation, including expression of the MLCK gene, MLCK activity, free Ca^{2+} concentration, and CaM activity. The following experiments were designed to access the effects of these factors on the role of the MLCK in MLC20 phosphorylation: 1) using siRNA to knockdown the MLCK expression in human bladder smooth muscle (hBSM) cells, 2) inactivating the MLCK by inhibitors of calmodulin and MLCK, and 3) depriving the calcium ion of cultured hBSM cells.

We first investigated the effect of alteration in MLCK expression on the MLC20 phosphorylation. To inhibit the MLCK expression, the siRNA against MLCK (from Santa Cruz Biotechnology Inc.) was transfected with Lipofectamine 2000 into cultured hBSM cells at a cell density of 40-50% confluence, which were isolated from the peri-cancer tissue of a bladder cancer patient. Only lower than seventh passage primary cells were used for experiments. Two days after transfection, cells were harvested for Western blotting. MLCK, *pan*-actin and phosphor MLC20 were detected by specific antibodies. Western blotting analysis showed that no significant alteration of MLC20 phosphorylation level was observed while the MLCK expression was knocked down (Figure 3A). Together with the data about the expression of MLCK and MLC20 phosphorylation from mice, this experiment provides further proof that MLC20 phosphorylation is independent of MLCK gene expression.

Equally vital to the MLCK gene expression, the MLCK activity or activation may contribute to its functions. Then, the subsequent experiments were designed to study the effects of the MLCK activity on the MLC20 phosphorylation. The MLCK and its upstream molecule, the CaM, were directly inactivated by the corresponding commercial inhibitory peptides (EMD Chemicals Inc., NJ) in both permeabilized or unpermeabilized hBSM cells (Figure 3B, C) and mouse smooth muscle extracts (Figure 3D). The MLC20 phosphorylation was examined by Western blotting. In addition, ethylene glycol tetraacetic acid (EGTA), a Ca^{2+} chelating agent, was added into the medium to prevent the formation of Ca^{2+}/CaM complex in cultured hBSM cells (Figure 3B and C). CaM and MLCK inhibitory peptides were added into mixtures of smooth muscle extracts 1 and 2, in which small amount of extract 2 was able to phosphorylate MLC20 of extract 1 (shown in figure 2). After one-hour incubation at 37°C, *pan*-actin and phosphor MLC20 were assayed by Western blotting. Likewise, none of the inhibitors of the MLCK and CaM, and EGTA inhibit the MLC20 phosphorylation in both cultured hBSM cells and tissue extracts.

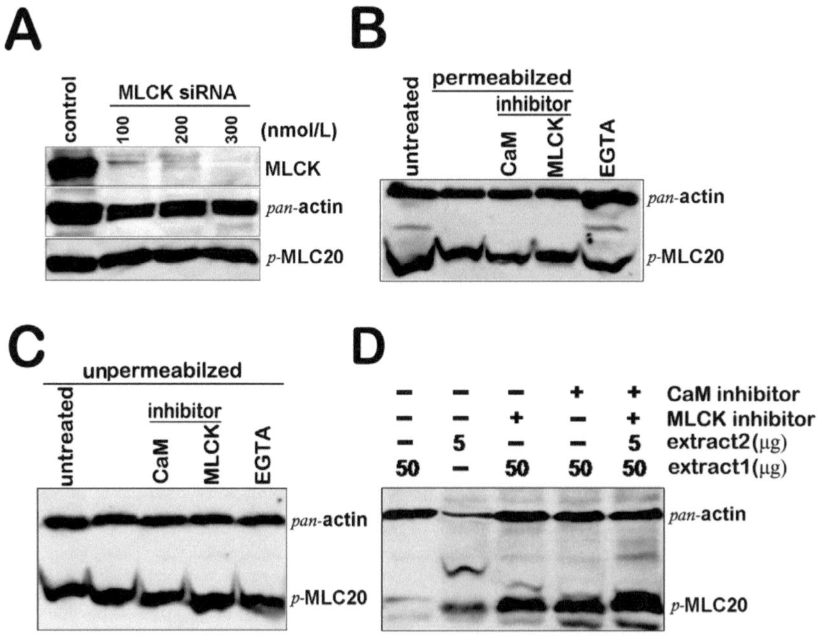

Figure 3. Inhibition of MLCK expression and activity does not block MLC20 phosphorylation. (A) siRNA-mediated MLCK knockdown did not inhibit MLC20 phosphorylation. (B and C) Inhibitors of CaM and MLCK and EGTA did not suppress MLC20 phosphorylation in permeabilized and unpermeabilized cells. (D) Neither CaM nor MLCK inhibitors blocked MLC20 phosphorylation in smooth muscle extracts.

Putting the data together, we firmly believe that an alternative mechanism plays an important part in the regulation of MLC20 phosphorylation. This potential regulatory mechanism is MLCK-independent.

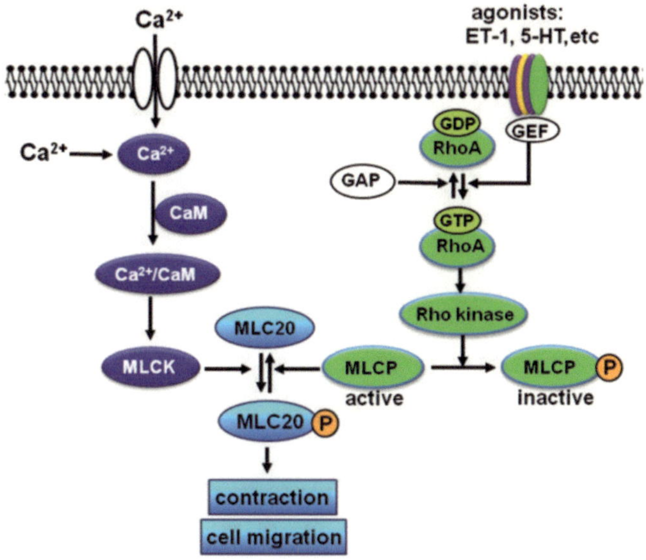

Figure 4. Two main pathways involved in regulation of smooth muscle cell contraction and migration. The contractile response is initiated by a rapid and transient rise in intracellular Ca^{2+}, followed by a Ca^{2+}-calmodulin interaction, the MLCK activation to initiate phosphorylation of the MLC20 and therefore contraction and cell migration (the blue). Agonists (such as ET-1) bind to serpentine receptors on the membrane of the smooth muscle cells to activate the RhoA/Rho kinase signaling pathway, which inactivates the MLC phosphatase (MLCP) by phosphorylation, and resulted in increased unphosphorylated MLCP and sustained smooth muscle contraction (the green). GAP, GTPase activating protein; GEF, GDP/GTP exchange factor.

6. MAP kinase pathways and functions in smooth muscle cells

After we verified the MLCK-independent regulation of MLC phosphorylation, we were interested in searching for the non-MLCK mechanism that may function as the trigger of MLC20 phosphorylation. The cell signaling pathways would be most likely to address this intractable issue. In recent years, a multitude of signaling pathways has been suggested to regulate the smooth muscle contractility. However, these pathways can be broken down into two major pathways (Figure 4): the calcium–calmodulin signalling pathway and Rho/Rho-kinase pathway. The former, the Ca^{2+}/CaM pathway, regulates smooth muscle contraction by the binding of increased intracellular calcium ion with CaM, which activates the MLCK. The MLCK phosphorylates the MLC20 at the neck of the myosin heavy chains. The phosphorylation increases the ATPase activity and thereby produces contraction. In the Rho/Rho-kinase pathway, the Rho A-GTP, activated by multiple stimuli, activates the Rho kinase (also known as Rok or Rock). The activated kinase phosphorylates the myosin phosphatase and therefore modulates the MLC20 phosphorylation.

In the following sections, we will describe a MAP kinase-mediated phosphorylative mechanism of MLC20.

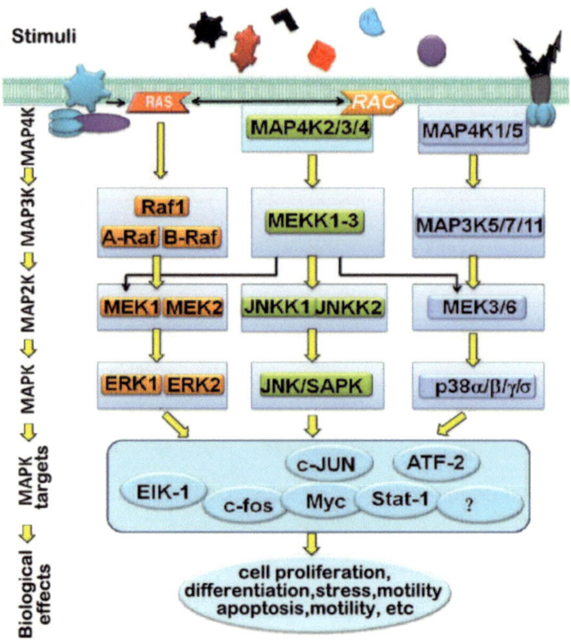

Figure 5. The main components of the three MAP kinase pathways. MAP kinase cascade is typically composed of four kinases that establish a sequential activation pathway comprising a MAP kinase kinase kinase kinase (MAP4K), MAP kinase kinase kinase (MAP3K), MAP kinase kinase (MAP2K), and MAP kinase (MAPK). Extracellular/intracellular stimuli activate MAP4K, MAP3K, MAP2K and MAPK sequentially, leading to the multiple biological effects.

MAP kinases (MAPKs) comprise a family of serine threonine kinases, which include three major sets of kinases: extracellular signal-regulated kinases (ERK1 and ERK2), c-Jun amino-terminal kinase/stress-activated protein kinases (JNKs), and p38 MAPKs (Figure 5). These kinases constitute three major discrete cascades and serve as focal points in response to a variety of extracellular stimuli. Members of all three MAPK families, the ERK, JNK, and the p38 MAPKs, are expressed in various types of tissues and cells, including smooth muscles.

6.1. ERK pathway

The best-characterized members of the MAPK family are ERK1 and ERK2, also known as p44 MAPK and p42 MAPK, respectively. The canonical ERK cascade comprises Ras, Rafs, MEK1/2, ERK1/2 and several MAPK-activating protein kinases (MAPKAPK, such as Elk-1, Sap1a and c-Fos). ERK1 and ERK2 are expressed to various extents in all tissues, including various types of smooth muscle. ERK1/2 is distributed throughout quiescent cells, but upon stimulation, a significant population of ERK1/2 accumulates in the nucleus. Immediate upstream kinases that activate ERK cascade are mitogen-activated protein kinase kinase 1/2

(MEK1/2). The ERK is activated by a large variety of diverse extracellular stimuli, including growth factors, cytokines, virus infection, transforming agents, and carcinogens. Activated ERK1/2 phosphorylate numerous substrates, including various membrane proteins (CD120a, Syk, and calnexin), nuclear substrates (SRC-1, Pax6, Elk-1, MEF2, c-Fos, c-Myc, and STAT3), cytoskeletal proteins, and several MAPK-activated protein kinases (MKs), and MAPK-interacting kinases(MNKs). As a consequence of the activation, ERK pathway regulates many distinct and even opposing cellular processing, including proliferation, differentiation, metabolism, morphology, survival and even apoptosis.

ERK pathway may be involved in regulation of contraction, migration and proliferation of smooth muscle cells. PD-098059, a specific inhibitor of MAP kinase kinase (MEK1/2, the upstream of ERK1/2), reduced vascular smooth muscle contraction and increase in blood pressure induced by Ang II and phenylephrine, perhaps by inhibiting ERK1/2 activation (Escano *et al.*, 2008) and airway smooth muscle contraction by isoprenaline (Lelliott *et al.*, 2012). Direct inhibition of ERK1/2 by U-0126 attenuates Ang II- and isoprenaline-induced contraction of bronchial smooth muscle (Sakai *et al.*, 2010; Lelliott *et al.*, 2012). Similar role of ERK1/2 are also observed in other smooth muscle (cells) (Jeong *et al.*, 2011). However, some reports show that the inhibitors of ERK and its upstream do not affect vascular smooth muscle contraction (Do *et al.*, 2009; Bauer *et al.*, 2011; Sathish *et al.*, 2011). Thus, this role of ERK in regulation of smooth muscle contraction is not universal, or even controversial.

The biological process of cell migration is similar to that of muscle cell contraction. The functional role of ERK in the regulation of smooth muscle cell migration is reported by a large number of articles (Aitken & Bagl, 2001; Kavurma & Khachigian, 2003). Many reports about the regulatory model of ERK1/2 in smooth muscle cell migration are consistent, even though it is not entirely clear how ERK activation promotes smooth muscle cell motility. For example, inhibition of ERK by inhibitor and antisense oligonucleotide blocks smooth muscle migration (Graf *et al.*, 1997; Gerthoffer, 2007). Our study confirmed role of ERK1/2 in smooth muscle cell migration by blocking movement by PD-098059(Deng *et al.*, 2011). In addition, Raf-1 kinase, the upstream molecule of MEK1/2 that activates ERK1/2, may be involved in regulation of smooth muscle contraction (Sathishkumar *et al.*, 2010).

6.2. p38 pathway

The p38 cascade is composed of MEKK3, MKK3/4/6, p38 and several MAPKAPKs (EIk-1, CHOP, ATF2 and MEF2A). The p38 family includes four splice variants: p38α, p38β, p38$\gamma$and p38δ. Like the rest of MAP kinases, p38 are ubiquitously expressed in various types of tissues, including various smooth muscle cells. Two main MAPKKs, MEK3 and MEK6, function as the upstream molecules of p38, which are known to activate p38. In addition, MEK4, an upstream kinase of JNK, can aid in the activation of p38α and p38δ in specific cell types. The p38 MAPK cascade is activated by numerous promigratory stimuli, including platelet-derived growth factor (PDGF), Ang II, S1P, and thrombin. This kinase is also responsive to a wide range of environmental stresses, such as ultraviolet irradiation, heat shock, osmotic shock, as well as response to inflammatory cytokines, but less by serum and growth factors. Substrates of p38 *in vivo* include transcription factors (ATF2, MEF2C), the RNA binding protein tristetraprolin, and several protein kinases, such as MAPKAP kinases 2, 3, and 5. The p38 signaling has been implicated in cellular responses including inflammation, cell cycle, cell death, development, cell differentiation, senescence, and tumorigenesis. Because p38 MAPKs are key regulators of inflammatory cytokine expression, they appear to be involved in many human diseases such as asthma and autoimmunity.

The p38 MAPKs may participate in multiple processes in smooth muscle cells including contraction, oxidative stress signaling, and cytokine synthesis . Contraction induced by some but not all agonists depends on p38 activity. SB203580, the specific inhibitor of p38, blocks spontaneous and agent-induced smooth muscle contraction (Lee *et al.*, 2007; Barona *et al.*, 2011). Besides, by using SB203580 and siRNA, researchers have demonstrated that p38 is involved in regulation of smooth muscle cell migration (Mugabe *et al.*, 2010).

6.3. JNK pathway

The JNK cascade comprises generally MAP-ERK kinase kinase-1 (MEKK1), MKK4/7, JNKs and MAPK-activated C-Jun, JunB and ATF2. JNK family, encoded by three genes, jnk1, jnk2, jnk3, includes three main spliced forms, JNK1, -2, and -3, also known as SAPKγ, SAPKα, and SAPKβ, respectively. JNK1 and JNK2, are believed to be expressed in every cell and tissue type, whereas the JNK3 protein is found primarily in brain (Bode & Dong, 2007). JNK activation is much more complex than that of ERK1/ERK2 owing to inputs by a greater number of MAP4Ks and MAP3Ks. This diversity of upstream messages allows a wide range of stimuli to activate this MAPK pathway, including UV and γ-irradiation, protein synthesis inhibitors (anisomycin), hyperosmolarity, toxins, ischemia/reperfusion injury in heart attacks, heat shock, anticancer drugs (cisplatinum, adriamycin, or etoposide), ceramide, peroxide, and inflammatory cytokines such as TNFα. The activated JNKs translocate to the nucleus where they phosphorylate the effector molecules. A well-known substrate for JNKs is the transcription factor c-Jun. Besides, several other transcription factors have been shown to be phosphorylated by the JNKs, such as ATF-2, NF-ATc1, HSF-1, c-Myc, p53, STAT3, DPC4/ SMAD4/MADH4. Some non-transcription factors can also be regulated by JNKs, such as Bcl-2, paxillin, and Bcl-xL. The dual role of JNK in both apoptotic and survival signaling pathways indicates that the functional role of JNK is complex. JNKs are involved in various physiological and pathological processes, such as apoptosis, neurodegeneration, neural development, cell differentiation and proliferation, inflammation, cytokine production and regulation of responses to stimuli.

Compared with ERK and p38 MAPKs, less is known about the role of JNK MAPKs in contraction and migration of smooth muscle cells. The chemical inhibitors of JNK family members are less selective than MEK (then ERK1/2) and p38 MAPK inhibitors and therefore somewhat less useful for definitive studies of kinase action in cells. However, recent evidences support an important role for JNK in regulation of airway smooth muscle contraction (Lei *et al.*, 2011). SP600125, a JNK inhibitor, attenuates vascular smooth muscle contraction and human prostate smooth muscle induced by norepinephrine and phenylephrine (Lee *et al.*, 2006; Strittmatter *et al.*, 2012). In addition, overexpression of an inactive mutant of JNK partially inhibits migration induced by PDGF-BB and Ang II (Ohtsu *et al.*, 2005). The mutant may attenuate natural JNK activity by interfering its binding to the up- and down-stream molecules. Besides, SP600125 inhibits smooth muscle migration (Kavurma & Khachigian, 2003).

7. Regulation of MAP kinase pathways on MLC20 phosphorylation

By using the MAPK inhibitors and adenoviral overexpression, we found a MAP kinase pathway-mediated regulatory mechanism of MLC20 phosphorylation.

7.1. Inhibition of MAP kinase pathways increases phosphorylation level of MLC20

Research on the role of the ERK, p38 MAPK, and JNK is facilitated by effective and relatively selective small molecule inhibitors. We first used the p38, JNK, and ERK inhibitors to access the effects of MAP kinase signaling on MLC20 phosphorylation in cultured hBSM cells (Figure 6A).

The SB203580, BIRB 796, SB 202190, and VX-702 have inhibitory effects for the p38 kinase but only the SB 203580, a specific inhibitor of p38α and p38β isoforms, is widely used for the p38 function-related studies. The half maximal inhibitory concentration (IC50) values recommended by the manufacturer are 50 and 500 nM for the p38$\alpha$and p38β, respectively. The concentration ranges used in published reports are 10–50 μM (Kramer et al., 1996; Fatima et al., 2001; Kim et al., 2002). The JNK inhibitors include the SP600125, JNK inhibitor I (a peptide), JNK inhibitor IX, and JNK inhibitor VIII. The SP600125, a JNK inhibitor for three isoforms of JNKs, is often used to inhibit JNKs in most studies. The recommended IC50 values are 40 nM for JNK-1 and JNK-2, and 90 nM for JNK-3. The concentration ranges used by most previous researchers are 10–30 μM (Renlund et al., 2008). The recently used concentration is 20 μM (Takahashi et al., 2011). The ERK inhibitors include the PD98059, ERK Inhibitor II ($C_{18}H_{13}N_7$), U0126, and inhibitory peptide. It is a 13-amino acid peptide that corresponds to the N-terminus of MEK1 and functions as a specific inhibitor of ERK activation. The IC50 recommended by the manufacturer is 2.5 μM. Different concentrations such as 10, 20, 40, and 50 μM were used in previous researches (Mathur et al., 2004; Monick et al., 2008). The concentration of 60 μM was used in our studies. Considering the difficulty to transport a peptide across a cell membrane, we permeabilized cells before the ERK inhibitor treatment, even though the inhibitor peptide is cell-permeable.

In this study, the cultured hBSM cells were exposed to 30 μM of SB203580 (a p38 inhibitor), SP600125 (a JNK inhibitor), and 60 μM of ERK inhibitory peptide for 40 min. Cells for ERK inhibition were previously permeabilized (Deng *et al.*, 2011). Proteins of cell lysates were separated by SDS-PAGE electrophoresis and then transferred to PVDF membranes. Actin, phosphor MLC20 and MLCK on the same membrane were detected. Inhibition of MAP kinase activity increased MLC20 phosphorylation (Figure 6A). To minimize the effects of the quantitative difference in protein loading on experimental judgment in Western blotting assay, we always incubated membranes with mixed primary antibodies against *pan*-actin and phosphor MLC20, followed by incubation with corresponding and different second antibody after PBS washing. The actin gene is a housekeeping gene, which is expressed invariably in various cells/tissues usually at different treatments. The *pan*-actin bands can show us how much protein is loaded in the wells of SDS-PAGE gels. Then, the membrane was re-probed with specific antibody against the MLCK. Unexpectedly, all of the three MAP kinase inhibitors appear to increase the phosphor MLC20 level (Figure 6A), especially treatments by the SB 203580 and SP600125 in lines 2–3. The MLCK expression was not altered when the phosphor MLC20 level increased due to the MAP kinase inhibitor treatment.

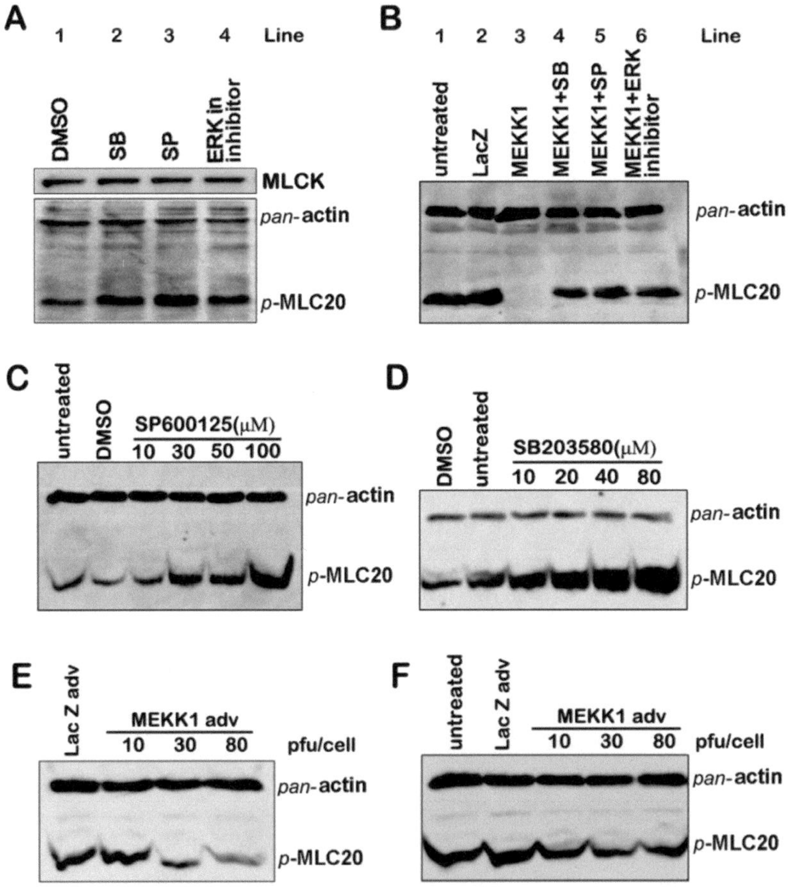

Figure 6. Negative regulation of MAP kinase pathway on MLC20 phosphorylation. (A) All of p38, JNK and ERK inhibitors raised phosphorylation level of MLC20. (B)Activation of MAP kinases by adenovirus-mediated overexpression of MEKK1 inhibited significantly MLC20 phosphorylation (line 3). MAPK inhibitors counteracted inhibition of MEKK1 on MLC20 phosphorylation (line4~6). (C, D) The level of MLC20 phosphorylation increased with JNK and p38 inhibitors in hBSM cells. (E, F) MLC phosphorylation decreased with MEKK1-viral doses. *P*-MLC20, phosphor MLC20; SB, SB203580; SP, SP600125; adv, adenovirus.

7.2. Activation of MAP kinase pathways inhibits MLC20 phosphorylation

Contrary to the MAP kinase-inhibition experiment (Figure 6A), the activity of the MAP kinases was to be increased in this experiment. The activity of the MAP kinase in quiescent cells is relatively low (Graf et al., 1997; Pizon & Baldacci, 2000; Hatton et al., 2003; Kuramochi et al., 2004; Mackeigan et al., 2005), and we observed low activation of MAP kinases in the hBSM cells used in this study. It is difficult to observe the alteration of the MAP kinase activity during a change of MLC20 phosphorylation. Thus, we increased the intracellular levels of the phosphor MAP kinases by adenovirus-mediated overexpression of active human MAP-ERK kinase kinase-1 (MEKK1), the upstream molecule of MAP kinases in hBSM cells.

The generation of MEKK1-containing adenovirus is described in brief as follows. To overexpress MEKK1, an HA-tagged 4.9-kb cDNA for active human wild-type MEKK1 was cloned at the downstream of cytomegalovirus (CMV) promoter of pAdtrack, a shuttle vector that transports target gene MEKK1 to the viral plasmid. To prepare the MEKK1-containing viral DNA, the shuttle vector was recombined with the adenoviral DNA in bacteria. Then, the recombinant adenoviral plasmid was used to produce the adenovirus in packaging cells following conventional protocol for adenovirus preparation. Two control adenoviruses were prepared in the same procedure but human MEKK1 was replaced by the β-galactosidase (LacZ) or green fluorescent protein (GFP). The control virus helped us to distinguish the MEKK1 effects from that of non-MEKK1 virus infection. After the amplification, the viruses were titrated by the agarose overlay plaque assay.

The viral infection efficiency in the cultured hBSM cells was examined by observing the percentage of the GFP-positive living cells under a fluorescence microscope. The viral expressive efficiency of MEKK1 was assayed in an independent experiment by Western blotting using an antibody against the HA-Tag (Deng et al., 2011). The exogenous MEKK1 activity expressed by the adenovirus was estimated by its capacity to activate the downstream molecules, including the JNK, ERK, and p38. These results show that the infection efficiency of the adenoviruses in the cultured hBSM cells reached up to 99% (the cell density was 70% when infected). The MEKK1 expressed by the adenovirus activated strikingly JNK1 and JNK2. In the meantime, it also activated the ERK1 and p38 kinase to a lower extent. Then, the MEKK1 adenovirus was used for the evaluation of the roles of the MAP kinase on MLC20 phosphorylation. When the cell density was between 60–70%, the hBSM cells in 6-well plates were infected with the MEKK1 and the control viruses at different doses (Deng et al., 2011). Three days after infection, the cells were harvested for Western blotting. The experimental results show that the MEKK1 adenovirus infection inhibited the MLC20 phosphorylation significantly (line 3 in Figure 6B). The inhibition of the MAP kinase pathway for the MLC20 phosphorylation is already hinted in Figure 6A but the extent of inhibition is beyond our expectation. Apparently, the inhibitory efficiency depends on viral infection efficiency, expressive capacity of MEKK1 adenovirus, and activating ability of MEKK1 for the downstream molecules.

In addition, we set up other treatments, using MAP kinase inhibitors to counteract the activation of the MEKK1 for the MAP kinases so that we could evaluate more accurately the role of the MEKK1-activated MAP kinase pathway in the regulation of the MLC20 phosphorylation. If it is true that the MEKK1 regulates negatively the MLC20 phosphorylation, the MAP kinase inhibitors would augment the MLC20 phosphorylation through suppression of the activity of the MEKK1 downstream molecules. As expected, the MAP kinase inhibitors neutralized the inhibition of the MEKK1 adenovirus on the MLC20 phosphorylation (lines 4–6 in Figure 6B). Likewise, the actin bands are detected to show a similar amount of loaded proteins. Obviously, the experimental results and even conclusions depend on the balance between the viral doses and inhibitor concentrations. A relatively low-concentration inhibitor may not affect the role of a relatively high-dose infection of MEKK1 adenovirus in MLC20 phosphorylation, and vice versa.

The dose-response relationship is a mathematic relationship between the dose and the organism's reaction to the dose. It describes the change in reaction or effect on an organism caused by differing levels of doses. Herein, we used dose-gradient methods to provide additional evidence that the MLC20 phosphorylation is regulated by the MAP kinase pathways. In the studies, the hBSM cells were exposed to different concentrations of the MAP kinase inhibitors for 40 minutes, followed by Western blotting assays for the phosphor MLC20. The results show that phosphor MLC20 level increases with the doses of inhibitors (Figure 6C and D, ERK inhibitor not shown) (Deng et al., 2011). A similar method was used to examine the effects of the MEKK1 at different infection doses (i.e., pfu/cell) on the MLC20 phosphorylation. The phosphor MLC20 decreased with doses of the MEKK1 adenovirus (Figure 6 E and F).

At this point, our experiments have demonstrated the negative regulation of MAP kinase pathway on the MLC20 phosphorylation.

8. Suggestions for the future investigation

The role of the MLC20 phosphorylation in contraction needs to be reassessed, even though it is accepted widely as a trigger of muscle contraction. A number of experiments have already verified the role of the MAP kinase pathway in the migration of different cell types and smooth muscle contraction. Therefore, according to current views, it is reasonable to assume that the MAP kinases would phosphorylate the MLC20, either via one or more enzymes. However, a series of experiments above showed the negative regulation of the MAP kinase pathways on the MLC20 phosphorylation. Because of conflict between our conclusion and current views about the mechanism of the MLC20 phosphorylation, we subsequently carried out a preliminary experiment to explore the synchrony between the cell migration and MLC20 phosphorylation. The migrating cells were fixed immediately and the phosphor MLC20 inside cells was immunostained with a specific antibody. However, we did not see a difference in the fluorescence intensity of the phosphor MLC20 between the migrating and quiescent cells (Deng et al., 2011).

In addition, when laboratory animals are killed, the muscle is usually relaxed, or rapidly relaxed in seconds (except for some smooth muscles). Because muscle contraction and relaxation are rapid, the MLC20 phosphorylation and dephosphorylation are also very fast, especially in the skeletal muscle. Even for smooth muscles, the muscle contraction and relaxation, such as the bladder emptying by detrusor and relaxation of tracheal/bronchial smooth muscle elicited by agents, may be fast. In one word, the MLC20 is phosphorylated in contracting muscle but unphosphorylated in relaxed muscle, and the transformation between the phosphorylation and nonphosphorylation is rapid. The muscle samples used in a lot of published articles for the MLC20 phosphorylation are almost in non–contracting status. However, those reports show that the phosphorylated MLC20 in various types of muscles are actually in relaxed state. Besides, a lot of data from different laboratories demonstrate that the MLC20 is phosphorylated strongly in cultured cells. As we all know, cells in culture dishes stop moving and are considered quiescent cells when they reach a high density like 90%. This suggests that the MLC20 is heavily phosphorylated in nonmigrating cells. Up to now, there is no direct evidence to support that the MLC20 phosphorylation triggers muscle contraction.

Author details

Maoxian Deng
Department of Animal Biology, School of Animal Husbandry and Veterinary Medicine, Jiangsu Polytechnic College of Agriculture and Forestry, Jurong, Jiangsu,China

Lixia Deng
West China Center of Medical Sciences, Sichuan University, Chengdu, Sichuan, China

Yarong Xue
School of Life Sciences, Nanjing University, Nanjing Jiangsu, China

9. References

Aitken K & Bagl DJ. (2001). Stretch-induced bladder smooth muscle cell (SMC) proliferation is mediated by RHAMM-dependent extracellular-regulated kinase (erk) signaling. *Urology* 57, 109.

Barona I, Fagundes DS, Gonzalo S, Grasa L, Arruebo MP, Plaza MA & Murillo MD. (2011). Role of TLR4 and MAPK in the local effect of LPS on intestinal contractility. *The Journal of pharmacy and pharmacology* 63, 657-662.

Bauer RM, Strittmatter F, Gratzke C, Gottinger J, Schlenker B, Reich O, Stief CG, Hedlund P, Andersson KE & Hennenberg M. (2011). Coupling of alpha1-adrenoceptors to ERK1/2 in the human prostate. *Urologia internationalis* 86, 427-433.

Bode AM & Dong Z. (2007). The functional contrariety of JNK. *Molecular carcinogenesis* 46, 591-598.

Cho YE, Ahn DS, Morgan KG & Lee YH. (2011). Enhanced contractility and myosin phosphorylation induced by Ca(2+)-independent MLCK activity in hypertensive rats. *Cardiovascular research* 91, 162-170.

Deng JT, Van Lierop JE, Sutherland C & Walsh MP. (2001). Ca^{2+}-independent smooth muscle contraction. a novel function for integrin-linked kinase. *The Journal of biological chemistry* 276, 16365-16373.

Deng M, Ding W, Min X & Xia Y. (2011). MLCK-independent phosphorylation of MLC20 and its regulation by MAP kinase pathway in human bladder smooth muscle cells. *Cytoskeleton (Hoboken, NJ* 68, 139-149.

Do KH, Kim MS, Kim JH, Rhim BY, Lee WS, Kim CD & Bae SS. (2009). Angiotensin II-induced aortic ring constriction is mediated by phosphatidylinositol 3-kinase/L-type calcium channel signaling pathway. *Experimental & molecular medicine* 41, 569-576.

El-Toukhy A, Given AM, Ogut O & Brozovich FV. (2006). PHI-1 interacts with the catalytic subunit of myosin light chain phosphatase to produce a Ca^{2+}-independent increase in MLC(20) phosphorylation and force in avian smooth muscle. *FEBS letters* 580, 5779-5784.

Emmert DA, Fee JA, Goeckeler ZM, Grojean JM, Wakatsuki T, Elson EL, Herring BP, Gallagher PJ & Wysolmerski RB. (2004). Rho-kinase-mediated Ca^{2+}-independent contraction in rat embryo fibroblasts. *American journal of physiology* 286, C8-21.

Escano CS, Jr., Keever LB, Gutweiler AA & Andresen BT. (2008). Angiotensin II activates extracellular signal-regulated kinase independently of receptor tyrosine kinases in renal smooth muscle cells: implications for blood pressure regulation. *The Journal of pharmacology and experimental therapeutics* 324, 34-42.

Fatima S, Khandekar Z, Parmentier JH & Malik KU. (2001). Cytosolic phospholipase A2 activation by the p38 kinase inhibitor SB203580 in rabbit aortic smooth muscle cells. *The Journal of pharmacology and experimental therapeutics* 298, 331-338.

Gerthoffer WT. (2007). Mechanisms of vascular smooth muscle cell migration. *Circulation research* 100, 607-621.

Graf K, Xi XP, Yang D, Fleck E, Hsueh WA & Law RE. (1997). Mitogen-activated protein kinase activation is involved in platelet-derived growth factor-directed migration by vascular smooth muscle cells. *Hypertension* 29, 334-339.

Harnett KM & Biancani P. (2003). Calcium-dependent and calcium-independent contractions in smooth muscles. *The American journal of medicine* 115 Suppl 3A, 24S-30S.

Hatton JP, Pooran M, Li CF, Luzzio C & Hughes-Fulford M. (2003). A short pulse of mechanical force induces gene expression and growth in MC3T3-E1 osteoblasts via an ERK 1/2 pathway. *J Bone Miner Res* 18, 58-66.

Jeong SI, Kwon OD, Kwon SC & Jung KY. (2011). Signalling pathways responsible for the methylisogermabullone-induced contraction of ileal longitudinal muscles. *The Journal of pharmacy and pharmacology* 63, 245-252.

Kim KY, Kim MY, Choi HS, Jin BK, Kim SU & Lee YB. (2002). Thrombin induces IL-10 production in microglia as a negative feedback regulator of TNF-alpha release. *Neuroreport* 13, 849-852.

Kavurma MM & Khachigian LM. (2003). ERK, JNK, and p38 MAP kinases differentially regulate proliferation and migration of phenotypically distinct smooth muscle cell subtypes. *Journal of cellular biochemistry* 89, 289-300.

Kohama K & K. S. (1995). Smooth Muscle Contraction: New Regulatory Modes (S. Karger AG, Basel). 1-159.

Kuramochi Y, Lim CC, Guo X, Colucci WS, Liao R & Sawyer DB. (2004). Myocyte contractile activity modulates norepinephrine cytotoxicity and survival effects of neuregulin-1beta. *American journal of physiology* 286, C222-229.

Lee HM, Won KJ, Kim J, Park HJ, Kim HJ, Roh HY, Lee SH, Lee CK & Kim B. (2007). Endothelin-1 induces contraction via a Syk-mediated p38 mitogen-activated protein kinase pathway in rat aortic smooth muscle. *Journal of pharmacological sciences* 103, 427-433.

Lee YR, Lee CK, Park HJ, Kim H, Kim J, Kim J, Lee KS, Lee YL, Min KO & Kim B. (2006). c-Jun N-terminal kinase contributes to norepinephrine-induced contraction through phosphorylation of caldesmon in rat aortic smooth muscle. *Journal of pharmacological sciences* 100, 119-125.

Lei Y, Cao Y, Zhang Y, Edvinsson L & Xu CB. (2011). Enhanced airway smooth muscle cell thromboxane receptor signaling via activation of JNK MAPK and extracellular calcium influx. *European journal of pharmacology* 650, 629-638.

Lelliott A, Nikkar-Esfahani A, Offer J, Orchard P & Roberts RE. (2012). The role of extracellular-signal regulate kinase (ERK) in the regulation of airway tone in porcine isolated peripheral bronchioles. *European journal of pharmacology* 674, 407-414.

Mackeigan JP, Murphy LO, Dimitri CA & Blenis J. (2005). Graded mitogen-activated protein kinase activity precedes switch-like c-Fos induction in mammalian cells. *Molecular and cellular biology* 25, 4676-4682.

Mathur RK, Awasthi A, Wadhone P, Ramanamurthy B & Saha B. (2004). Reciprocal CD40 signals through p38MAPK and ERK-1/2 induce counteracting immune responses. *Nature medicine* 10, 540-544.

McFawn PK, Shen L, Vincent SG, Mak A, Van Eyk JE & Fisher JT. (2003). Calcium-independent contraction and sensitization of airway smooth muscle by p21-activated protein kinase. *Am J Physiol Lung Cell Mol Physiol* 284, L863-870.

Monick MM, Powers LS, Barrett CW, Hinde S, Ashare A, Groskreutz DJ, Nyunoya T, Coleman M, Spitz DR & Hunninghake GW. (2008). Constitutive ERK MAPK activity regulates macrophage ATP production and mitochondrial integrity. *J Immunol* 180, 7485-7496.

Mugabe BE, Yaghini FA, Song CY, Buharalioglu CK, Waters CM & Malik KU. (2010). Angiotensin II-induced migration of vascular smooth muscle cells is mediated by p38 mitogen-activated protein kinase-activated c-Src through spleen tyrosine kinase and epidermal growth factor receptor transactivation. *The Journal of pharmacology and experimental therapeutics* 332, 116-124.

Niiro, N., Ikebe, M. Zipper-interacting protein kinase induces Ca^{2+}-free smooth muscle contraction via myosin light chain phosphorylation. *The Journal of Biological Chemistry* 276, 29567-29574

Ohtsu H, Mifune M, Frank GD, Saito S, Inagami T, Kim-Mitsuyama S, Takuwa Y, Sasaki T, Rothstein JD, Suzuki H, Nakashima H, Woolfolk EA, Motley ED & Eguchi S. (2005). Signal-crosstalk between Rho/ROCK and c-Jun NH2-terminal kinase mediates migration of vascular smooth muscle cells stimulated by angiotensin II. *Arteriosclerosis, thrombosis, and vascular biology* 25, 1831-1836.

Oishi K, Takano-Ohmuro H, Minakawa-Matsuo N, Suga O, Karibe H, Kohama K & Uchida MK. (1991). Oxytocin contracts rat uterine smooth muscle in Ca^{2+}-free medium without any phosphorylation of myosin light chain. *Biochemical and biophysical research communications* 176, 122-128.

Pizon V & Baldacci G. (2000). Rap1A protein interferes with various MAP kinase activating pathways in skeletal myogenic cells. *Oncogene* 19, 6074-6081.

Ratz PH, Miner AS & Barbour SE. (2009). Calcium-independent phospholipase A2 participates in KCl-induced calcium sensitization of vascular smooth muscle. *Cell calcium* 46, 65-72.

Renlund N, Pieretti-Vanmarcke R, O'Neill FH, Zhang L, Donahoe PK & Teixeira J. (2008). c-Jun N-terminal kinase inhibitor II (SP600125) activates Mullerian inhibiting substance type II receptor-mediated signal transduction. *Endocrinology* 149, 108-115.

Sakai H, Nishizawa Y, Nishimura A, Chiba Y, Goto K, Hanazaki M & Misawa M. (2010). Angiotensin II induces hyperresponsiveness of bronchial smooth muscle via an activation of p42/44 ERK in rats. *Pflugers Arch* 460, 645-655.

Sathish V, Yang B, Meuchel LW, VanOosten SK, Ryu AJ, Thompson MA, Prakash YS & Pabelick CM. (2011). Caveolin-1 and force regulation in porcine airway smooth muscle. *Am J Physiol Lung Cell Mol Physiol* 300, L920-929.

Sathishkumar K, Yallampalli U, Elkins R & Yallampalli C. (2010). Raf-1 kinase regulates smooth muscle contraction in the rat mesenteric arteries. *Journal of vascular research* 47, 384-398.

Strittmatter F, Walther S, Gratzke C, Gottinger J, Beckmann C, Roosen A, Schlenker B, Hedlund P, Andersson KE, Stief CG & Hennenberg M. (2012). Inhibition of adrenergic human prostate smooth muscle contraction by the inhibitors of c-Jun N-terminal kinase, SP600125 and BI-78D3. *British journal of pharmacology*, Feb 24. doi: 10.1111/j.1476-5381.2012.01919.x. [Epub ahead of print].

Takahashi S, Ebihara A, Kajiho H, Kontani K, Nishina H & Katada T. (2011). RASSF7 negatively regulates pro-apoptotic JNK signaling by inhibiting the activity of phosphorylated-MKK7. *Cell death and differentiation* 18, 645-655.

Chapter 6

Two Guanylylcyclases Regulate the Muscarinic Activation of Airway Smooth Muscle

Marcelo J. Alfonzo, Fabiola Placeres-Uray, Walid Hassan-Soto,
Adolfo Borges, Ramona González de Alfonzo and Itala Lippo de Becemberg

Additional information is available at the end of the chapter

http://dx.doi.org/10.5772/48684

1. Introduction

Muscarinic activation of Airway smooth muscle (ASM) is of major importance to the physiological and patho-physiological actions of acetylcholine, which induces bronchoconstriction, airway smooth muscle thickening, and the modulation of cytokine and chemokine production by these cells as described in [1]. The parasympathetic nervous system is the dominant neuronal pathway in the control of airway smooth muscle tone. Stimulation of cholinergic nerves causes bronchoconstriction, and mucus secretion as in [2]. The human airways are innervated via efferent and afferent autonomic nerves, which regulate many aspects of airway function. It has been suggested that neural control of the airways may be abnormal in asthmatic patients, and that neurogenic mechanisms may contribute to the pathogenesis and pathophysiology of asthma. Although abnormalities of the cholinergic innervation have been suggested in asthma, thus far the evidence for cholinergic dysfunction in asthmatic subjects is not convincing as mentioned in references [1,2].

Acetylcholine (Ach) is the predominant parasympathetic neurotransmitter, acting as an autocrine or paracrine hormone in the airways and its role in the regulation of bronchomotor tone and mucus secretion from airway submucosal glands in the respiratory tract is well established as described in [2]. More recent findings suggest that acetylcholine regulates additional functions in the respiratory tract, may promoting airway inflammation and remodelling, including airway smooth muscle thickening in inflammatory lung diseases as reported in references [3-5]. Another source of ACh is the non-neuronal cells and tissues, particularly inflammatory cells and the airway epithelium as described in references [6-8].

Collectively, these findings indicate that acetylcholine, derived from the vagal nerve and from non-neuronal origins such as the airway epithelium, may induce cell responses associated with airway wall remodelling and trigger proinflammatory Cytokines as IL-6 and

IL-8 release in [9] by structural cells of the airways, including the airway smooth muscle itself. In addition, muscarinic receptors regulate proliferative and proinflammatory functions of the airway smooth muscle as mentioned in references [9,10].

Acetylcholine acting on muscarinic receptors (mAChRs) anchored at the airway smooth muscle sarcolemma are involved in the generation of a number of signal transducing cascades allowing the activation of the smooth muscle machinery as described in [11]. Airway smooth muscle muscarinic activation is mediated through mAChRs, which are members of the so called G protein-coupled receptors (GPCR) family, which are cell surface receptors, that activate intracellular responses by coupling G proteins as described in [12] to specific effectors in [13]. Molecular cloning studies have revealed the existence of five mammalian subtypes of muscarinic receptors (m1-m5) as described in [14]. In trachea, smooth muscle expresses mRNAs coding for both m2 and m3 receptors as reported in [15]. Airways smooth muscarinic receptors have been identified as a mixed population of M_2 and M_3 subtypes roughly in a 4:1 ratio using pharmacological ligand binding studies in [16,17] being the M_2 subtype, the most abundant muscarinic receptor in tracheal plasma membranes as reported in references [16,18,19]. In addition, a muscarinic antagonist heterogeneity associated with the M_3AchR subtype present in plasma membrane fractions from tracheal smooth muscle has been described in [20].

It has been claimed that M_3AChR represents a primary target of acetylcholine in the airways, involved in the regulation of bronchoconstriction as stated in references [17,18,21,22]. Classically, M_3AChRs in ASM are coupled to phospholipase C (PLC)/protein kinase C (PKC) pathway via pertussis toxin (PTX)-insensitive G proteins of the Gq/11 family. The contractile response evoked by M_3AChRs stimulation is attributed to the formation of inositol trisphosphate (IP3), the subsequent release of Ca^{2+} from intracellular stores, the additional influx of extracellular calcium, and the Ca^{2+}-sensitizing effect of PKC as mentioned in [21,23,24].

On the other hand, the stimulation of M_2 muscarinic receptors (M_2AChRs) in ASM inhibits adenylyl cyclase via activation of PTX-sensitive G proteins of the Gi/o family in [25,26] and therefore M_2AChRs are thought to counteract relaxation as reported in [27]. Experimental evidence has been provided that M_2AChRs participate directly in ASM contraction but the molecular mechanisms by which the M_2AChRs in ASM induce contraction is, still unknown.

Recently, it has been shown that M_2AChRs stimulate Gi/o proteins to released $\beta\gamma$ dimer, which inhibit the Large Conductance Ca^{2+}-activated K^+ Channel Activity (BK channels) as described in [28]. The inhibition of BK channel activity favors contraction of ASM and these BK channels are opposed to the M_2AChR-mediated depolarization and activation of calcium channels by restricting excitation–contraction coupling to more negative voltage ranges as mentioned in [29].

In addition, the influence of M_2AChRs to modulate the relaxant effects of atrial natriuretic peptide (ANP) has been reported. Thus, the stimulation of M_2AChRs suppresses ANP-induced activation of particulate guanylyl cyclase via a PTX-sensitive G protein as reported in [30].

More recently, we showed that muscarinic agonists, via M_2AChR induced a massive and selective $\alpha 1\beta 1$-NOsGC migration from cytoplasm to plasma membranes in a dose-dependent manner. Such migration was blocked by PTX, suggesting the involvement of Go/Gi proteins in [31].

Some of the signal cascades activated by mAChRs at TSM are linked to the generation of second messengers such as cyclic nucleotides: cAMP and cGMP as described in reference [32]. In this review, we address the cGMP generation as a product of the muscarinic activation of ASM. It is well known that in mammalian cells, cGMP is produced by the action of two distinctive guanylyl cyclases, named the NO-sensitive soluble guanylylclyclases (NO-sGC) as reported in [33] and the single membrane-spanning guanylyl cyclases as published in references [33-35].

2. Body: Cyclic GMP signals during muscarinic activation

The muscarinic activation of tracheal smooth muscle (TSM) fragments associated with smooth muscle contraction, involves the generation of two cGMP signals, at 20-s and 60-s as described in reference [36] and a kinetic behavior is shown in Figure 1.

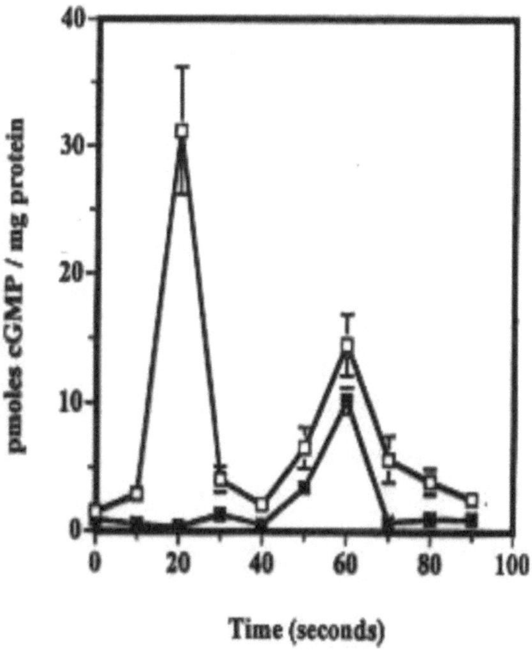

Effect of ODQ on the time course of cGMP signals induced by agonist muscarinic carbachol ($1x10^{-5}$ M) in TSM. Tracheal smooth muscle strips were pre-incubated for 30 min with ODQ (1H-[1, 2,4] Oxadiazolo[4,3-a]quinoxalin-1-one) (■) (100 nM) selective inhibitor of NO-sGC, following the Procedure 2 as described in reference 36. Control experiments (□) without drug were run simultaneously under the same conditions. Each value is the mean ± SE of 4 different tracheas and the cGMP determinations were carried out by triplicate as described in reference [36].

Figure 1. Time course of muscarinic agonist action on cGMP levels from TSM in the presence of ODQ. Taken from reference [36].

Interestingly, the 20-s signal is linked to the onset and the 60s signal is related to the plateau of the smooth muscle contraction as described in [36]. The 20-s signal is associated with the activity of a Soluble guanylyl cyclases, which are nitric oxide stimulated guanylyl cyclase (NO-sGC), which are described in [33] and the second, the 60-s signal, is linked to membrane-bound natriuretic peptide receptor guanylyl cyclase (NPR-GC) in references [33-35], which has been previously characterized at TSM in [37,38].

The details of the activation of an NO-sGC, generating the early 20-s cGMP signal, presents in the plasma membrane of BTSM, has been previously established in reference [31].

Soluble guanylyl cyclases are nitric oxide stimulated guanylyl cyclase (NO-sGC) due to the fact that the primary and best-studied endogenous activator is nitric oxide (NO). NO-sGC is a heterodimeric hemoprotein formed by two different subunits, α- and β-subunits, which exist in four types (α1, α2, β1, and β2), each the product of a separate gene as described in [39,40]. Structurally, each subunit consists of N-terminal H-NOX domain, a central domain related to the dimerization, and a C-terminal consensus nucleotide catalytic cyclase domain as described in [41,42]. For the formation of a catalytically active enzyme, both α- and β-subunits are required as reported in [43]. Although the α1β1 isoform is ubiquitous, the α2β1 isoform is less broadly distributed described in [39-43].

The best-characterized heterodimers are the α1/β1 and the α2/β1 isoforms as mentioned in [44] being the first ones, relevant in our studies as in [31]. At molecular level, His-105 at the amino terminus of the β1 subunit of NO-sGC is the axial ligand of the pentacoordinated reduced iron center of heme, which is required for NO activation of the enzyme. Thus, NO activates sGC by binding to the sixth position of the heme ring, which breaks the bond between the axial histidine and iron to form a 5-coordinated ring with NO in the fifth position as reported in [45].

By using several experimental approaches as biochemical, pharmacological and molecular biology methods, we established that the first 20-s signal is a product of NO-sGC being sensitive to ODQ, as shown in Figure 2, which is translocated from cytoplasm to the inner face of the airway smooth muscle sarcolemma under muscarinic activation.

In addition, there is a coupling mechanism between M_2AChRs and NO-sGC involving a Go/Gi heterotrimeric proteins as previously demonstrated in [31]. Thus, in intact smooth muscle fragments and isolated plasma membranes from bovine tracheal smooth muscle, we showed that the heterodimer of NO-sGC involved in such translocation is the α1β1sGC. The experimental evidence using Western blotting with specific antibodies against all subunits of NO-sGC demonstrate that under muscarinic activation, the α1β1NO-sGC-heterodimer isoform is translocated from cytoplasm to plasma membranes of BTSM. These experimental data are shown in Figure 3.

Since the capability of this α1β1-sGC to migrate to plasma membranes under muscarinic activation, a purification procedure and further identification of this α1β1-sGC heterodimer was also performed. Such NO-sGC translocation involves a M_2AChR subtype as previously described in [31].

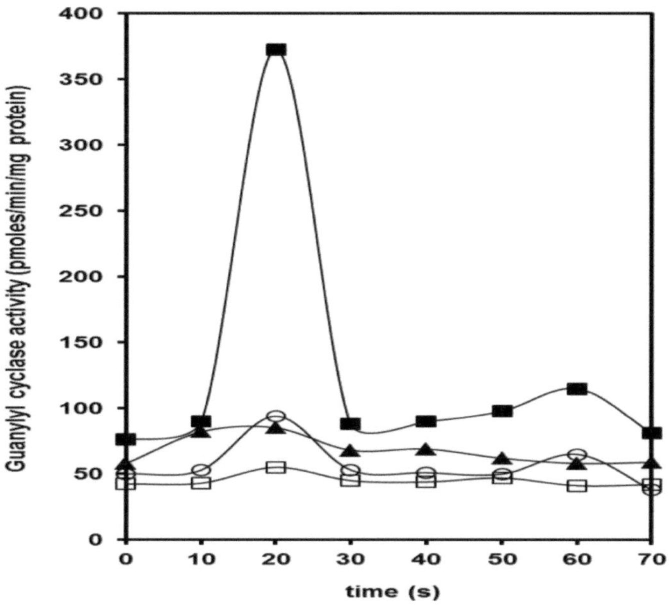

Isolated BTSM strips were incubated in KRB at 37C in the presence and absence of muscarinic agonist Cch (1 x 10^{-5} M) for 70 sec in [36]. After each 10 s, the BTSM strips were removed and immediately frozen in liquid N2 and processed to prepare crude membranes fraction as described in [54]. In these membrane sediments, GC assays were performed in duplicate as described previously in [53]. Empty symbols represent GC assays performed by duplicate in the presence of 3 mM MnCl2, 0.2M GTP, 5 mM creatine phosphate, and 10 IU phospho-creatine kinase in 0.01% defatted BSA. Full symbols represent GC activity in the presence of 100 μM SNP. Basal (□), muscarinic agonist CC 1 x 10^{-5} M (○), basal plus SNP (▲), and CC+SNP (■). Each value is the mean of three different BTSM strips.

Figure 2. Time-course of guanylyl cyclase activity in crude plasma membrane fractions isolated from BTSM strips under muscarinic agonist action. Taken from reference [31].

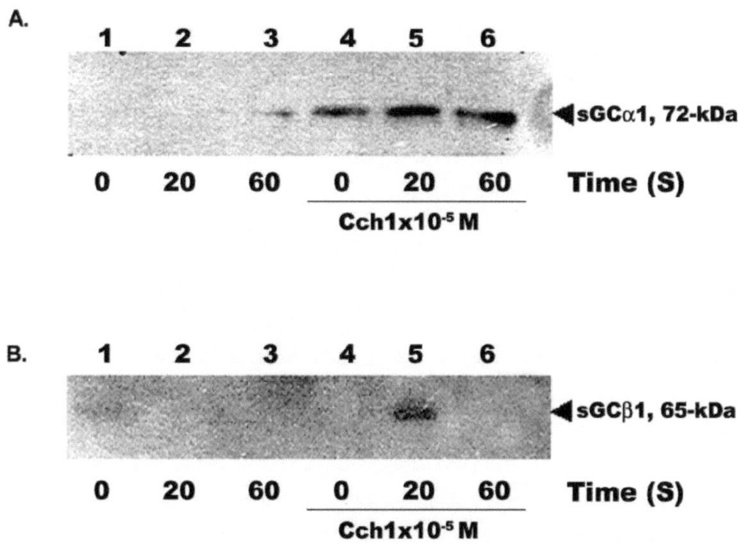

Figure 3. Western blotting of α1β1 NO-sGC-heterodimer from plasma membranes of BTSM under muscarinic activation as described in reference [31].

Identification of $\alpha1/\beta1$ sGC heterodimer in membrane fractions isolated from BTSM strips under muscarinic agonist exposure. Isolated BTSM were incubated in KRB at 37°C in the presence and absence of muscarinic agonist CC (1 x10^{-5} M) during 0, 20 s and 60 s. The samples were immediately frozen and pulverized with a mortar in liquid nitrogen, following the protocols described to isolate crude plasma membranes fraction as described in [31,54]. The final membranes sediments were electrophoresed in 12% PAGE-SDS, transferred to nitrocellulose membranes, and probed with specific antibodies against the $\alpha1$, $\alpha2$, (A) $\beta1$ and $\beta2$ (B) of NO-sGC subunits. The specific antibodies against $\alpha2$ and $\beta2$ produced negative results. This experiment was performed three times with similar results.

Thus, we demonstrate that the first 20-s cGMP signal is a product of a novel signaling cascade involving M_2AChR coupled to Go/i proteins, which facilitates the $\alpha1\beta1$-sGC isoform migration from cytoplasm to the BTSM sarcolemma. These experimental evidences support a model for the M_2AChR, Gi/o, and heterodimer of $\alpha1\beta1$ of NO-sGC novel signal transducing cascade as illustrated in Figure 4.

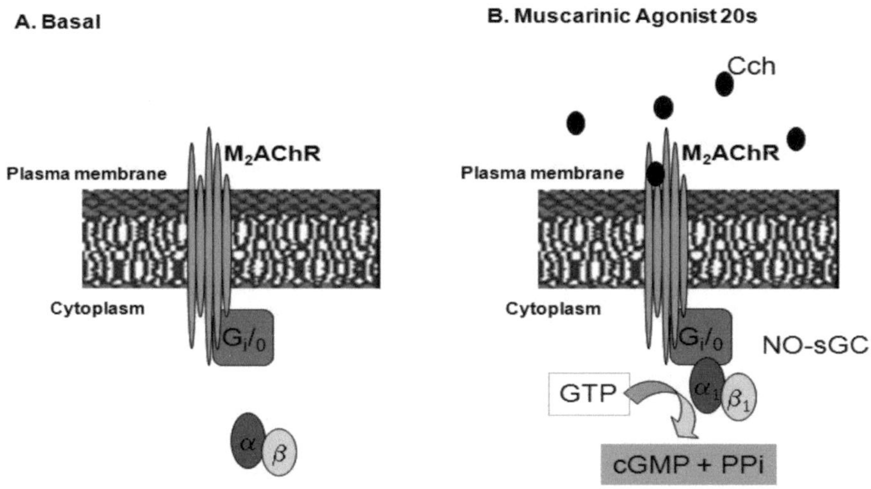

Figure 4. Model for the M_2AChR, Gi/o, and heterodimer of $\alpha1\beta1$ of NO-sGC novel signal transducing cascade in mammalian cells taken from reference [31].

Proposed model for a novel signal transducing cascade in mammalian cells involving three distinct molecular entities: M_2AChR, Gi/o protein, and heterodimer of $\alpha1\beta1$ of NO-sGC. (A) Basal condition: The hetero-dimer ($\alpha1\beta1$NO-sGC) is located in the cytoplasm and the M_2AChR and Gi/o proteins are macromolecules spanning and associated respectively with the plasma membrane bilayer. (B) Under muscarinic exposure, the agonist (Cch) binds at the extracellular domains of M_2AChR causing the activation of M_2AChR. This induces conformational changes at the M_2AChR cytoplasm domains stimulating a PTX-sensitive G protein (Gi/o), which may induce a migration and further activation of the $\alpha1\beta1$ heterodimer NO-sGC from cytoplasm to the plasma membrane resulting in a fast rise in cGMP production, which is related to the 20-s cGMP signal generated during the muscarinic activation of tracheal smooth muscle cell.

In the other hand, the 60-s cGMP signal is a product of a Natriuretic Peptide Receptor Guanylylcyclase-B (NPR-GC-B), which was previously identified at TSM, using biochemical as reported in [38] and molecular biology approaches as described in [37]. The NPR-GC-B (GC-B) is a membrane-spanning homodimer form, which contains an extracellular ligand-binding, trans-membrane, kinase homology, dimerization and carboxyl-terminal catalytic domains that was published in references [46,47]. NPR-GC-B, which is also called NPR-B or NPR2, is activated by CNP, which exists in 22 and 53 amino acid forms that are structurally similar to ANP and BNP as reported in references [48,49].

Furthermore, this GC-B from TSM is a novel G-protein coupled guanylyl cyclase presents in isolated plasma membranes fractions from this smooth muscle subtype as described in [53]. This G-protein-coupled NPR-GG-B showed complex kinetics and regulation, which is summarized in a proposed model as shown in Figure 5 as illustrated in reference [38]. This NPR-B was activated by Natriuretic Peptides (CNP-53>CNP-22>ANP-28) at the Ligand Extracellular Domain, stimulated by Gq-protein activators such as mastoparan, and inhibited by a chloride sensitive-Gi/o, interacting at the Juxtamembrane Domain. The Kinase Homology Domain was evaluated by the ATP inhibition of Mn^{2+}-activated-NPR-B, which was partially reversed by mastoparan. The Catalytic Domain was studied by its kinetics of Mn^{2+}/Mg^{2+} and GTP, and the catalytic effect with GTP analogs with modifications of the $\beta\gamma$ phosphates and ribose moieties. Most NPR-B biochemical properties remained after detergent-solubilization but the mastoparan-activation and chloride-inhibition of NPR-B disappeared. This NPR-GC-B is a highly regulated nano-machinery with domains acting at crosstalk points with other signal transducing cascades initiated by G-Protein Coupled Receptors (GPCR) and affected by intracellular ligands such as chloride, Mn^{2+}, Mg^{2+}, ATP and GTP. In addition, this model contains a novel GPRM domain, a G-protein-regulatory site, which was identified and characterized as the binding domain for the G proteins subunits as described in [38].

GC-B is abundantly expressed in brain, lung, bone, heart and ovary tissue as reported in [50]. GC-B contains three intramolecular disulfide bonds and is highly glycosylated on asparagine residues. Moreover, GC-B is highly phosphorylated and dephosphorylation is associated with receptor inhibition as described in [51]. ATP increases the enzymatic activity of GC-B by reducing the Michaelis-Menten constant for GTP, an order of magnitude and this nucleotide seems to be essential for maximal activity as demonstrated in reference [52].

On the other hand, this Natriuretic Peptide Receptor Guanylylcyclase–B (NPR-GC-B), which has been previously identified as the predominant functional TSM membrane-bound GC subtype by using molecular biology in [37] and biochemical approaches as described in [38]. Nucleotide sequence of membrane-bound GC transcripts retrieved by RT-PCR correspond to the bovine GC-B isoform, indicating the predominance of this NPR-sensitive GC-B over GC-A and GC-C, which are also expressed in TSM in reference [37]. Both ANP and CNP stimulates the TSM membrane-bound GC activity in a concentration-dependent manner, but the CNP effect (10^{-8} to 10^{-5} M) was significantly higher than that exerted by ANP, whereas guanylin (a GC-C activator) showed no effect. Given that CNP specifically activates GC-B receptors and that it activates GC-A receptors only marginally, the significant CNP effect on TSM cGMP production indicates that NPR-GC-B is the most abundantly expressed tracheal membrane-spanning GC isoform, further confirming the RT-PCR results in [37].

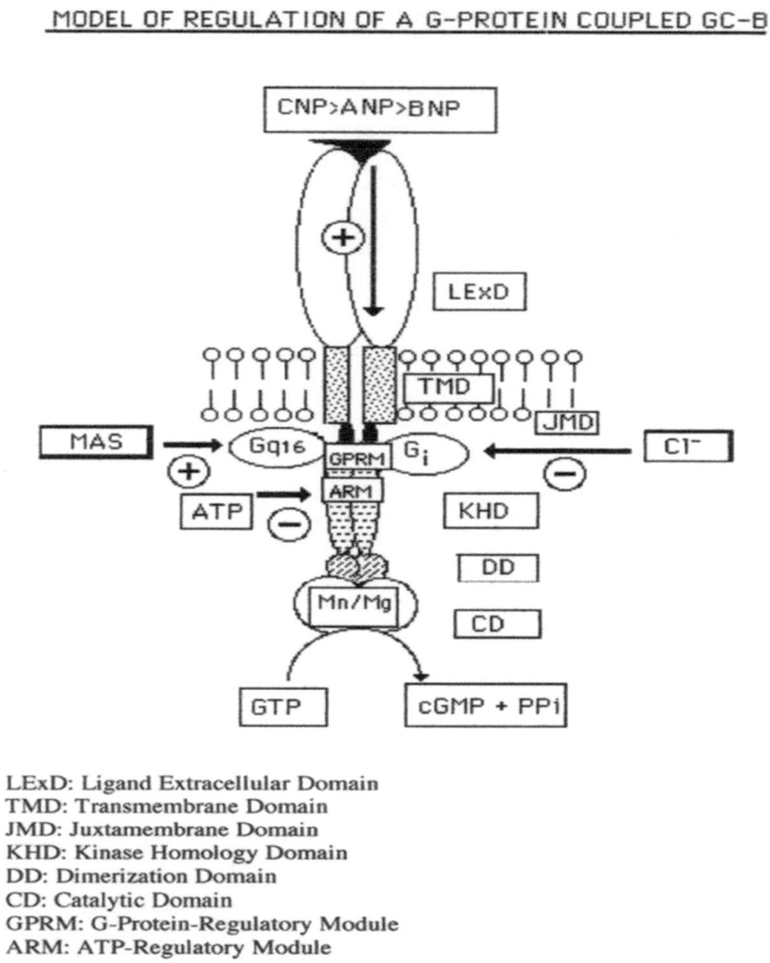

Figure 5. Proposed model of the G-protein coupled NPR-GC-B as illustrated in reference [38].

This NPR-GC-B is regulated in an opposite way by two MAChRs signaling cascades, acting the M_2AChR as an inhibitor and the M_3AChRs as an activator of this NPR-GC-B as mentioned in [55].

This novel signal transducing system is illustrated as a model in Figure 6. The relevant feature of these two opposite signal cascades that regulate this NPR-GC-B is the coupling of two MAChRs and distinctive G proteins. Thus, this activation system coupled a M_3AChR to a Gq16 being stimulated by mastoparans (tetradecapeptides from wasp venoms), whereas, on the inhibitor cascade of NPR-GC-B, via M_2AChR coupled to a Go/Gi protein has been previously described in references [35,38,56].

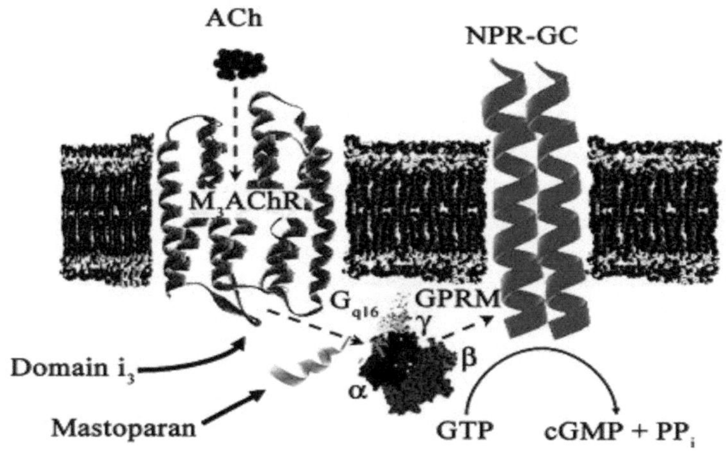

Figure 6. Schematic model of signal cascade involving m3AChR/Gq16/NPR-GC-B at BTSM in [56].

A model for coupling M_3AChR, via $Gq\alpha_{16}\beta\gamma$ to activate NPR-GC-B in plasma membranes from BTSM. This model is composed of three separate and different molecular entities, M_3AChR, a GPCR seven transmembrane receptor, a heterotrimeric G protein and homodimeric NPR-GC (cGMP producing enzyme) as the effector. The drawings do not take into account the actual structural biology (molecular mass) of these entities; it is a scheme to suggest the flow of information in this novel signal transducing cascade, which is indicated by the dashed lines. Thus, a muscarinic agonist (ACh) binds at extracellullar domains of M_3AChR inducing conformational changes at the cytoplasmic i_3M_3AChR domain, which stimulates the $Gq\alpha_{16}\beta\gamma$, to release its active subunits that interact with NPR-GC. Mastoparan and its active analogues may act at the interactions between i_3M_3AChR domain and the $Gq\alpha_{16}\beta\gamma$ protein as indicated in the scheme.

Mastoparans are tetradecapeptides isolated from wasp venom as reported in [57,58] and they are well known as heterotrimeric G-protein activators as described in [59,60]. Moreover, mastoparan is able to activate the NPR-GC-B associated with plasma membranes fractions from TSM in [56]. Following these results, we evaluate the effect of these peptides on TSM contraction. It was found that mastoparan and analogs were able to inhibit in a selective manner the BTSM contraction induced by muscarinic agonist as carbachol (CC) as shown in Figure 7. Thus, our original findings indicated that mastoparans in the nM range decreased the contractile maximal responses induced by CC without changing its EC_{50}. One explanation for the decrement on the contractile maximal responses by mastoparans may be related to the ability of these tetradecapeptides to disturb the function and the contractile machinery of the BTSM type through cytotoxic mechanisms, which have been described in other biological models, specifically in the μM concentration range as described in [61,62].

This assumption is not supported by our results on the effects of classic TSM spasmogens as 5-HT in as described in reference [63], which produced potent contractions even in the presence of mastoparan (nM) as shown in Figure 8. These findings can be explained since this bioactive amine has been claimed to exert their physiological effects on TSM through specific GPCRs. These receptors are the 5-HT2A, which induced activation of the Gq/11 protein and its downstream effector phospholipase C (PLC) leading to intracellular phosphatidylinositol turnover and Ca^{2+}mobilization as reported in [64]. These Gq/11 proteins are mastoparan-insensitive ones. The latter facts can explain the mastoparan-insensitivity of the serotoninergic transducing cascades at TSM. Furthermore, our results demonstrated that mastoparan inhibits selectively the muscarinic activation without altering other spasmogens transducing cascades at TSM.

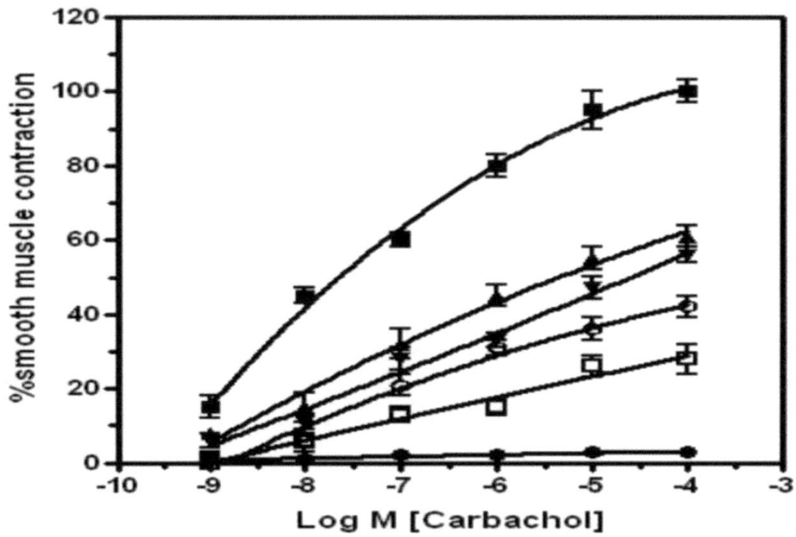

The BTSM strips contractile activity was measured using Procedure 1 as previously described in [36]. Carbachol cumulative responses using concentrations from $1x10^{-9}$ M to $1x10^{-4}$ M were measured in recording period until 3 min. Mastoparan was pre-incubated for 10 min before the muscarinic agonist (carbachol) addition. Experimental conditions: (■) Control, (▲) 0.5nM (▼) 1 nM (○) 5 nM (□) 10 nM (●) 50nM. The maximal contractile activity was considered as 100% (3.2 ± 0.2 g) and this value was used to estimate other contractile responses. Each value is the mean ± SEM of three different tracheas assayed in duplicate.

Figure 7. Carbachol cumulative concentration curves responses from BTSM, pre-treated with mastoparan.

From the data above mentioned, mastoparan can affect specifically the signal cascades associated with the muscarinic activation at TSM sarcolemma, which are initiated with a mAChRs coupled to heterotrimeric G-proteins in reference [65]. These results indicate that the most like candidates implicated in mastoparan effects are the heterotrimeric G-proteins coupled to these mAChRs. G protein involvement on the mastoparan inhibition on TSM muscarinic activation is supported by the ability of this G-protein activator to alter the generation of the two GMPc signals at 20-s and 60-s as previously described in [36]. Mastoparan (50 nM) induced a potent inhibition of these cGMP signals, as shown in Figure 8. After mastoparan pre-incubation, the kinetics of cGMP intracellular levels at TSM was evaluated following the muscarinic agonist exposure. Interestingly, the first cGMP signal (20-s) decreased in more than 60% and the second signal peak (60-s) completely vanished.

The disappearance of the second signal of cGMP (60-s) correlates well with a significant reduction on the contractile maximal responses as here described in Figure 7. This 60-s cGMP signal is a product of NPR-GC-B as above mentioned in references [35,54,55]. Recently, we further recognized that muscarinic agonist and mastoparan activations of NPR-GC-B in isolated BTSM plasma membranes fraction involve this mastoparan-sensitive-Gq16 protein in [54]. Thus, the pre-exposure of TSM strips to mastoparan stimulated the Gq16 protein, disrupting this signal cascade inducing the failure of this muscarinic-dependent TSM contraction. However, the serotoninergic-dependent TSM contraction remained functionally active indicating that the smooth muscle machinery remains fully active.

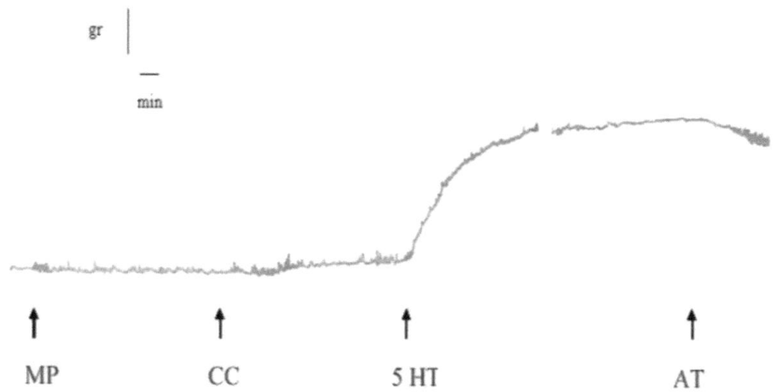

BTSM fragments were stabilized for 1 h using a procedure described previously in [36]. Later, mastoparan (50 nM) was added for 10 min, followed by carbachol (CC) ($1x\ 10^{-4}$ M), serotonin or 5-hydroxytryptamine (5-HT) ($1x\ 10^{-4}$ M) and Atropine (AT) ($1x10^{-4}$M) was added. Values between parentheses are the final concentration of drugs in the incubation media. This trace is representative of 3 experiments performed with 3 different tracheas fragments.

Figure 8. Serotonin (5HT) induced contractile responses in mastoparan-treated BTSM strips.

Until now, the BTSM muscarinic activation is unique biological system that involves two cGMP signals as second messengers in references [35,36]. In addition, this activation is a highly regulated biological process, which starts with M_2/M_3AChRs coupled to two different heterotrimeric G proteins, leading to a fine time regulation and stimulation of two distinctive guanylyl cyclases that accomplish the generation of these two (20-s and 60-s) cGMP signal peaks.

Based in our results, it can be postulated that the first cGMP signal (20-s) is a product of the activation of one M_2AChR subtype coupled to a PTX-sensitive-Gi/o protein that leads to the translocation and further activation of the heterodimer $\alpha1\beta1$ NO-sGC isoform, anchored to plasma membrane. Recently, it has been suggested that a plasma membrane-bound GC, may provide a localized pool of cGMP in [68], which seems to be in a similar trend suggested by our work. The second signal peak is a product of a the activation of M_3AChR coupled to the stimulation of mastoparan-sensitive Gq16 as reported in reference [54] to turn on a transmembrane-homodimer as NPR-GC-B as described in [35,55].

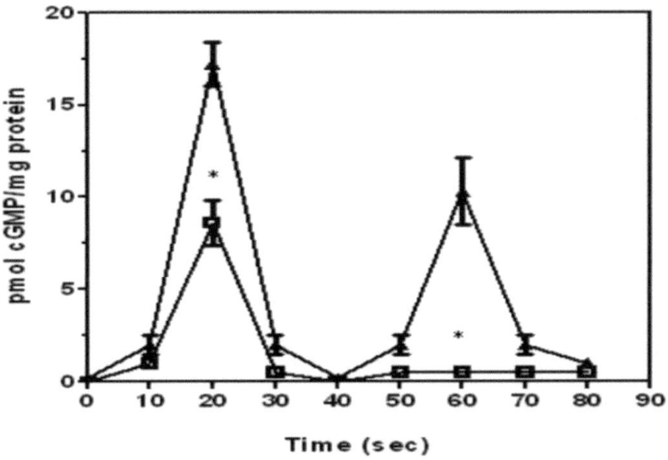

BTSM strips were assayed using following Procedure 2 as described in [36]. Pre-incubation for 10 min with mastoparan (□) (50nM) was performed. Control experiments (▲) without drugs were run simultaneously under the same conditions. Each value is the mean X ± SEM of four different tracheas, and the cGMP determinations were carried out in triplicate as described in references [36,53]. Statistically significant difference between the Control with respect to mastoparan as indicated with asterisk (* $p<0.05$).

Figure 9. Mastoparan effect on cGMP signal peaks induced by muscarinic agonist (carbachol) in TSM.

Thus, a dysfunction of these M_2/M_3AChR signal transducing cascades has been implied in the pathophysiological mechanisms of bronchial asthma as mentioned in references [1,3,4] and Chronic Obstructive Pulmonary Disease (COPD) in [3]. In this sense, we attempted to evaluate these novel signal transducing cascades involving guanylyl cyclase activities above discussed, in an experimental asthma model in rats as described in reference [70]. In addition, it has been claimed that excessive NO production that occurs in asthma induces a down-regulation of NO-sGC as reported in [69].

Following these rationale, we used cultured Airway Smooth Muscle Cells (ASMC) from Control and an experimental asthma model (Ovoalbumin exposed rats or OVA-ASMC), which were sensitized to Ovoalbumin using a procedure described in [70,71]. In these ASMC, we evaluate the cGMP production by NPR-GC-B stimulated by muscarinic agonists, mastoparans, natriuretic peptides (ANP, CNP) and for NO-sGC, a classic NO donor as Sodiun NitroPrusside (SNP) and a selective inhibitor for the NO-sGC as 1H-[1, 2,4] Oxadiazolo[4,3-a]quinoxalin-1-one (ODQ) were used.

All ASMC exposed to a NO-donor compound as SNP and muscarinic agonist as CC increased cGMP intracellular levels, which were inhibited by ODQ, suggesting that NO-sGC is present and this activity is partially responsible for cGMP production in these ASMC. However, OVA-ASMC showed low basal cGMP production compared to CONTROL ASMC in reference [71] as shown in Figure 10 possibly due to substantial reduced NO-sGC expression reflected in decrease the steady state levels of NO-sGC subunit mRNAs and protein level expression, as described in intact lung tissue from OVA-sensitized mice described elsewhere in reference [69].

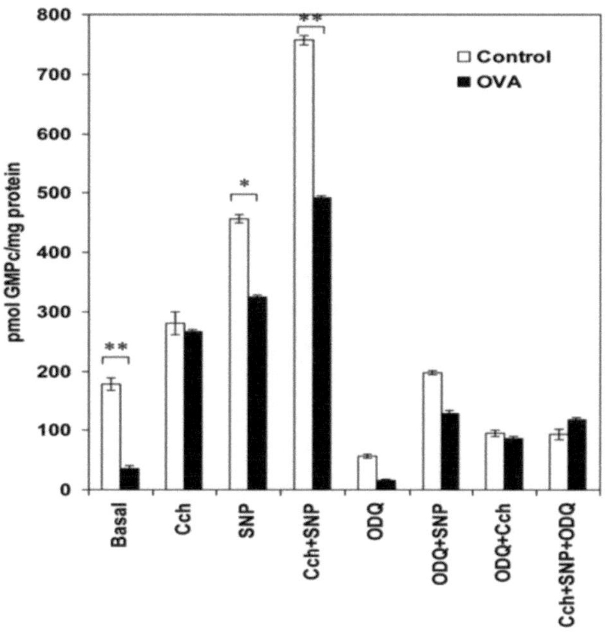

Cyclic GMP (cGMP) production in airway smooth muscle cells (ASMC) from CONTROL and OVA- sensitized rats. The ASMC were incubated for 15 min in the presence of IBMX as described in reference [70]. Cyclic GMP production was determined in the presence of Cch, SNP and ODQ as described in Methods. The cGMP produced was estimated by duplicate using a radioimmunoassay kit from Amersham as described in [70]. Each value is the mean ± of 5 different experiments. In both groups, the amounts of cell cultures plates were obtained for a pool of 5 rats. The stadistical significance between CONTROL vs OVA was established as $p<0.05$ (*) and $p< 0.001$ (**).

Figure 10. The effect of NO-sGC activators and inhibitors on total GC activity from Control and OVA ASMC. Taken from reference [70].

Trying to understand the role of the NPR-GC-B in the ASMC, we studied the effect of several activators of this NPR-GC such as natriuretic peptides (ANP and CNP) and mastoparans as previously described in BTSM in references [37,38,71]. Thus, ASMC from Control and OVA-exposed rats were cultured and exposed to these NPR-GC-B activators. Thus, in Figure 11, the total GC activity at OVA-ASMC was stimulated by CNP, ANP and muscarinic agonist (Cch) indicating that NPR-GC-B was present in Control ASMC, which are similar results described in intact BTSM strips and isolated plasma membranes from BTSM in references [37,38,53,55]. In addition, the OVA-ASMC showed a more significant stimulation by these NPR-GC activators. Interestinly, the combination of CC plus ANP and CNP increased in more than 6 times the GC activity. As expected, mastoparan, an activator of NPR-GC-B, via Gq16 dramatically increased in 5 times, the GC activity as reported in [72]. These data showed for the first time that there is an hyperstimulation of M_3AChR/Gq16/NPR-GC-B cascade in ASMC from a experimental asthma model.

Taking together all these results indicate that the ASMC from OVA-sensitized rats express a reduced NO-sGC activity and an increased in the NPR-GC-B activity. This imbalance between these two guanylyl cyclases can contribute to airway hyperreactivity and might be implicated in the hyperplastic smooth muscle responses and remodeling present in asthma.

All these previous experimental data unravels some of complex molecular mechanisms associated with the muscarinic activation of ASM. This muscarinic activation is the most physiological, pathophysiological and pharmacological relevant mechanisms because Ach is the neurotransmitter-linked stimulation of ASM in asthma and COPD. Thus, this works opens new trends for the pathophysiological and pharmacological mechanisms, which may lead to new therapeutic approaches for the treatment of these chronic respiratory diseases, such as asthma and COPD, in which, the ASM is involved.

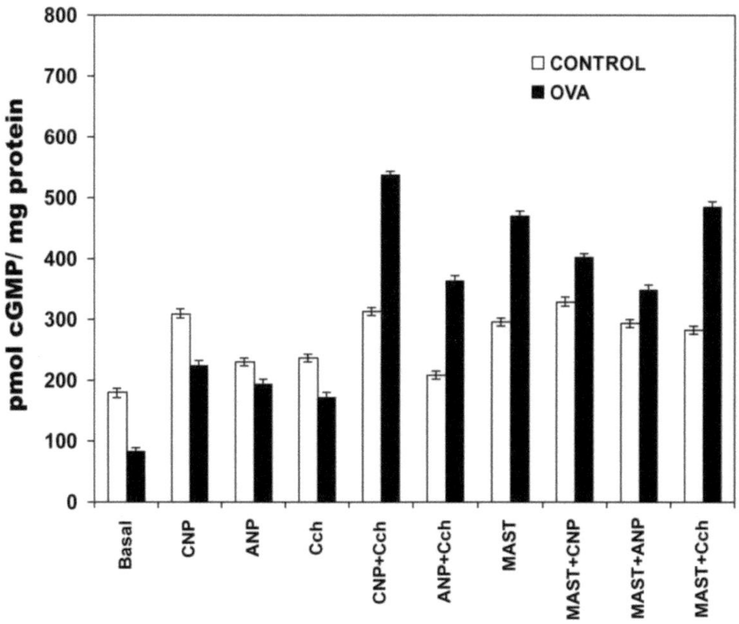

The effect of NPR-GC activators on total GC activity in Control and OVA ASMC. The OVA and Control ASMC were incubated for 15 min in the presence of 100 μM IBMX (non-selective PDEs inhibitor). Cyclic GMP production was determined in the presence of Cch ($1x10^{-5}$ M), Mastoparan ($1x10^{-7}$M), CNP ($1x10^{-7}$M) and ANP ($1x10^{-7}$M). The reaction was stopped by removing the medium immediately and freezing with liquid N_2 was later 500 μl of TCA 6%, mixed vigorously and centrifuged for 1,500 g x 15 min. The acid extract was treated 2 times with ether saturated with water and the water phase was then lyophilized and suspended in 0.150 ml of water. The cGMP produced by these cells were determined using a radioimmunoassay kit as previously described in reference [70].

Figure 11. The effect of NPR-GC activators on total GC activity in Control and OVA ASMC.

3. Conclusion

In this review, we exposed the recent experimental evidences, in relation to the generation of cGMP, on the muscarinic activation of ASM, which is the essential element in the bronchoconstriction presents in asthma.

Moreover, this muscarinic activation seems to be involved in the remodelling and functional changes of ASM described in asthma and COPD as described in references [1-6,73]. We discussed the existence of two cGMP signals, at 20-s and 60-s in [36], which are products of two distinctive guanylyl cyclases, the NO-sensitive soluble guanylyl cyclases (NO-sGC) and

Natriuretic Peptide Receptor Guanylyl Cyclase-B (NPR-GC-B). The 20-s cGMP signal is linked to the activation of a M_2AChR coupled to Go/Gi proteins inducing a massive and transient $\alpha 1\beta 1$-NO-sGC translocation from cytoplasm to plasma membranes of ASM as reported in [31].

The 60-s cGMP signal is associated with a NPR-GC-B in [37], a novel G-protein coupled NPR-GC-B as described in references [38,53,55], which is nano-machine regulated by GPCR and also modulated, in an opposite way, by an activator M_3AChRs coupled to Gq16 to activate NPR-GC-B (M_3AChR/Gq16/NPR-GC-B cascade) that is stimulated by mastoparan as reported in [56] and an inhibitor M_2AChR signal cascade that was partially characterized.

Mastoparan inhibited in a selective manner the muscarinic-dependent ASM contraction and affected the two cGMP signals. The ASM muscarinic activation is unique biological system involving two cGMP signals and a dysfunction of these M_2/M_3AChR cascades has been implied in asthma and COPD as published in references [1-6].

We used another mammalian model as ASM from rats, which are more prone to develop experimental asthma using Ovoalbumin leading to the OVA-sensitized rats model as described in [70,71]. In isolated and cultured Airway Smooth Muscle Cells (ASMC) from Control and OVA, we evaluated the cGMP production in these ASM cultured cells. All ASMC showed cGMP increments by a classic NO donor as SNP being ODQ-sensitive, indicating that NO-sGC is present, but OVA-ASMC showed low basal cGMP production compared to Control ASMC described in [71], which confirmed the molecular biology results reported elsewhere in reference [69]. Moreover, NPR-GC-B is present in Control ASMC but OVA ASMC displayed an hyperstimulation of M_3AChR/Gq16/NPR-GC-B cascade, which is an original experimental findings as reported in [72].

These results indicate that the OVA-ASMC express a reduced NO-sGC and increased NPR-GC-B activities. This imbalance between these two GCs can contribute to airway hyperreactivity and might be implicated in the abnormal ASM responses present in asthma and COPD.

Author details

Marcelo J. Alfonzo, Fabiola Placeres-Uray, Walid Hassan-Soto, Adolfo Borges,
Ramona González de Alfonzo and Itala Lippo de Becemberg
Sección de Biomembranas, Instituto de Medicina Experimental Facultad de Medicina,
Universidad Central de Venezuela, Apdo, Sabana Grande, Caracas, Venezuela

Acknowledgement

This work was supported by grants from CDCH-UCV # PG -09-7401-2008/2 (RGA) and CDCH-UCV # PI -09-7726.2009/2 (ILB) and financial support for the publication of this book chapter. WHS is a Graduate student at Ph-D program of Curso de Postgrado en Ciencias

Fisiológicas. Facultad de Medicina, Universidad Central de Venezuela (UCV). The authors thank to Dr. Marcelo Alfonzo-González for the editing process of this manuscript.

4. References

[1] Belmonte KE. Cholinergic pathways in the lungs and anticholinergic therapy for chronic obstructive pulmonary disease. Proc Am Thor Soc. 2005; 2: 297-304.
[2] Van der Velden VH, Hulsmann AR. Autonomic innervation of human airways: structure, function, and pathophysiology in asthma. Neuroimmunomodulation 1999; 6: 145-59.
[3] Gosens R, Zaagsma J, Meurs H, Halayko AJ. Muscarinic receptor signaling in the pathophysiology of asthma and COPD. Resp Res 2006; 7: 73-86.
[4] Racké K, Matthiesen S. The airway cholinergic system: physiology and pharmacology. Pulm Pharmac Ther. 2004; 17: 181-198.
[5] Racké K, Juergens UR, Matthiesen S. Control by cholinergic mechanisms. Eur J of Pharmacol. 2006; 533: 57-68.
[6] Proskocil BJ, Sekhon HS, Jia Y, Savchenko V, Blakely RD, Lindstrom J, Spindel ER. Acetylcholine is an autocrine or paracrine hormone synthesized and secreted by airway bronchial epithelial cells. Endocrinology 2004; 145: 2498-2506.
[7] Wessler I, Kirkpatrick CJ. Acetylcholine beyond neurons: the non-neuronal cholinergic system in humans. Br J Pharmacol. 2008; 154:1558-1571.
[8] Wessler IK, Kirkpatrick CJ. The non-neuronal cholinergic system: an emerging drug target in the airways. Pulmonary Pharmacol Ther. 2001; 14: 423–434.
[9] Gosens D, Rieks H, Meurs DK, Ninaber, KF, Rabe K, J. Nanninga J, Kolahian S, Halayko AJ, Hiemstra PS, Zuyderduyn S. Muscarinic M_3 receptor stimulation increases cigarette smoke-induced IL-8 secretion by human airway smooth muscle cells. Europ Resp. 2009; J 34: 1436-1443
[10] Tliba O, Panettieri RA. Noncontractile functions of airway smooth muscle cells in asthma. Ann. Rev. Phys. 2009; 71: 509-535.
[11] Challiss RA, Adams D, Mistry R, Boyle JP. Second messenger and ionic modulation of agonist-stimulated phosphoinositide turnover in airway smooth muscle. Biochem Soc Trans. 1993; 21:1138-1145.
[12] Oldhman WM, Hamm HE. Heterotrimeric G protein activation by G-protein-coupled receptors. Nat Rev Mol Cell Biol. 2008; 9: 60-71.
[13] Kotenis E, Zeng FY, Wess J. Structure-function of muscarinic receptors and their associated G proteins. Life Sci. 1999; 64: 335-362.
[14] Caufield MP. Muscarinic receptors-characterization, coupling and function. Pharmacol Ther. 1993; 58: 319-379.
[15] Maeda A, Kubo T, Mishina M, Numa S. Tissue distribution of mRNAs encoding muscarinic acetylcholine receptor subtypes. FEBS Lett. 1998; 239: 339-342.
[16] Lucchesi PA, Scheid CR, Romano FD, Kargacin ME, Mullikin-Kilpatrick D, Yamaguchi H, Honeyman TW. Ligand binding and G protein coupling of muscarinic receptors in airway smooth muscle. Am J Physiol. 1990; 258: C730-738.

[17] Eglen RM, Hedge SS, N. Watson N. Muscarinic receptor subtypes and smooth muscle function. Pharmacol Rev.1996; 48: 531-565.

[18] Roffel, AF, Elzinga CRS, Van Amsterdam RGM, De Zeueuw RA, Zaagsma J. Muscarinic receptors in bovine tracheal smooth muscle: Discrepancies between binding and function. Eur J Pharmacol. 1988; 153:73-82

[19] Misle AJ, Bécemberg IL, Alfonzo RG, Alfonzo MJ. Methoctramine binding sites sensitive to alkylation on muscarinic receptors from tracheal smooth muscle. Biochem Pharmacol. 1994; 48: 191-195.

[20] Misle A, Bruges G, Herrera VN, Alfonzo MJ, Becemberg IL, Alfonzo, RG. G-protein-dependent antagonists binding in M3 AChR from tracheal smooth muscle. Arch Venez Farm Terap. 2001; 20:144-152.

[21] Meurs H, Roffel AF, Postema JB, Timmermans A, Elzinga, CRS, Kauffman, HF, Zaagsma J. Evidence for a direct relationship between phosphoinositide metabolism and airway smooth muscle contraction induced by muscarinic agonists. Eur J Pharmacol.1988; 156: 271-274.

[22] Roffel AF, Elzinga CR, Zaagsma J. Muscarinic M_3 receptors mediate contraction of human central and peripheral airway smooth muscle. Pulm Pharmacol. 1990; 3: 47–51

[23] Grandordy BM, Cuss FM, Sampson AS, Palmer JB, Barnes PJ. Phosphatidylinositol response to cholinergic agonists in airway smooth muscle: relationship to contraction and muscarinic receptor occupancy. J Pharmacol Exp Ther. 1986; 238:273-279.

[24] Roffel AF, Meurs H, Elzinga CR, Zaagsma J. Characterization of the muscarinic receptor subtype involved in phospho-inositide metabolism in bovine tracheal smooth muscle. Br J Pharmacol. 1990; 99: 293-296.

[25] Jones CA, Madison JM, Tom-Moy M, Brown JK. Muscarinic cholinergic inhibition of adenylate cyclase in airway smooth muscle. Am J Physiol. 1987; 253:C97-104.

[26] Sankary RM, Jones CA, Madison JM, Brown JK. Muscarinic cholinergic inhibition of cyclic AMP accumulation in airway smooth muscle. Role of a pertussis toxin-sensitive protein. Am Rev Respir Dis. 1988; 138:145-150.

[27] Fernandes LB, Fryer AD, Hirshman CA. M2 muscarinic receptors inhibit isoproterenol-induced relaxation of canine airway smooth muscle. J. Pharmacol Exp Ther. 1992; 262, 119-126.

[28] Zhou XB, Wulfsen I, LutzS, Utku E, Sausbier U, Ruth P, Wieland T, Korth M M2 Muscarinic Receptors Induce Airway Smooth Muscle Activation via a Dual, G $\beta\gamma$-mediated Inhibition of Large Conductance Ca2+ -activated K Channel Activity. J Biol Chem. 2008; 283: 21036–21044.

[29] Semenov I, Wang B, Herlihy JT, Brenner R. BK channel $\beta 1$ subunits regulate airway contraction secondary to M_2 muscarinic acetylcholine receptor mediated depolarization. J Physiol. 2011; 589.7: 1803–1817.

[30] Nakahara T, Yunoki M, Mitani A, Sakamoto K, Ishii K. Stimulation of muscarinic M_2 receptors inhibits atrial natriuretic peptide-mediated relaxation in bovine tracheal smooth muscle. Naunyn Schmiedebergs Arch Pharmacol. 2002; 366: 376-379.

[31] Uray FP, de Alfonzo RG, de Becemberg IL, Alfonzo MJ. Muscarinic agonists acting through M2 acetylcholine receptors stimulate the migration of an NO-sensitive guan-

ylyl cyclase to the plasma membrane of bovine tracheal smooth muscle. J Recept Signal Transduct Res. 2010, 30:10-23.

[32] Katsuki S, Murad F. Regulation of adenosine cyclic 3',5'-monophosphate and guanosine cyclic 3',5'-monophosphate levels and contractility in bovine tracheal smooth muscle. Mol Pharmacol. 1977;13: 330-341.

[33] Potter LR. Guanylyl cyclase structure, function and regulation. Cellular Signalling 2011; 23:1921-1926.

[34] Potter LR, Abbey-Hosch S, Dickey DM. Natriuretic peptides, their receptors, and cyclic guanosine monophosphate-dependent signaling functions. Endocr Rev. 2006, 27: 47-72.

[35] Alfonzo MJ, Guerra LG., Villarroel SS, Toba GF, Misle A, Herrera VN, Alfonzo RG , Lippo IB. Signal transduction pathways through mammalian guanylyl cyclases New Advances in Cardiovascular Physiology and Pharmacology 1998; 147-175.

[36] González LG, A. Misle A, G. Pacheco G, Herrera VN, Alfonzo RG, Bécemberg IL, and M.J. Alfonzo MJ. Effects of 1H-[1, 2, 4]Oxadiazolo[4, 3, α] quinoxalin-1-one (ODQ) and N^{ω}(6)-nitro-L-arginine methylester (NAME) on cyclic GMP levels during muscarinic activation of tracheal smooth muscle. Biochem Pharmacol. 1999; 58: 563-569.

[37] Borges A, de Villarroel SS, Winand NJ, de Bécemberg IL, Alfonzo MJ, de Alfonzo RG. Molecular and biochemical characterization of a CNP-sensitive guanylyl cyclase in bovine tracheal smooth muscle. Am J Respir Cell Mol Biol. 2001; 25: 98-103.

[38] Alfonzo MJ, de Aguilar EP, de Murillo AG, de Villarroel SS, de Alfonzo RG, Borges A, de Becemberg IL. Characterization of a G-protein coupled guanylyl cyclase-B receptor from Bovine Tracheal Smooth Muscle. J Recept Signal Transduct Res. 2006; 26: 269-297.

[39] Sharina IG, Krumenacker JS, Martin E, Murad F. Genomic organization of α1 and β1 subunits of the mammalian soluble guanylyl cyclase genes. Proc Natl Acad Sci (USA) 2000; 97:10878-10883.

[40] Jiang Y, Stojilkovic SS. Molecular cloning and characterization of soluble of alpha1-soluble guanylyl cyclase gene promoter in rat pituitary cells. J Mol Endocrinol. 2006; 37:503-515.

[41] Pyriochou A, Papapetropoulos A. Soluble guanylyl cyclase: more secrets revealed. Cell Signal 2005; 17:407-413.

[42] Wedel B, Harteneck C, Foerster J, Friebe A, Schultz G, Koesling D. Functional domains of soluble guanylyl cyclase. J Biol Chem. 1995; 270: 24871-24875.

[43] Buechler WA, Nakane M, Murad F. Expression of soluble guanylate cyclase activity requires both enzyme subunits. Biochem Biophys Res Commun. 1991;174: 351-357.

[44] Wagner C, Russwurm M, Jäger R, Friebe A, Koesling D. Dimerization of nitric oxide-sensitive guanylyl cyclase requires the alpha 1 N terminus. J Biol Chem. 2005; 280:17687-17693.

[45] Russwurm M, Koesling D. NO activation of guanylyl cyclase. EMBO J. 2004; 10: 4443-4450.

[46] Sunahara RK, Beuve A, Tesmer JJ, Sprang SR, Garbers DL, Gilman AG. Exchange of substrate and inhibitor specificities between adenylyl and guanylyl cyclases. J Biol Chem. 1998; 273:16332-16338.

[47] Tucker CL, Hurley JH, Miller TR, Hurley JB. Two amino acid substitutions convert a guanylyl cyclase, RetGC-1, into an adenylyl cyclase. Proc Natl Acad Sci (USA). 1998; 95: 5993-5997.

[48] Suga S, Nakao K, Hosoda K, Mukoyama M, Ogawa Y, Shirakami G, Arai H, Saito Y, Kambayashi Y, Inouye K. Receptor selectivity of natriuretic peptide family, atrial natriuretic peptide, brain natriuretic peptide, and C-type natriuretic peptide. Endocrinology 1992; 130: 229-239.

[49] Koller KJ, Lowe DG, Bennett GL, Minamino N, Kangawa K, Matsuo H, Goeddel DV. Selective activation of the B natriuretic peptide receptor by C-type natriuretic peptide (CNP). Science 1991; 252:120-123.

[50] Nagase M, Katafuchi T, Hirose S, Fujita T. Tissue distribution and localization of natriuretic peptide receptor subtypes in stroke-prone spontaneously hypertensive rats. J Hypertens. 1997; 15:1235-43.

[51] Potter LR, Hunter T. Identification and characterization of the major phosphorylation sites of the B-type natriuretic peptide receptor. J Biol Chem. 1998; 273:15533-15539.

[52] Antos LK, Potter LR. Adenine nucleotides decrease the apparent Km of endogenous natriuretic peptide receptors for GTP. Am J Physiol Endocrinol Metab. 2007; 293:E1756-1763.

[53] Lippo de Bécemberg I, Correa de Adjounian MF, Sánchez de Villaroel S, Peña de Aguilar E, González de Alfonzo R, Alfonzo MJ. G-protein-sensitive guanylyly cyclase activity associated with plasma membranes. Arch Biochem Biophys. 1995; 324:209-215.

[54] Alfonzo RG, Becemberg IL, Alfonzo MJ. A Ca2+/CAM protein kinase associated with Ca2+ transport in sarco(endo)plasmic vesicles from tracheal smooth muscle. Life Sci. 1996; 58:1403-1412.

[55] Alfonzo MJ, de Bécemberg IL, de Villarroel SS, de Herrera VN, Misle JA, de Alfonzo RG Two opposite signal transducing mechanisms regulate a G protein coupled guanylyl cyclase. Arch Biochem Biophys. 1998; 350: 19-25.

[56] Bruges G, Borges A, Sánchez de Villarroel S, Lippo de Bécemberg I, Francis de Toba G, Pláceres F, González de Alfonzo R, Alfonzo MJ. Coupling of M3 acetylcholine receptor to Gq16 activates a natriuretic peptide receptor guanylyl cyclase. J Recept Signal Transduct Res. 2007; 27:189-216.

[57] Hirai Y, Yasuhara T, Yoshida H, Nakajima T, Fujino M, Kitada C. A new mast cell degranulating peptide "mastoparan" in the venom of Vespula lewisii. Chem Pharm Bull (Tokyo) 1972; 27: 1942-1944.

[58] Higashijima T, Uzu S, Nakajima T, Ross E. Mastoparan a peptide toxin from wasp venom, mimics receptors by activating GTP-binding regulatory proteins (G proteins). J Biol Chem. 1988; 263; 6491-6494.

[59] Higashijima T, Burnier J, Ross EM. Regulation of Gi and Go by mastoparan, related amphiphilic peptides, and hydrophobic amines. Mechanism and structural determinants of activity. J Biol Chem. 1990; 265:14176-14186.

[60] Shpakov AO, Pertseva MN. Molecular invertebrates mechanisms for the effect of mastoparan on G proteins in tissues of vertebrates and in vertebrates. Bull Exp Biol Med. 2006; 141:302-306.

[61] Jones S, Howl J. Charge delocalization and the design of novel mastoparan analogues: enhanced cytotoxicity and secretory efficacy of [Lys5, Lys8, Aib10]MP. Regul Pept. 2004;15:121-128.

[62] Sugama J, Yu JZ, Rasenick MM, Nakahata N. Mastoparan inhibits betaadrenoceptor-G(s) signaling by changing the localization of Galpha(s) in lipid rafts. Cell Signal 2007; 19:2247-2254.

[63] Shi J, Damjanoska KJ, Singh RK, Carrasco GA, Garcia F, Grippo AJ, Landry M, Sullivan NR, Battaglia G, Muma NA. Agonist induced-phosphorylation of Galpha11 protein reduces coupling to 5-HT2A receptors. J Pharmacol Exp Ther. 2007; 323:248-56.

[64] Shi J, Zemaitaitis B, Muma NA. Phosphorylation of Gq11 Protein Contributes to Agonist-Induced Desensitization of 5-HT2A Receptor Signaling. Mol Pharmacol. 2007; 71:303-313.

[65] Murthy KS, Makhlouf GM. Differential coupling of muscarinic m2 and m3 receptors to adenylyl cyclases V/VI in smooth muscle. Concurrent M2-mediated inhibition via Galphai3 and m3-mediated stimulation via Gbetagammaq. J Biol Chem. 1997; 272: 21317-21324.

[66] Agulló L, Garcia-Dorado D, Escalona N, Ruiz-Meana M, Mirabet M, Inserte J, Soler-Soler J. Membrane association of nitric oxide-sensitive guanylyl cyclase in cardiomyocytes. Cardiov Res. 2005; 68:65-74.

[67] Bidmon HJ, Mohlberg H, Habermann G, Buse E, Zilles K, Behrends S. Cerebellar localization of the NO-receptive soluble guanylyl cyclase subunits-alpha(2)/beta (1) in non-human primates. Cell Tissue Res. 2006; 326:707-714.

[68] Belligham M, Evans TJ. The $\alpha2\beta2$ isoform of guanylyl cyclase mediates plasma membrane localized nitric oxide signaling. Cellular Signalling 2007; 19: 2183-2193.

[69] Papapetropoulos A, Simoes DC, Xanthou G, Roussos C, Gratziou C. Soluble guanylyl cyclase expression is reduced in allergic asthma. Am J Physiol Lung Cell Mol Physiol. 2006; 290: L179-L184.

[70] Hjoberg J, Shore S, Kobzik L, Okinaga S, Hallock A, Vallone J, Subramaniam V, De Sanctis GT, Elias JA, Drazen JM, Silverman ES. Expression of nitric oxide synthase-2 in the lungs decreases airway resistance and responsiveness. J Appl Physiol. 2004; 97: 249-259.

[71] Placeres-Uray F, de Alfonzo RG, de Becemberg IL, Alfonzo MJ. Soluble guanylyl cyclase is reduced in airway smooth muscle cells from a murine model of allergic asthma. World Allergy Organiz J. 2010; 3: 271-276.

[72] [Placeres-Uray F, de Alfonzo RG, Alfonzo MJ, de Becemberg IL. Hypersensitivity of the M_3AChR/ Gq16 protein/NPR-GC-B coupling mechanism associated to muscarinic activation of airway smooth muscle cells in a rat asthma model. World Allergy Organization 2011 Annual Meeting. Boston. USA. (Abstract 81).

[73] Kolahian S, Gosens R. Cholinergic Regulation of Airway Inflammation and Remodelling. J Allergy (Cairo) 2012: 2012:681258.

Chapter 7

Melanophores: Smooth Muscle Cells in Disguise

Saima Salim and Sharique A. Ali

Additional information is available at the end of the chapter

http://dx.doi.org/10.5772/48256

1. Introduction

Melanophores are specialized cells derived from the neural crest that contain membrane bound vesicles called melanosomes. Melanosomes are filled with melanin, a dark, non-fluorescent pigment that plays a principal role in physiological color adaptation of animals. Melanophores regulate melanosome trafficking on cytoskeletal filaments to generate a range of striking chromatic patterns. The mechanism of physiological color change by these melanophores encompasses both physical and biochemical aspects of melanosome dynamics. Melanophores aggregate or disperse their melanosomes when the host requires changing its color in response to the environmental cues (e.g., social interactions or camouflage). Interestingly, morphological, embryological and physiological evidence has revealed that melanophores of fish, amphibian and reptiles are functionally modified smooth muscle cells [1]. Moreover, during contraction of melanophore there is an aggregation of melanin granules which is to be considered the visible counterpart of an aggregation of colloidal particles that occur during smooth muscle contraction. The biochemical events underlying melanosome dispersion is analogous to those for smooth muscle relaxation: both processes result from increase in cAMP levels and require the presence of extracellular Ca2+ for their action.

1.1. Regulation of smooth muscle contraction

Ligands like neurotransmitters and hormones bind to specific receptors to activate contraction in smooth muscle. Subsequent to this binding there is an increase in cellular phospholipase C activity via coupling to a G protein resulting into the activation of a membrane phospholipid phosphatidylinositol 4, 5-biphosphate (PIP_2). Phospholipase C (PLC) produces two potent second messengers from PIP_2: Diacylglycerol (DAG) and inositol 1, 4, 5-triphosphate (IP3). IP3 binds to specific receptors, causing release of calcium (Ca^{2+}). Later in the cascade DAG activates PKC, which further phosphorylates specific target proteins. In smooth muscles PKC promotes phosphorylation of Ca^{2+}

channels that regulates cross-bridge cycling. Ca^{2+} binds to a calcium modulated protein called calmodulin, resulting into the activation of MLC kinase, which is a myosin light chain protein [2]. This step initiates the shortening of the muscle cell together with actin via cross bridge cycling. The state of contraction is maintained by a Ca^{2+} sensitizing mechanism that is brought about by another protein, Rho kinase also known as ROCK. ROCK increases the activity of the motor protein myosin II by two different mechanisms: Firstly, phosphorylation of the myosin light chain (MLC) increases the myosin II ATPase activity. Thus several bundled and active myosins, which are asynchronously active on several actin filaments, move actin filaments against each other resulting in the net shortening of actin fibres. Secondly by inactivating MLC phosphatase, leading to increased levels of phosphorylated MLC [2, 3, 100]. (Figure 1)

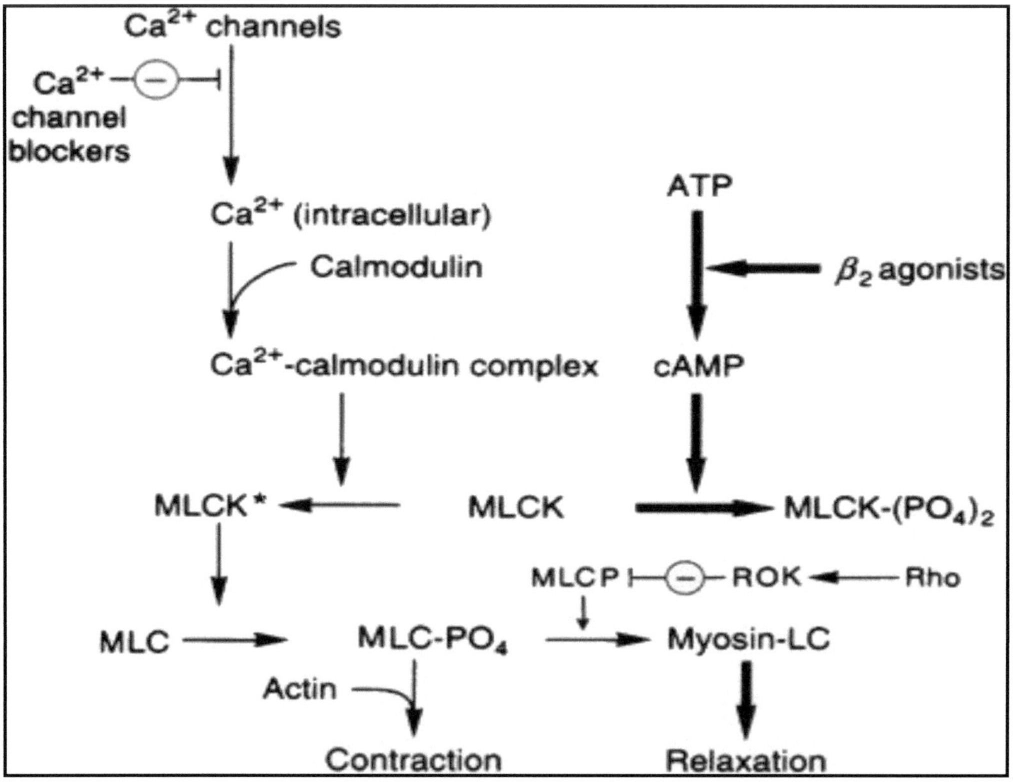

Figure 1. Regulation of Smooth Muscle Contraction and Relaxation. Contraction is triggered by the influx of calcium through the transmembrane channels (ligand-gated (noradrenaline) and the voltage-gated Ca^{2+}-channels). The calcium combines with calmodulin to form a complex that converts myosin light chain kinase to its active form (MLCK*). The latter phosphorylates the myosin light chains (MLC), thereby initiating the interaction of myosin with actin. Activation of RhoA leads to the activation of Rho-kinase (ROCK), which in turn phosphorylates the regulatory myosin-binding subunit of myosin phosphatase (MLCP), resulting into the inhibition of the enzyme. Substances that increase the cAMP may cause relaxation in smooth muscle by accelerating the inactivation of MLCK and facilitating the expulsion of calcium from the cell. (Reproduced with permission from Macmillan Publishers Ltd: [100] Internat. Journ. of Impot. Res. 16: 459–469, Copyright © 2004, Nature Publishing Group.

1.2. Regulation of smooth muscle relaxation

The process of relaxation requires a decreased intracellular Ca^{2+} concentration and increased myosin light chain (MLC) phosphatase activity. The plasma membrane contain Ca Mg-ATPases along with Na^{+}/ Ca^{2+} exchangers that remove Ca^{2+} from the cytosol. Also the voltage operated Ca^{2+} channels in the plasma membrane close to the entry of Ca^{2+} in the cell resulting into relaxation [2].

In short the contractile activity in smooth muscle is initiated by a Ca^{2+}-calmodulin interaction to stimulate phosphorylation of the light chain of myosin. A Ca2+ sensitization of the contractile proteins is signaled by the RhoA/Rho kinase pathway to inhibit the dephosphorylation of the light chain by myosin phosphatase, maintaining force generation. Removal of Ca2+from the cytosol and stimulation of myosin phosphatase initiate the process of smooth muscle relaxation [2,100] (Figure 1).

2. Historical and comparative perspective of melanophores with reference to smooth muscles

The comparison of melanophores to smooth muscle cells dates back as early as 1917 when Spaeth provided evidence for the similarity of melanophores with smooth muscle cells. Rhythmical pulsations in the isolated scale melanophores of *Fundulus heteroclitus* were first observed by Spaeth [1]. The observation followed by the in depth analysis of the behavior and responses of melanophores to various stimuli and their parallelism to smooth muscle cells led to the conceptualization of melanophore as a modified derivative of smooth muscle. Rhythmical movements are, of course, intimately associated with this latter tissue and their persistence for considerable periods of time in isolated strips of the esophagus and stomach of frogs and other vertebrates is well known. The rhythmical movements in isolated scale melanophores are comparable and perhaps similar to these movements in smooth muscle. This discovery opens a new field for research that might contribute in understanding the ramifications of inter and intra-cellular dynamics.

The biochemical events underlying the melanosome dispersion are analogous to smooth muscle relaxation since both these processes result from an increase in cyclic AMP (cAMP) levels [102]. Unlike cardiac muscle, increased cAMP in smooth muscle causes relaxation. The reason for this is that cAMP normally inhibits myosin light chain kinase (MLCK), the enzyme that is responsible for phosphorylating smooth muscle myosin and causing contraction. It is known that the increase in cAMP initiates the influx of Ca^{2+} from the cytoplasm to the endoplasmic reticulum and/or mitochondria or out of the smooth muscle cell. Whereas the melanosome dispersion progresses in the absence of Ca^{2+} and presence of extracellular Ca^{2+} (from the action of MSH) [4]. The crucial role of Ca^{2+} and MSH in the bidirectional movement of melanophores has been shown in the figure 2. Extracellular stimuli (either hormonal; adrenaline or mechanical stimulus) could result in inward movement of Ca^{2+} from the extracellular matrix. This is followed by activation of phospholipase C (PLC) promoting 1) inositol trisphosphate (IP3)-dependent release of Ca^{2+}

from intracellular stores and 2) diacylglycerol (DAG)-sensitive activation of protein kinase C (PKC), which facilitates Ca^{2+} entry through voltage-dependent calcium channels, resulting into pigment aggregation. Stimulation of alpha MSH receptor that couples to Gsα proteins stimulates adenylyl cyclase and increase intracellular cAMP resulting into pigment dispersion by activating a calcium pump that drives the expulsion of calcium out from the cells resulting into pigment dispersion. In the case of pigment cells the action of Melanocyte concentrating hormone (MCH) is enhanced by the absence of intracellular Ca^{2+} [5, 101] which in the case of smooth muscle may be an apparent counterpart triggered by β adrenergic receptors (NE or noradrenaline). The actual process of how melanosomes translocate from the concentrated to dispersed state and vice versa is not completely unfurled. However the mechanism is highly coordinated and requires a multitude of factors and entities.

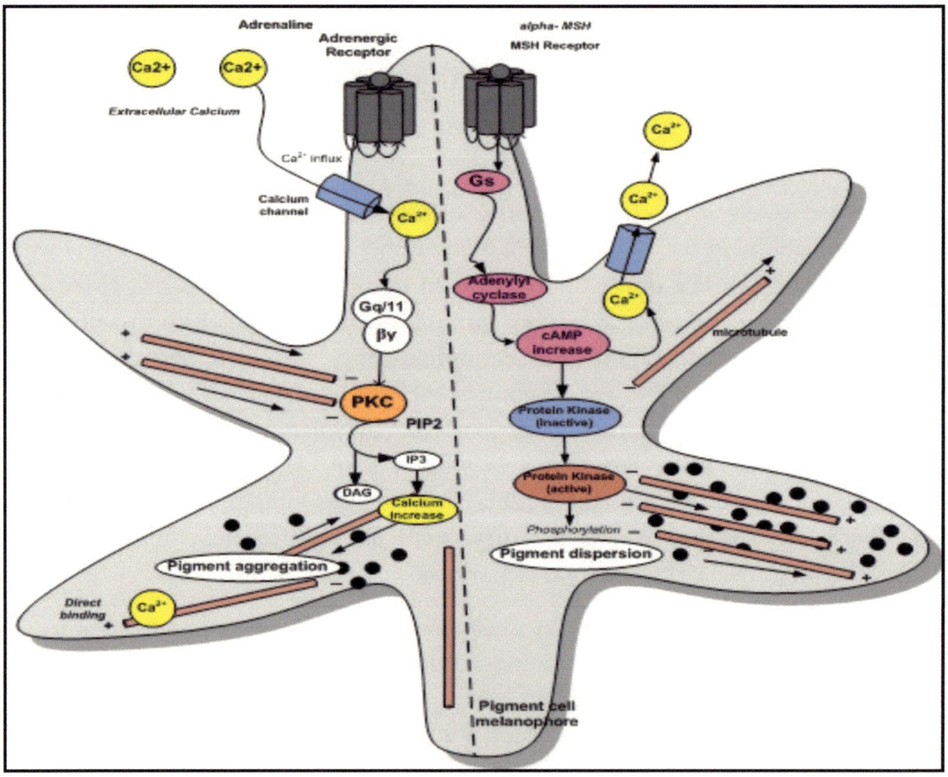

Figure 2. Model for the crucial role of Ca^{2+} and MSH in the bidirectional movement of melanophores. Extracellular stimuli (either hormonal; adrenaline or mechanical stimulus) could result in inward movement of Ca^{2+} from the extracellular matrix. This is followed by activation of phospholipase C (PLC) promoting 1) inositol trisphosphate (IP3)-dependent release of Ca^{2+} from intracellular stores and 2) diacylglycerol (DAG)-sensitive activation of protein kinase C (PKC), which facilitates Ca^{2+} entry through voltage-dependent calcium channels, resulting into pigment aggregation. Stimulation of alpha MSH receptor that couples to Gsα proteins stimulates adenylyl cyclase and increase intracellular cAMP resulting into pigment dispersion by activating a calcium pump that drives the expulsion of calcium out from the cells resulting into pigment dispersion

3. Experiments

Since all muscle cells are "excitable cells," i.e., they are capable of responding to appropriate stimulation. The response to this is contraction, however the stimulus can vary. It may be an electrical signal or depolarization produced at the surface of the muscle cell by the activity of neurons. Some muscle cells (particularly smooth muscle) respond not only to neuronal signaling, but to direct chemical or mechanical stimulation [6]. Hormonal signals can cause contraction in some situations. Interestingly, melanophores have an instinctive ability to quickly reposition pigment granules within the cells on given appropriate stimulus. The responses of melanophores to external stimuli are highly coordinated; where external signals are received and integrated, thereby eliciting a concerted and appropriate response [7, 8]. This cellular communication depends largely on the transmission of signal couriers (i.e. "ligands/agonist") which are received via cell surface and intracellular recognition molecules (i.e. "receptors") on the recipient cells, resulting into remarkably coordinated bidirectional movement of pigment granules within the cells [7, 8].

By far the most striking resemblance between the melanophores and smooth muscle cells has been furnished on physiological grounds [9, 1]. Also the morphological, as well as embryological evidence has been brought to light that supports the idea that color changes in the three groups of lower vertebrates; fishes, amphibians and reptiles along with crustaceans and cephalopods are brought about by responses of specialized functional smooth muscle cells [10, 11, 12].

The most striking evidence of resemblance between smooth muscles and melanophores in terms of physiological basis was provided as early as 1906 by Franz [13]. He recorded the similarity between the physiological responses of the sphincter papillae in *Acanthias* and the dermal melanophores of frog. A number of investigators have demonstrated the involvement of nervous system and direct innervations of chromatophores. This view has been widely accepted by a number of workers and a significant amount of reports have been published [14, 15, 16, 17].

3.1. Innervation

There is a certain resemblance between the mechanisms coordinating the mechanism of smooth muscles with that of vertebral melanophores in terms of innervation. Both are controlled by sympathetic nervous system and influenced by hormonal secretion during nervous excitement [16, 18, 8]. In the case of smooth muscles, however there is an autonomic innervation of antagonistic fibers that are morphologically distinct. Interestingly, the neural control of skeletal muscles differs significantly from that of smooth muscles. A skeletal muscle fiber has only one junction with a somatic nerve fiber, and the receptors for the neurotransmitter are located only at the neuromuscular junction [19]. By contrast, the entire surface of smooth muscle cells contains neurotransmitter receptor proteins. Neurotransmitter molecules (norepinephrine) are released along a stretch of an autonomic nerve fiber that is located some distance from the smooth muscle cells and stimulate a number of smooth muscle cells [20].

The innervation of the melanophores has been demonstrated physiologically in several species of teleosts [22, 23, 24, 1]. As early as 1876, Pouchet [21] reported that in teleost fishes, the sectioning of peripheral nerves or the electrical stimulation of spinal nerves led respectively to the darkening or paling in definite areas, claiming the innervation to be of sympathetic nature. Interestingly, the surface of melanophores contains numerous receptors of neurotransmitters as well as hormones including adrenoceptors [8]. It is now generally accepted that the melanophores are sympathetically innervation and the catecholamines released from the adrenergic fibers are responsible for causing pigment aggregation.

3.2. Effect of light

Interestingly it has been demonstrated that the excised iris of certain fishes (elasmobranchs and teleosts) and amphibians responds to illumination by contraction of sphincter papillae. Also a light induced stimulation in chromatophores of cephalopods, Octopus and Loligo have been demonstrated [25], which was a direct result of contraction and relaxation of radially arranged smooth muscles. In another report the smooth muscle cells of the frog, (*Rana pipiens*) iris sphincter also showed light-evoked contraction [26]. In general, melanosome movements in fish melanophores are controlled by adrenergic nerves. That means that the melanosomes aggregate in the presence of epinephrine and the aggregation is inhibited by α-adrenergic blocking drugs [104,105]. The light-sensitive cultured melanophores also aggregate on application of epinephrine (Wakamatsu, 1978). Dibenamine (α-adrenergic blocker) inhibited the response to epinephrine, but did not inhibit the light-induced melanosome aggregation (Wakamatsu, unpublished observation). These facts suggest that light controls the melanosome movements without mediation of the adrenergic receptors and acts directly on phosphodiesterase enzyme systems in the light-sensitive melanophores.

3.3. Mechanical stimulation

It has been reported that gentle stretching or pinching of excised pieces of fish and frog stomach and esophagus induced powerful contractions, provided the stimulus has not been too violent. Similarly, melanophores from the portions of skin of Loligo resulted into wide expansion. Single aggregated melanophores from isolated scales of the angelfish, *Pterophyllum scalare* were submitted to mechanical compression with forces ranging from 50–320 μ p. With increasing force melanophores disperse their pigment, the degree of dispersion being proportional to the intensity of the force [27].

3.4. Electrical stimulation

The responses of several types of smooth muscles as well as the melanophores from the three groups of lower vertebrates show considerable similarity in their contraction to faradic stimulation.

It has been reported that sphincter papillae of eel and frog (*Rana*) contracted when stimulated electrically. Likewise the radial muscles of the chromatophores of cephalopods also respond to electrical stimulation by contracting [15]. Such contractions have been seen in strips of frog stomach and the digestive tube of several species of teleosts. Such parallelism in responses to melanophores on stimulation with induction currents of sufficient intensity and durations have been observed as well. *Fundulus* melanophores contract invariably when proper strength and duration of current and salt concentration of mounting medium was selected [1]. Similar results were observed with melanophores of *Rana esculenta* and *Hyla arborea* [28]. The reports by Bert and Krukenberg [29, 30] were in line with the original observation of Brücke that the direct faradic stimulation of excised chameleon skin pieces produced paling or lightening by contraction of melanophores. Since the nervous system of lower vertebrates like fish has evolved to allow faster chromatic adaptation, the sympathetic division of the autonomic nervous system has been shown to be involved in aggregation of pigment in melanophores.

3.5. Chemical and pharmacological stimulation

Effect of various chemicals and pharmacological compounds has been extensively studied on pigment as well as muscle cells. The responses to various biogenic amines/ compounds and neurotransmitters have been analyzed and it is confirmed that the effector cells contain several types of receptors for different neurotransmitters and hormones [8]. The extracellular signals are translated by the receptor system into an increase or a decrease in levels of intracellular second messenger. Until recently three kinds of second messengers were known to take part in the motile responses of chromatophores, namely cyclic adenosine mono phosphate (cAMP), Calcium (Ca^{2+}), and inositol 1,4,5 tri-phosphate (IP3). Interestingly the roles of these second messengers in smooth muscle contraction have been described [31, 32].

Calcium is one of the most important cations in terms of diversity of function and has a crucial role in muscle contraction and pigment translocation respectively (due to the extent of depth only the role of Ca^{2+} is discussed). The role of Ca^{2+} in the process of muscle contraction and relaxation has been reviewed by Karaki et al., in quite detail [103].

In smooth muscle, the influx of calcium leads to depolarization. There calcium binds to calmodulin, causing the calmodulin-caldesmon complex to change its configuration and pull the caldesmon away from the myosin binding sites on the actin strand. The myosin heads bind to actin resulting into muscle contraction. Calcium ions are required in the medium for the full darkening action of melanocyte- stimulating hormone. It has been reported that frog (*Rana pipiens*) skin the action of melanocyte-stimulating hormone involves the production of pigment granule movements as a result of the interaction of calcium ions with intracellular microfilaments and possibly by the breakdown of microtubules [33, 34] In relation to the nervous system controlling fish chromatophores [35] first reported that Ca^{2+} is required for catecholamine release from the sympathetic nerve terminals in the goby *Chasmichthys gulosus.* It has been suggested that Ca^{2+} act as the coupling agent in muscle excitation and contraction. Fish chromatophores provide an ideal cell model for studying the regulation of directional, microtubule-based organelle trafficking [36, 37, 38]. Pigment granules in these cells exhibit distinct movements, which are closely associated with and dependent upon a dense radial array of microtubules [39, 40, 41]. This transport is either completely retrograde (inward aggregation toward the "minus'-ends of microtubules at the cell center) or completely anterograde (outward dispersion toward the "plus'-ends at the cell periphery). Thus, the cellular events that signal a specific directional organelle movement can be manipulated and studied. The data supporting the role of Ca^{2+} in the regulation of pigment movement is quite convincing. Studies using ionophores [42, 43] or lysed cell models [44] have implicated a Ca^{2+} based regulatory system in erythrophores. Raising Ca^{2+} in these two models causes aggregation, and lowering Ca^{2+} causes dispersion, suggesting that intracellular Ca^{2+} is the exclusive regulatory signal for both aggregation and dispersion.

Another signaling molecule Nitric oxide (NO) plays a significant role in a number of cellular processes like the relaxation of blood vessels, sperm motility, and polymerization of actin. It has been found that the signal transduction by NO can be mediated through the activation of soluble guanylate cyclase (sGC), which leads to increased production of cGMP and activation of cGMP-dependent kinases (PKG) [108]. Studies show that melanosome aggregation depends on synthesis of NO, and NO deprivation causes dispersion indicating a crucial role of NO and cGMP in regulation of melanosome translocation. Similar results have been reported by other workers on teleostean melanophores [109]. Interestingly the parallelism of NO action in terms of smooth muscles can be drawn since increased intracellular cGMP by NO, inhibits calcium entry into the cell, thereby decreasing intracellular calcium concentrations and causing smooth muscle relaxation. NO also activates K^+ channels, which leads to hyperpolarization and relaxation. Finally, NO acting through cGMP can stimulate a cGMP-dependent protein kinase that activates myosin light chain phosphatase that dephosphorylates myosin light chains, which leads to relaxation (Figure 3).

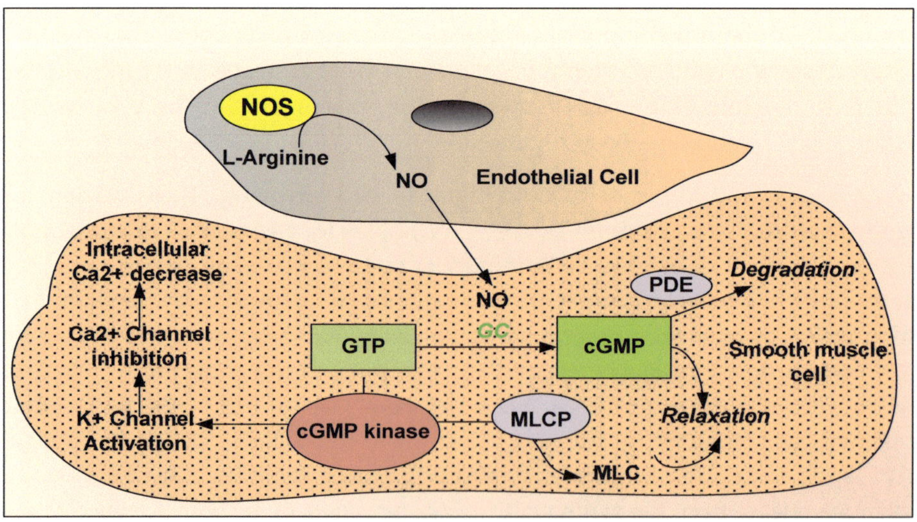

Figure 3. NO induced smooth muscle Relaxation, NO; nitric oxide, NOS; nitric oxide synthase, GC; guanylyl cyclase, PDE cyclic GMP-dependent phosphodiestrase, MLC lyosin light chain

4. Immunocytochemical experiments to reveal the involvement of similar intercellular proteins in melanophores and smooth muscle

Immunocytochemical experiments conducted on the melanophores, xanthophores, and iridophores from the skins of the two Antarctic fish species *Pagothenia borchgrevinki* and *Trematomus bernacchii* tested for a variety of muscle proteins confirmed the presence of actin, myosin, and calmodulin, for all three chromatophores types of the two fishes [46]. Interestingly, the presence of caldesmon and calponin that are both characteristic proteins of smooth muscle fibers have also been found. It has been suggested that calponin's role, in the presence of Ca^{2+} and calmodulin, is that of a modulator and that caldesmon, a molecule that competes with calponin for actin binding sites, is in a position in which it can switch on and off Ca^{2+}-dependent contractility and relaxation. It is not known at this stage whether these proteins occur also in the chromatophores of other fishes and are not restricted to Antarctic species. Since the control of pigment translocation in fish chromatophores and the regulation of smooth muscle tension both involve the sympathetic nervous system, the presence of similar target proteins in both the systems is quite convincing. Also the fact that none of the chromatophores tested positive for troponin (a characteristic regulatory protein in skeletal and cardiac muscle but not in smooth muscle) shows that there is no close relationship between pigment cells and striated muscle. Also the presence of alpha-actinin (a typical actin binding protein) in melanophores and xanthophores unlike iridophores could be most likely due to greater degree of pigment translocation within the former class of cells [45, 46].

5. Implication of motor proteins and their dynamics in pigment translocation within melanophores

The cytoskeleton is now no longer considered to be a rigid scaffold, but instead is viewed as a complex and dynamic network of protein filaments that can be modulated by internal and external cues. Three cytoskeletal polymers- actin filaments, microtubules and intermediate filaments cooperate to maintain the physical integrity of eukaryotic cells and, together with molecular motors, allow cells to move themselves and their intracellular organelles like melanosomes. Pigment cells provide an excellent model to study organelle transport as they specialize in the translocation of pigment granules in response to defined chemical cues. Pigment cells of lower vertebrates have traditionally been used as a model for these studies as they transport pigment organelles in a highly concerted and coordinated fashion. These cells can be cultured and transfected, are ideal for biochemical and in vitro studies. Changes in pigment translocation can be easily monitored under light microscopy. Many important mechanism of organelle transport like the regulation and interactions of cytoskeletal filaments (actin and microtubules) and motor proteins have been studied using pigment cells of lower vertebrates. Genetic studies of mouse melanocytes allowed the discovery of essential elements involved in organelle transport including the myosin-Va motor and its receptor and adaptor molecules on the organelle surface [47]. Future studies of pigment cells will contribute in unraveling the mechanisms of regulation of microtubule motors and other related mechanisms that may involve the cooperation of motor proteins.

5.1. Molecular mechanism of intracellular translocation of pigment granules

The most interesting example of organelle transport is within melanophores of lower vertebrates where the cytoskeletal actin and microtubules act in cooperation. The melanophores contain a radial array of ~1000–2000 microtubules [48, 40, 49]. The bidirectional movement of pigments occurs along the radially-organized microtubule cytoskeleton of these cells and also transported along filamentous actin [50]. Microtubules act as tracks for the transport of several intracellular organelles, including pigment vesicles, melanosomes. The trafficking of melanosomes is controlled by two classes of microtubule-associated motor proteins, kinesins, and cytoplasmic dyneins. Kinesins power the plus-end-directed microtubule-based motility, while cytoplasmic dyneins drive the minus end motility [51, 52]. Dyneins and kinesins also have well-established roles in retrograde (aggregation) and in anterograde (dispersion) transport of melanosomes [53, 54, 55, 56]. In microtubule-poor regions of the cell, pigment granules are transported along microfilaments powered by a myosin motor.

5.2. *Xenopus laevis* melanophore system: A fascinating example of organelle transport

Xenopus laevis melanophore system represents one of the most interesting mechanisms of organelle transport. The pigment transport is regulated by the intracellular levels of cAMP (Figure 4). Up-regulation (induces dispersion) and down regulation (induces aggregation) of cAMP levels are triggered by Melanocyte stimultating hormone (MSH) and Melatonin respectively [57]. This antagonistic action of hormones that regulates pigment translocation within the cells is carried by kinase and phosphatase activities [58].

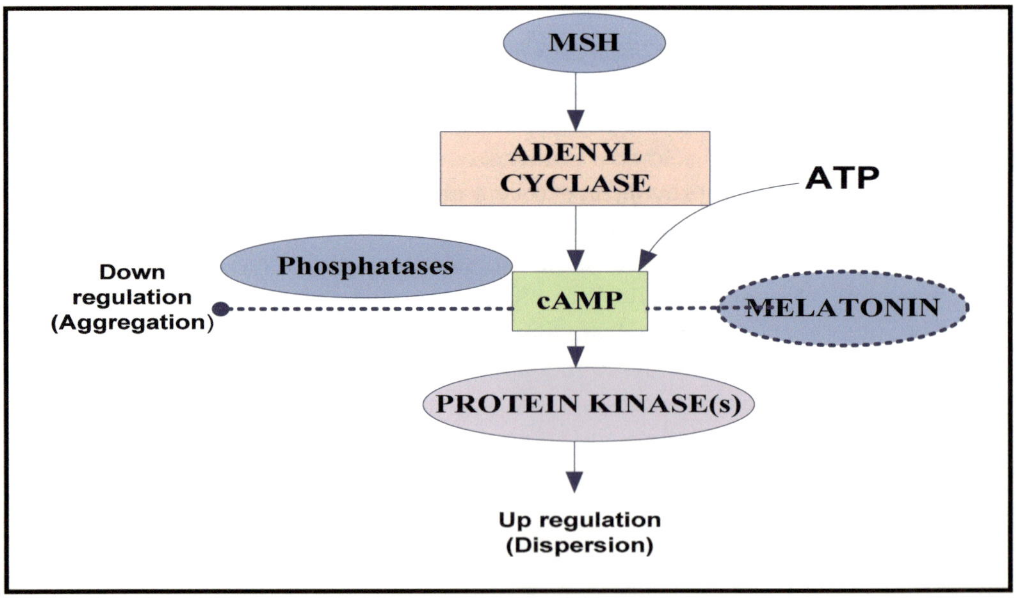

Figure 4. Involvement of cAMP in melanosome movement

5.2.1. Cytoskeletal tracks: Microtubule and actin filaments

Although actin filaments and microtubules differ in origin and structure, their shared features reflects extensive convergence of function. The microtubules and actin filaments show tremendous interaction by binding of several motors to the same cargo at the same time, and movement of pigment granules along both types of cytoskeleton tracks. Studies have demonstrated that the microtubule motor activity is coordinated while the microtubule and actin-based transport are competitive [59]. The direction of melanosome movement is determined by the orientation of plus and minus ends of microtubules and actin filaments within the melanophores [60]. As in most cells, melanophore microtubules are oriented with the minus ends located at the nucleus and their plus ends toward the periphery near the plasma membrane [61, 47]. In melanophores, melanosome aggregation and dispersion are achieved by minus- and plus-end directed movement respectively [59]. In contrast to microtubule organization which is typically in the center of the cells, actin filaments are randomly oriented in a form of a meshwork close to the plasma membrane. The filaments consist of globular subunits arranged head-to-tail into double helical polymers, giving the structure polarity [62]. Actin filaments are required for complete melanosome dispersion but also for maintenance of the dispersed state [34].

5.2.2. Kinesin and dynein

Kinesin and Dynein are key entities of microtubule-based motion and are termed as Microtubule Associated Proteins (MAPs). These two families of motor proteins transport membrane-bounded vesicles, proteins, and organelles along microtubules. Purified *Xenopus* melanosomes have been shown to possess bound kinesin and cytoplasmic dynein and to move on microtubules in vitro [63]. Nearly all kinesins move cargo toward the (+) end of microtubules (anterograde transport) also called plus-end directed motors, whereas dyneins transport cargo toward the (−) end (retrograde transport), oriented towards the cell center, also called "minus-end directed motors". The structure of molecular motors consists of two parts: a motor domain that reversibly bind to the cytoskeleton converting chemical energy contained in ATP into mechanical energy (motion) and the tail which is attached to the cargo. The divergent tail domains allow the motor to bind various types of organelles and particles via interactions with receptor proteins [64]. *Xenopus* melanophore plus-end directed movement (dispersion) has been shown to be driven by kinesin-II, a heterotrimeric kinesin of the kinesin super family which is found in many species. Kinesins typically contain two heavy chains with motor heads which move along microtubules via a pseudo-processive asymmetric walking motion. It is formed by two homologous motor subunits of 85 and 95 kD and a non-motor subunit of 115 kD, called kinesin-associated protein (KAP), which is thought to mediate cargo binding [65] (Figure 5).

Cytoplasmic dynein is responsible for the aggregation of pigment organelles in melanophores and consist of two heavy chains containing the motor domains as well as various intermediate, light intermediate and light chains [66, 64]. Like kinesin, cytosolic dynein is a two-headed molecule, with two nearly identical heavy chains forming the head domains. However, unlike kinesin, dynein cannot mediate transport by itself. Rather, dynein-related motility requires a large complex of *microtubule-binding proteins* that link pigment carrying melanosomes to microtubules but by themselves do not exert force to cause movement. Dynein has been shown to interact with many of its cargos, including pigment organelles, via dynactin [67, 64, 68]; which is a hetero-complex of at least ten subunits, including a 150,000-MW protein called *Glued or p150 Glued),* the actin-capping protein Arp1, and dynamatin or p50 [68]. The P150 *Glued* subunit plays a key role, in dynactin-dynein interaction by binding to the cytoplasmic dynein. In addition the p150Glued subunit of dynactin interact with the KAP subunit of kinesin II, thus serving as a receptor for kinesin II during dispersion of *Xenopus* melanosomes [68] (Figure 6).

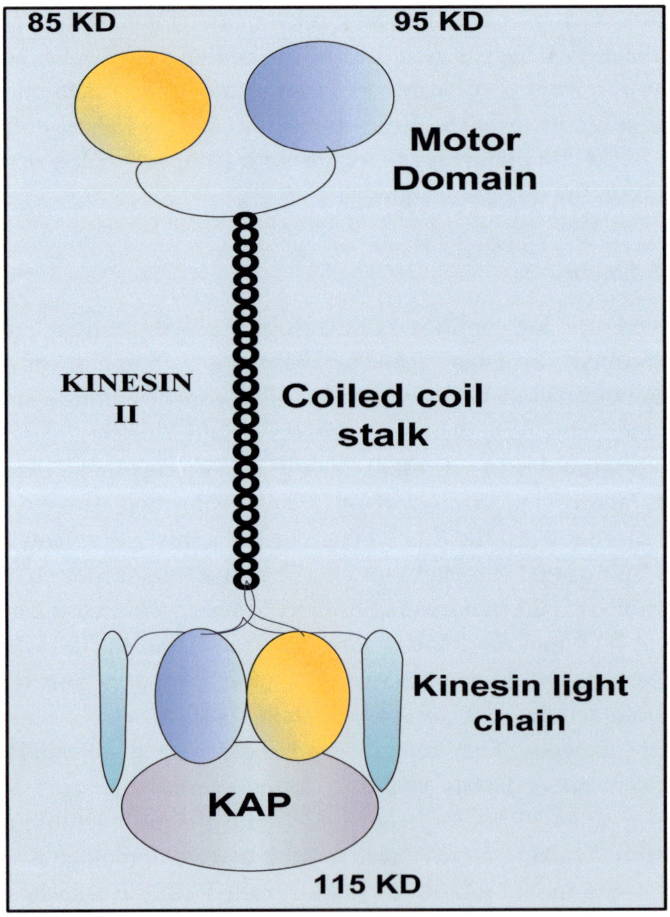

Figure 5. Kinesin II is a heterotrimeric motor protein. Motor heads shown in blue (95 KD) and orange (85 KD) and a non motor subunit shown in grey (115 KD) called kinesin-associated protein (KAP).The motor domains contain binding sites for cytoskeleton. The coiled-coil domain connects the motor domain to the tail domain. Kinesin light chain (KLC) binds through the globular tail domain of kinesin heavy chain (KHC).

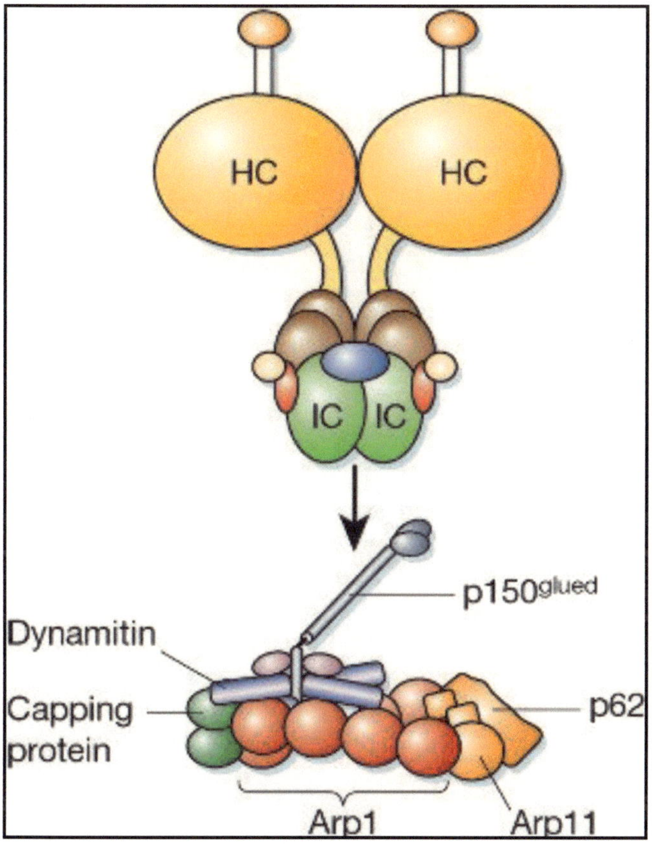

Figure 6. The dynein molecule, itself a complex of heavy (HC), intermediate (IC) and light chains, interacts with the p150glued subunit of the dynactin complex through its intermediate chains (arrow), although the precise mode of interaction is not known. The most prominent component of the dynactin complex is a short filament of the actin-related protein Arp1 (Reproduced with permission from Macmillan Publishers Ltd [107], Nature, Copyright © 2003, Nature Publishing Group.

5.2.3. Myosin V is responsible for melanosome transport along actin filaments in xenopus

Myosin V is the unconventional class V of the myosin family which is found in many different organisms [69,70]. It is composed of:

- The *head domain* that binds the filamentous actin, and uses ATP hydrolysis to generate force and to "walk" along the filament towards the barbed (+) end.
- The *neck domain* acts as a linker and as a lever arm for transducing force generated by the catalytic motor domain. The neck domain can also serve as a binding site for myosin *light chains* which are distinct proteins that form part of a macromolecular complex and generally have regulatory functions.
- The *tail domain* generally mediates interaction with cargo molecules and/or other myosin subunits. In some cases, the tail domain may play a role in regulating motor activity.

Myosin Va binds to melanosomes through a complex of melanophilin and Rab27a, a small GTP-binding Ras-like GTPase (Figure 6) [71,47]. Rab27a first binds to the melanosome and then recruits melanophilin, who's N-terminal interact with the GTP- bound Rab27a. Subsequently myosin Va, requiring binding of both Rab27a and melanophilin to the melanosome, is able to interact with the C-terminal portion of melanophilin [70, 71] (Figure 7).

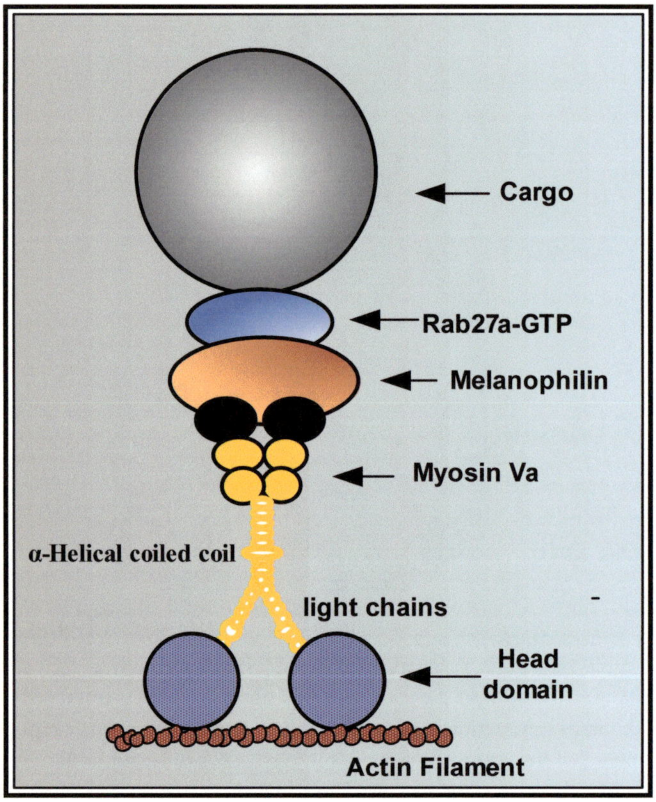

Figure 7. Mysoin Va binds to melanosomes through a complex of GTP-bound Rab27a and melanophilin. Rab27a binds to the melanosome first and then recruits melanophilin. Myosin Va requires the binding of both Rab27a and melanophilin before being able to interact with melanophilin.

During pigment translocation across the cells microtubules and actin filaments interact with several motors that bind to the same cargo (melanosome) at the same time. In *Xenopus laevis* melanophores the retrograde transport of melanosomes (aggregation) is driven by the molecular motor dynein along microtubules and anterogate movements (dispersion) by kinesin II and myosin Va along both microtubules and actin filaments. The actin filaments trap melanosomes at the cell periphery assuming an even distribution of pigment throughout the cell. It has been demonstrated [59] that the microtubule motor activity is coordinated while microtubule-actin based transport is competitive. During aggregation the myosin Va dependent transport is down regulated whereas the kinesin based transport predominate. The net movement of melanosomes results from the combined action of these three motors (Figure 8).

However there are two different mechanisms postulated to explain the unidirectional (either aggregation or dispersion) transport of melanosomes attached to two opposing motor proteins. In the first model, the motors are involved in a competitive manner called the "tug of war" with the stronger motor determining the direction of motion at any given time. On the other hand the second model, there is coordination, so that when the melanosome moves in one direction the opposing motors are inactive.

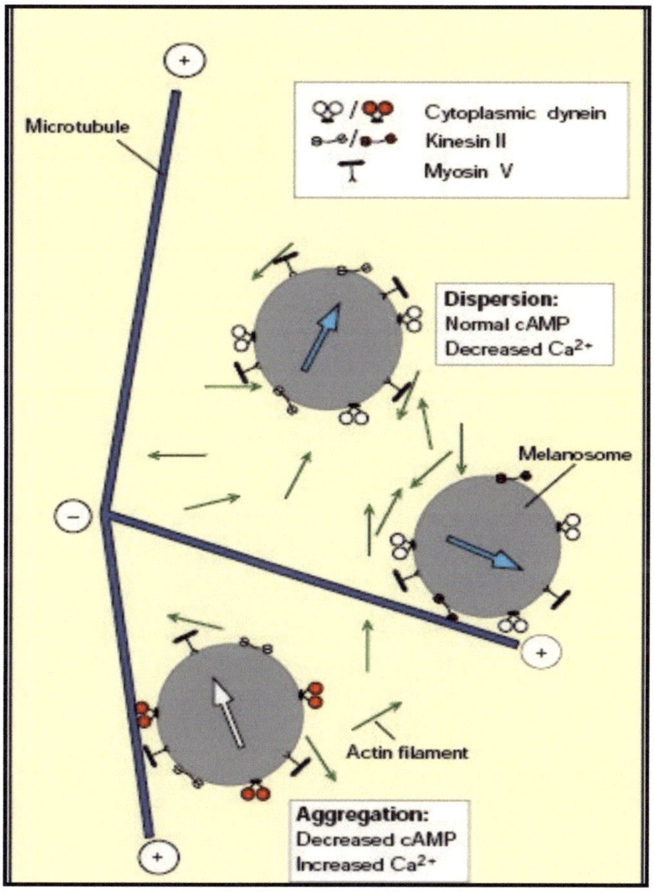

Figure 8. The external stimulation of pigment cells result into a highly coordinated bidirectional movement of pigment granule that either result into aggregation (movement towards the minus ends of microtubules, indicated by white arrow) or dispersion (movement towards the plus ends of microtubules and the periphery of the cell, indicated by a blue arrow). Melanosomes that are in the process of dispersion, move off microtubule tracks and then move along short, randomly-oriented actin filaments. 'Active' microtubule motors are indicated by red globular motor domains; 'inactive' microtubule motors are indicated by white motor domains. (Figure from reference [97]; Reproduced with permission from Elsevier, Copyright © 1998 Elsevier Science Ltd.

The process of pigment translocation within the pigment cells as evident from above discussion is a remarkably regulated phenomenon that involves the coordination of numerous cellular entities. Biochemical studies of melanosomes purified from aggregated and dispersed states have indicated that motor proteins remain attached to the melanosome even when their activity is not required; for example, kinesin II remains associated with melanosomes in an inactive state during melanocortin-induced pigment aggregation [58]. In contrast to mammalian cells, the motor activity in melanophores is acutely altered in a cAMP- and protein kinase A (PKA)-dependent manner. Elevated cAMP results in increased kinesin activity, which leads to melanosome dispersion, whereas reduced cAMP results in increased dynein activity, which gives rise to aggregation [72]. PKA directly associates with melanosomal kinesin and dynein and might be recruited to melanosomes by the small GTPase Rab32 [73, 74].

Studies show that one to three dynein molecules can transport each melanosome in the minus end direction. The transport in the plus end direction is driven by one –two copies of kinesin II. The number of dyneins transporting a melanosome increases during aggregation, whereas the number of active Kinesin II stays the same during aggregation and dispersion. Thus the direction of net melanosome transport is regulated by the number of active dynein molecules [111]. Kinesin-II and dynein compete for the same binding site on p150Glued of dynactin [68]. During aggregation, the release of myosin V helps the dynein-mediated movement to 'win' over the kinesin-II mediated movement [59]. Also studies show that the long-range, bidirectional, microtubule-dependent melanosome movements, coupled with actomyosin Va–dependent capture of melanosomes in the periphery, is the predominant mechanism responsible for the centrifugal transport and peripheral accumulation of melanosomes in mouse melanocytes [71] (Figure 9).

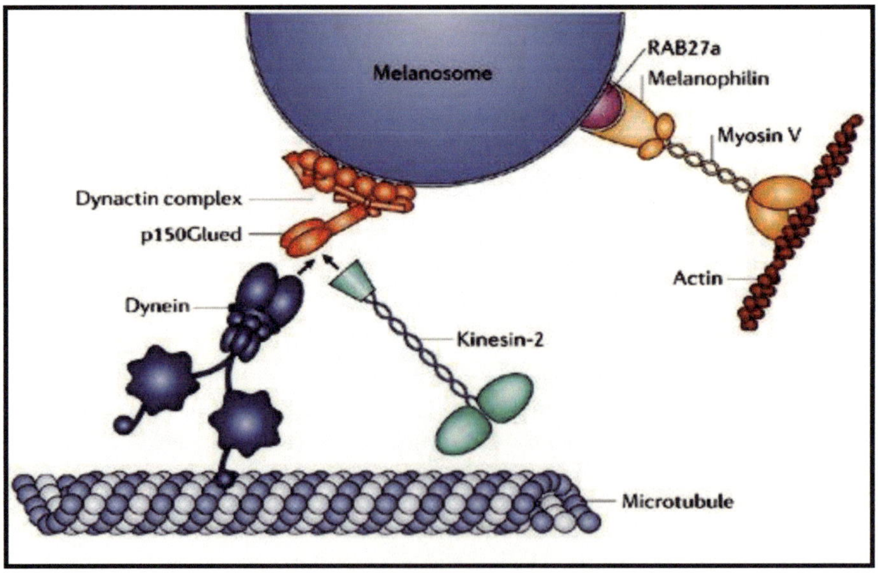

Figure 9. Coordinating several motors: the melanosome paradigm. (Reproduced with permission from Macmillan Publishers Ltd [110], Nature, Copyright © 2006, Nature Publishing Group).

Gross *et al.*,[59] (Figure 10) have presented a very interesting model (Figure 10) that reflects how these motor proteins interact with the cytoskeletal tracks during melanosome transport. They have proposed that the pigment organelle can be moved in three different ways, along the microtubule by kinesin II (A and D) or by Dynein (B and E), or along the actin filaments by means of myosin V (C and F). It is proposed that the motion at any given moment is dominated by the two transport system i.e either action or microtubules. It is now known that the microtubule- and actin-based transport is not coordinated but rather competitive like in a "tug-of-war". It has been observed that there has been a significant stimulation of microtubule-based motion due to a loss of myosin V function [71, 75] that clearly support that the microtubule- and actin-based transport are not coordinated but rather in a tug-of-war. , the motors of the other transport system are transiently active and interacting with their substrate (i.e., actin or microtubules). In their model they postulated that the motors of the other transport system are transiently active and interacting with their substrate (i.e., actin or microtubules) and during dispersion, these transient interactions (indicated by T) are significant and allow myosin V to reduce the velocity of microtubule-based transport (A and B). The interactions also play a role in the switch between the two transport systems, that is, between B and C. Since myosin V activity predominantly decreases the length of minus end microtubule-based motion, the switch from microtubule- to actin-based transport occurs only between state B and C (from dynein movement to myosin V movement) and those transitions from kinesin II movement (A) to myosin V movement (C) are rare. Because kinesin II but not dynein appears to win in tug-of-wars with myosin V, we suggest that the C to A transition is possible but not the C to B transition; however, we have no direct evidence on this point. During aggregation, myosin V activity is decreased, which results in reduced weak transient interactions (indicated by weak T) in D and E, and myosin V is no longer able to interfere with microtubule-based motion. Due to the weakness of the interactions, any time the microtubule motors are in contact with the microtubules they win the tug-of-war with myosin V and there is a transfer (F to D or F to E). Similarly, the reverse transfer from microtubules to actin-based transport (D to F or E to F) does not occur. M-V, myosin V; K-II, kinesin II; Dy, dynein; Ms, melanosome.

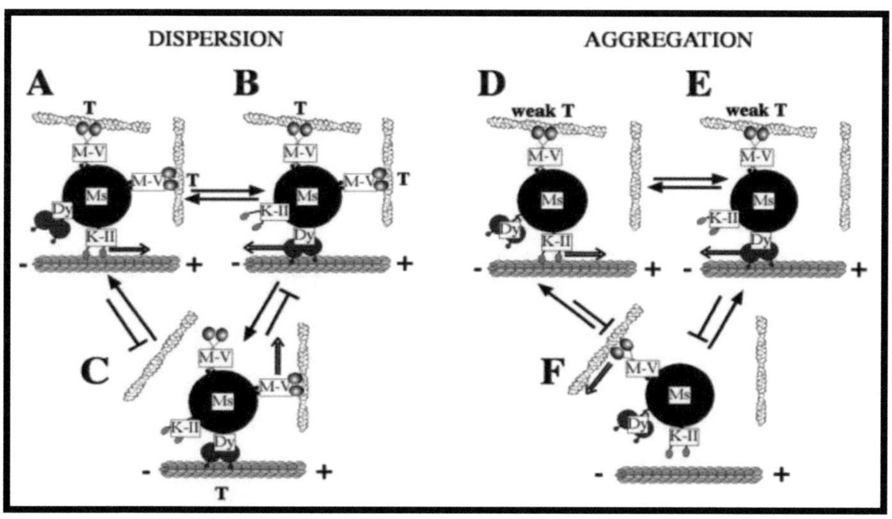

Figure 10. Model for transport of melanosomes by actin- and microtubule-based motors. Figure originally published in reference [59]; Gross et al., (2002) Interactions and regulation of molecular motors in Xenopus melanophores. J Cell Biol. 4;156(5):855-65.Copyright © 2002.The Rockefeller University Press.

6. Melanophores as potential model for drug discovery

The field of drug discovery is enormous with the development of novel and sensitive pharmaceutics and drugs that exert a distinguished response on targeted sites. On these grounds the noteworthy contribution of pigment bearing cells, melanophores cannot be overlooked. The fact that melanophores are highly specialized in bidirectional and coordinated translocation of pigment granules on given appropriate stimulus has conferred them as exceptional model system for pharmacological and physiological assays. The movement of pigment granules on external cues has been a highly coordinated phenomenon that requires an orchestrated and coordinated work of numerous cellular entities. This cellular connection depends largely on the transmission of signal couriers which are received via cell surface and intracellular recognition molecules (*i.e.* "receptors") on the recipient cells, resulting into remarkably coordinated bidirectional movement of pigment granules within the cells. Extensive pharmacological studies have devised the presence of various cell surface receptors (GPCR; G-protein coupled receptors) on these melanophores. Selective screening of these receptors has been a subject of investigation amongst researchers. The recognition of various ligands (agonists and antagonists) and characterizing their effects on pigment motility within the cells has answered many baffling questions regarding the complexed phenomenon of cellular motility and the signaling pathways that trigger pigment translocation.

Also the striking chromatic changes that occur are dependent on the cytoskeletal platforms that form the intricate networks within the cells and that these pigment motions are clearly microtubule dependent, since any disruption of microtubules blocks pigment dispersion and prevents directed pigment aggregation [39, 40, 77]. In addition, it is now known that melanosome aggregation may be mediated by the retrograde microtubule-dependent motor protein dynein [76, 78] and dispersion is supported by the anterograde motor protein kinesin [80]. Therefore pigment motion within the melanophores could be a used as a revolutionary tool to study the possible relationship between microtubule dynamics and intracellular transport. Nonetheless the search is still on to bridge in the gap that still exists.

6.1. Melanophores as bio-sensory opoids

Selected G-protein coupled receptors can be functionally expressed in cultured frog melanophores. It has been demonstrated that the recombinant frog melanophores can be used as a biosensor for the detection of opoids. Transfection of melanophores with selected receptors enables the creation of numerous melanophore biosensors, which respond selectively to certain substances. The successful generation of in vitro cultures of these cells [51] has facilitated further characterization of melanophore cell signaling pathways and has made possible their use as a cell based reporter system [81]. The melanophore biosensor has potential use for measurement of substances in body fluids such as saliva, blood plasma and urine. Since the pigment granules, termed melanosomes, may be stimulated to undergo rapid dispersion throughout the melanophores (cells appear dark), or aggregation to the center of the melanophores (cells appear light). This simple physiological response, which can easily be measured in a photometer, has been used in a sensitive biosensor for catecholamine in blood plasma and pertussis toxin in saliva, based on melanophores in isolated fish scales [82, 83]. The melanophores are also an attractive model for studies in pharmacology, for drug design and in cytotoxicity screening [84]. Odorant, pheromone, and gustatory receptors all belong to the GPCR family, suggesting that melanophore biosensors, expressing the appropriate receptor, may also be a new principle for odor, pheromone, and taste sensing [86]. In an interesting report by De-Camp et al. [87] a 7-pass trans-membrane protein "Smoothened" was investigated for its ability to act as a G-protein-coupled receptor in immortalized *Xenopus laevis* melanophores. Cells expressing the protein showed a phenotype of persistent pigment aggregation, a hallmark of constitutive Galpha(i) activation. The findings demonstrated that the human Sonic hedgehog receptor complex can be functionally reconstituted in melanophores and that it is capable of trans-membrane signaling by utilizing endogenous Galpha(i).

According to Karlsson *et al.*, [85] the melanophores of *Xenopus laevis* show a fast response initially, although it takes about 1h to get maximal aggregation response. However, detection of melanosome movement can be noticed after a few minutes. If an even faster response is desired, development of fish cell lines might be an interesting alternative. For example, fish melanophores aggregate within 5s after sympathetic nerve stimulation [88] and within 30s after light stimulation [89]. Owing to the super sensitivity of fish melanophores, the detection of signaling cues or compounds may be detected on a faster pace. However, a fish melanophore cell line has to be developed to allow efficient introduction of foreign receptor genes. The zebra fish stem cell line might be an attractive way to achieve above-mentioned goals [90]. Alternatively, the melanophores of frog can be manipulated to increase the speed of pigment aggregation [86].

6.2. Zebrafish melanophore model

Zebrafish model system has emerged as one of the most reliable systems for studies related to developmental gene function and disease processes in nervous system. During the past 20 years zebrafish has served as an excellent model for understanding normal development and birth defects based on its powerful genetics and exquisite embryology. For example, recent breakthroughs made in zebrafish include the isolation of a human skin color gene, the development of a melanoma model, and the isolation of a chemical that can correct cardiovascular defects. As vertebrates they possess a brain structure similar to that found in mammals. The intracellular signaling downstream of hormone stimulation and the biomechanical processes involved in zebrafish pigment translocation, has confirmed the importance of cyclic adenosine monophosphate (cAMP) as a mediator of pigment translocation and the presence of intact microtubules essential for both melanin dispersion and aggregation, has rendered it as an experimental model for studying both physiological color change and the molecular basis of pigment translocation [91]. Also in the recent years zebrafish melanophore model has been used to study melanocyte biology and melanoma [92]. Due to its prolific reproduction and the external development of the transparent embryo, the zebrafish serve as a cutting edge tool for genetic and developmental studies, as well as research in toxicology and genomics. Zebrafish melanophores are externally visible, and single cells can be visualized in a living animal. Zebrafish melanocytes retain melanin unlike mammals where the melanin pigment-containing melanosomes are transported to neighboring keratinocytes. For this reason zebrafish melanocytes can serve as a reliable and useful cell-type marker. Furthermore, the characteristic pattern of pigment cells in the zebrafish skin, combined with newly developed techniques molecular genetics, makes the zebrafish an ideal experimental system for the evaluation of melanogenic regulatory compounds and the mechanism of pigmentation pattern formation [93, 95, 96].

Since a growing number of diseases, including neuropathology and developmental disorders, are thought to result from disrupted transport of organelles, the regulation of molecular motor proteins by applying genetics to the problem of melanosome transport in zebrafish melanophores could be used as a promising tool in investigating such problems. This will be accomplished by screening zebrafish mutants specifically for alterations in melanosome dynamics. Mutants of interest can be examined by isolating their melanophores via live cell imaging and characterizing their defects in pigment transport [94].

7. Future directions

Our earlier impression of cell has been completely reconditioned with the novel concepts that has brought cell to be a dynamic rather than a rigid entity. The regulation of normal responsiveness of a pigment cell result from dynamic interaction of cytoskeletal processes, molecular motor proteins and associated regulatory processes that are viewed as complex and integrated array of events. How do cells regulate this complex array of motor proteins and their interaction to control the trafficking of same cargo across the cells? Future investigations of melanosome dynamics promise to answer questions about how motor proteins and their motility are controlled and coordinated. The ability to isolate biochemically-defined melanosomes in large quantities from cultured melanophores, coupled with in vitro motility assays to test the activity of specific motors under different conditions, provide a great opportunity to gain insights into how these complex interactions are regulated [97]. Fish melanophores are a classical example [98] of a cell highly specialized in intracellular microtubule-dependent transport, and it is worth considering whether microtubule polymerization and de-polymerization in these cells may be directly coupled to the translocation of pigment organelles [50]. The direct observation and manipulation of cultured melanophores will provide ways to draw conclusions from the in vitro experiments and expand our understanding of this mechanism. These experiments would provide deeper insights into the control and coordination of motility for a variety of cargos in different eukaryotic cell systems [97]. It is known that several genetic disorders in mice and humans are linked to disrupted intracellular organelle transport. The recent characterization of genes defective in these diseases has drawn immense interest in the melanosome as a model system for understanding the molecular mechanisms that underlie intracellular membrane dynamics [99]. The regulation of normal cellular morphology is a highly concerted system that must integrate the temporal-spatial control of multiple cellular processes, including cytoskeletal and membrane dynamics. Since cells interact with other cells through their surfaces, therefore cellular phenotypes are a dynamic process that reflects the influence of other cells. On these grounds we emphasize upon the role of pigment cells and their dynamics that might help to better comprehend the dynamics of other cellular entities like the muscle cells and vice versa. The study of these cellular systems would significantly contribute towards comprehension of both physiological and pathological events related to muscle as well as pigment cells.

Author details

Saima Salim
Pigment Biology, Saifia College Bhopal, India

Sharique A. Ali*
Physiology and Department of Biotechnology at Saifia College Bhopal, India

8. References

[1] Spaeth RA. Evidence proving melanophores as disguised type of Smooth muscle cells. Journ of Exp Zool. 1917; 20: 2,193-215.
[2] Webb CR. Smooth Muscle contraction and Relaxation. Adv Physiol Educ.2003; 27; 4, 201-206.
[3] Riento K, Ridley AJ. "Rocks: multifunctional kinases in cell behaviours". Nat Rev Mol Cell Biol. 2003; 4, 6: 446–456.
[4] Vesely DL, Hadley ME Ionic requirements for melanophore stimulating hormone (MSH) action on melanophores. Comp. Biochem. Physiol. 1979; 62A:501-507.
[5] Hadley ME. Endocrinology, 5th Edition, Printice Hall. 1988.
[6] Zhang F, Wang Li-P, Boyden ES, Deisseroth K.Channelrhodopsin-2 and optical control of excitable cells. Nature Methods, 2006; 3:10, 785-792.
[7] Aspengren S, Hedberg D,Wallin M.Melanophores: A model system for neuronal transport and exocytosis? Journ of Neuro Res. Sp Issue Intraneural Transport. 2006;85: 12, 2591-2600.
[8] Salim S, Ali SA. Vertebrate melanophores as potential model for drug discovery and development: a review. Cell Mol Biol Lett. 2010; 6,1:162-200.
[9] Redfield AC.The Physiology of the melanophores of the horned toad Phrynosoma. Journ of Exp Zool. 1918; 26:2, 275-333.
[10] Smith DC. Melanophore Pulsations in the isolated scales of *Fundulus heteroclitus*. Proc. Nat. Am Soc. 1930; 16:381-385.
[11] Parker GH. Animal Color changes and their Neurohumors. Quat. Rev of Biol. 1943; 18:3, 205-227.
[12] Messenger JB. Cephalopod chromatophores: neurobiology and natural history. Biol. Rev. 2001;76: 473-528.

* Corresponding author

[13] Franz V. Beobachtungen am lebenden Selachierauge.Jenaische Zeitschr. f. Naturwiss., 41(Nf.Bd. 34) 1906; 429-471.

[14] Pouchet G. Color changes in crustaceans and fishes. J. Anat. Physiol. 1876; 12:1-90, 113-116.

[15] Brücke E. Untersuchungen uber den Farbenwechsel des afrikanischen Chamaleons. Denschr. Akad. Wiss. Wien, Mathnat. Kl, 1852; 4:179.

[16] Parker GH. Animal colour changes and their neurohumors. Cambridge Univ. Press, Cambridge. (U.K) 1948.

[17] Fujii, R. Cytophysiology of fish chromatophores. Int Rev Cytol. 1993; 143: 191-255.

[18] Canning BJ. Reflex regulation of airway smooth muscle tone. Jour of App. Physiol. 2006; 101, 3: 971-985.

[19] Sherwood L. Fundamentals of Human Physiology 4/E. The Muscle Physiology, pp 185-186. Cengage learning, 2011.

[20] Van De Graaff, Fox SI. Concepts of human anatomy and physiology. W.C Brown.Cornell University.1986

[21] Pouchet G. Des changements de coloration sous I'innuence des neK J. Anat Paris 1876; 12: 1-90, 113-165.

[22] Miyata S Yamada K. Innervation pattern and responsiveness of melanophores in tail fins of teleosts. Journ of Experim Zool. 1987; 241: 1, 31-39.

[23] Pye JD. Nervous control of chromatophores in teleost fishes. I I. T he influence of certain drugs in the minnow (Phoxinus phoxinus (L.)). J. Exp. Biol.1964; 41, 535—41-

[24] Frisch VK. Beitrage zur Physiologic der Pigmentsellen in der Fischhaut. Pflugers Arch.Gesante Physiol. Menschen Tiere. 1911; 138: 319-387.

[25] Hertel. Ztschr. F.allag. Physiol. 1907; 6,44.

[26] Kargacin GJ, Detwiler PB. Light-evoked Contraction of the Photosensitive Iris of the Frog' The Journ. of Neurosc. 1985; 5: 11,3081-3087.

[27] Schliwa M, Bereiter-Hahn. Pigment movements in fish melanophores: Morphological and physiological studies. Cell and Tiss. Res. 1975; 158:1, 61-73

[28] Winkler.Arch. F. Derm. U. Syphilis. 1910; 101, 255.

[29] Bert P. Compt. Rend., 1975; 81, 938.

[30] Krukenberg CF. I-Vergleich, Physiol.Studien, 1 Reihe, 1.1881.

[31] Kapoor BG Khanna B. Integument: Dermal Skeleton Coloration and Pigment Cells. 65-73. Ichthyology Handbook, Springer.2004.

[32] Thaler CD, Haimo LT. Control of organelle transport in melanophores: Regulation of Ca^{2+} and cAMP levels. Cell Motility and Cytoskeleton. 2005; 22:175-184.

[33] Novales RR . The effect of the divalent cation ionophore A23187 on amphibian melanophores and iridophores. J Invest Dermatol. 1977; 69:446-50.

[34] Tuma CM, Gelfand VI. Molecular mechanism of Pigment Transport in melanophores. Pigm. Cell Res. 1999;12:5, 283-294.

[35] Fujii R, Fujii Υ. Mechanism of nervous control of fish melanophores. IL Role of divalent ions in the transmission at the melanin-aggregating nerve endings; Zool. Mag. 1965; 74–351.

[36] McNiven MA, Porter KR. Chromatophores-models for studying cytomatrix translocations. J. Cell Biol. 1984; 99:152s-158s.

[37] Schliwa, M. Review article: permeabilized cell models for the study of granule transport in pigment cells. Pigm. Cell Res.1978;1:65-68.
[38] Haimo LT, Rozdzial, MM. Lysed chromatophores: a model system for the study of bidirectional organelle transport. Methods Cell Biol.1989; 31: 3-24.
[39] Schliwa M, Bereiter-Hahn J. Pigment movement in fish melanophores. III. The effects of colchicine and vinblastine. Z. Zellforsch. 1973; 147:127-148.
[40] Murphy DG, Tilney LG..The role of microtubules in movement of pigment granules in teleost melanophores.J. Cell Biol. 1974; 61:757-779.
[41] Obika M, Negishi S. Effects of hexylene glycol and nocodazole on microtubules and melanosome translocation in melanophores of the Medaka, Oryzias latipes. J. Exp. Zool. 1985; 235: 55 – 63.
[42] Luby-Phelps K, Porter KR The control of pigment migration in isolated erythrophores of Holocentrus ascensionis (Osbeck). II. The role of calcium. Cell. 182; 29:441-450.
[43] Oshima N, Suzuki M, Yamaji N, Fujii.R Pigment aggregation is triggered by an increase in free calcium ions within fish chromatophores. Comp. Biochem. Physiol. 1988; 9 1A: 27- 32
[44] McNiven M.A, Ward JB. Calcium regulation of pigment transport in vitro. J. Cell Biol.1988; 106:111-125.
[45] Rochow VBM, Royuela M. Calponin, Caldesmon and chromatophores: the smooth muscle connection. Microscopy Res. and Technique. Sp Issue:Biol of Pig Cells in fish.2002; 58:6, 504-513.
[46] Rochow VBM, Royuela, Fraile B, Paniagua R Smooth muscle proteins as intracellular components of the chromatophores of the Antarctic fishes *Pagothenia borchgrevinki* and *Trematomus bernacchii* (Nototheniidae). Protoplasma , 2002; 218, : 1-2, 24-30.
[47] Nascimento AA, Roland JT, Gelfand VI. Pigment cells: a model for the study of organelle transport. Annu Rev Cell Dev Biol. 2003; 19:469-91.
[48] Bikle D, Tilney LG, Porter KR. Microtubules and pigment migration in the melanophores of *Fundulus heteroclitus*. Protoplasma.1966; 61, 322–345.
[49] Schliwa, M. Mechanisms of intracellular transport. Cell and Muscle Motility (Shay, J. W. ed) 1984; 5: 1– 81.
[50] Rodionov VI, Nadezhdina ES, Borisy GG. Centrosomal control of microtubule dynamics. Proc. Natl. Acad. Sci. USA 1999; 96, 115–120.
[51] Vallee RB, Wall JS, Paschal BM. The role of dynein in retrograde axonal transport, TINS, 1989; 12:66-70.
[52] Schnapp BJ, Vale RD, Sheetz MP, Reese TS. Single microtubules from squid axoplasm support bidirectional movements of organelles, Cell. 1985; 40: 455-462.
[53] Byers HR, Yaar M, Eller MS, Jalbert NL, Gilchrest BA. Role of cytoplasmic dynein in melanosome transport in human melanocytes. J. Invest. Dermatol.2000; 114, 990-997.
[54] Hara M, Yaar M, Byers HR, Goukassian D, Fine R E, Gonsalves J, Gilchrest BA. Kinesin participates in melanosomal movement along melanocyte dendrites. J. Invest. Dermatol. 2000; 114: 438-443.
[55] Wu XS, Tsan GL, Hammer JA, 3rd. Melanophilin and myosin Va track the microtubule plus end on EB1. J. Cell Biol 2005. 171; 201-207.

[56] Vancoillie G, Lambert J, Mulder A, Koerten HK, Mommaas AM, Van Oostveldt P, Naeyaert JM. Cytoplasmic dynein colocalizes withmelanosomes in normal human melanocytes. Br. J. Dermatol. 2000; 143:298-306.

[57] Daniolos A, Lerner AB, Lerner MR. Action of light on frog pigment cells in culture. Pigment Cell Res.1990; 3, 38-43.

[58] Reilein AR, Tint IS, Peunova NI, Enikolopov GN, Gelfand VI. Regulation of organelle movement in melanophores by Protein Kinase A (PKA), Protein Kinase C and Protein Phosphatase 2A (PIP2). JCB, 1998; 142:3, 803-813.

[59] Gross SP, Tuma CM, Deacon SW, Serpinskaya AS, Reilein AR, Gelfand VI. Interactions and regulation of molecular motors in Xenopus melanophores. J. Cell Biol. 2002; 156, 855-865.

[60] DePina AS, Langford GM. Vesicle transport: the role of actin filaments and myosin motors. Microsc. Res. Tech. 1999; 47, 93-106.

[61] Gundersen GG, Cook TA. Microtubules and signal transduction. Curr. Opin. Cell Biol.1999; 11, 81-94.

[62] Pollard TD, Borisy GG.Cellular motility driven by assembly and disassembly of actin filaments. Cell 2003;.112:453-65.

[63] Rogers SL, Tint LS, Fanapour PC, Gelfand VI. Regulated bidirectional motility of melanophore pigment granules along microtubules in vitro. Proc. Natl. Acad. Sci. USA.1997; 94:3720–3725.

[64] Karcher RL, Deacon SW, Gelfand VI. Motor-cargo interactions: the key to transport specificity. Trends Cell Biol. 2002;12, 21–27.

[65] Tuma MC, Zill A, Le Bot N, Vernos I, Gelfand V. Heterotrimeric kinesin II is the microtubule motor protein responsible for pigment dispersion in *Xenopus* melanophores.J. Cell Biol.1998; 143:1547–1558.

[66] Nilsson, H, Wallin M. Evidence for several roles of dynein in pigment transport in melanophores. Cell Motil. Cytoskeleton.1997; 38:397–409.

[67] Hirokawa N. Kinesin and dynein superfamily proteins and the mechanism of organelle transport. Science.1998; 279 :519-26.

[68] Deacon SW, Serpinskaya AS, Vaughan PS, Lopez Fanarraga M, Vernos I, Vaughan KT, Gelfand VI.. Dynactin serves as a receptor for kinesin II on *Xenopus laevis* melanosomes. J. Cell Biol. 2003; 160:297–301.

[69] Rogers SL, Gelfand VI. Myosin cooperates with microtubule motors during organelle transport in melanophores. Curr. Biol. 1998; 8:161–164.

[70] Fukuda M, Kuroda TS, Mikoshiba K.. Slac2-a/melanophilin, the missing link between Rab27 and myosin Va: implications of a tripartite protein complex for melanosome transport. J. Biol. Chem. 2002; 277, 12432–12436.

[71] Wu X, Wang F, Rao K, Sellers JR Hammer JA 3d. Rab27a is an essential component of melanosome receptor myosin Va. Mol Biol. Cell 2002; 13, 1735-1749.

[72] Rodionov V, Yi J, Kashina A, Oladipo A, Gross SP. Switching between microtubule- and actin-based transport systems in melanophores is controlled by cAMP levels. Curr. Biol.2003; 13, 1837-1847.

[73] Kashina AS, Semenova IV, Ivanov PA, Potekhina ES, Zaliapin I, Rodionov VI. Protein kinase A, which regulates intracellular transport, forms complexes with molecular motors on organelles. Curr. Biol. 2004; 14, 1877-1881.
[74] Park M, Serpinskaya AS, Papalopulu N, Gelfand VI. Rab32 regulates melanosome transport in Xenopus melanophores by protein kinase A recruitment. *Curr. Biol.* 2007; 17, 2030-2034.
[75] Bridgman, PC. Myosin Va movements in normal and dilute-lethal axons provide support for a dual filament motor complex. J. Cell Biol. 1999; 146:1045–1060.
[76] Beckerle MS, Porter KR. Inhibitors of dyne in activity block intracellular transport in erythrophores. Nature (Lond.) 1982; 295:701-703.
[77] Beckerle MC, Porter KR. Analysis of the role of microtubules and actin in erythrophore intracellular motility. J. Cell Biol. 1983; 96:354-362.
[78] Ogawa KH, Hosoya E, Yokota T, Kobaya shi, Wakamatsu Y, Ozato K (1987) Melanoma dynein: evidence that dyne in is a general "motor" for microtubule-associated cell motilities. Fur. J. Cell Biol 1987;43: 3-9.
[79] Rodionov VI, Gyoeva FK, Gelfand VI. Kinesin is responsible for centrifugal movement of pigment granules in melanophores. Proc. Natl.Acad. Sci. USA. 1991; 88:4956-4960.
[80] Jayawickreme CK, Kost TA. Gene expression systems in the development of high-throughput screens. Curr. Opin. Biotechnol. 1991; 8: 629–634.
[81] Lundström I. Gustafsson A., Ödman, S. Karlsson JOG , Andersson RGG, .Grundström, Sundgren N, Elwing H (1990). Fish scales as biosensors. Sensors & Actuators: B. Chemical 1990; 1-6: 533-536.
[82] Karlsson JOG, Grundsrom, Elwing H, Andersson RGG. The fish pigment cell; An alternative model in biochemical research. ATLA, 1990; 18:201-224.
[83] Lundstrom I, Svensson S. Biosensing with G-protein coupled receptor systems. Biosens and Bioelectron. 1998;13, 689 – 695.
[84] Karlsson AM, Lerner MR, Unett D, Lundström I, Svensson SP. Melatonin-induced organelle movement in melanophores is coupled to tyrosine phosphorylation of a high molecular weight protein. Cell Signal.; 2000; 12:469-74.
[85] Karlsson AM, Bjuhr K, Testorf M, Oberg PA, Lerner E, Lundstrom I, Svensson SPS. Biosensing of opioids using frog melanophores. Biosens and Bioelectron.2002; 17: 331-335.
[86] DeCamp DL, Thompson TM, de Sauvage FJ, Lerner MR Smoothened activates Galphai-mediated signaling in frog melanophores. J. Biol Chem; 2000; 275:26322-26327.
[87] Svensson SP, Adolfsson PI, Grundström N, Karlsson JO. Multiple alpha 2-adrenoceptor signalling pathways mediate pigment aggregation within melanophores.Pigment Cell Res.1997; 10:395-400.
[88] Oshima N, Nakata E, Ohta M, Kamagata S. Light-induced pigment aggregation in xanthophores of the medaka, Oryzias latipes. Pigment Cell Res. 1998;11, 362–367.
[89] Ma C, Fan L, Ganassin R, Bols N, Collodi P. Production of zebrafish germ-line chimeras from embryo cell cultures. Proc. Natl. Acad. Sci. USA. 2001; 98, 2461–2466.
[90] Logan DW, Burn SF, Jackson IJ. Regulation of pigmentation in zebrafish melanophores. Pigment Cell Res.2006; 19:206-13.
[91] Ceol CJ, Houvras Y, White RM, Zon LI. Melanoma biology and the promise of zebrafish. Zebra fish. 2008; 5:247–255.

[92] Choi TY, Kim JH, Ko DH, Kim C-H, Hwang J-S, Ahn S, Kim SY, Kim C-D, Lee J-H, Yoon T-J. Zebrafish as a new model for phenotype-based screening of melanogenic regulatory compounds. Pigment Cell Res. 2007; 20:120–127.

[93] Lavinia Sheets. Regulation of Molecular Motors in Zebrafish.2007 http://www.researchgrantdatabase.com/g/5F31GM071198-03/Regulation-of-Molecular-Motors-in-Zebrafish/

[94] Takahashi G, Kondo S. Melanophores in the stripes of adult zebrafish do not have the nature to gather, but disperse when they have the space to move. Pigment Cell Melanoma Res. 2008; 21:677–686.

[95] Newman M, Wilson L, Camp E, Verdile G, Martins R, Lardelli M. A zebrafish melanophoremodel of amyloid beta toxicity. Zebrafish.. 2010; 7:155-159.

[96] Kelleher JF and Margaret AT. Intracellular motility: How can we all work together?. Current Biology, 1998; 8:R394–R397.

[97] Bikle D, Tilney LG, Porter K.R. Microtubules and pigment migration in the melanophores of Fundulus heteroclitus. *Protoplasma. 1966;* 61:322-345.

[98] Marks MS, Seabra MC. The melanosome: membrane dynamics in black and white. Nature Reviews Molecular Cell Biology, 2001; 2, 10: 738-748.

[99] Ghalayini IF. Nitric oxide–cyclic GMP pathway with some emphasis on cavernosal contractility. Internat. Journ. of Imp. Res. 2004; 16: 459–469.

[100] Castrucci AM, Hadley ME, Lebl M, Zechel C, Hruby VJ. Melanocyte stimulating hormone and melanin concentrating hormone may be structurally and evolutionarily related. Regulatory Peptides. 1989; 24: 27-35.

[101] Negi CN. " Regulation of Pigmentation: Pars Intermedis". Introduction to Endocrinology. PHI learning Ltd. New Delhi. 2009;. 113-115.

[102] Karaki H, Ozaki H, Hori M, Mitsui-Saito M, Amano K, Harada K, Miyamoto S, Nakazawa H, Won KJ, Sato K.. Calcium movements, distribution, and functions in smooth muscle. Pharmacol Rev 1997; 49:157-230.

[103] Fujii R, Novales R. Cellular aspects of the control of physiological color changes in fishes. Integ and Comp Biol. 1969; 9: 453-463.

[104] Bagnara JT, Hadley ME. Chromatophores and color changes. Prentice-Hall, Inc, Englewood Cliffs, N. J. 1973.

[105] Wakamatu Y. Light-sensitive fish melanophores in culture. J. exp. Zool. 1978. 204:299-304.

[106] Schliwa M, Woehlke G. Molecular Motors. Nature. 2003.422: 759-765.

[107] Nilsson HM, Karlsson AM, Loitto VM, Svensson SP, Sundqvist T. Nitric oxide modulates intracellular translocation of pigment organelles in *Xenopus laevis* melanophores. Cell Motil Cytoskeleton 2000;.47:209-218.

[108] Hayashi H, Fujii R.. Possible involvement of nitric oxide in signaling pigment dispersion in teleostean melanophores. Zoolog Sci.2001;18:1207-1215.

[109] Soldati T, Schliwa M. Powering membrane traffic in endocytosis and recycling. Nature Rev. Molec Cell Biol.2006; 7: 897-908.

[110] Levi V, Serpinskaya A, Gratton E, Gelfand V. Organelle Transport along Microtubules in *Xenopus* Melanophores: Evidence for Cooperation between Multiple Motors. Biophys J 2006; 90: 318–327.

Section 2

Pathological Aspects of Cardiac and Smooth Muscle Cells

Chapter 8

Cardiomyocyte and Heart Failure

Shintaro Nakano, Toshihiro Muramatsu,
Shigeyuki Nishimura and Takaaki Senbonmatsu

Additional information is available at the end of the chapter

http://dx.doi.org/10.5772/47772

1. Introduction

In recent years, outcome of therapy in patients with heart failure is going up. Many clinical trials have demonstrated that renin angiotensin aldosterone system inhibitors and β-blockers have functional roles in stabilizing and /or reversing cardiac remodeling via suppression of the excessive activation of renin angiotensin aldosterone and the adrenergic nervous system. Additively, the cardiac resynchronization therapy and ventricular assist device therapy also achieve remarkable success in heart failure therapy. Conversely, in many counties that come up against an elderly society, heart failure is a looming public health problem. Therefore, much further advancement of heart failure therapy and decrement of patients with heart failure are one of most important assignments in the medical services. In this chapter, we describe the recent topics of heart failure including 1,molecular basis of cardiomyocyte, 2,mechanisms of progression in heart failure, 3,renin angiotensin aldosterone system and heart failure, 4,β-adrenergic receptor and heart failure, 5, non-drug treatment and heart failure, 6,heart transplantation and heart failure, 7,Cardiac regeneration and heart failure.

2. Molecular basis of cardiomyocyte

The heart is a highly organized tissue and consists of ventricular or atrial cardiomyocytes, pace maker cells, Purkinje cells, vasculature, and connective tissue. The ventricular cardiomyocytes are columnar shaped cells of 20μm in diameter and 60-140μm in length, while the atrial cardiomyocytes are ellipsoidal shaped cells of 5μm in diameter and 10-20μm in length (Table 1). The ventricular cardiomyocytes occupies approximately 50% of the heart weight, and 2-4 billion of them make up the human left ventricle. Approximately 50% of the cell volume in an individual contracting cardiomyocyte is made up of myofibrils and 25% of the cell volume is occupied by mitochondria. The remainder consists of nucleus, sarcoplasmic reticulum (SR), and the cytosol (Fig 1). Myofibril is the rodlike bundle that

Characteristics of cardiac cells

	Ventricular myocyte	Atrial myocyte	Purkinje cells
Shape	columnar	ellipsoidal	Long and broad
Length μm	60-140	10-20	150-200
Diameter μm	20	5	35-40
T- tubules	Plentiful	Rare or none	Absent
	Mitochondria and sarcomeres abundant		

Table 1. Characteristics of cardiac cells

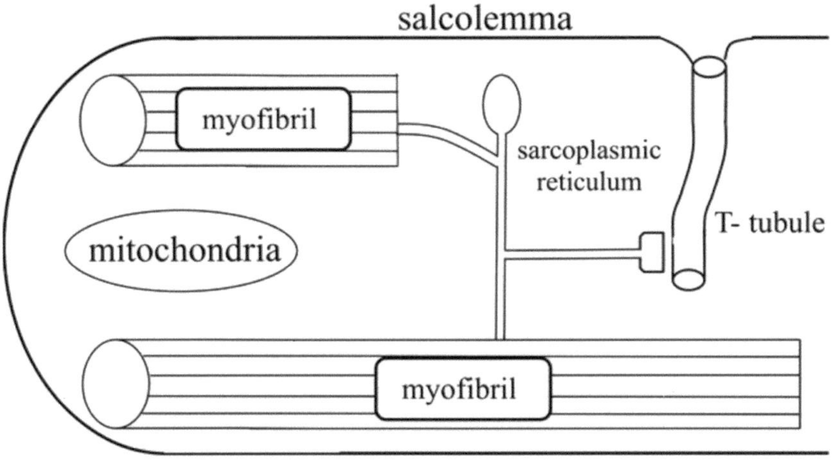

Figure 1. Ventricular cardiomyocyte

forms the contractile elements within cardiomyocytes. As one of the specialized structures of the cardiomyocyte, there is the sarcolemma, which is a coalescence of the plasma membrane proper and the basement membrane. The sarcolemma iscomposed of a lipid bilayer, which contains hydrophilic heads and hydrophobic tails. This structural fabric allows the sarcolemma to regulate the interactions with the intracellular and extracellular environment. The transverse tubular system (T- tubules) is specialized organo-parts of cardiomyocyte in the sarcolemma. The T- tubules are invagination of the sarcolemma into the cardiomyocyte, and they form a barrier between the intracellular and extracellular space. When electrical action potential reaches T-tubules, the wave of depolarization induces Ca^{2+} influx into the cardiomyocyte through the voltage-sensitive L-type Ca^{2+} channel of the T-tubules. This leads to Ca^{2+} discharge of the sarcoplasmic reticulum into cytosol resulting in contraction of the heart. Thus, the T-tubules are important structural components in the excitation-contraction coupling system described later. Myofibril is composed of actin thin filament, myosin thick filament and titin, which stabilizes myosin at the Z-line (Fig 2). The cardiomyocyte has aggregation of myofibrils and the fundamental contractile unit within the cardiomyocyte is the sarcomere, which has a length of 1.8 μm in the systole and 2.2 μm in the diastole. Other than myofibril, the contractile apparatus contains tropomyosin, the troponin complex. Myosin has a filamentous tail and a globular head region that contains the site for actin binding. Actin has 2 forms G and F. F-actin is the backbone of the thin filament, while G-actin works as a stabilizing protein. Using ATP, the G-actin interacts with the myosin globular head leading to the crossbridge formation and sarcomere shorting. Tropomyosin lies on the side of actin for rigidity to thin filament. The troponin complex, also present in the thin filament, is composed of troponin C, I and T. These proteins regulate crossbridge formation. In the systole, an increased Ca^{2+} binding to the troponin C leads to the actin-myosin interaction resulting in initiating crossbridge formation. The troponin I and T suppress actin-myosin interaction in decreased Ca^{2+} of the diastole. The previous report indicates that cTnT1, isoform of troponin T that is not expressed under normal heart, is induced expression level in heart failure [1]. Ca^{2+} is the fundamental ion for evoking the excitation-contraction coupling complex (Fig 3). Upon the wave of depolarization, the voltage-sensitive L-type Ca^{2+} channel of the T-tubules opens and allows Ca^{2+} influx. This rapid but small Ca^{2+} influx causes activation of large amounts of Ca^{2+} release from the ryanodine receptor (RyR2) on the sarcoplasmic reticulum. Finally, cytosolic Ca^{2+} level changes from 100 nmol/L to 10 μmol/L in concentration. Ten μmol/L of Ca^{2+} also binds to the troponin C. Active relaxation of the cardiomyocyte is dependent on the function of the sarcoplasmic reticulum Ca^{2+}-ATPase (SERCA2a in the heart). For each 1 mol of ATP hydrolyzed, 2 mol of Ca^{2+} is transported back into the sarcoplasmic reticulum. Phospholamban (PLB) regulates the function of SERCA2a. Additionally, the Na^+/Ca^{2+} exchanger on the plasma membrane removes Ca^{2+} from cytosol. Human heart excretes 1 ton of blood in a day. Therefore, cardiomyocytes are required to maintain high level of ATP. Usually, the heart produces 6kg of ATP in a day. To produce high level of ATP, fatty acid and glucose are expended as substrates of ATP.

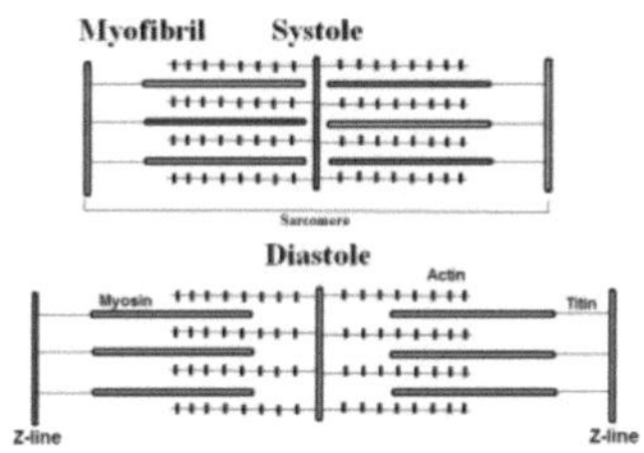

Figure 2.

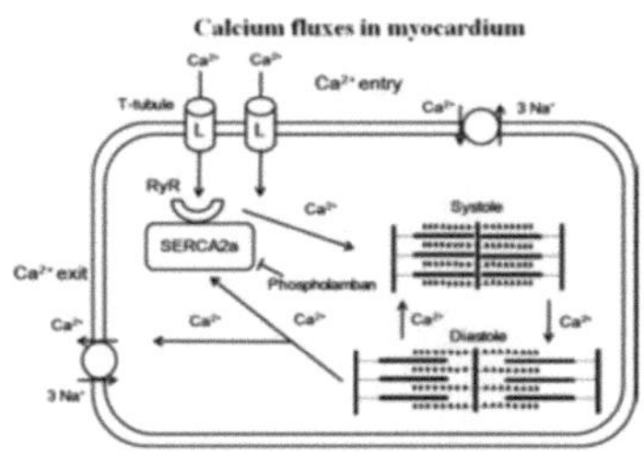

Figure 3. Calcium fluxes in myocardium

3. Mechanisms of progression in heart failure

Heart failure is observed as a progressive disorder that is initiated after an index event. This index event contains myocardial infarction, sustained hypertension, severe arrhythmia, viral infection, stressed environment, or a genetic disease. Finally, the index event damages the cardiomyocytes resulting in loss of function or collapses in the pumping of the heart. Heart failure is clearly a major clinical and a public health problem. Despite the recent innovations in treating heart failure and its predisposing conditions, it still remains highly prevalent and lethal due to increasing life spans across the cultures. It is estimated that nearly 23 million people have heart failure worldwide. Elderlies consist of 80% of the total heart failure population, and the morbidity prevalence of heart failure in the elderly is over 1%. This epidemiological study clearly indicates that human heart failure is an age-related disorder. Heart failure evokes the overexpression of biologically active molecules that are capable of exerting deleterious effects on the heart and circulation [2]. Under this pathological environment, the compensatory mechanisms induce activation of the adrenergic nervous system and renin angiotensin system, which is termed "neurohormonal alternation" in heart failure. These systems are responsible for maintaining cardiac output through increased retention of salt and water, peripheral arterial vasoconstriction, and increased contractility and activation of inflammatory mediators, which are responsible for cardiac repair and remodeling. Although sustained neurohormonal alternation is interpreted to be the key to disease progression, there is an increasing clinical evidence to suggest against it. Cardiac hypertrophy has two basic patterns to response to hemodynamic overload (Fig 4). Pressure overload induces concentric hypertrophy, which shows a thick appearance, whereas volume overload induces eccentric hypertrophy, which displays an elongated appearance. Cardiac hypertrophy induces alterations in the biological phonotype of the cardiomyocyte, which in turn reactivates fetal genes that are normally not expressed [3]. The reactivation of these fetal genes is associated with a decreased expression of a number of genes that are normally expressed in the adult normal heart. This may contribute to the contractile dysfunction that develops in the failing heart. During heart failure, the progressive cardiomyocyte loss may also contribute to cardiac dysfunction and left ventricular remodeling through necrotic, apoptotic or autophagic cell death pathways.

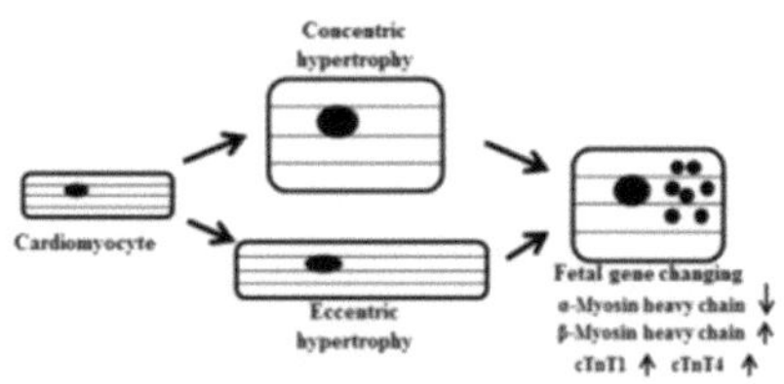

Figure 4. Process of ventricular remodeling

3.1. Heart failure with a normal ejection fraction

Now heart failure with a normal ejection fraction (HFnlEF) is a common term of cardiologists, because it is possible that the prevalence of HFnlEF has increased over time, leading to more widespread recognition. However, in the 20th century, existence of such patients with HFnlEF had not been considered. The term HFnlEF has been used in current management guidelines. Although consensus of HFnlEF seems to be building toward use of EF higher than 50% to designate HFnlEF, the approach to patients with borderline reduction in EF (EF of 40 to 50%) adds to the complexity of the classification [4]. Numerous epidemiologic studies and national registers have defined the prevalence of HFnlEF in various heart failure populations and have documented a prevalence of 50% to 55% [5]. The prevalence of heart failure increases with age and is similar in men and women. The prevalence of heart failure with reduced EF increases with age but is more common in the men than in women at any age, whereas the prevalence of HFnlEF increases even more dramatically with age more than heart failure with a reduced EF and is much more common in women than in men at any age [6]. Most large contemporary studies have suggested that all-cause mortality for HFnlEF is similar to that of heart failure with reduced EF [13]. Meanwhile, there are minimal differences in heart failure readmission rates between morbidity of patients with HFnlEF and with heart failure with a reduced EF [7]. Patients with HFnlEF have been shown to have pathophysiologic characteristics similar to those of heart failure patients with a reduced EF, including severely reduced exercise capacity, neuroendocrine activation, and impaired quality of life [8]. Since LV structure and function are altered by age, gender, and cardiovascular disease in absence of heart failure, understanding of the pathophysiologic mechanisms in HFnlEF dictates a clear understanding of LV diastolic and systolic function and the manner under physiological and pathological conditions. So there are wide-ranging abnormalities in extracardiac, whole heart, extracellular matrix, cardiomyocyte and myofilaments as mechanisms of particular current or emerging clinical interests in HFnlEF.

4. Renin angiotensin aldosterone system and heart failure

The renin–angiotensin system (RAS) plays pivotal roles in the regulation of the cardiovascular system under normal and pathological conditions (Fig 5) [9]. Renin is released from the juxtaglomerular cells in the kidney, and cleaves the N-terminal end of circulating angiotensinogen, which is synthesized in the liver, to form the biologically inactive decapeptide angiotensin I (Ang I). Angiotensin-converting enzyme (ACE) cleaves 2 amino acids from Ang I to the biological active octapeptide Angiotensin II (Ang II). Ang II binds to two major G-protein coupled receptor (GPCR) subtypes, AT_1 and AT_2. Although both the AT_1 and AT_2 receptors are expressed in the human myocardium, expression level of the AT_2 receptor is less than half the level of the AT_1 receptor. Cellular localization of the AT_1 receptor in the heart is most abundant in nerves distributed in the myocardium. The AT_2 receptor is localized more highly in the fibroblasts and the interstitium. Activation of the AT_1 receptor evokes vasoconstriction, cell growth, aldosterone secretion, and catecholamine release with strong effects on cardiac hypertrophy and congestive heart failure (Table 2). In contrast, accumulating evidences show that the function of the AT_2 receptor is vasodilation, the inhibition of cell growth, and bradykinins release (Table 2) [17]. However, the opposite functions of the AT_2 receptor against the AT_1 receptor have not yet reached the consensus. Senbonmatsu et al. reported that the AT_2 receptor binds to promyelo cyticleukemia zinc finger protein (PLZF), which is a transcription factor, and its subsequent translocation into the nucleus, where it up-regulates the p85α regulatory subunit of phosphoinositide 3-kinase (PI3K) resulting in the development of cardiac hypertrophy similar to the AT_1 receptor (Fig

6) [10,11]. Since PLAF selectively expresses in the heart but not in the kidney or the vascular, the AT_2 receptor may have dual effects depending on the cell components.

It has been thought that RAS plays as the dual manners. One way is that RAS works as the neuroendocrine system and thus acts on the heart in an endocrine manner, which is termed "the circulating RAS" (Fig 5, Right side). The other way is that Ang II is synthesized directly within the myocardium and thus acts in an autocrine and paracrine manner, which is termed "the tissue or local RAS" (Fig 5, Left side). The accumulating evidences suggested that the pathologic states may be mediated by mainly the local RAS [12]. However, the local RAS still remains an enigma because renin is secreted from only the juxtaglomerular cells in the kidney. What supplies renin in the local RAS? Plasma concentration of prorenin, which is a precursor of renin, is about 10 folds of that of renin because of expressions in various tissues. However, prorenin does not display protease activity in the plasma because the enzymatic cleft is covered by the prosegment, and is not converted to active renin in the plasma. Recently, the (pro)renin receptor ((P)RR) was discovered [13]. (P)RR binds both renin and prorenin [14]. Although the binding of renin to (P)RR may increase its catalytic activity, the binding affinity between (P)RR and renin is lower than that of (P)RR and prorenin [15]. The binding of prorenin to (P)RR evokes conformational change of prorenin resulting in the renin activity without removal of its prosegment (Fig 7). This nonproteolytic activation of prorenin may contribute to the activation of the local RAS. In addition to the enzymatic activity, prorenin has been shown to provide other (P)RR-mediated effects. The binding of prorenin to (P)RR induces the activation of intracellular signaling, including the p38 MAP kinase-HSP27 cascade, the PI3K pathway and the ERK 1/2 pathway; these effects occur independently of Ang II [16]. Coincidentally, the direct renin inhibitor, aliskiren is available in clinics and basic scientific experiments.

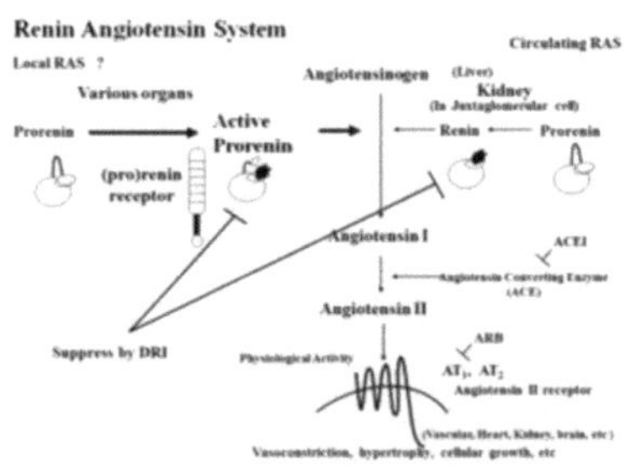

Figure 5. Renin Angiotensin System

RAS is activated in patients with heart failure. The presumptive mechanisms for RAS activation in heart failure include renal hypoperfusion; decreased filtered sodium reaching the macula densa in the distal tubule; and increased sympathetic stimulation of the kidney, leading to increased renin (Fig 8) [17]. RAS has several important actions that are critical for the maintenance of circulatory homeostasis. However, sustained activation of RAS is maladaptive and leads to fibrosis of the heart, kidney and other organs. Activated RAS also leads to worsening neurohormonal activation by enhancing the release of norepinephrine (NE) and stimulating the adrenal cortex to produce aldosterone. The sustained expression of aldosterone also exerts harmful effects by provoking hypertrophy and fibrosis within the vasculature and the myocardium. Thus prolonged activation of RAS contributes to reduced vascular compliance and increased ventricular stiffness. Hence, the drugs, which counteract the excessive activation of RAS and the adrenergic nervous system, hold potential for a power to relieve the symptoms of heart failure with a depressed left ventricular function by stabilizing and/or reversing cardiac remodeling. From the last decade of the 20th century, many clinical trials have been performed for evidence of efficacy of RAS inhibitors against patients with heart failure.

Physiological Function and Regulation of Angiotensin Receptors

	AT_1	AT_2
Affinity	Ang II > Ang III > Ang I	Ang III > Ang II > Ang I
Antagonist	ARBs	PD123319, CGP42112A
Structure	359 amino acids, GPCR 42 kDa	363 amino acids, GPCR 41 kDa
G-Protein	Gq/i	Gi, Gs?, Not?
Signaling	PLC activation RAS/ERK JAK2/STAT	PLZF activation Phosphatase activation Kinin/NO/cGMP
Function	Growth Stimulation	Growth Stimulation? Suppression?

Table 2. Physiological Function and Regulation of Angiotensin Receptors

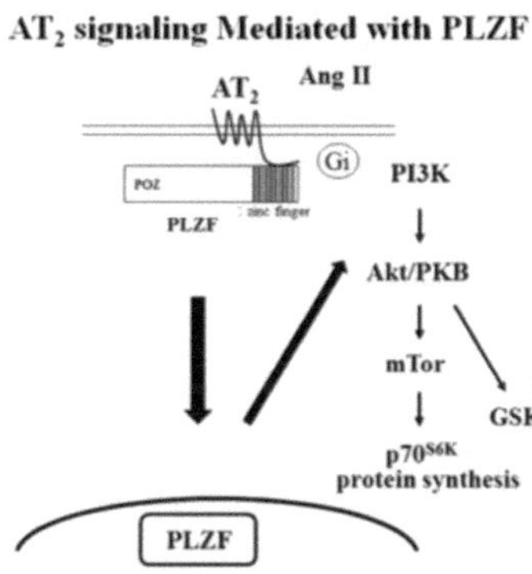

Figure 6. AT_2 signalling Mediated with PLZF

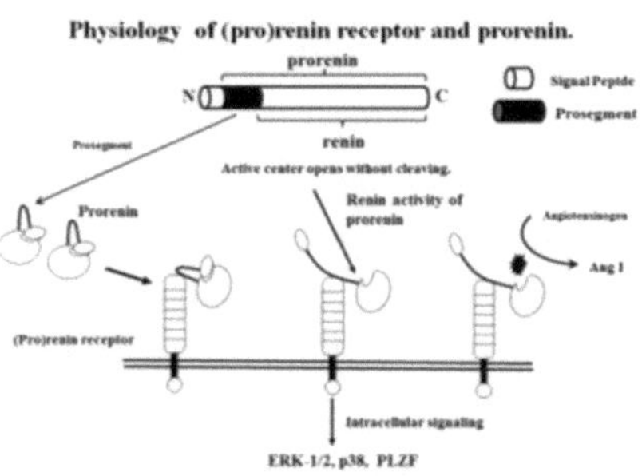

Figure 7. Physiology of (pro)renin receptor and prorenin

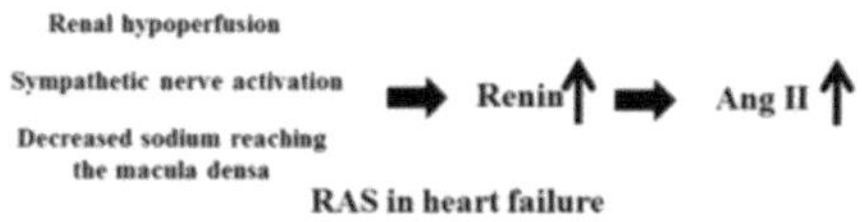

Figure 8.

4.1. Angiotensin converting enzyme inhibitor and heart failure

ACEIs should be used in symptomatic and asymptomatic patients with reduced left ventricular function, because there is overwhelming evidence of ACEI to heart failure. ACEIs suppress the production of Ang II through inhibition of ACE. ACEIs also have diverse effects independent of RAS inhibition in contrast to other RAS inhibitors. This is because ACEIs cleave carboxyl-terminal dipeptides of various oligopeptides such as angiotensin (Ang) I, kinins, Ang (1-7) or matrix metalloproteases (MMPs) (Fig 9). In Blood Pressure Lowering Treatment Traialists' Collaboration (BPLTTC) suggested that ACEIs but not ARBs hold evidence of blood pressure-independent effects on the risk of major coronary disease events [18]. Thus it is thought that ACEIs have superior benefits to other RAS inhibitors due to their cardioprotective effects. The Cooperative North Scandinavian Enalapril Survival Study (CONSENSUS), which recruits patients with New York heart association (NYHA) class IV heart failure shows that ACEIs treatment is tremendously advantageous in severe heart failure [19]. ACEIs also exhibit efficacy for patients with mild to moderate heart failure [20, 21]. In the Vasodilator in heart failure II (V-HeFT-II) trial, enalapril had significantly lower mortality than that of the combination of hydralazine plus isosorbide dinitrate, which does not directly suppress neurohormonal system, despite weaker blood pressure lowering the effects of enalapril [22]. These observations of clinical trials support that ACEIs have the power to improve the natural history in a patient with broad range of reduced left ventricular function through several mechanisms including blood pressure lowering, suppression of neurohormonal system, and RAS independently cardioprotective effects. ACEIs should be initiated in low doses, followed by increments in each dose if lower doses have been well tolerated. Usually, titration is achieved by doubling the dosage every 3 to 5 days. The dose of ACEIs should be increased until the doses used are similar to those that have been shown to be effective in clinical trials or permissibly maximum dosage in each country. Higher doses of ACEIs are more effective than lower doses in preventing hospitalization because of suppression of the sustained activated RAS in patients with heart failure. ACEIs should keep being used for patients with reduced left ventricular function for reasons other than severe hypotension, severe renal dysfunction or high potassium retention associated with ACEIs treatment. The side effects of ACEIs that are related to kinin potentiation include a nonproductive cough, which is in about 10% of patients, and angioedema, which is in 1% of patients. For patients who cannot tolerate ACEIs taking because of the cough or angioedema, ARBs are the next recommended line of therapy. Patients intolerant to ACEIs because of hyperkalemia or renal insufficiency are likely to experience the same side effects with ARBs. The combination of hydralazine and an oral nitrate should be considered to the latter patients.

4.2. Angiotensin II receptor blocker and heart failure

ARBs are well tolerated in patients who are intolerant of ACEIs treatment because of the development of nonproductive cough, angioedema or skin rash. Under such conditions, ARBs should be used in symptomatic and asymptomatic patients with reduced left ventricular function who are ACEI-intolerant for reasons other than hyperkaremia or renal insufficiency. Although the target of ACEIs and ARBs is the inhibition of the AT_1 receptor, their mechanisms are different. ACEIs suppress Ang II production, while ARBs interfere the activation of the AT_1 receptor leading to an unlocking of the negative feedback of RAS, which results in an increment of the RAS peptides. The increased renin, Ang I, Ang II may evoke an unblocked AT_1 receptor by ARB. Therefore, high-dose ARBs appear to be better than low-dose ARBs for treating patients with heart failure. The question of high-dose versus low-dose ARB clinical outcomes was evaluated in the Heart Failure Endpoint Evaluation of Angiotensin II Antagonist Losartan (HEAAL) trial [23]. However, this study showed that treatment with high-dose losartan was not associated with a significant reduction in the primary endpoint of all-cause death or admission for heart failure when compared to that of low-dose losartan. Although ARBs are as effective as ACEIs in some clinical trials, ARBs does not cap ACEIs in a direct comparison of ACEIs versus ARBs trails. In the Losartan Heart failure Survival Study (ELITE-II), losartan was not associated with improved survival in older heart failure patients when compared to captopril, but was significantly better tolerated [24]. In the Valsartan in Acute Myocardial Infarction Trail (VALIANT), losartan was not as effective as captopril on all-cause mortality in post myocardial infarction patients who developed left ventricular dysfunction associated with signs of heart failure, while valsartan was shown to be non-inferior to captopril on all-cause mortality [25]. Hence, the general consensus is that ACEIs remain as the first-line drug for the treatment of systolic heart failure, while ARBs are strongly recommended for ACE-intolerant patients.

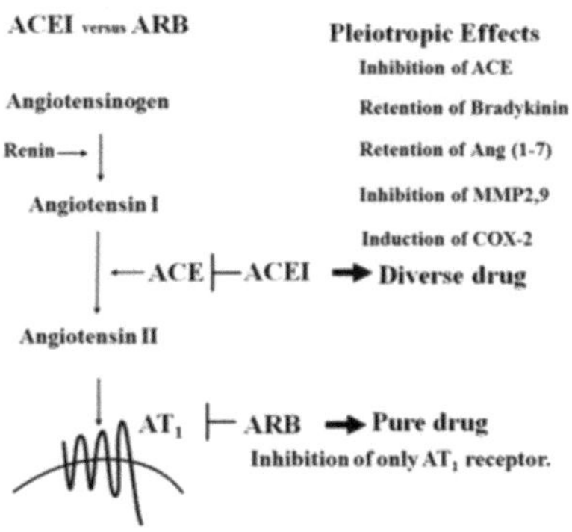

Figure 9.

4.3. Direct renin inhibitor and heart failure

Direct renin inhibitor, aliskiren, is the 3rd RAS inhibitor and it is available in clinics since the 21th century. Aliskiren is an orally active renin inhibitor and is a competitively non-peptide inhibitor that binds to the active site in cleft of renin instead of angiotensinogen. Since renin is the limiting protease of RAS, aliskiren may be a rationalized RAS inhibitor in three RAS inhibitors. In the Aliskiren Observation of Heart Failure Treatment (ALOFT) study in patients with NYHA class II to IV heart failure. NT-pro BNP was significantly lower in patients who were randomized to aliskiren when compared to placebo [26].

4.4. Aldosterone blocker and heart failure

We already described that ACEIs is the first-line drug for patients with heart failure. Although ACEIs may transiently decrease aldosterone secretion, long-term usage of ACEIs rapidly return of aldosterone to levels similar to those before ACEIs. This is termed "aldosterone breakthrough". The predictable mechanism of aldosterone breakthrough is that RAS takes a detour through the tissue chymases but not ACE. The results of the Eplerenone Post-Acute Myocardial Infarction Heart Failure Efficacy and Survival Study (EMPHASIS-HF) study, which recruits patients with NYHA class II heart failure and an ejection fraction of no more than 35% to receive eplerenone (up to 50 mg daily) or placebo, in addition to recommended therapy, displays that the administration of an aldosterone blocker is an available drug in patients with severe heart failure [27]. The dose of aldosterone blocker should be increased until the doses used are similar to those that have been shown to be effective in clinical trials or permissibly maximum dosage in each country. Patients should be counseled to avoid high potassium-containing foods. Potassium levels and renal function should be rechecked within 3 days and again, 1 week after initiation of an aldosterone blocker.

5. β-adrenergic receptor and heart failure

In the cardiomyocyte, β-adrenergic receptors dominate, and NE evokes increment of heart rate and contractile force, while in the arterioles, NE has predominantly vasoconstrictive effects acting through postsynaptic α_1-receptor. In addition, NE stimulates presynaptic α_2-receptors to invoke feedback inhibition of its own release, thereby modulating excess release of NE. Predominant β-adrenergic receptors are β_1 subtype in the cardiomyocyte, while most non-cardiac receptors are β_2. The left ventricle of the human heart also expresses β_2-receptors that are about 20% of the total β-receptor population, whereas the atria express β_2-receptors about 40% of total population. The cardiac β_1-receptors are colligated stimulatory G protein G_s, which is a component of the G protein-adenylyl cyclase system. However on the contrary, the cardiac β_2-receptors are colligated both G_s and the inhibitory G protein G_i. Therefore, the intracellular signaling of β_2-receptors remains controversially. Hypothetically, β_2-receptors are more strongly coupled to G_s under normal conditions, but this coupling is weakened and the coupling to G_i is strengthened under heart failure. The percentage of β_2-receptors in the left ventricle during heart failure is up to double because of β_1-receptor downregulation. The β_2-receptors may modulate the total valance of the adrenergic receptor system. Upon NE stimulation, the activation of G_s-adenylyl cyclase system is initiated as the positive inotropic effects in the cardiomyocyte. NE stimulation induces the molecular change in β_1-receptors, leading to the binding of GTP to α_s subunit of G_s. The dissociated GTP-α_s subunit of G_s from β_s, γ_s subunits stimulates adenylyl cyclase resulting in the formation of cAMP from ATP. cAMP activates cAMP-dependent protein kinase A (PKA). PKA plays important roles as phosphorylation of various key proteins and enzymes. PKA is locally bound to A-kinase anchoring protein (AKAP), which induces phosphorylation of a sarcolemmal protein p27 leading to increased entry of calcium ion through increased opening of the voltage-dependent L-type calcium channels in the sarcolemma. This small influx of calcium ion through the L-type calcium channels is a trigger of phosphorylation of the ryanodine receptor resulting in greater and more rapid rise of intracellular free calcium ion concentration. High concentration of the intracellular calcium ion increases calcium-troponin C interaction with deinhibition of tropomyosin effect on actin-myosin interaction. Thereby, increased rate and number of cross bridges interacting with increased myosin ATPase activity are amplified. Finally the heart procures increased rate and peak of force development. The increased relaxant effect is the consequence of increased PKA-mediated phosphorylation of phospholamban. Increased phosphorylation of troponin I also help desensitize the contractile apparatus to calcium ions. Sustained β receptor stimulation rapidly induces the activity of the β-agonist receptor kinase (βARK1), G protein-coupled receptor kinase (GRK2). βARK1- GRK2 increases the affinity of the β receptor for another protein family, arrestins, which cause the dissociation. β-arrestin is scaffolding and signaling protein that links to one of the cytoplasmic loops of the GPCR coupled β adrenergic receptor, lessening activation of adenylyl cyclase to inhibit the function of this receptor. Furthermore, β-arrestin switches the agonist coupling from G_s to G_i [28].

In heart failure, activation of the sympathetic nervous system is one of the most important adaptations. This occurs early in the course of heart failure. This activation is accompanied by a concomitant withdrawal of parasympathetic tone. This imbalance results in a resultant loss of heart rate and variability and increased peripheral vascular resistance in patients with heart failure. As a result of the increase in sympathetic tone, there is an increase in circulating levels of NE, a potent adrenergic neurotransmitter. The elevated levels of circulating NE result from a combination of increased release of NE from adrenergic nerve endings, and its consequent "spillover" into the plasma, with reduced uptake of NE by adrenergic nerve endings. In patient with moderate heart failure, the coronary sinus NE concentration exceeds the arterial concentration, indicating increased adrenergic stimulation of the heart. However, as heart failure progresses, there is a significant decrease in the myocardial concentration of NE. The mechanism responsible for cardiac NE depletion in severe heart failure is not clear and may relate to an exhaustion phenomenon resulting from the prolonged adrenergic activation of the cardiac adrenergic nerves in heart failure. For this reason, β-blocker therapy represents a major advance in the treatment of heart failure patients with reduced left ventricular function. Although there are a number of potential benefits to blocking all three receptors that are β_1, β_2 and α_1, the blocking of β_1-adrenergic receptor display most of the deleterious effects of sustained sympathetic activation. Three β blockers have been shown to be effective in reducing the risk of death in patients with chronic heart failure [29-31]. Sustained –release metoprolol succinate and bisoprolol both competitively block the β_1-adrenergic receptor, and carvedilol competitively blocks the α_1-, β_1- and β_2-adrenergic receptor. β-blockers should be initiated in low doses followed by gradual increments if low doses have been well tolerated. The dose of β-blockers should be increased until the doses used are similar to those that have been shown to be effective in clinical trials or permissibly maximum dosage in each country. However, the dose titration of β-blockers should proceed no sooner than at 2-week intervals, because the initiation and/or increased dosing of these agents may lead to worsening fluid retention because of the abrupt withdrawal of adrenergic support to the heart and circulation. Therefore, it is important to optimize the dose of diuretics before starting of β-blockers.

6. Non-drug treatment and heart failure

Other than internal and surgical therapies, there are implantable devices including the cardiac resynchronization therapy (CRT) or left ventricular assist device (LVAD) for the management, monitoring and assisted circulation in heart failure. Patients with severe heart failure may require the non-drug treatments for the purpose of surviving or facilitating the process of heart transplantation. 6-1, Cardiac resynchronization therapy (CRT) and heart failure.

Delays in interventricular or intraventricular electrical activation cause marked abnormalities in the sequence of global and segmental right and left ventricular activation, and impair mechanical performance. In patients with moderate to severe heart failure colligating wide QRS, a significant improvement was demonstrated an increase in exercise duration, and quality of life [32]. CRT was associated with reverse remodeling of left ventricular resulting in improved EF, dimensions and volume, and reduced mitral regurgitation. Moreover, CRT reduced the risk of complications ant death among patients with moderate or severe heart failure owing to left ventricular systolic dysfunction and cardiac dysynchrony, and this effect was not limited to ischemic heart disease. The combination of Implantable cardiac defibrillator (ICD) and CRT (CRT-ICD) in addition of optimal medical therapy has resulted in a 39% reduction in heart failure hospitalization and a 36% reduction in mortality in comparison with ICD alone [33]. CRT also has led to a degree of improvement in left ventricular volume and EF in patients with mild heart failure similar to that in patients with severe heart failure [34]. CRT reduced mortality and hospitalizations among asymptomatic or mildly symptomatic heart failure patients [52]. Hence, recent clinical trials are directed toward focus on delaying progression of heart failure in asymptomatic or less symptomatic patients.

6.1. Ventricular assist device (VAD) and heart failure

Timely referral for mechanical circulatory support (MCS) evaluation and appropriate implantation depends on familiarity with recent advances in pump design and clinical outcomes. The expansion of durable left ventricular assist device (LVAD) options for patients with advanced heart failure came just as the significant shortage of donor hearts was becoming apparent. In the U.S., according to the Centers for Medicare and Medicaid Services, implant strategies are divided into four groups; such as bridge to transplant (BTT), bridge to candidacy (BTC), destination therapy (DT), and Bride to recovery (BTR). In contemporary thinking, the dichotomous decision of either a bridge to transplantation or destination therapy is no longer tenable, and one could consider mechanical circulatory in the context of a "bridge to decision"[35]. Evolving pump design has driven clinical progress. After the invention of a smaller high-speed, rotary impeller pump with a single moving part, continuous-flow VADs with enhanced durability and near-silent operation became available. The transition from pulsatile technology toward continuous flow has been remarkably swift, and this rapid rise of continuous flow has made improved survival and performance [36, 37]. Pump complications, such as stroke, bleeding and infection, remain substantial risks. Embolic strokes appear more common than hemorrhagic strokes with all device designs. The Heatmate II has relatively low thrombotic risk provided patients are on an anticoagulation regimen that features an antiplatelet agent such as aspirin along with warfarin with an international normalized ratio (INR) goal of 1.5 to 2.0 [38]. Infection related to LVAD is reported 11-20%. The importance of infections in the VAD patient prompted the creation of a comprehensive set of guidelines and definitions [39].

Another pump development is miniaturization along with less invasive surgery. INTERMACS profiles have been developed to define clinically important differences in the severity of disease among patients with advanced heart failure [40]. Sicker subset of INTERMACS profile has been consistently associated with higher perioperative mortality. This trend will prompt the application of implantation of mechanical circulatory support to less sick heart failure patients in earlier stage. Adequate right ventricular function is necessary for proper LVAD function. Right heart failure after LVAD implant results in up to a 6-hold increase risk of death and is a major contributing factor in prolonged hospitalizations [41]. Right ventricular failure (RVF) results in persistently elevated venous pressure and insufficient LVAD preload, which occurs 6 to 35% of LVAD recipients [41]. In DT setting, in addition to a right ventricular assist device (RVAD) support, biventricular ventricular assist device (BiVAD) support with two continuous flow devices has been reported [42]. However, if RVF persists and long-term RV support is required, then the total artificial heart (TAH) is an option for those patients who are eligible for transplant. The TAH offers full circulatory replacement therapy for patients with irreversible biventricular failure. Freedom Driver, one of the smaller-sized TAH may allow discharge from hospital, and is undergoing investigation [43].

7. Heart transplantation and heart failure

Heart transplantation (HT) is indicated for those with chronic progressive heart failure despite optimal therapy, or with cardiogenic shock requiring mechanical support or high-dose inotropes. Heart failure patients with adult congenital heart disease are also taken into consideration for HT [44]. Various organizations for HT in the world have updated the waiting list of HT candidates to ensure an equitable system of donor organ allocation under the shortage of donor hearts. Cardiopulmonary exercise (CPX) is routinely used in the determination of candidacy for cardiac transplantation [45, 46]. In the presence of beta-blocker, a cutoff for peak VO2 of <14ml/kg/min should be used to guide listing (Class I) [47]. Right ventricular failure (RVF) and pulmonary hypertension (PH) are factors that prompt to reconsider suitability for waiting list. PH and elevated pulmonary vascular resistance (PVR) should be considered as relative contraindications to cardiac transplantation when the PVR is greater than 5 Woods units or the pulmonary vascular index is 6 or the transpulmonary gradient exceeds 16 to 20 mm Hg. If the systolic pulmonary arterial pressure exceeds 60 mmHg in conjunction with any of the aforementioned three variables, the risk of RVF and early death is increased [48]. For those with irreversible pulmonary pressures, a combined heart-lung transplant is a therapeutic choice.

Advances in post-transplant care have improved outcomes in older patients. A follow-up of HT recipients >65 years of age demonstrated survival rates comparable to those of younger patients [49]. Although the Patients older than 70 years have also been reported to have acceptable outcome with presumably less donor organ rejection, usually alternate-type program or permanent mechanical support should be pursued [50]. Active or recent malignancy is a contraindication to HT due to limited survival rates. However, pre-existing neoplasms may be treatable with chemotherapy to induce remission. Therefore it is essential to assess each patient as to their risk of tumor recurrence.

Diabetes with end-organ damage other than nonproliferative retinopathy or poor glycemic control with glycosylated hemoglobin (HbA1C) greater than7.5 despite optimal effort is a relative contraindication for transplant. It is reasonable to consider the presence of irreversible renal dysfunction (eGFR greater than 40ml/min) as a relative contraindication for HT. Obese patients with BMI > 30 kg/m^2 demonstrated nearly twice the 5-year mortality [51]. Therefore for this population, weight loss should be mandatory before listing for HT. Other comorbidity includes cirrhosis, peripheral vascular disease, addictions (tabacco, excessive alcohol) [52]. Psychosocial evaluation is mandatory before listing-up for HT. Immunologic evaluation is also needed. Immunocompatibility testing including ABO blood group typing, human leukocyte antigen and antibody screening should be completed. Panel-reactive antibody (PRA) test, which can identify the presence of circulating anti-human leukocyte antigen (HLA), and should be performed preferably by flow cytometry [53]. In France, single center data reported that actuarial survival rates were 75%, 58%, and 42% at 5, 10, and 15 years, respectively [54]. In Netherland, comparable survival rate was reported with the overall 1-, 5-, 10- and 15-year survival was respectively 77%, 67%, 53% and 42% [55]. Recent advance in HT technology along with surrounding circumstances has disclosed further issues to revise. The proposed challenges in this regard include optimization and individualization of immunosuppressive therapies, expansion and optimization of the donor and recipient candidate population, characterization of comorbidities, and understanding of antibody mediated rejection [56]. Late outcomes in the HT population remain poor with a median cardiac allograft survival of 11 years, a statistic that has not improved in over a decade [57]. The major causes of late morbidity and mortality are chronic kidney disease, cardiac allograft vasculopathy (CAV), and malignancy [46]. The dosing of calcineurin inhibitor (CNI), cyclosporine or tacrolimus, a purine synthesis inhibitor such as mycophenolate mofetil, and corticosteroids, which have a narrow therapeutic index, is typically based on the weight and renal function of a patient. A key research priority should be to develop clinical trials that evaluate how CNI sparing and elimination approaches (CNI-free immunosuppression). Better understanding of individualized immunologic characteristics is a key component to perform proper immunosuppressive therapy.

8. Cardiac regeneration and heart failure

Since usually heart failure results from deficiency of the cardiomyocyte, heart regeneration may become the prospective therapeutic technology of heart failure through regenerating lost cardiomyocytes to recovery of cardiac function. However, from the 19th century to the early 20th century, there had been the consensus that indicates that the heart is an organ incapable of regeneration [58]. Ventricular hypertrophy had been cause by enlargement rather than proliferation of the cardiomyocyte. From '60s, the investigators have opened up the milestone articles that display the evidence of heart regeneration of the human adult heart [59]. Pathologically hypertrophied heart demonstrates the evidence of cardiomyocyte proliferation when the heart weight exceeds 450g that contains about 210g of myocardium [60]. To evaluate cardiomyocyte proliferation, biochemical measurement of tissue DNA content and fluorescent analysis of individual nuclear DNA content associated with histopathology have been employed [61]. Most human cardiomyocyte nuclei are polyploid by the onset of puberty. In response to pathological overloads, human cardiomyocytes commonly reinitiate DNA synthesis without nuclear division [62]. Human cardiomyocytes seems to remain mononucleated throughout life. Thus, DNA synthesis is common in the adult human heart. Although this cannot be equated to cardiomyocyte proliferation, the measurement of cardiomyocyte DNA content is useful for investigation of heart proliferation. Using these methods, researchers have displayed that the cardiomyocyte nuclear number is steady at ~2 billion, which is reached at about 2 months of age, in the range of heart weight from 50g to 350g [63]. However, there is a linear increase in nuclear number, reaching 4 billion cardiomyocyte nuclei in hypertrophied hearts, which are weighting 700-900g. Since the number of non-cardiomyocytes such as fibroblasts and vascular cells increases linearly with heart weight throughout life, these results indicate that cardiomyocyte renewal occurs during pathological hypertrophy in the adult human heart [60, 64]. In 2009, there was definitive evidence of regeneration of the human heart. Employing ^{14}C, generated by nuclear bomb tests during the Cold War, infiltrate nuclear and label the DNA of dividing cells, the age of the cardiomyocyte composing the human heart was performed [65]. Mathematical modeling suggested that cardiomyocyte renewal was age-dependent, 1% of human cardiomyocytes were renewed at the age of 20, and this rate was reduced to 0.45% at the age of 75. About 45% of the cardiomyocytes would be predicted to be renewed over a normal human life on the basis of this kinetics. Most of the cardiac regeneration studies focused on the proliferation of existing cardiomyocytes, and were not designed to detect cardiomyocytes formed from progenitor cells or not. To determine whether such progenitor cells contribute to cardiomyocyte renewal, the genetic fate-mapping experiment was performed using transgenic mice [66]. This system allowed the authors to distinguish between cardiomyocyte renewal from existed cardiomyocytes via proliferation and cardiomyocyte renewal from progenitor cells. The adult mammalian heart shows that heart regeneration depends on replenishment by cardiomyogenic progenitor cells than on replacement by cardiomyocyte proliferation. Thus, these human and rodent heart studies provide strong evidence for plasticity in the adult human heart. Although actually cardiomyocyte regeneration from progenitor cells probably occurs in the human heart, it seems to be a very slow process different from that of the zebrafish, which rapidly promotes cardiac regeneration through cardiac proliferation, besides ageing is associated with the loss of ~ 1g /of myocardium per year in the absence of specific heart disease [67].

Stem cell biology is one of frontier areas of biomedical research including regeneration medicine. In the latter part of 20th century, bone marrow stem cell (BMCs) transplantation was gotten a lot of attention as a next regeneration medicine, however, the accumulating evidences indicate that BMC do not work by directly differentiating into new cardiomyocytes. In the 21th century, the existence of several types of cardiac stem cells has been reported. Cardiac stem cells display cell surface markers as c-kit positive, Sca-1 positive, Abcg2 positive, cardiospere-drived cells (CDCs) positive and islet-1 positive respectively [68]. These cells can be isolated and differentiated into fully mature cardiomyocytes that express contractile proteins, generation of calcium transients and respond to β-adrenergic stimulation. However, their abundant presence in the adult human heart and their capacity to engraft, regenerate myocardium leading to improving of cardiac function does not reach the sufficient evidence as the consensus. In fact, clinical trials using CDCs and c-kit positive cells are underway in California, Louisville and Kentucky respectively. Embryonic stem cells (ESC) and induced pluriopotent stem cells (iPS) are able to generate any cell type in our body. They have a tremendous potential for regeneration associated with obvious problems such as immune rejection, the carcinogenic potential. Therefore, they are a potentially inexhaustible supply of the human cardiomyocytes. IPS was originally generated by the reprograming of adult somatic cells by the forced expression of up to four stem cell related transcription factors, which is termed "Yamanaka factors". So the cardiomyocytes from any pluripotent stem cell type are immature and lack the expression profile, morphology and function of the adult ventricular cardiomyocyte. Therefore, the cardiomyocytes from patient-derived iPS cells may play a normal cardiac function. Human ESC-derived cardiomyocyte express early cardiac transcription factors such as NKX2.5, as well as the expected sarcomeric proteins, ion channels, connexins and calcium-handling proteins. They show similar functional properties to those reported for cardiomyocytes in the developing heart, and undergo comparable mechanisms of excitation- contraction coupling and neurohormonal signaling [69, 70]. Human iPS-derived cardiomyocytes show a very similar phenotype [71, 72]. Furthermore, these cells have shown to engraft in infarct mouse, rat, guinea pig and pig heart, forming islands of nascent, proliferating human myocardium within the scar zone [73, 74]. Furthermore, two research groups achieved directly reprogrammed cardiomyocyte from somatic cells [75, 76]. These results may be one of most important evidences of cardiac regeneration employing pluripotent stem cell. Final goal of these biochemical tools will depend on the long-term engraftment of regenerative cells.

9. Summary

We described recent topics of the heart failure in basic and clinical field. To materialize applicable conditions responding to an elderly society, therapeutic, economic or Social security problems associated with heart failure have to be gotten fixed. Thereby, the research system close linkage between basic and clinic is important to prevent and remedy heart failure in the elderly societies.

Author details

Shintaro Nakano, Toshihiro Muramatsu, Shigeyuki Nishimura and Takaaki Senbonmatsu
Division of Cardiology, International Medical Center, Saitama Medical University, Saitama, Japan

Takaaki Senbonmatsu
Department of Pharmacology, Saitama Medical University, Saitama, Japan

Acknowledgement

We thank Ayumi Hara for secretarial assistance.

10. References

[1] Nassar R, Malouf NN, Mao L, et al. (2005) cTnT1, a cardiac troponin T isoform, decreases myofilament tension and affects the left ventricular pressure waveform. Am J Physiol Heart Circ Physiol. 288: H1147-56.

[2] Mann DL, Bristow MR. (2005) Mechanisms and models in heart failure: the biomechanical model and beyond. Circulation. 111: 2837-2849.

[3] Lowes BD, Gilbert EM, Abraham WT, et al. (2002) Myocardial gene expression in dilated cardiomyopathy treated with beta-blocking agents. N Engl J Med. 346: 1357-1365.

[4] Redfield MM, Jacobsen SJ, Burnett JC Jr, et al. (2003) Burden of systolic and diastolic ventricular dysfunction in the community: appreciating the scope of the heart failure epidemic. JAMA. 289: 194-202.

[5] Hogg K, Swedberg K, McMurray J. (2004) Heart failure with preserved left ventricular systolic function; epidemiology, clinical characteristics, and prognosis. J Am Coll Cardiol. 43: 317-327.

[6] Ceia F, Fonseca C, Mota T, et al. (2002) Prevalence of chronic heart failure in Southwestern Europe: the EPICA study. Eur J Heart Fail. 4: 531-539.

[7] Bhatia RS, Tu JV, Lee DS, et al. (2006) Outcome of heart failure with preserved ejection fraction in a population-based study. N Engl J Med. 355: 260-269.

[8] Kitzman DW, Little WC, et al. (2002) Pathophysiological characterization of isolated diastolic heart failure in comparison to systolic heart failure. JAMA. 288: 2144-2150.

[9] de Gasparo M, Catt KJ, Inagami T, et al. (2000) International union of pharmacology. XXIII. The angiotensin II receptors. Pharmacol Rev. 52: 415-472.

[10] Senbonmatsu T, Saito T, Landon EJ, et al. (2003) A novel angiotensin II type 2 receptor signaling pathway: possible role in cardiac hypertrophy. EMBO J. 22: 6471-6482.

[11] Wang N, Frank GD, Ding R, et al. (2012) Promyelocytic Leukemia Zinc Finger Protein Activates GATA4 Transcription and Mediates Cardiac Hypertrophic Signaling from Angiotensin II Receptor 2. ProS One. In press.
[12] Iwai N, Shimoike H, Kinoshita M. (1995) Cardiac renin-angiotensin system in the hypertrophied heart. Circulation. 92: 2690-2696.
[13] Nguyen G, Delarue F, Burckle C, et al. (2002) Pivotal role of the renin/prorenin receptor in angiotensin II production and cellular responses to renin. J Clin Invest. 109: 1417-1427.
[14] Ichihara A, Hayashi M, Kaneshiro Y, et al. (2004) Inhibition of diabetic nephropathy by a decoy peptide corresponding to the "handle" region for nonproteolytic activation of prorenin. J Clin Invest. 114: 1128-1135.
[15] Batenburg WW, Krop M, Garrelds IM, et al. (2007) Prorenin is the endogenous agonist of the (pro)renin receptor. Binding kinetics of renin and prorenin in rat vascular smooth muscle cells overexpressing the human (pro)renin receptor. J Hypertens. 25: 2441-2453.
[16] Saris JJ,'t Hoen PA, Garrelds IM, et al. (2006) Prorenin induces intracellular signaling in cardiomyocytes independently of angiotensin II. Hypertension. 48: 564-571.
[17] Timmermans PB, Wong PC, Chiu AT, et al. (1993) Angiotensin II receptors and angiotensin II receptor antagonists. Pharmacol Rev. 45: 205-251.
[18] Turnbull F, Neal B, Pfeffer M, et al. (2007) Blood pressure-dependent and independent effects of agents that inhibit the renin-angiotensin system. J Hypertens. 25: 951-958.
[19] (1987) Effects of enalapril on mortality in severe congestive heart failure. Results of the Cooperative North Scandinavian Enalapril Survival Study (CONSENSUS). The CONSENSUS Trial Study Group. N Engl J Med. 316: 1429-1435.
[20] Pfeffer MA, Braunwald E, Moye LA, et al. (1992) Effect of captopril on mortality and morbidity in patients with left ventricular dysfunction after myocardial infarction. Results of the survival and ventricular enlargement trial. The SAVE Investigators. N Engl J Med. 327: 669-677.
[21] Torp-Pedersen C, Kober L. (1999) Effect of ACE inhibitor trandolapril on life expectancy of patients with reduced left-ventricular function after acute myocardial infarction. TRACE Study Group. Trandolapril Cardiac Evaluation. Lancet. 354: 9-12.
[22] Rector TS, Johnson G, Dunkman WB, et al. (1993) Evaluation by patients with heart failure of the effects of enalapril compared with hydralazine plus isosorbide dinitrate on quality of life. V-HeFT II. The V-HeFT VA Cooperative Studies Group. Circulation. 87: VI71-VI77.
[23] Konstam MA, Neaton JD, Dickstein K, et al. (2009) Effects of high-dose versus low-dose losartan on clinical outcomes in patients with heart failure (HEAAL study): a randomised, double-blind trial. Lancet. 374: 1840-1848.
[24] Pitt B, Poole-Wilson PA, Segal R, et al. (2000) Effect of losartan compared with captopril on mortality in patients with symptomatic heart failure: randomised trial--the Losartan Heart Failure Survival Study ELITE II. Lancet. 355: 1582-1587.

[25] Pfeffer MA, McMurray JJ, Velazquez EJ, et al. (2003) Valsartan, captopril, or both in myocardial infarction complicated by heart failure, left ventricular dysfunction, or both. N Engl J Med. 349: 1893-1906.

[26] McMurray JJ, Pitt B, Latini R, et al. (2008) Effects of the oral direct renin inhibitor aliskiren in patients with symptomatic heart failure. Circ Heart Fail. 1: 17-24.

[27] Pitt B, Remme W, Zannad F, et al. (2003) Eplerenone, a selective aldosterone blocker, in patients with left ventricular dysfunction after myocardial infarction. N Engl J Med. 348: 1309-1321.

[28] Baillie GS, Sood A, McPhee I, et al. (2003) beta-Arrestin-mediated PDE4 cAMP phosphodiesterase recruitment regulates beta-adrenoceptor switching from Gs to Gi. Proc Natl Acad Sci U S A. 100: 940-945.

[29] Waagstein F, Bristow MR, Swedberg K, et al. (1993) Beneficial effects of metoprolol in idiopathic dilated cardiomyopathy. Metoprolol in Dilated Cardiomyopathy (MDC) Trial Study Group. Lancet. 342: 1441-1446.

[30] CIBIS-II Investigators and Committees. (1999) The Cardiac Insufficiency Bisoprolol Study II (CIBIS-II): a randomised trial. Lancet. 353: 9-13.

[31] Poole-Wilson PA, Swedberg K, Cleland JG, et al. (2003) Comparison of carvedilol and metoprolol on clinical outcomes in patients with chronic heart failure in the Carvedilol Or Metoprolol European Trial (COMET): randomised controlled trial. Lancet. 362: 7-13.

[32] Abraham WT, Fisher WG, Smith AL, et al. (2002) Cardiac resynchronization in chronic heart failure. N Engl J Med. 346: 1845-1853.

[33] Adabag S, Roukoz H, Anand IS, et al. (2011) Cardiac resynchronization therapy in patients with minimal heart failure: a systematic review and meta-analysis. J Am Coll Cardiol. 58: 935-941.

[34] Bleeker GB, Holman ER, Steendijk P, Boersma E, van der Wall EE, Schalij MJ, Bax JJ. (2006) Cardiac resynchronization therapy in patients with a narrow QRS complex. J Am Coll Cardiol. 48: 2243-2250.

[35] Felker GM, Rogers JG. (2006) Same bridge, new destinations rethinking paradigms for mechanical cardiac support in heart failure. J Am Coll Cardiol. 47: 930-932.

[36] Kirklin JK, Naftel DC, Kormos RL, et al. (2011) Third INTERMACS Annual Report: the evolution of destination therapy in the United States. J Heart Lung Transplant. 30: 115-123.

[37] Slaughter MS, Rogers JG, Milano CA, et al. (2009) Advanced heart failure treated with continuous-flow left ventricular assist device. N Engl J Med. 361: 2241-2251.

[38] Boyle AJ, Russell SD, Teuteberg JJ, et al. (2009) Low thromboembolism and pump thrombosis with the HeartMate II left ventricular assist device: analysis of outpatient anti-coagulation. J Heart Lung Transplant. 28: 881-887.

[39] Hannan MM, Husain S, Mattner F, et al. (2011) Working formulation for the standardization of definitions of infections in patients using ventricular assist devices. J Heart Lung Transplant. 30: 375-384.

[40] Stevenson LW, Pagani FD, Young JB, et al. (2009) INTERMACS profiles of advanced heart failure: the current picture. J Heart Lung Transplant. 28: 535-541.

[41] Fitzpatrick JR 3rd, Frederick JR, Hsu VM, et al. (2008) Risk score derived from preoperative data analysis predicts the need for biventricular mechanical circulatory support. J Heart Lung Transplant. 27: 1286-1292.

[42] Kirklin JK, Naftel DC, Kormos RL, et al. (2010) Second INTERMACS annual report: more than 1,000 primary left ventricular assist device implants. J Heart Lung Transplant. 29: 1-10.

[43] Jaroszewski DE, Anderson EM, Pierce CN, et al. (2011) The SynCardia freedom driver: a portable driver for discharge home with the total artificial heart. J Heart Lung Transplant. 30: 844-845.

[44] Simmonds J, Burch M, Dawkins H, et al. (2008) Heart transplantation after congenital heart surgery: improving results and future goals. Eur J Cardiothorac Surg. 34: 313-317.

[45] Mudge GH, Goldstein S, Addonizio LJ, et al. (1993) 24th Bethesda conference: Cardiac transplantation. Task Force 3: Recipient guidelines/prioritization. J Am Coll Cardiol. 22: 21-31.

[46] Costanzo MR, Augustine S, Bourge R, et al. (1995) Selection and treatment of candidates for heart transplantation. A statement for health professionals from the Committee on Heart Failure and Cardiac Transplantation of the Council on Clinical Cardiology, American Heart Association. Circulation. 92: 3593-3612.

[47] Mehra MR, Kobashigawa J, Starling R, et al. (2006) Listing criteria for heart transplantation: International Society for Heart and Lung Transplantation guidelines for the care of cardiac transplant candidates--2006. J Heart Lung Transplant. 25: 1024-1042.

[48] Butler J, Stankewicz MA, Wu J, et al. (2005) Pre-transplant reversible pulmonary hypertension predicts higher risk for mortality after cardiac transplantation. J Heart Lung Transplant. 24: 170-177.

[49] Zuckermann A, Dunkler D, Deviatko E, et al. (2003) Long-term survival (>10 years) of patients >60 years with induction therapy after cardiac transplantation. Eur J Cardiothorac Surg. 24: 283-291.

[50] Blanche C, Blanche DA, Kearney B, et al. (2001) Heart transplantation in patients seventy years of age and older: A comparative analysis of outcome. J Thorac Cardiovasc Surg. 121: 532-541.

[51] Lietz K, John R, Burke EA, et al. (2001) Pretransplant cachexia and morbid obesity are predictors of increased mortality after heart transplantation. Transplantation. 72: 277-283.

[52] Radovancevic B, Poindexter S, Birovljev S, et al. (1990) Risk factors for development of accelerated coronary artery disease in cardiac transplant recipients. Eur J Cardiothorac Surg. 4: 309-313.

[53] Kobashigawa J, Mehra M, West L, et al. (2009) Report from a consensus conference on the sensitized patient awaiting heart transplantation. J Heart Lung Transplant. 28: 213-225.

[54] Roussel JC, Baron O, Perigaud C, et al. (2008) Outcome of heart transplants 15 to 20 years ago: graft survival, post-transplant morbidity, and risk factors for mortality. J Heart Lung Transplant. 27: 486-493.

[55] Tjang YS, van der Heijden GJ, Tenderich G, et al. (2008) Survival analysis in heart transplantation: results from an analysis of 1290 cases in a single center. Eur J Cardiothorac Surg. 33: 856-861.
[56] Shah MR, Starling RC, Schwartz Longacre L, et al. (2012) Heart transplantation research in the next decade-a goal to achieving evidence-based outcomes: national heart, lung, and blood institute working group. J Am Coll Cardiol. 59: 1263-1269.
[57] Stehlik J, Edwards LB, Kucheryavaya AY, et al. (2010) The Registry of the International Society for Heart and Lung Transplantation: twenty-seventh official adult heart transplant report--2010. J Heart Lung Transplant. 29: 1089-1103.
[58] Karsner HT, Saphir O, Todd TW. (1925) The State of the Cardiac Muscle in Hypertrophy and Atrophy. Am J Pathol. 1: 351-372.1.
[59] LINZBACH AJ. (1960) Heart failure from the point of view of quantitative anatomy. Am J Cardiol. 5: 370-382.
[60] Adler CP, Costabel U. (1975) Cell number in human heart in atrophy, hypertrophy, and under the influence of cytostatics. Recent Adv Stud Cardiac Struct Metab. 6: 343-355.
[61] Herget GW, Neuburger M, Plagwitz R, et al. (1997) DNA content, ploidy level and number of nuclei in the human heart after myocardial infarction. Cardiovasc Res. 36: 45-51.
[62] Adler CP, Friedburg H. (1986) Myocardial DNA content, ploidy level and cell number in geriatric hearts: post-mortem examinations of human myocardium in old age. J Mol Cell Cardiol. 18: 39-53.
[63] Adler CP. (1975) Relationship between deoxyribonucleic acid content and nucleoli in human heart muscle cells and estimation of cell number during cardiac growth and hyperfunction. Recent Adv Stud Cardiac Struct Metab. 8: 373-386.
[64] Grajek S, Lesiak M, Pyda M, et al. (1993) Hypertrophy or hyperplasia in cardiac muscle. Post-mortem human morphometric study. Eur Heart J. 14: 40-47.
[65] Bergmann O, Bhardwaj RD, Bernard S, et al. (2009) Evidence for cardiomyocyte renewal in humans. Science. 324: 98-102.
[66] Hsieh PC, Segers VF, Davis ME, et al. (2007) Evidence from a genetic fate-mapping study that stem cells refresh adult mammalian cardiomyocytes after injury. Nat Med. 13: 970-974.
[67] Kikuchi K, Holdway JE, Werdich AA, et al. (2010) Primary contribution to zebrafish heart regeneration by gata4(+) cardiomyocytes. Nature. 464: 601-605.
[68] Carvalho AB, de Carvalho AC. (2010) Heart regeneration: Past, present and future. World J Cardiol. 2: 107-111.
[69] Kehat I, Kenyagin-Karsenti D, Snir M, et al. (2001) Human embryonic stem cells can differentiate into myocytes with structural and functional properties of cardiomyocytes. J Clin Invest. 108: 407-414.
[70] Zhu WZ, Santana LF, Laflamme MA. (2009) Local control of excitation-contraction coupling in human embryonic stem cell-derived cardiomyocytes. PLoS One. 4: e5407.
[71] Zhang J, Wilson GF, Soerens AG, et al. (2009) Functional cardiomyocytes derived from human induced pluripotent stem cells. Circ Res. 104: e30-e41.

[72] Zwi L, Caspi O, Arbel G, et al. (2009) Cardiomyocyte differentiation of human induced pluripotent stem cells. Circulation. 120: 1513-1523.

[73] Laflamme MA, Chen KY, Naumova AV, et al. (2007) Cardiomyocytes derived from human embryonic stem cells in pro-survival factors enhance function of infarcted rat hearts. Nat Biotechnol. 25: 1015-1024.

[74] Fernandes S, Naumova AV, Zhu WZ, et al. (2010) Human embryonic stem cell-derived cardiomyocytes engraft but do not alter cardiac remodeling after chronic infarction in rats. J Mol Cell Cardiol. 49: 941-949.

[75] Ieda M, Fu JD, Delgado-Olguin P, et al. (2010) Direct reprogramming of fibroblasts into functional cardiomyocytes by defined factors. Cell. 142: 375-386.

[76] Efe JA, Hilcove S, Kim J, et al. (2011) Conversion of mouse fibroblasts into cardiomyocytes using a direct reprogramming strategy. Nat Cell Biol. 13: 215-222.

Chapter 9

Implication of MicroRNAs in the Pathophysiology of Cardiac and Vascular Smooth Muscle Cells

Valérie Metzinger-Le Meuth, Eléonore M'Baya-Moutoula, Fatiha Taibi, Ziad Massy and Laurent Metzinger

Additional information is available at the end of the chapter

http://dx.doi.org/10.5772/47770

1. Introduction

In the last 10 years, microRNAs (miRNAs) have emerged as critical regulators of numerous physiological and pathological mechanisms [1-2], including cardiac and vascular smooth muscle cell (VSMC) plasticity [3-5]. These small molecules (approx. 20 to 25 nucleotides) comprise a novel and abundant class of endogenous interfering RNAs. More than 1 500 miRNAs are now listed by dedicated internet databases such as miRBase, Tarbase, MicroRNA.org or miRdb (See sub-chapter II for URL adresses). They are transcribed and matured, in a process known as miRNA biogenesis [6] which starts with the transcription of a larger RNA product, called pri-miRNA, by the RNA polymerase II in the vast majority of cases. Pri-miRNA, which is a few hundred to a few thousand nucleotides long, is then submitted to cleavage in the nucleus by a specific RNase III (Drosha) and its protein partner, DiGeorge syndrome critical region 8 (DGCR8), near the base of the miRNA hairpin stem. This process releases a pre-miRNA hairpin (of approx. 60 to 70 nt). The pre-miRNA is then released in the cytoplasm where it is recognized and cleaved within its stems by the Dicer RNase III and its protein partners. This results in a double stranded RNA, known as the miRNA/miRNA* duplex (approx. 22 bp). This complex is unwound to single strands. One strand (guiding strand / mature miRNA) is incorporated in the RNA-induced-silencing complex (*RISC*) that contains Argonaute 2 (Ago2), another endonuclease, the other strand is usually rapidly degraded. Finally, the RISC complex carries the mature miRNA to its target messenger RNAs (mRNAs), which results in gene silencing, in a post-transcriptional manner [7]. Figure 1 shows a representative example of the biogenesis of miR-143 and miR-145, which are the main miRNAs expressed in smooth muscle cells.

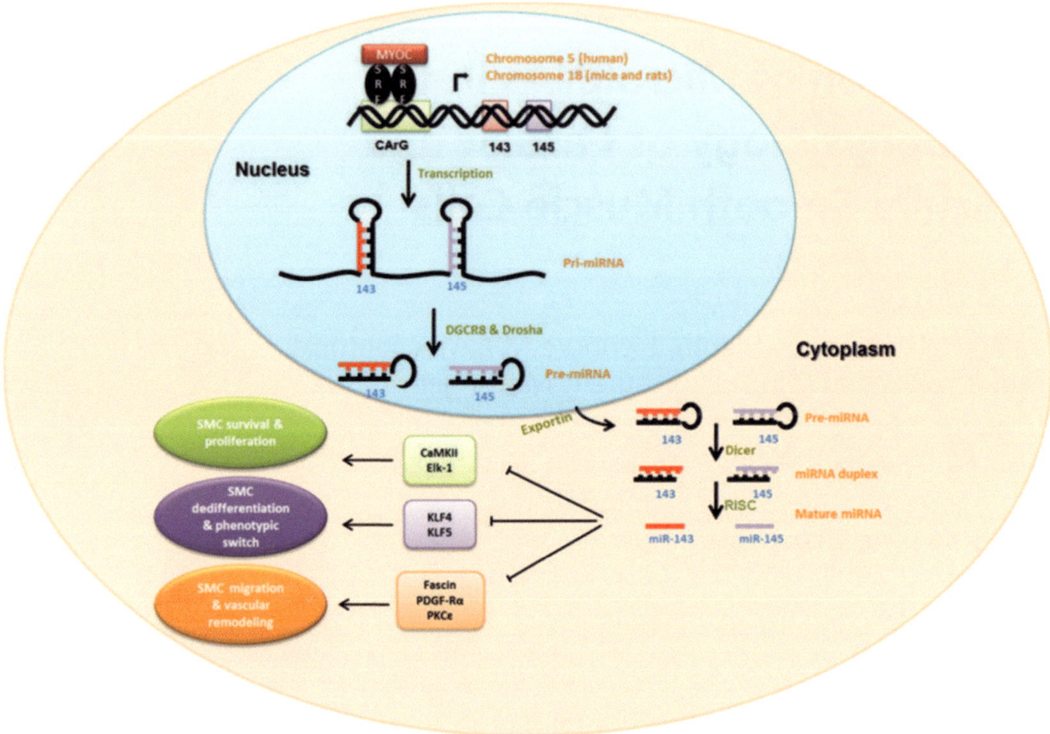

Figure 1. Schematic representation of the miR-143/145 cluster biogenesis, which expresses the main miRNAs in vascular smooth muscle. The transcription factor Myocardin in complex with its cofactor Serum Response Factor (SRF) binds CArG box-containing promoters, which then activates the transcription of the miR-143/145 cluster. MiR-143 and 145 are cotranscribed as a single pri-miRNA transcript. The Drosha and DGCR8 (DiGeorge Syndrome Critical Region 8) complex processes the pri-miRNA into a hairpin-structured pre-miRNA. The pre-miRNAs corresponding to miR-143 and 145 are exported by a nucleocytoplasmic shuttle protein known as exportin, from the nucleus to the cytoplasm and are then cleaved by the Dicer complex into miRNA duplexes. The miRNA duplexes are finally unwound to obtain the mature miR-143 and miR-145, which are incorporated into the RNA-induced silencing complex (RISC) and bind various mRNA targets in their 3′ untranslated region. In vascular smooth muscle cells (VSMCs), miR-143 and 145 inhibit the expression of CaMKII (Ca^{2+}/calmodulin-dependent protein kinases II) and Elk-1 (E twenty-six (ETS)-like transcription factor 1), which induces a repression of VSMC survival and proliferation. Also, the miR-143 and 145 inhibit the expression of Fascin, PDGF-Rα (Platelet-derived growth factor receptor α) and PKCε (Protein Kinase Cε) which induces a repression of VSMC migration and of the vascular remodeling. Finally, miR-143 and 145 also inhibit the expression of KLF4 and 5 which are repressors of Myocardin activity. As a consequence, VSMC differentiation is maintained.Adapted from [3].

Determining how the RISC complex carries a specific miRNA to its target mRNAs, and thus regulates gene expression, remains an intense field of research. The most important feature in the miRNA sequence is a short but critical region called the seed sequence, which is only 7 nts long, and most of the times, located in nucleotides in positions 2 to 8 of the miRNA. The base-pairing uses canonical Watson-Crick complementarity. Conversely, this small stretch of nucleic acids is a useful tool to classify miRNA into families based on shared seed sequences. Since the miRNA seed is so short, each miRNA can potentially bind hundreds of target mRNAs when one considers the large number of possible binding sites in mRNA regulatory sequences. This is one of the reasons why one single miRNA can regulate the expression of multiple target genes, by binding as many as several hundreds of mRNA targets. This clearly shows the role of these small RNAs in the intricate tapestry of gene regulations. Another, foremost reason for this complexity is that miRNAs, outside of the seed region, bind their mRNA targets mostly as imperfect complements. Similarly, one mRNA can be regulated by several miRNAs. All this further explains how the 1 500 miRNAs known to date are able to regulate the expression of approximately one third of the human genes. Thus, microRNAs are likely to impact multiple mechanisms of gene regulation and developmental pathways, by using extensive regulatory gene expression networks [1-2].

The current paradigm states that miRNAs act mostly by inhibiting the translation of their target mRNAs, rather than by inducing their degradation. Bartel's team has however recently challenged that view by showing that, in a vast majority of cases, mammalian microRNAs act by destabilizing their target mRNAs and by decreasing their levels [8]. No matter the case, a wide consensus agrees that miRNAs are posttranscriptional regulators which bind to their target mRNAs, mostly in their 3′ Untranslated Region (UTR). Note, however, that recent unbiased studies have shown that, in some particular cases, miRNAs bind the coding region or 5′UTR of respective target mRNAs ([9-11].

The miRNA nomenclature is remarkably standardized and straightforward [12]: the prefix "mir" is followed by a dash and an assigned number, reflecting the prevalence of discovery, to experimentally confirmed miRNAs. Gene is referred to in italic, eg *mir-143*. The pre-miRNA is designated by the suffix mir-, eg mir-143. Finally, the mature miRNA is abbreviated miR-, eg miR-143. MiRNAs with similar sequences / structure except for 1-2 nts, will be assigned a supplementary letter, eg miR-29a, miR-29b, miR-29c and often originate from the same gene. Different loci will produce different pre-miRNAs, but yield a mature miRNA with the precise same sequence. In this case, the nomenclature miR-1-1, miR-1-2 etc will be used. In some cases, one pre-miRNA will result in two different mature miRNAs, one from the 5-prime stem and one originating from the 3-prime stem. They will be designated by the suffixes miR-142-5p and miR-142-3p. Species can also be taken into consideration: hsa-miR refers to *Homo sapiens* miRNAs, mmu-miRNAs to *Mus musculus*, etc.

Mature miRNA sequences and pathways are remarkably preserved throughout phylogeny. Also, the evolutionary complexity of multicellular organisms positively correlates with the number of miRNA genes, their expression, and the diversity of their targets [13]. This remarkable conservation is used by most miRNA target prediction algorithms. These dedicated softwares use a standardized method to evaluate interactions between miRNAs and their specific target mRNAs based on (1) complementarity between mRNA 3′ UTR and miRNA seed sequence and (2) the degree of conservation of miRNA across species (see subchapter 2 for url adresses). The end result is straightforward: the higher the conservation, the higher the score given.

2. Current methods for studying miRNAs: Functional analysis

Various techniques have been developed over the last decade to quantify the expression of miRNAs and study their function (extensively reviewed in [14]). Figure 2 exposes the most widely used methods. Northern blotting was first used to quantify miRNAs, but this tedious technique, relying on radioactive labeling, was rapidly replaced by the much more convenient qRT-PCR. Two interchanging chemistries can be used for qPCR: (1) Sybr Green, which is often associated with a modified oligo(dT) technique, and the adjunction of an universal primer for reverse transcription which enables reverse transcription of all transcripts within an RNA sample; therefore, target miRNA and normalizing mRNA can be analyzed from the same RT reaction; or (2) Taqman chemistry, in association with the use of stem-loop miRNA-specific RT primers to produce cDNA, with the advantage of additional specificity. Both give accurate results. Each method has its own assets and setbacks. Technically speaking, the modified oligo(dT) method requires only a single RT reaction to reverse transcribe both miRNA and its target mRNAs and is less time consuming. Both techniques have been optimized and developed to screen most of known miRNAs in one experiment. Indeed, analysis of high-throughput miRNA expression remains a challenge since the number of miRNAs continues to increase with *in silico* prediction and experimental verification. Oligonucleotide microchip (microarray) was first widely used for high throughput miRNA screening, but a novel miRNA expression profiling approach, quantitative RT-PCR array (qPCR-array) is now rapidly gaining ground. Comparison between microarray and qPCR-array indicated a superior sensitivity and specificity of qPCR-array [15], and qPCR arrays are now rapidly becoming the method of choice.

Over-expressing or inhibiting miRNA activity, using RNA constructs, and examining the resulting phenotypic effects is crucial for understanding microRNA involvement, both *in vitro* and *in vivo*. For that, solutions are readily available for *in vitro* use [16]. Knock-in is most of the time induced by the addition of pre-miR precursors of a particular miRNA, often nicknamed "mimics". Alternatively, viral vectors have been developed by various industrials, most of them relying on modified *lentivirus* or adeno-associated virus (AAV). On the other hand, knock-down is classically induced by transfection of so-called "antagomiRs", which have been developed to interfere with expression of a specific miRNA. These synthetic RNA inhibitors incorporate the reverse complement of the mature miRNA (which represents here the target site). They are chemically modified to enhance binding affinity and decrease nucleolytic cleavage by the RISC complex and degradation by other RNAses. The last years have also seen the development of miRNAs sponges, which are equivalent to the endogenous sponges described below (see sub-chapter V). These long transcripts contain repeated regions complementary to specific miRNAs which will bind them instead of their dedicated mRNA targets, thus resulting in miRNA silencing. All these techniques can also be implemented *in vivo,* in various animal models.

Finally, classic knock-out genetic strategies can be applied to miRNA genes [17]. MiRNA-processing proteins such as Dicer, Dgcr8, Drosha or Ago2 are essential for viability in mice. Knock-out mice individually lacking these key miRNA-processing genes die during early gestation with severe developmental defects, including in vessels and heart. However, conditional murine knockout of these genes have been developed recently and offer valuable tools to study miRNA importance, including in the cardiac organ. Also, one can obtain animals models totally devoid of the miRNA of interest. We describe below the interesting example of mice knock-outed of the most widely expressed vascular miRNAs, miR-143 and miR-145 (see subchapter III, [18-20]).

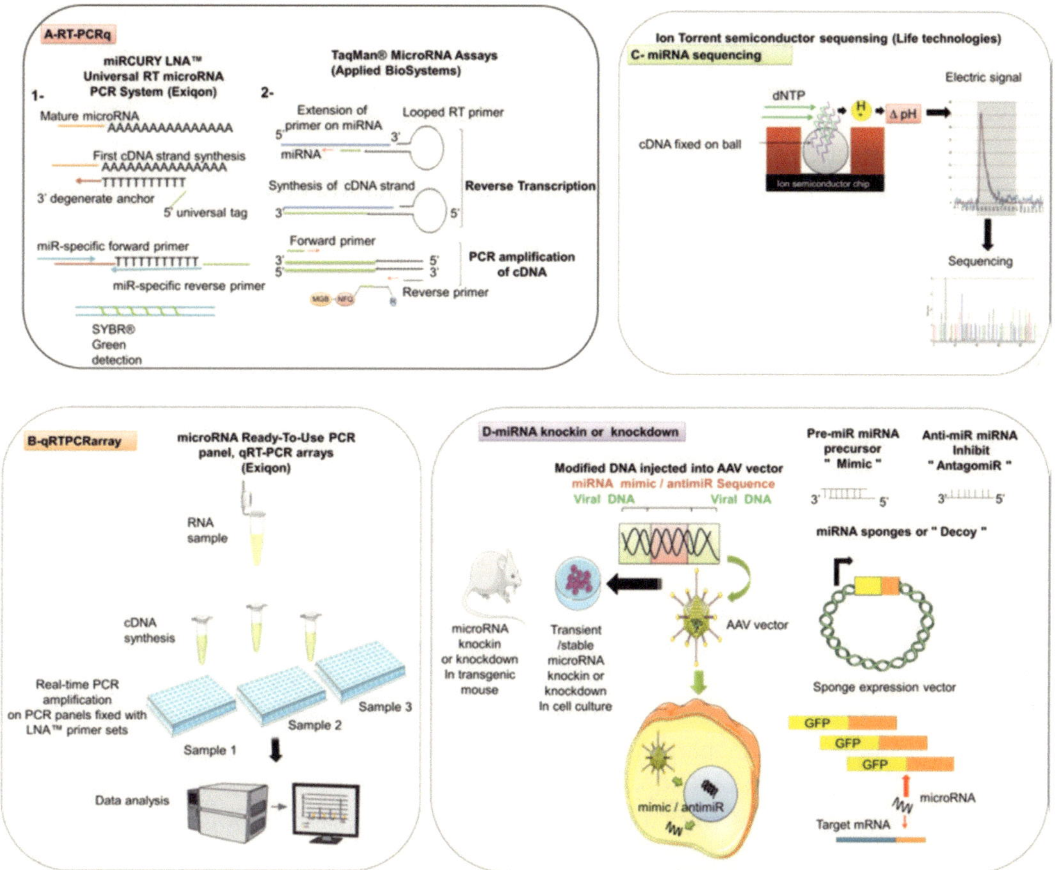

A- Schematic outline of microRNA RT-qPCR Systemq. (1) A poly-A tail is added to the mature microRNA template. cDNA is synthesized using a poly-T primer with a 3′ degenerate anchor and a 5′ universal tag. The cDNA template is then amplified using microRNA-specific and LNA™-enhanced forward and reverse primers. SYBR® Green is used for detection . (2) Taqman chemistry with the use of specific stem loop RT primers, and of an internal Taqman probe for qPCR quantitation.

B-qRT-PCR arrays: The microRNA PCR array protocol is a two-part protocol consisting of, 1- First-strand cDNA synthesis, 2- Real-time PCR amplification . Two 384-well plates enable to quantitate more than 700 miRNA in one experiment.

C-High throughpout sequencing. Nucleotides flow sequentially over Ion semiconductor chip– Direct detection of natural DNA extension– A few seconds per incorporation

D- miRNA sponges interfere with miRNA function. Sponges are ectopically expressed or artificial RNAs that contain multiple miRNA target sites. These target sites compete miRNAs away from their natural mRNA targets. miRNA sponges are suitable for use in a variety of experimental systems, including cultured cells and transgenic animals.

Figure 2. Methods currently used to study miRNA expression and function.

Websites are dedicated to identification of microRNA gene targets, and experimental validation of these *in silico* data: miRBase (http://www.mirbase.org/), Tarbase (diana.cslab.ece.ntua.gr/DianaToolsNew/index.php?r=tarbase/index), MicroRNA.org (http://www.microrna.org/microrna/home.do) or miRdb (http://mirdb.org/miRDB). They were described above. Unfortunately, as of today, the results given by the various websites are often very different from each other and the informations must therefore be subject to careful scrutiny. An interesting complement is provided by Patrocles (http://www.patrocles.org/). This software attends to the referencing of polymorphisms (Single nucleotide polymorphisms, SNP, mostly) and the interactions between target genes and relevant miRNAs in seven vertebrate species. Significant progress will certainly come from this sort of research efforts in the next few years since it is now increasingly clear that non-coding SNPs provide a potential mechanism for transmission of phenotypes and diseases. Finally, biochemical approaches, using affinity purification, are also being developed for direct empirical detection of miRNA associating with the 3'UTR of mRNA targets [21].

3. MicroRNAs are implicated in the pathophysiology of vascular smooth muscle cells

VSMCs are not terminally differentiated cells like skeletal and cardiac muscle cells. They have a remarkable plasticity which allows them to undergo phenotypic modulation inducing a switch from a "synthetic" to a "contractile" phenotype, in response to physiological and pathological environmental cues. On one hand, vascular injury or growth factors like PDGF provoke VSMC dedifferentiation which, as a consequence, transdifferentiates the cells into a highly migratory and proliferative ("synthetic") phenotype necessary for vascular repair or angiogenesis. On another hand, Transforming Growth Factor-β (TGF-β) and its related family member Bone Morphogenetic Protein 4 (BMP4) promote differentiation into a less migratory and less proliferative phenotype known as the "contractile" phenotype. This VSMC phenotypic modulation, called transdifferentiation, is characterized by significant changes in cellular gene expression pattern. In particular, high expression of VSMC-specific genes, such as smooth muscle α-actin (SMαA), calponin1 (CNN), and SM22α (SM22) are associated with the contractile phenotype. Transcription of contractile genes is regulated by SRF through a DNA sequence motif known as the CArG box (CC(A/T)$_6$GG) which is present in the promoter of VSMC-specific genes. A coactivator (and binding partner) of SRF, myocardin, activates VSMC expression of key contractile genes ([22-23].

The recent emerging role of miRNAs in gene expression regulation *via* gene silencing (through mRNA degradation or translation inhibition) suggests a role for these small nucleic acids in VSMC phenotypic regulation. Indeed, numerous publications have documented their importance through *in vitro* and *in vivo* studies in the cardiac and vascular biology fields and their related diseases [24]. This important role of miRNAs in VSMC development, differentiation, and related pathologies has been emphasized by two independent teams [25-26] that independently showed that knock-outing the miRNA processing enzyme Dicer in murine VSMCs provokes severe vascular abnormalities, resulting in embryonic lethality. Among vascular miRNAs, miR-143 and miR-145 are the most documented to date, and will be explored in greatest detail. They are the most highly expressed miRNAs in smooth muscle cells, and their down-regulation is directly associated with a phenotypic switch from contractile, *i.e.* fully differentiated to synthetic, i.e. proliferative, VSMCs. Other miRNAs such as miR-21, miR-221 and miR-222 also have demonstrated roles in VSMC differentiation. Their functions in smooth muscle will also be described. See Figure 3A for a pictorial representation of their roles in VSMCs.

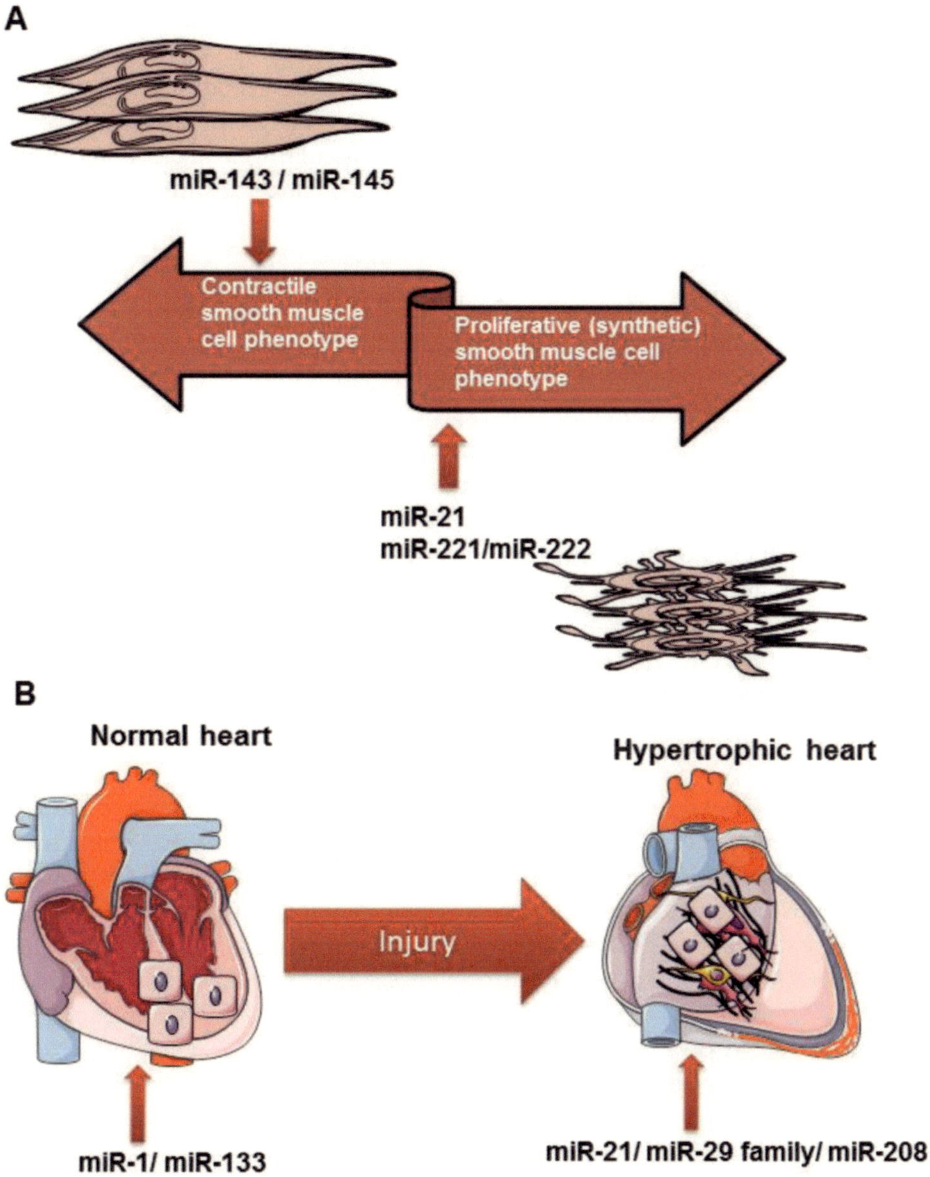

Figure 3. (A) Main micro-RNAs involved in the regulation of smooth muscle cell phenotype, including transdifferentiation from synthetic to contractile phenotype and (B) main micro-RNAs involved in cardiac muscle cell physiopathology.

3.1. The miR-143/145 cluster

The bicistronic unit which encodes miR-143 and miR-145 is critical for maintaining the VSMC contractile phenotype. For example, miR-143 and miR-145 are down-regulated in synthetic VSMCs [19-20] when VSMC dedifferentiation is induced by PDGF and during neointimal formation [27]. On the opposite, TGFβ1 (Transforming Growth Factor 1), a strong activator of VSMC differentiation, stimulates both miRNAs expression in a dose- and time-dependent manner [27]. The transcription of miR-143/145 is under the control of two independent signaling pathways: SRF/myocardin/Nkx2.5 and Jag-1/Notch signaling [28]. The expression of miR-143/145 is drastically reduced in several models of vascular disease: carotid artery ligation injury in mouse, carotid balloon-injury in rat, and ApoE Knock-out mice [27]. In miR-143 or miR-145 KO mice, abnormal vascular tone and reduced contractile activity have been detected but VSMCs are functional [29]. Moreover, miR-143/145 levels are decreased in aortas from patients with aortic aneurism and lower circulating levels are detected in the serum of patients with coronary artery disease [20-22]. Overexpression of miR-145 induced lower neointima formation in balloon-injuried arteries [27].

Dimmeler and colleagues showed recently an example of vesicle-mediated miRNA transfer between human vascular endothelial cells and human aortic SMCs [30]. Blood vessels exposed to laminar blood flow undergo high shear stress. It is known that under shear stress conditions vascular endothelial cells overexpress the transcription factor Krüppel-like factor 2 (KLF2) which in turn induces up-regulation of miR-143 and miR-145. MiR-143/145 are transported in extracellular vesicles such as exosomes and they reduced the expression of miR-143 and miR-145 specific targets in co-cultured VSMCs. Additionally, the authors showed that extracellular vesicles derived from KLF2-expressing endothelial cells decrease atherosclerotic lesion formation in the ApoE KO mice kept on a high fat diet [30].

3.2. MiR-21, miR-221, and miR-222

In contrast to miR-143/145, miR-21, miR-221, and miR-222 are up-regulated in neointimal lesions.

TGF-β and its related family member BMP4 promote contractile gene expression and VSMC differentiation. Interestingly, they induce the transcription of the miR-143/145 cluster and promote also an increased expression of miR-21 post-transcriptionally. The critical target of miR-21 that is down-regulated in this process is programmed cell death 4 [22]. As a consequence, miR-21 induces VSMCs transdifferentiation to the contractile phenotype in response to BMP4 and TGFβ [31]. Additionally, miR-21 promotes VSMC proliferation and reduces apoptosis [32]. These miR-21 actions were confirmed in balloon-injured rat carotid arteries [33]. Moreover, knock-down of mir-21 using antisense oligonucleotides (antogomiRs) in the rat decreases vascular remodeling following balloon injury in carotid arteries [32]. Although an increase in differentiation is usually coupled to a decrease in proliferation, this is not necessarily the case in VSMCs. MiR-21 indeed targets a diverse set of genes and mediates differential biological outcomes depending on the cellular context [22].

Mir-221 and *miR-222* genes are clustered on the X chromosome and share a common seed sequence. Some reports indicate that they are transcribed from a common promoter [34].

Mir-221 and miR-222 contribute to VSMC dedifferentiation from the differentiated / contractile to the undifferentiated / synthetic phenotype and thus to increased cellular proliferation. Indeed, miR-221 and miR-222 are strongly elevated *in vivo* in VSMC following balloon injury of the vessel. Knock-down of mir-221 and miR-222 in the vessel reduced VSMC proliferation and neointimal lesion formation after angioplasty. Mir-221 and miR-222 are important for PDGF-cell mediated proliferation, by repressing tyrosine kinase c-kit, $p57^{Kip2}$ and the cyclin-dependent kinase inhibitor $p27^{Kip1}$. Interestingly, inhibition of c-kit reduced the expression of myocardin [32,35]. Overexpression of miR-221 induces an important decrease of myocardin expression, even if this miRNA does not target myocardin directly. Instead, it is the down regulation of c-Kit that is responsible for the up-regulation of myocardin. MiR-221 overexpression also increases during VSMC migration but the targets are still unknown. All these findings provide an example of the potential of one miRNA to mediate various cellular outcomes by regulating multiple targets [22,32].

4. MicroRNAs are implicated in the pathophysiology of cardiac muscle cells

Cardiovascular pathologies represent the prevalent causes of human morbidity and mortality in the Western hemisphere. As a consequence, a vast number of research groups consider that studying heart molecular and cellular characteristics is a major step in order to develop novel diagnostic and therapeutic strategies and to counteract cardiovascular diseases. It is now clear that miRNAs are an important part of the complex transcriptional and posttranscriptional regulatory circuit essential for the homeostasis of the cardiac tissue. They are powerful modulators in virtually all aspects of cardiac biology, from cardiac development to cardiomyocyte survival and hypertrophy, which we will now describe in more detail in this subchapter.

In the recent literature, more than a hundred microRNAs have been described as stably expressed in the cardiac tissue [36-38]. However, the vast majority (90%) of these miRNAs are represented by no more than 18 miRNAs in the mature murine organ. Even more remarkable is the fact that all these 18 miRNAs show an altered expression in pathological conditions, including coronary artery diseases and cardiomyopathies. Interestingly, it has been shown that a strong characteristic of these various models of cardiovascular disorders is the re-expression of a fetal cardiac miRNA program. This miRNA expression will finally trigger the over-expression of several fetal proteins, such as the atrial and brain natriuretic factor genes and the fetal isoform of the β-Myosin Heavy Chain gene (βMHC). Exploring further how miRNAs regulate gene expression in the heart will thus provide us with unique mechanistic insights into cardiac diseases. We will now describe in further details the miRNAs which have been the most implicated in the process. See Figure 3B for a pictorial representation of their roles in cardiac muscle.

4.1. Anti-hypertrophic miRNAs

Muscle miRNAs, such as miR-1 and miR-133, are integrated into myogenic regulatory networks: their expression is under the transcriptional and posttranscriptional control of myogenic factors, and they in turn have widespread control of the muscle gene expression program. Recent studies demonstrated that both miR-1 and miR-133 are significantly downregulated in hypertrophic and failing hearts. They play major roles in the development of cardiac hypertrophy, and have thus been nicknamed anti-hypertrophic miRNAs. In addition, miR-1 and the related miRNA miR-133 arise from a common precursor RNA which is regulated by the transcription factors Serum Response Factor (SRF) and Myocyte Enhancer Factor 2 (MEF2, [37]), which clearly suggest their importance in an intricate cardiac regulatory network.

The mature miR-1 transcript is the product of two genes, miR-1-1 and miR1-2, and it is now proven that its elevation induces arrhythmia in cardiac disease states. Its expression is specific for both cardiac and skeletal muscle. Overexpression of mature miR-1 in rat exacerbates cardiac arrhythmia whereas its knock-down by an antagomiR in the same animal, in the infarcted heart relieves arrhythmogenesis [39]. MiR-1 is also overexpressed in individuals with coronary artery disease. Part of miR-1 action is mediated by down regulation of connexin 43 and the inward rectifier K channel (Kir2.1, [39]). Another important role of miR-1 is to modulate cardiac excitation–contraction coupling by selectively increasing phosphorylation of the L-type and RyR2 channels *via* disrupting localization of PP2A activity to these channels [40]. Determining plasma levels of miR-1, using qPCR techniques (see subchapters II and VI) can be used as a sensitive biomarker for myocardial infarction and its expression is strongly down-regulated in hearts from patients afflicted with myocardial infraction compared to healthy adult hearts [41].

There are three known *mir-133* genes: *mir-133a-1, mir-133a-2* and *mir-133b* found on chromosomes 18, 20 and 6 respectively. In the human genome, all three genes encode miRNA with identical mature sequence [42]. Actually, miR-133a-1 and miR-133a-2 are each expressed bicistronically with miR-1-1 and miR1-2. Knockdown experiments of miR-133 gave at first puzzling results: *in vitro* overexpression of miR-133 or miR-1 inhibited cardiac hypertrophy while infusing mice with antagomiRs against the mature miR-133 sequence induced cardiac hypertrophy [43]. On the other hand, genetic models with knockout of either *mir-133a* gene did not display significant cardiac pathologies, or actually any phenotype. However, deleting both *mir-133a* genes resulted in a drastic phenotype: ectopic expression of cardiac-specific markers genes in VSMCs, embryonic lethality, and aberrant proliferation of cardiac muscle cells [37]. The phenotypic difference between mice treated by antagomiRs at the adult stage, and genetic models which have been deprived from miRNAs from conception clearly shows the limits of both models. The sum of these studies however clearly emphasizes the role of the miR-133 mature sequence in cardiac muscle biology. On a mechanistic side, Horie et al. [44] have shown a direct role of miR-133 in cardiomyocyte glucose transport: overexpression of the miRNA decreased levels of the glucose transporter GLUT4 and reduced insulin-induced glucose uptake. Additionally, this increase of miR-133 reduced Krüppel-like transcription factor 15 (KLF15) expression, which induces GLUT4 expression.

4.2. Pro-hypertrophic miRNAs

The role of mir-21 in cardiac modeling and pathophysiology is clearly controversed. MiR-21 inhibition by antagomir strategies was first reported as causing an alleviation of murine cardiac hypertrophy [38,45]. A divergence arose when the first team attributed this pro-hypertrophic effect to an effect on cardiomyocytes [38] whereas the second team claimed the primary site of miR-21 action was actually cardiac fibroblasts [45]. In contrast to both team results, Cheng et al. [46] reported that miR-21 was indeed increased by fourfold in hypertrophic mouse hearts, but that modulating miR-21 *via* antisense depletion had a significant negative effect on cardiomyocyte hypertrophy. Finally, Patrick et al. [47] found that a genetic deletion of the *mir-21* gene results in mice with a normal phenotype that did not respond differently to normal littermates when exposed to cardiac stress conditions. Also, in the same study, LNA-modified antagomiRs specific for miR-21 did not block a remodeling response of the murine heart to stress conditions. The authors concluded that, although miR-21 is highly up-regulated during cardiac remodeling, it is not essential for cardiac hypertrophy, a disease state associated with fibrosis in response to heart injury. Nonetheless, miR-21 is a miRNA of interest in the cardiac field, at least as an innovative biomarker, since it is almost undetectable in the healthy heart, but is strongly over-expressed in cardiac pathologies.

The human miR-29 family of microRNAs is encoded by two gene clusters. As a consequence, three matures members exist: miR-29a, miR-29b, and miR-29c. In this instance, the miR-29 family has been shown to be expressed in both cardiac fibroblasts and cardiomyocytes [48]. In these cell types, sixteen of their targets are extracellular matrix genes. This clearly shows a striking example of a single microRNA mature sequence which is capable to target a large group of functionally related genes. As a consequence, miR-29 expression induces strong antifibrotic effects in heart and other tissues. MiR-29s have also been shown to be pro-apoptotic and involved in cell differentiation. Acute myocardial infarction due to coronary artery occlusion also results in a decrease of the expression of the miR-29 family in the region of the fibrotic scar [48]. Using up- and down-regulation of miR-29, the same authors showed that this miRNA regulates the expression of collagens, and as a result the fibrotic response. Finally, the miR-29 family has also been shown to down-regulate elastin and other extracellular matrix (ECM) genes implicated in elastogenesis [49]. Jones et al. [50] have examined miRNA expression using qPCR in aortic tissue collected from patients with ascending thoracic aortic aneurysm and shown that miR-29a expression is correlated with cardiac tissue proteolytic degradation and aortic size. These last results show the interest of determining specific miRNA levels in human diagnostic.

Another miRNA of interest in the cardiac field of investigation is miR-208a, which is expressed strictly in the heart. Mir-208a overexpression in transgenic mice induces hypertrophic growth of the cardiac muscle and induces arrhythmias. This hypertrophic growth is concomitant with fibrosis and a decrease of contractility, which results from down-regulation of the faster isoform, α-myosin heavy chain (α-MHC) and up-regulation of the fetal specific, slower isoform, β-MHC [51-52]. Thus, cardiac-specific overexpression of miR-208a induces cardiac remodeling and regulates the expression of hypertrophic proteins, including β-MHC. Conversely, the same authors showed that genetic deletion of miR-208a in mouse induces a decrease of β-MHC. Additionally, miR-208 targets other proteins, such as thyroid hormone-associated protein 1 and myostatin 2, which are both inhibitors of muscle growth and hypertrophy, with for final consequence hypertrophic cardiac growth. MiR-208a is also strongly up-regulated in the diseased human heart as detected in biopsies from patients afflicted with myocardial infarction [41]. Also, miR-208a is not detected in plasma from healthy patients, but is raised to a detectable level as soon as 1 h after coronary artery occlusion [53]. This result is important, since it clearly shows that, at least for this miRNA, plasma levels reflect tissue amounts, and thus that miRNA are strong candidates as non-invasive biomarkers (see subchapter VI for more information on this topic).

5. The miRNA regulators: Who watches the watchmen?

Expressional patterns of miRNAs vary according to organs and their developmental stages. Indeed, several transcriptional and post-transcriptional processes control the levels of mature miRNAs. The study of these regulatory mechanisms is still in its very early steps. We will review here what is known, starting with other non-coding RNAs which are as long as miRNAs are short, we will proceed with transcription factors, and finally explore how cells are able to communicate with each other, using microRNAs.

In the last few years, it has been shown that long non-coding RNAs act as miRNA sponges and/or competing endogenous RNAs, are able to regulate miRNA function by binding complementarily with them, thus competing with their dedicated mRNAs targets, and thereby to impose an additional level of post-transcriptional regulation. Little is known about the mechanisms of action of these exciting new regulatory RNAs in muscle cells. A groundbreaking result came from Cesana et al [54]: they have identified a long non coding RNA, called linc-MD1, in the skeletal muscle. linc-MD1 is stably expressed in mouse and human myoblasts, and controls the myogenesis program by binding, and thus "sponging" two instrumental miRNAs, miR-133 and miR-135, which in turn regulate transcription factors that activate muscle-specific gene expression. By "inhibiting the inhibitors", linc-MD1 accelerates myogenesis. Interestingly, this RNA's expression is strongly reduced in Duchenne muscular dystrophy, a genetic disorder which is characterized by a drastic reduction of myoblasts. No equivalent of linc-MD1 has yet been described in smooth muscle biogenesis or in cardiogenesis but one can strongly guess that they will be identified in the forseeable future.

Several transcription factors have been characterized as miRNA regulators in smooth, cardiac and skeletal muscle cells, thus revealing novel mechanisms underlying VSMC differentiation. Myocardin is the best characterized in VSMCs and cardiac cells [55]. Myocardin, with its co-activator Serum Response Factor (SRF), is a cardiac- and muscle specific trans-acting protein, and a master regulator of the smooth muscle phenotype. It has been shown that this transcription factor regulates several miRNAs in VSMCs. It induces miR-1 expression, which in turn inhibits VSMC proliferation, and increases their differentiation, by targeting Pim-1, a serine/threonine kinase [56]. Similarly, myocardin represses versican, a chondroitin sulfate proteoglycan of the extra-cellular matrix that is produced by synthetic VSMCs and promotes VSMC migration and proliferation, by inducing the expression of miR-143, a miR instrumental in VSMC differentiation [57]. SRF has been shown to regulate the expression of several mIRs, including miR-1, miR-133a and miR-21 [58]. Interestingly, myocardin co-activator SRF regulates microRNA biogenesis, specifically the transcription of pri-microRNA, thereby affecting the mature microRNA level, by binding to the proximal promoter region of miR genes. Transforming Growth Factor-β1 (TGFβ1), another known stimulus capable of inducing VSMC differentiation, has also been shown to induce both miR-143 and miR-145 in human coronary artery SMCs [59]. We have already discussed the importance of transcription factors SRF and MEF4 in the regulation of miR-1 and miR-133 (see subchapter 5). Finally, although skeletal muscle is outside of the topic of this chapter, it is interesting to note that another important muscle-specific transcription factor, MyoD, impacts miR-1 and miR-206 expressions, with strong consequences on myoblast apoptosis levels [60]. Although it has not been shown yet, one can speculate that similar systems exist in smooth muscle and cardiac muscle cells.

Decay mechanisms affecting miRNAs in order to regulate their expression are not well understood. However, new notions have recently been put to the forefront: it seems that changes in cellular density and cell adhesion mechanisms affect rapidly miRNAs expression [61]. When cells are grown at low density or after cell splitting, some miRNAs are rapidly degraded while others remain unaffected. This rapid, and yet unexplained, degradation of persistent regulatory molecules such as miRNAs may facilitate cellular plasticity and remodeling in response to various stresses.

Wang et al. [62] have recently shown that several human cell lines from various origins (glioblastoma, hepatocytes, lung bronchial epithelium, pulmonary fibroblasts, alveolar basal epithelial cells) actively release miRNAs in a short time period of time (approx. 1 h) after serum deprivation. Thus, one can hypothesize that, at least some, exported miRNAs are used for cell-to-cell communication. More studies will of course be needed to determine exactly how miRNAs are specifically targeted to relevant target cells, and what information is transduced. It will also be important to determine why evolution has selected several different means of transportation for miRNAs: protein complexes, exosomes, Microvesicles, High Density Lipoprotein (HDL) or apoptotic bodies (Figure 4). For example, miRNA complexed with proteins, could be targeted to specific cell surface receptors, and miRNAs inside vesicles to others targets. All these recent results ask important questions about cell-to-cell communication mediated by miRNAs, and raise the possibility that a yet undiscovered biological information transduction system exists, and could be important to explain many biological processes including development, differentiation, and stress response. In cardiovascular diseases, for example, the general decrease in circulating miRNAs detected in patients with CAD might be caused by a disregulation of this miRNA trafficking system in atherosclerotic lesions or in the infarcted myocardium.

6. MiRNAs: New biomarkers in vascular and cardiac diseases

Being instrumental players in the fine-tuning of gene regulation networks, microRNAs have significant diagnostic and prognostic value, as biomarkers of disease etiology and progression. Until recently, however, miRNA quantitation and usefulness as a biomarker was dependent on the availability of the pathological tissue. This was not a major setback for the diagnosis of cancer, where biopsies were readily available in most cases, but proved to be serious concern when dealing with heart or vascular pathologies. Very recently, these concerns were alleviated when several teams showed that miRNAs can be detected and precisely measured in human blood (eg [63,64]). These papers, showing a stable presence of miRNA in human plasma, came as a surprise. Indeed, any researcher having experience with RNA work considers ribonucleic acids as fragile and unable to survive in a liquid like serum which contains a wealth of specific and non-specific degrading enzymes.

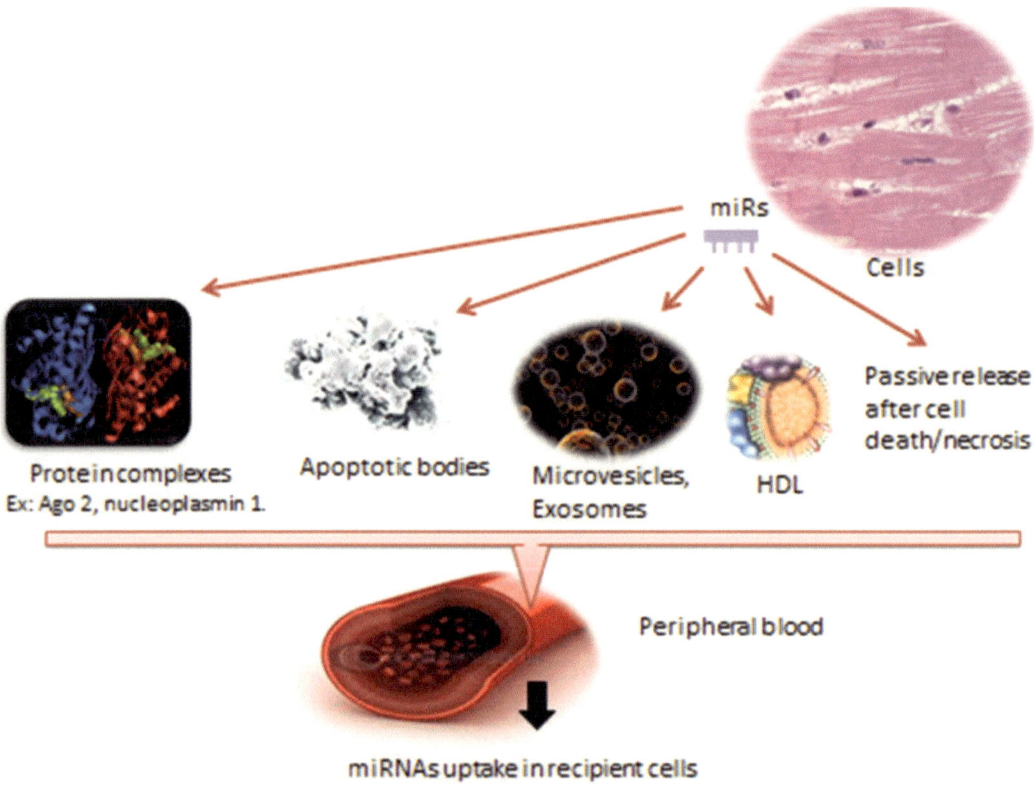

Figure 4. The various mechanisms of miRNA release from cells in the peripheral blood circulation and their uptake in recipient cells. Ago2 Argonaute 2; HDL High density Lipoprotein.

Stephanie Dimmeler's pioneer studies show that miRNA can be detected in the serum of patients with coronary artery diseases (CAD) and that their levels are altered in patient's serum when compared to healthy counterparts [65] [65-66]. MiRNAs are thus prime candidates as novel non-invasive biomarkers in cardiovascular diseases, which can be measured in routine clinical diagnosis. The main question here is which endogenous referent genes to use as this is instrumental in qPCR. In cardiovascular disease studies, various endogenous circulating miRNAs (eg miR-17-5p, miR-454, U6 or RNU6b) have been used for normalization of circulating miRNAs, but the use of spiked-in miRNAs, *i.e.* adding a known amount of exogenous non-human miRNA, (eg synthetic *Caenhorabditis elegans* miR-

39) is now increasingly common as the experimenter knows clearly the amount of the referent miRNA, and no further experimental bias is added [66].

An important question is the exact localization of miRNAs in the bloodstream. They exist in a highly stable, extracellular form and are remarkably persistent in the RNase-rich environment of blood. The first model postulates that circulating miRNAs are protected by encapsulation in membrane-bound vesicles such as exosomes, phagosomes, apoptotic bodies…[65-66] Arroyo et al. [67] have however recently challenged this view: they used a combination of differential centrifugation and size-exclusion chromatography and showed that circulating miRNAs cofractionate mostly with protein complexes rather than with vesicles, in human plasma and serum. Even more surprising was the fact that the main miRNA binding partner was Ago2, the key effector protein of miRNA-mediated silencing, which is considered as a cytosolic protein. Ago2 seems to be one of the factors protecting circulating miRNAs from plasma RNases, since purified miRNAs, devoid of protein partners, were sensitive to RNase treatment. Figure 4 summarizes the different hypotheses that have been put forward to explain how miRNAs can be secreted and exported in human blood.

Goren et al. [68] have very recently published that four miRNAs, miR-22, miR-92b, miR-320a and miR-423-5p were significantly increased in the serum of patients with heart failure. By relying on a signature derived from the expression of these miRNAs, the authors were able to discriminate between systolic heart failure patients and healthy controls with a sensitivity and specificity of 90%. Moreover, there was a significant correlation with important clinical prognostic parameters such as an elevated serum natriuretic peptide and a wide QRS. Other recent papers have highlighted the interest of determining their levels in serum and other body fluids, including urine, feces and saliva [69-70]. This forebodes well for the increasing usefulness of specific miRNA expression as non-invasive biomarkers.

7. Potential of miRNAs as innovative drug targets

In the last decade, the increasing interest in small RNAs has triggered the arrival of innovative drug targets on the pharmaceutical market. Among these small RNAs, miRNAs provide perhaps the most promising new opportunities for developing new compounds, especially with the recent advances in anti-miRNA chemistry. On one hand, therapeutic nucleic acids can be administered using lentivirus-mediated antagomir expression, which induces a stable knockdown phenotype for a specific miRNA [71-72]. On the other hand, the vast majority of anti-miRs used in trials are in fact altered locked nucleic acids (LNA), also known as inaccessible RNA (reviewed in [73]). The canonic nucleic acid ribose sugar backbone is modified with an extra bridge connecting the 2' oxygen and 4' carbon [74]. This conformation enhances base stacking and backbone pre-organization, which will significantly enhance the hybridization properties for the compounds. These poly-anionic molecules tend to distribute broadly but also to accumulate in liver, kidney and phagocytes. They are highly hydrophilic, with a molecular weight ranging from 2 to 6 kD. For the moment, routes of administration used are essentially intravenous and subcutaneous injections [75-76]. One has however got to keep in mind that developing innovative drugs is risky, represents a tremendous cost in resources, time, etc., and should not be undertaken lightly.

A first clinical trial, in this post-genome era, is under way in human patients affected by viral hepatitis C. Phase IIa results of this promising trial, focusing on the liver-specific miR-122 [77], aims to develop the related drug called Miravirsen, developed by Santaris Pharma A/S as a LNA antagomir, and thus to antagonize miR-122, which is instrumental for Hepatitis C virus C (HCV) infection. Langford et al. [78] had already shown in primates that a LNA-specific for miR-122 was able to suppress HCV viremia, with no evidence of viral resistance or side effects in the treated animals. These promising results are confirmed in humans and show that using an antagomir approach induces a decrease of the patient's viral load, and that this revolutionary treatment is less toxic and more effective than current medicine (http://www.santaris.com/news/2011/11/05/santaris-pharma-phase-2a-data-miravirsen-shows-dose-dependent-prolonged-viral-reduct), perhaps due to the specificity brought by RNA strand complementarity.

Concerning cardiovascular diseases, we will now focus on the expanding interest of miRNAs in cardiovascular molecular medicine, and the various studies that have been undertaken, on animal models, for the time being.

The important role of the miR-29 family of microRNAs has already been evoked in this chapter (see subchapter V). A promising study dealt with their effects in two murine models of abdominal aortic aneurysm (AAA) (porcine pancreatic elastase [PPE] infusion model in C57BL/6 mice and the AngII infusion model in ApoE-/- mice). Antagomirs against miR-29b was administered *in vivo* under the form of LNA. This resulted in an increase of collagen expression [79], which resulted in an early fibrotic response in the aortic wall and an actual reduction of AAA progression in both models. Conversely, overexpression of miR-29b using lentiviral vectors resulted in an aggravation of AAA, and a premature rupture of the aortic wall. This miRNA is thus a promising target for creating an innovative treatment for AAA.

Matkovich et al. [80] have shown that over-expression of miR-133a in the heart of transgenic mice prevented TAC-associated miR-133a downregulation and improved myocardial fibrosis and diastolic function. In another, more exotic, model, Yin et al. [81] have shown that miR-133 restricts injury-induced cardiomyocyte proliferation.

Very recently, miR-33, although not specific for cardiac or vascular tissues, has gained a lot of attention in atherosclerosis treatment [82]. Both miR-33a and miR-33b target the adenosine triphosphate-binding cassette transporter A1 (ABCA1), an important regulator of high-density lipoprotein (HDL) synthesis and reverse cholesterol transport in a murine model. Inhibiting miR-33 using two different methods (overexpression of dedicated lentiviral particles or injection of LNA antagomiRs) led to an up-regulation of ABCA1 and importantly to an increase of cholesterol influx and concomitant increase in the levels of HDL, and thus of atheroprotective effects. These authors show thus clearly that increasing HDL levels in Mouse via miR-33-specific antagomiRs promotes reverse cholesterol transport and suggest that it may be a promising strategy to induce atherosclerosis regression. Several months later, the same team published similar results in primates, more precisely African green monkeys [83]. In addition to the beneficial effects already detected in Mouse, the authors showed a strong decrease of plasma levels of very-low-density lipoprotein (VLDL)-associated triglycerides. This difference can tentatively be attributed to the presence of miR-33b in the SREBF1 gene of medium and large mammals and its absence in rodents. Pharmacological use of antagomiRs specific for miR-33a and miR-33b is thus able to markedly raise plasma HDL and lower VLDL triglyceride levels and therefore a promising therapeutic strategy to treat dyslipidaemias, and their induced cardiac consequences in human patients.

8. In conclusion: MicroRNAs, a bright future?

In the last decade, many advances have been made to decipher miRNA roles in cardiovascular development and pathogenesis. New methods have been developed in order to use them as innovative biomarkers in diagnostics, and as groundbreaking drugs in pharmacological treatments. A first, promising, clinical trial in humans is in progress right now. However, many questions remain still to be answered. Each miRNA targets up to one hundred mRNA targets, which poses significant challenges to the identification, and specific targeting, of the mRNAs that are relevant to a particular pathological process. On the other hand, this problem could also become a solution, since it is now clear that a particular family of miRNAs is associated with the same disease type [84]. So it could prove more efficient to target a predefined network of related miRNAs rather than a single one [85].

With the current pace of evolution in understanding the basic ways of miRNA action in cardiovascular development and disease, one can safely trust that these small molecules will still amaze us with more revelations in the near and not so near future.

Author details

Valérie Metzinger-Le Meuth, Eléonore M'Baya-Moutoula,
Fatiha Taibi, Ziad Massy and Laurent Metzinger
INSERM U-1088, Amiens, France

Valérie Metzinger-Le Meuth, Eléonore M'Baya-Moutoula,
Fatiha Taibi, Ziad Massy and Laurent Metzinger
Faculty of Pharmacy and Medicine, University of Picardie Jules Verne, Amiens, France

Valérie Metzinger-Le Meuth
Université Paris 13, UFR SMBH, Bobigny, France

Ziad Massy
Division(s) of Pharmacology and Nephrology, Amiens University Hospital, Amiens, France

Laurent Metzinger*
Biochemistry Laboratory, Amiens University Hospital, Amiens, France

---

* Corresponding Author

Acknowledgement

This work was funded by grants from the Picardie Regional Council (MARNO-MPCC and *Modulation des calcifications cardiovasculaires*), including a PhD fellowship for FT and a post-doctoral fellowship for EMM.

9. References

[1] Bartel DP. MicroRNAs: target recognition and regulatory functions. Cell 2009;136(2) 215-233.

[2] Gommans WM, Berezikov E. Controlling miRNA regulation in disease. Methods Mol Biol 2012;822 1-18.

[3] Rangrez AY, Massy ZA, Metzinger-Le Meuth V, Metzinger, L. miR-143 and miR-145: Molecular keys to switch the phenotype of vascular smooth muscle cells, Circulation Cardiovascular Genetics 2011;4(2) 197-205.

[4] Small EM, Frost RJ, Olson EN. MicroRNAs add a new dimension to cardiovascular disease. Circulation 2010;121(8) 1022-1032.

[5] Thum T, Galuppo P, Wolf C, Fiedler J, Kneitz S, van Laake LW, Doevendans PA, Mummery CL, Borlak J, Haverich A, Gross C, Engelhardt S, Ertl G, Bauersachs J. MicroRNAs in the human heart: a clue to fetal gene reprogramming in heart failure. Circulation 2007;116(3) 258-267.

[6] Kim VN. MicroRNA biogenesis: coordinated cropping and dicing. Nature Reviews Molecular and Cellular Biology 2005;6(5) 376-385.

[7] Starega-Roslan J, Koscianska E, Kozlowski P, Krzyzosiak WJ. The role of the precursor structure in the biogenesis of microRNA. Cellular and Molecular Life Sciences 2011;68(17) 2859-2871.

[8] Guo H, Ingolia NT, Weissman JS, Bartel DP. Mammalian microRNAs predominantly act to decrease target mRNA levels. Nature 2010;466(7308) 835-40.

[9] Forman JJ, Coller HA. The code within the code: microRNAs target coding regions; Cell Cycle 2010;9(8) 1533-1541.

[10] Lee I, Ajay SS, Yook JI, Kim HS, Hong SH, Kim NH, Dhanasekaran SM, Chinnaiyan AM, Athey BD. New class of microRNA targets containing simultaneous 5'-UTR and 3'-UTR interaction sites, Genome Research 2009;19(7) 1175-1183.

[11] Shin C, Nam JW, Farh KK, Chiang HR, Shkumatava A, Bartel, DP. Expanding the microRNA targeting code: functional sites with centered pairing. Molecular Cell 2010;38(6) 789-802.

[12] Griffiths-Jones S. miRBase: microRNA sequences and annotation. Current Protocols in Bioinformatics 2010 Chapter 12 Unit 12(9) 1-10..

[13] Berezikov E. Evolution of microRNA diversity and regulation in animals. Nature Reviews in Genetics 2011;12(12) 846-860.

[14] Bernardo BC, Charchar FJ, Lin RC, McMullen JR. A microRNA guide for clinicians and basic scientists: background and experimental techniques. Heart Lung Circulation 2012;21(3) 131-42.

[15] Chen Y, Gelfond JA, McManus LM, Shireman PK. Reproducibility of quantitative RT-PCR array in miRNA expression profiling and comparison with microarray analysis. BMC Genomics 2009;10 407.

[16] Xie J, Ameres SL, Friedline R, Hung JH, Zhang Y, Xie Q, Zhong L, Su Q, He R, Li M, Li H, Mu X, Zhang H, Broderick JA, Kim JK, Weng Z, Flotte TR, Zamore PD, Gao G. Long-term, efficient inhibition of microRNA function in mice using rAAV vectors. Nature Methods 2012;9(4) 403-409.

[17] Albinsson S, Skoura A, Yu J, DiLorenzo A, Fernandez-Hernando C, Offermanns S, Miano JM, Sessa WC. Smooth muscle miRNAs are critical for post-natal regulation of blood pressure and vascular function, PLoS One 2011;6(4) e18869.

[18] Boettger T, Beetz N, Kostin S, Schneider J, Kruger M, Hein L, Braun T. Acquisition of the contractile phenotype by murine arterial smooth muscle cells depends on the Mir143/145 gene cluster. Journal of Clinical Investigation 2009;119(9) 2634-2647.

[19] Cordes KR, Sheehy NT, White MP, Berry EC, Morton SU, Muth AN, Lee TH, Miano JM, Ivey KN, Srivastava D. miR-145 and miR-143 regulate smooth muscle cell fate and plasticity. Nature 2009;460(7256) 705-710.

[20] Elia L, Quintavalle M, Zhang J, Contu R, Cossu L, Latronico MV, Peterson KL, Indolfi C, Catalucci D, Chen J, Courtneidge SA, Condorelli G. The knockout of miR-143 and -145 alters smooth muscle cell maintenance and vascular homeostasis in mice: correlates with human disease. Cell Death Differentiation 2009;16(12) 1590-1598.

[21] Vo NK, Dalton RP, Liu N, Olson EN, Goodman RH. Affinity purification of microRNA-133a with the cardiac transcription factor, Hand2. Proceedings of the National Academy of Sciences U S A 2010;107(45) 19231-19236.

[22] Davis-Dusenbery BN, Chan MC, Reno KE, Weisman AS, Layne MD, Lagna G, Hata. A. Down-regulation of Kruppel-like factor-4 (KLF4) by microRNA-143/145 is critical for modulation of vascular smooth muscle cell phenotype by transforming growth factor-beta and bone morphogenetic protein 4. Journal of Biological Chemistry 2011;286(32) 28097-28110.

[23] Torella D, Iaconetti C, Catalucci D, Ellison GM, Leone A, Waring CD, Bochicchio A, Vicinanza C, Aquila I, Curcio A, Condorelli G, Indolfi C. MicroRNA-133 controls vascular smooth muscle cell phenotypic switch in vitro and vascular remodeling in vivo. Circulation Research 2011;109(8) 880-893.

[24] Miano JM, Small EM. MicroRNA133a: a new variable in vascular smooth muscle cell phenotypic switching. Circulation Research 2011;109(8) 825-827.

[25] Harfe BD, McManus MT, Mansfield JH, Hornstein E, Tabin CJ. The RNaseIII enzyme Dicer is required for morphogenesis but not patterning of the vertebrate limb. Proceedings of the National Academy of Sciences U S A 2005;102(31) 10898-10903.

[26] Kanellopoulou C, Muljo SA, Kung AL, Ganesan S, Drapkin R, Jenuwein T, Livingston DM, Rajewsky K. Dicer-deficient mouse embryonic stem cells are defective in differentiation and centromeric silencing. Genes Development 2005;19(4) 489-501.

[27] Cheng B, Liu HW, Fu XB, Sun TZ, Sheng ZY. Recombinant human platelet-derived growth factor enhanced dermal wound healing by a pathway involving ERK and c-fos in diabetic rats. Journal of Dermatological Science 2007;45(3) 193-201.

[28] Boucher JM, Peterson SM, Urs S, Zhang C, Liaw L. The miR-143/145 cluster is a novel transcriptional target of Jagged-1/Notch signaling in vascular smooth muscle cells. Journal of Biological Chemistry 2011;286(32) 28312-28321.

[29] Xin M, Small EM, Sutherland LB, Qi X, McAnally J, Plato CF, Richardson JA, Bassel-Duby R, Olson E.N. MicroRNAs miR-143 and miR-145 modulate cytoskeletal dynamics and responsiveness of smooth muscle cells to injury. Genes Development 2009;23(18) 2166-2178.

[30] Hergenreider E, Heydt S, Treguer K, Boettger T, Horrevoets AJ, Zeiher AM, Scheffer MP, Frangakis AS, Yin X, Mayr M, Braun T, Urbich C, Boon RA, Dimmeler S. Atheroprotective communication between endothelial cells and smooth muscle cells through miRNAs. Nature Cell Biology 2012;14(3) 249-256.

[31] Davis BN, Hilyard AC, Nguyen PH, Lagna G, Hata A. Induction of microRNA-221 by platelet-derived growth factor signaling is critical for modulation of vascular smooth muscle phenotype. Journal of Biological Chemistry, 2009;284(6) 3728-3738.

[32] Ji R, Cheng Y, Yue J, Yang J, Liu X, Chen H, Dean DB, Zhang C. MicroRNA expression signature and antisense-mediated depletion reveal an essential role of MicroRNA in vascular neointimal lesion formation. Circulation Research 2007;100(11) 1579-1588.

[33] Zampetaki A, Mayr M. MicroRNAs in vascular and metabolic disease. Circulation Research 2012;110(3) 508-522..

[34] Le Sage C, Nagel R, Egan DA, Schrier M, Mesman E, Mangiola A, Anile C, Maira G, Mercatelli N, Ciafre SA, Farace MG, Agami R. Regulation of the p27(Kip1) tumor suppressor by miR-221 and miR-222 promotes cancer cell proliferation, Embo Journal 2007;26(15) 3699-3708.

[35] Liu X, Cheng Y, Zhang S, Lin Y, Yang J, Zhang C. A necessary role of miR-221 and miR-222 in vascular smooth muscle cell proliferation and neointimal hyperplasia. Circulation Research 2009;104(4) 476-487.

[36] Bauersachs J, Thum, T. Biogenesis and regulation of cardiovascular microRNAs. Circulation Research 2011;109(3) 334-347.

[37] Liu N, Bezprozvannaya S, Williams AH, Qi X, Richardson JA, Bassel-Duby R, Olson EN. microRNA-133a regulates cardiomyocyte proliferation and suppresses smooth muscle gene expression in the heart, Genes Development 2008;22(23) 3242-3254.

[38] Tatsuguchi M, Seok HY, Callis TE, Thomson JM, Chen JF, Newman M, Rojas M, Hammond SM, Wang DZ. Expression of microRNAs is dynamically regulated during cardiomyocyte hypertrophy. Journal of Molecular and Cellular Cardiology 2007;42(6) 1137-1141.

[39] Yang B, Lin H, Xiao J, Lu Y, Luo X, Li B, Zhang Y, Xu C, Bai Y, Wang H, Chen G, Wang Z. The muscle-specific microRNA miR-1 regulates cardiac arrhythmogenic potential by targeting GJA1 and KCNJ2. Nature Medicine 2007;13(4) 486-491.

[40] Terentyev D, Belevych AE, Terentyeva R, Martin MM, Malana GE, Kuhn DE, Abdellatif M, Feldman DS, Elton TS, Gyorke S. miR-1 overexpression enhances Ca(2+) release and promotes cardiac arrhythmogenesis by targeting PP2A regulatory subunit B56alpha and causing CaMKII-dependent hyperphosphorylation of RyR2, Circulation Research 2009;104(4) 514-521.

[41] Bostjancic E, Zidar N, Stajer D, Glavac D. MicroRNAs miR-1, miR-133a, miR-133b and miR-208 are dysregulated in human myocardial infarction. Cardiology 2010;115(3) 163-169.

[42] Ivey KN, Muth A, Arnold J, King FW, Yeh RF, Fish JE, Hsiao EC, Schwartz RJ, Conklin BR, Bernstein HS, Srivastava D. MicroRNA regulation of cell lineages in mouse and human embryonic stem cells. Cell Stem Cell, 2008;2(3) 219-229.

[43] Carè A, Catalucci D, Felicetti F, Bonci D, Addario A, Gallo P, Bang ML, Segnalini P, Gu Y, Dalton ND, Elia L, Latronico MV, Hoydal M, Autore C, Russo MA, Dorn G.W. 2nd., Ellingsen O, Ruiz-Lozano P, Peterson KL, Croce CM, Peschle C, Condorelli G. MicroRNA-133 controls cardiac hypertrophy, Nature Medicine 2007;13(5) 613-618.

[44] Horie T, Ono K, Nishi H, Iwanaga Y, Nagao K, Kinoshita M, Kuwabara Y, Takanabe R, Hasegawa K, Kita T, Kimura T. MicroRNA-133 regulates the expression of GLUT4 by targeting KLF15 and is involved in metabolic control in cardiac myocytes. Biochemical and Biophysical Research Communications 2009;389(2) 315-320.

[45] Thum T, Gross C, Fiedler J, Fischer T, Kissler S, Bussen M, Galuppo P, Just S, Rottbauer W, Frantz S, Castoldi M, Soutschek J, Koteliansky V, Rosenwald A, Basson MA, Licht JD, Pena JT, Rouhanifard SH, Muckenthaler MU, Tuschl T, Martin GR, Bauersachs J, Engelhardt S. MicroRNA-21 contributes to myocardial disease by stimulating MAP kinase signalling in fibroblasts. Nature 2008;456(7224) 980-984.

[46] Cheng Y, Ji R, Yue J, Yang J, Liu X, Chen H, Dean DB, Zhang C. MicroRNAs are aberrantly expressed in hypertrophic heart: do they play a role in cardiac hypertrophy? American Journal of Pathology 2007;170(6) 1831-1840.

[47] Patrick DM, Montgomery RL, Qi X, Obad S, Kauppinen S, Hill JA, van Rooij E, Olson EN. Stress-dependent cardiac remodeling occurs in the absence of microRNA-21 in mice. Journal of Clinical Investigation 2010;120(11) 3912-3916.

[48] van Rooij E, Sutherland LB, Thatcher JE, DiMaio JM, Naseem RH, Marshall WS, Hill JA, Olson EN. Dysregulation of microRNAs after myocardial infarction reveals a role of miR-29 in cardiac fibrosis. Proceedings of the National Academy of Sciences U S A 2008;105(35) 13027-13032.

[49] Ott CE, Grunhagen J, Jager M, Horbelt D, Schwill S, Kallenbach K, Guo G, Manke T, Knaus P, Mundlos S, Robinson PN. MicroRNAs differentially expressed in postnatal aortic development downregulate elastin via 3' UTR and coding-sequence binding sites. PLoS One 2011;6(1) e16250.

[50] Jones JA, Stroud RE, O'Quinn EC, Black LE, Barth JL, Elefteriades JA, Bavaria JE, Gorman JH 3rd, Gorman RC, Spinale FG, Ikonomidis JS. Selective microRNA suppression in human thoracic aneurysms: relationship of miR-29a to aortic size and proteolytic induction. Circulation Cardiovascular Genetics 2011;4(6) 605-613.

[51] Callis TE, Pandya K, Seok HY, Tang RH, Tatsuguchi M, Huang ZP, Chen JF, Deng Z, Gunn B, Shumate J, Willis MS, Selzman CH, Wang DZ. MicroRNA-208a is a regulator of cardiac hypertrophy and conduction in mice. Journal of Clinical Investigation 2009;119(9) 2772-2786.

[52] van Rooij E, Sutherland LB, Qi X, Richardson JA, Hill J, Olson EN. Control of stress-dependent cardiac growth and gene expression by a microRNA. Science 2007;316(5824) 575-579.

[53] Wang K, Zhang S, Weber J, Baxter D, Galas DJ. Export of microRNAs and microRNA-protective protein by mammalian cells. Nucleic Acids Research 2010;38(20) 7248-7259.

[54] Cesana M, Cacchiarelli D, Legnini I, Santini T, Sthandier O, Chinappi M, Tramontano A, Bozzoni I. A long noncoding RNA controls muscle differentiation by functioning as a competing endogenous RNA. Cell 2011;147(4) 358-69.

[55] Wang Z, Wang DZ, Hockemeyer D, McAnally J, Nordheim A, Olson EN. Myocardin and ternary complex factors compete for SRF to control smooth muscle gene expression. Nature 2004;428 (6979) 185-189.

[56] Chen J, Yin H, Jiang Y, Radhakrishnan SK, Huang ZP, Li J, Shi Z, Kilsdonk EP, Gui Y, Wang DZ, Zheng XL. Induction of microRNA-1 by myocardin in smooth muscle cells inhibits cell proliferation, Arterioscleris Thrombosis Vascular Biology 2011;31(2) 368-75.

[57] Wang X, Hu G, Zhou J. Repression of versican expression by microRNA-143. Journal of Biological Chemistry 2010;285(30) 23241-23250.

[58] Zhang X, Azhar G, Helms SA, Wei JY. Regulation of cardiac microRNAs by serum response factor. Journal of Biomedical Science 2011;18(1) 15.

[59] Long X, Miano JM. Transforming growth factor-beta1 (TGF-beta1) utilizes distinct pathways for the transcriptional activation of microRNA 143/145 in human coronary artery smooth muscle cells. Journal of Biological Chemistry, 2011;286(34) 30119-30129.

[60] Hirai H, Verma M, Watanabe S, Tastad C, Asakura Y, Asakura A. MyoD regulates apoptosis of myoblasts through microRNA-mediated down-regulation of Pax3. Journal of Cell Biology 2010;191(2) 347-365.

[61] Kim YK, Yeo J, Ha M, Kim B, Kim VN. Cell adhesion-dependent control of microRNA decay. Molecular Cell 2011;43(6) 1005-1014.

[62] Wang GK, Zhu JQ, Zhang JT, Li Q, Li Y, He J, Qin YW, Jing Q. Circulating microRNA: a novel potential biomarker for early diagnosis of acute myocardial infarction in humans, European Heart Journal, 2010;31(6) 659-666.

[63] Lodes MJ, Caraballo M, Suciu D, Munro S, Kumar A, Anderson B. Detection of cancer with serum miRNAs on an oligonucleotide microarray. PLoS One 2009;4(7) e6229.

[64] Wang K, Zhang S, Marzolf B, Troisch P, Brightman A, Hu Z, Hood LE, Galas DJ. Circulating microRNAs, potential biomarkers for drug-induced liver injury, Proceedings of the National Academy of Sciences U S A 2009;106(11) 4402-4407.

[65] De Rosa S, Fichtlscherer S, Lehmann R, Assmus B, Dimmeler S, Zeiher AM. Transcoronary concentration gradients of circulating microRNAs. Circulation 2011;124(18) 1936-1944.

[66] Fichtlscherer S, De Rosa S, Fox H, Schwietz T, Fischer A, Liebetrau C, Weber M, Hamm CW, Roxe T, Muller-Ardogan M, Bonauer A, Zeiher AM, Dimmeler S. Circulating microRNAs in patients with coronary artery disease. Circulation Research 2010;107(5) 677-684.

[67] Arroyo JD, Chevillet JR, Kroh EM, Ruf IK, Pritchard CC, Gibson DF, Mitchell PS, Bennett CF, Pogosova-Agadjanyan EL, Stirewalt DL, Tait JF, Tewari M. Argonaute2 com-

plexes carry a population of circulating microRNAs independent of vesicles in human plasma. Proceedings of the National Academy of Sciences U S A 2011;108(12) 5003-5008.

[68] Goren Y, Kushnir M, Zafrir B, Tabak S, Lewis BS, Amir O. Serum levels of microRNAs in patients with heart failure. European Journal of Heart Failure 2012;14(2) 147-154.

[69] Koga Y, Yasunaga M, Takahashi A, Kuroda J, Moriya Y, Akasu T, Fujita S, Yamamoto S, Baba H, Matsumura Y. MicroRNA expression profiling of exfoliated colonocytes isolated from feces for colorectal cancer screening, Cancer Prevention Research (Phila), 2010;3(11) 1435-1442.

[70] Wang G, Tam LS, Li EK, Kwan BC, Chow KM, Luk CC, Li PK, Szeto CC. Serum and urinary free microRNA level in patients with systemic lupus erythematosus. Lupus 2011;20(5) 493-500.

[71] Scherr M, Venturini L, Battmer K, Schaller-Schoenitz M, Schaefer D, Dallmann I, Ganser A, Eder M. Lentivirus-mediated antagomir expression for specific inhibition of miRNA function. Nucleic Acids Research 2007;35(22) e149.

[72] Surdziel E, Eder M, Scherr M. Lentivirus-mediated antagomir expression. Methods in Molecular Biology 2010;667 237-48.

[73] Lennox KA, Behlke MA. Chemical modification and design of anti-miRNA oligonucleotides. Gene Therapy 2011;18(12) 1111-1120.

[74] Obika S, Nanbu D, Hari Y, Morio KI, In Y, Ishida T, Imanishi T. Synthesis of 2′-O,4′-C-methyleneuridine and -cytidine. Novel bicyclic nucleosides having a fixed C3'-endo sugar puckering. Tetrahedron Letters 1997;38(50) 8735-8738

[75] Baker M. RNA interference: Homing in on delivery. Nature 2010;464(7292) 1225-1228.

[76] Gambari R, Fabbri E, Borgatti M, Lampronti I, Finotti A, Brognara E, Bianchi N, Manicardi A, Marchelli R, Corradini R. Targeting microRNAs involved in human diseases: a novel approach for modification of gene expression and drug development, Biochemical Pharmacology 2011;82(10) 1416-1429.

[77] Jopling C. Liver-specific microRNA-122: Biogenesis and function, RNA Biology 2012;9(2) [Epub ahead of print].

[78] Lanford RE, Hildebrandt-Eriksen ES, Petri A, Persson R, Lindow M, Munk ME, Kauppinen S, Orum H. Therapeutic silencing of microRNA-122 in primates with chronic hepatitis C virus infection. Science, 2010;327(5962) 198-201.

[79] Maegdefessel L, Azuma J, Toh R, Merk DR, Deng A, Chin JT, Raaz U, Schoelmerich AM, Raiesdana A, Leeper NJ, McConnell MV, Dalman RL, Spin JM, Tsao PS. Inhibition of microRNA-29b reduces murine abdominal aortic aneurysm development. Journal of Clinical Investigation 2012;122(2) 497-506.

[80] Matkovich SJ, Wang W, Tu Y, Eschenbacher WH, Dorn LE, Condorelli G, Diwan A, Nerbonne JM, Dorn GW 2nd. MicroRNA-133a protects against myocardial fibrosis and modulates electrical repolarization without affecting hypertrophy in pressure-overloaded adult hearts. Circulation Research 2010;106(1) 166-75.

[81] Yin VP, Lepilina A, Smith A, Poss KD. Regulation of zebrafish heart regeneration by miR-133. Developmental Biology 2012;365(2) 319-327.

[82] Rayner KJ, Sheedy FJ, Esau CC, Hussain FN, Temel RE, Parathath S, van Gils JM, Rayner AJ, Chang AN, Suarez Y, Fernandez-Hernando C, Fisher EA, Moore KJ. Antag-

onism of miR-33 in mice promotes reverse cholesterol transport and regression of atherosclerosis. Journal of Clinical Investigation 2011;121(7) 2921-2931.

[83] Rayner KJ, Esau CC, Hussain FN, McDaniel AL, Marshall SM, van Gils JM, Ray TD, Sheedy FJ, Goedeke L, Liu X, Khatsenko OG, Kaimal V, Lees CJ, Fernandez-Hernando C, Fisher EA, Temel RE, Moore KJ. Inhibition of miR-33a/b in non-human primates raises plasma HDL and lowers VLDL triglycerides. Nature 2011;478(7369) 404-407.

[84] Xiao Y, Xu C, Guan J, Ping Y, Fan H, Li Y, Zhao H, Li X. Discovering dysfunction of multiple microRNAs cooperation in disease by a conserved microRNA co-expression network, PLoS One 2012;7(2) e32201.

[85] Kasinski AL, Slack FJ. Epigenetics and genetics. MicroRNAs en route to the clinic: progress in validating and targeting microRNAs for cancer therapy. Nature Reviews Cancer, 2011;11(12) 849-864.

Chapter 10

Cardiovascular Lesions of Kawasaki Disease: From Genetic Study to Clinical Management

Ho-Chang Kuo and Wei-Chiao Chang

Additional information is available at the end of the chapter

http://dx.doi.org/10.5772/48341

1. Introduction

Kawasaki disease (KD) is an acute febrile systemic vasculitis that was first described by Kawasaki *et al.*(1) in 1967 in Japanese(2) and in 1974 in English. Currently, it is the leading cause of acquired heart disease in children in developed countries; however, its etiology remains unknown.(3-5) KD mainly affects children less than 5 years of age, especially those in Asian countries. In Japan, Korea, and Taiwan, the incidence ranges from 69 to 218 cases per 100,000 children less than 5 years of age.(6-9) The incidence of KD in Taiwan has increased from 66 to 69 cases per 100,000 children aged less than 5 years.(9-12) Its incidence worldwide is increasing, especially in Japan, where, in 2010, Nakamura *et al.* reported the country's highest rate of 239.6 cases per 100,000 children aged 0–4 years.(13) An epidemiologic survey of KD in Taiwan spanning 2003–2006 found that 1.5% of all cases was recurrent (having a second episode of KD and receiving intravenous immunoglobulin [IVIG] treatment).(9) In Taiwan, KD occurs most frequently in the summer (April to June) and least frequently in the winter; for unknown reasons, its seasonal occurrence varies in other countries. The most serious complication of KD is the development of coronary artery lesions (CAL), including myocardial infarction, coronary artery fistula formation,(14) coronary artery dilatation, and coronary artery aneurysm.(15)

The most commonly used definition of CAL (also known as coronary artery abnormality [CAA] or CAL) is based on the Japanese Ministry of Health criteria: maximum absolute internal diameter > 3 mm in children younger than 5 years of age or >4 mm in children 5 years and older, or a segmental diameter 1.5 times greater than that of an adjacent segment, or the presence of luminal irregularity.(16-21) Coronary arteries should be corrected relative to body surface area (if available) and expressed as standard deviation units from the mean (Z scores).(22) Several studies analyzed CAL, including aortic root dimension,(23) and transient CAL (the definition of "transient" varies among studies, from 30 days to 6–8 weeks after diagnosis of disease). Thus, KD patients with coronary artery ectasia or dilatation that disappears within the first 8 weeks after disease onset are defined as having transient ectasia or dilatation (transient CAL). Kuo *et al.* reported the serious CAL analysis that comprised 341 KD patients;(24) 35% of KD patients had dilatation during the acute phase of admission, 17.2% still had dilatation 1 month after disease onset, 10.2% had dilatation at 2 months follow-up, and 4.1% had persistent CAL for more than 1 year.(25, 26) Ectasia or transient dilatations are somewhat considered to be a risk for a subsequent cardiovascular event or inflammation duration, rather than normal status.(27)

Although the clinical features of KD are recognizable, its underlying immunopathogenetic mechanisms are still under investigation, particularly the agent responsible for the development of CAL. KD is regarded as an autoimmune disorder rather than an infectious disease.(23) Kuo *et al.* reported that persistent monocytosis after IVIG treatment is associated with CAL formation.(28) Eosinophils in KD patients were also higher than that in age-matched febrile controls. In addition, IVIG treatment significantly increased eosinophils in KD patients. This increase of eosinophils after IVIG treatment is inversely correlated with IVIG treatment failure in KD.(29) Further studies have shown that eosinophil changes after IVIG treatment were positively correlated with changes in interleukin (IL)-5 levels. An increase in eosinophils and IL-5 levels after IVIG treatment is inversely correlated with CAL formation.(30) Recently, we found that incidence of allergic diseases (asthma and allergic rhinitis) after onset of KD were higher than that in age and sex-matched controls in a population cohort in Taiwan.

2. Allergy potential of Kawasaki disease patients

Brosius *et al.*(31) showed that the incidence of atopic dermatitis among children with KD was 9 times greater than that of controls. Burns *et al.*(32) reported associations of KD with atopic dermatitis and allergy, elevated serum IgE levels, and eosinophilia and that increased circulating numbers of monocytes/macrophages expressing the low-affinity IgE receptor (FCεR2) may be related to the effects of IL-4. Liew *et al.*(33) reported that KD may be a risk factor for subsequent allergic disease and postulated that KD occurs more frequently in children at risk of immune disequilibrium, with an initial abnormal inflammatory response, and subsequently, more allergic manifestations. Currently, Webster *et al.*(34) also reported that KD patients were more likely to have been admitted at least once with asthma/allergy than controls were. From our previous reports, we found that the T-helper (Th) type 2 immune response was elevated in the acute stage of KD, including eosinophils,(29) IL-4, IL-5,(35) and eotaxin. The eosinophil changes were correlated to changes of IL-5 levels but not to eosinophil cationic protein (ECP) levels, suggesting a Th2 immune reaction in KD. There are several lines of evidence pointing to an abnormal Th1/Th2 balance in KD patients.(28, 29, 35-39) Lin *et al.*(40) reported the comparison of eosinophils in KD and enterovirus (EV) patients with IVIG treatment and demonstrated a more significant eosinophil increase in KD patients. EV patients also had elevated eosinophil levels after IVIG therapy, but not as high as that of the KD patients after IVIG treatment. This may indicate an imbalance of the Th1/Th2 immune response, with a skewed Th2 response in KD.

3. Clinical phenotype and presentation of Kawasaki disease

As shown in Figures 1–8, the clinical characteristics of KD patients include fever lasting longer than 5 days, diffuse mucosal inflammation, bilateral non-purulent conjunctivitis, dysmorphic skin rashes, indurative angioedema over the hands and feet, and cervical lymphadenopathy. In addition to the diagnostic criteria, there is a broad range of non-specific clinical features, including irritability, uveitis, aseptic meningitis, cough, vomiting, diarrhea, abdominal pain, gallbladder hydrops, urethritis, arthralgia, arthritis, hypoalbuminemia,(5) liver function impairment, and heart failure.(4, 29, 41)

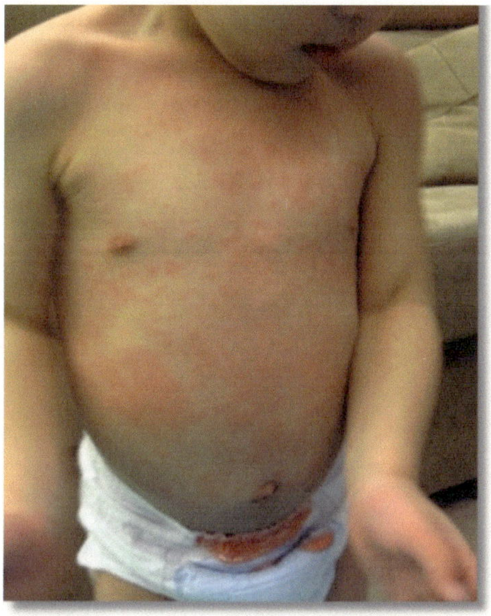

Figure 1. Dysmorphic skin rash of Kawasaki disease

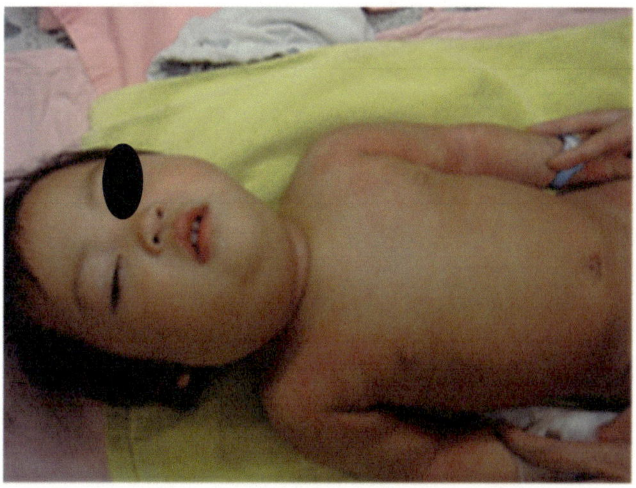

Figure 2. Skin rash and neck lymphadenopathy (right side, >1.5 cm in diameter)

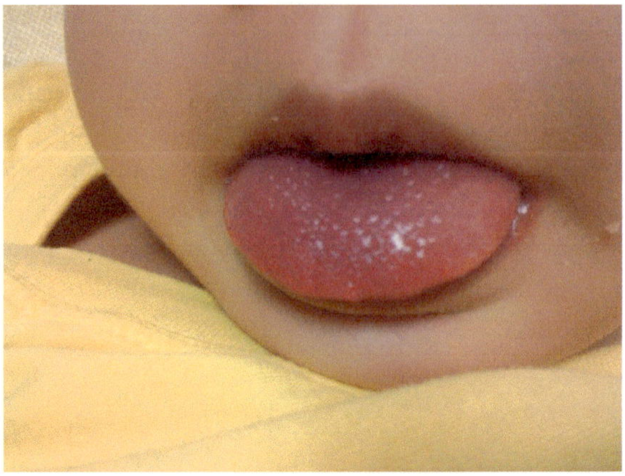

Figure 3. Strawberry tongue

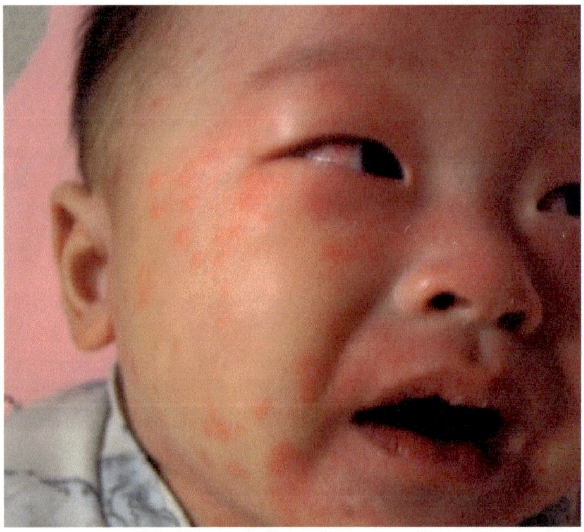

Figure 4. Face of Kawasaki disease patient exhibiting conjunctivitis, fissured lips, and skin rashes

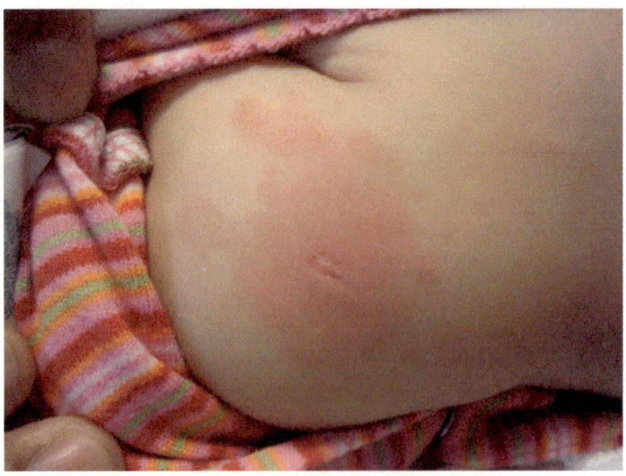

Figure 5. BCG injection site indurations

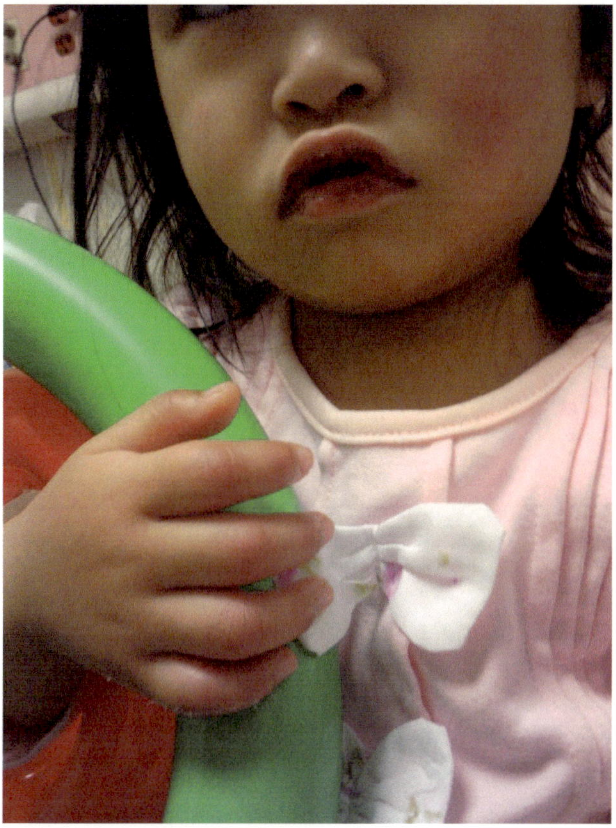

Figure 6. Fissured lips and swelling of finger joints

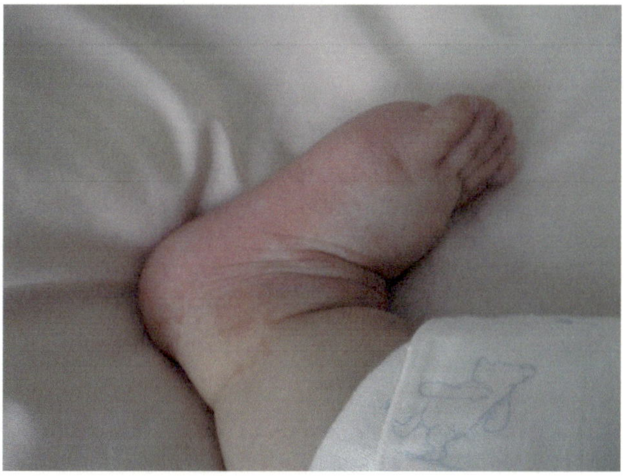

Figure 7. Induration change over foot

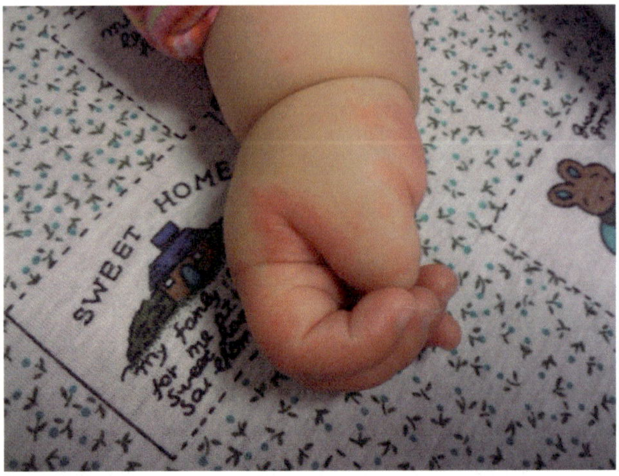

Figure 8. Induration change over palm

3.1. Diagnosis of Kawasaki disease

To date, there is no specific diagnostic laboratory test for KD. Diagnosis is based on the clinical phenotype, i.e., presence of fever lasting longer than 5 days and fulfillment of 4 of 5 specific clinical criteria. In Japan, at least 5 of 6 criteria (fever and 5 other clinical criteria) should be fulfilled for a diagnosis of KD. However, patients with 4 of the principal clinical features can be diagnosed when coronary aneurysm or dilatation is identified.(42) From the Japanese Circulation Society Joint Working Groups criteria (JCS 2008, Guidelines for Diagnosis and Management of Cardiovascular Sequelae in Kawasaki Disease),(43) KD can be diagnosed even when fever lasts less than 5 days. However, according to the American Heart Association (AHA) criteria,(15) fever lasting more than 5 days is essential for the diagnosis of KD.

Some patients who do not fulfill the criteria have been diagnosed with "incomplete" or "atypical" KD, a diagnosis often based on echocardiographic identification of CAL. The term "incomplete" may be preferable to "atypical" because these patients have insufficient criteria instead of atypical presentation.(15)

In countries with a bacillus Calmette-Guérin (BCG) vaccine policy (i.e., Taiwan and Japan), KD with erythematous induration or even ulceration of the BCG scar has been observed in one-third to half of KD patients (the incidence of BCG site induration is higher than that of neck lymphadenopathy in these countries).(3) Uehara *et al.*(44) reported that redness or the formation of a crust at the BCG inoculation site is a useful diagnostic sign for KD in children aged 3–20 months. Even if patients exhibit 4 or fewer signs of the clinical criteria for KD, physicians should consider the redness or crust formation at the BCG inoculation site as a possible indicator of KD.

Incomplete cases of KD are not uncommon (up to 15–20%). The incidence of CAL in patients exhibiting 4 principal symptoms of KD is slightly higher than that in patients with 5 to 6 principal symptoms.(45) Presentation of a small number (<4) of principal symptoms does not indicate a milder form of the disease. Patients with at least 4 principal symptoms require the same treatment as patients with complete (typical) presentation of KD, and those with 3 or fewer principal symptoms should be treated similarly when they meet the supplementary criteria. Herein, common supplementary criteria for the diagnosis of incomplete KD are introduced.

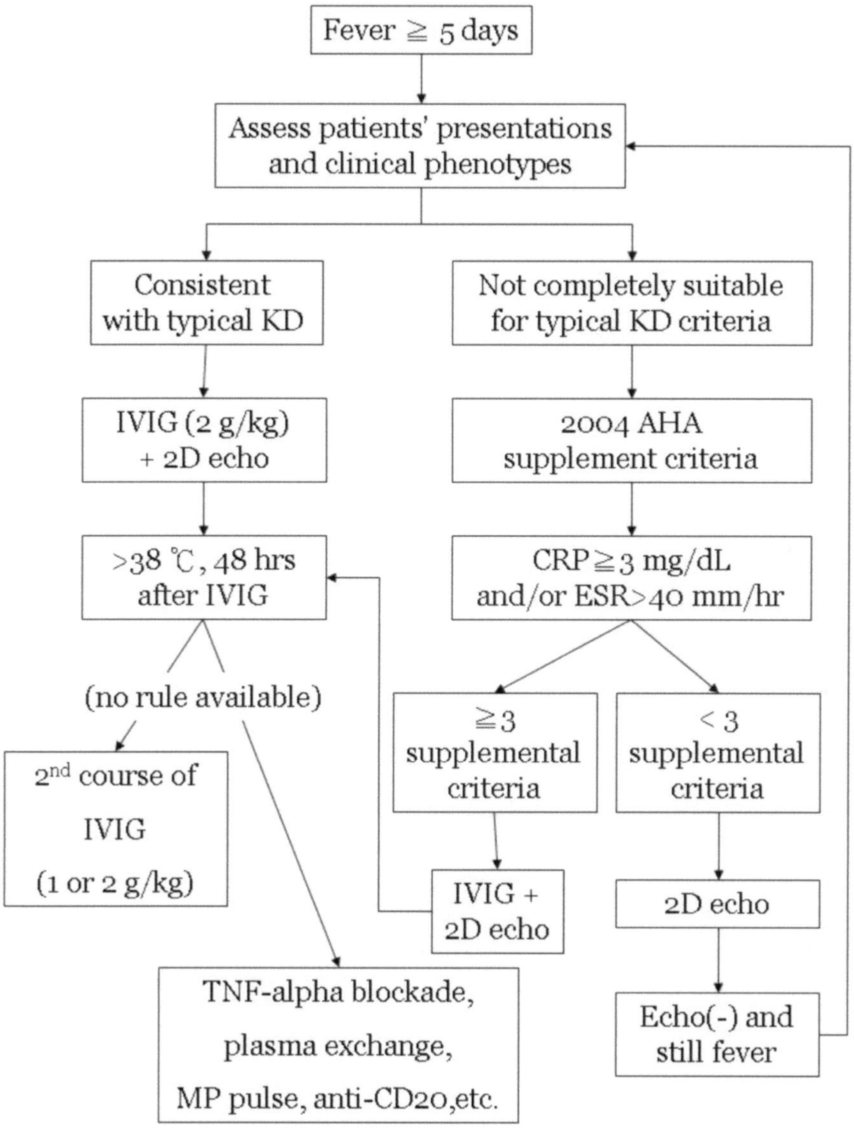

Figure 9. Flowchart of Kawasaki disease management

Incomplete KD is more common in young infants than in older children, making accurate diagnosis and timely treatment especially important in these young patients, who are at substantial risk of developing coronary abnormalities.(46, 47) The incidence of KD is actually higher than that previously reported worldwide, partly because earlier reports did not take incomplete forms into account. The AHA criteria (2004), which incorporate suggestions for laboratory tests and early echocardiography, are helpful for diagnosing incomplete KD.(41, 48) Consultation with an expert (cardiologist, immunologist, or rheumatologist) should be sought whenever assistance in making a diagnosis is needed. Patients with fever for 5 days or more (with 2 or 3 principal clinical features for KD) without other causes should undergo laboratory testing, and if there is evidence of systemic inflammation, an echocardiogram should be obtained even if the patient does not fully meet the clinical criteria for KD. Likewise, infants 6 months or younger with fever for 7 days or more without other causes should undergo laboratory testing, and if evidence of systemic inflammation is found, an echocardiogram should be obtained even if the infant fulfills no clinical criteria for KD.(15)

The 2004 AHA supplemental laboratory criteria include (1) albumin ≤ 3.0 g/dL; (2) anemia for age; (3) elevation of alanine aminotransferase (ALT); (4) platelets after 7 days ≥ 450,000/mm^3; (5) white blood cell count ≥ 15,000/mm^3; and (6) urine ≥ 10 white blood cells/high-power field.(15) If a patient has more than 3 supplementary criteria, incomplete KD is diagnosed and IVIG should be prescribed before performing echocardiography.(15) The flowchart for incomplete KD diagnosis and treatment are depicted in Figure 9.

4. Treatment for Kawasaki disease

The standard treatment for KD is IVIG (2 g/kg) infusion for 8–12 hours with high-dose aspirin (80–100 mg/[kg·day]).(20, 24, 29) The most serious complication of KD is the development of CAL, including myocardial infarction, coronary artery dilatation, coronary artery aneurysms, and coronary fistula formation.(25, 30, 49) Coronary artery aneurysms occur as a sequela of the vasculitis in 20–25% of untreated children. There are several risk factors for developing coronary arteritis, such as low serum albumin, age younger than 1 year, and long duration of the fever before treatment. Young patients with low albumin run a very high risk for CAL and IVIG treatment resistance.(5, 29) Although the introduction of IVIG therapy has greatly decreased the rate of coronary aneurysm to 3–10% of patients still develop some type of CAL. Durongpisitkul *et al.* showed that 11.6% patients are unresponsive to initial IVIG (2 g/kg) treatment. The worst prognosis occurs in children with so-called "giant aneurysms of the coronary arteries" (those with a maximal diameter of >8 mm), as thrombosis is promoted both by sluggish blood flow within the massively dilated vascular space and by the frequent development of stenotic lesions later. The treatment for KD are reviewed and introduced as follows.

4.1. Aspirin

Aspirin has been used in the treatment of KD for many years, even before the usage of IVIG. Although aspirin has important anti-inflammatory (high dose) and anti-platelet (low dose) effects, it does not appear to reduce the frequency of CAL formation. During the acute phase of the illness, aspirin is administered in 4 doses of 80–100 mg/kg per day (30–50 mg/[kg·day] in Japan)(50) with IVIG. High-dose aspirin and IVIG appear to possess additive anti-inflammatory effects.

Practices regarding the duration of high-dose aspirin administration vary across countries and centers, many of which reduce the aspirin dose when the patient is afebrile. When high-dose aspirin is discontinued, low-dose aspirin (3–5 mg/[kg·day]) is administered until there is no evidence of CAL and inflammatory markers (including platelets, C-reactive protein [CRP], and erythrocyte sedimentation rate [ESR]) have returned to normal levels, which usually occurs 6–8 weeks after disease onset. For children who develop CAL, low-dose aspirin (or other anti-platelet agents) is continued indefinitely until the inflammatory markers return to the normal range and the echocardiogram does not display abnormalities. Hsieh *et al.*(50) reported that regardless of timing (before or after day 5 of the illness), single-infusion, high-dose (2 g/kg) aspirin in the acute stage of KD had no effect on the response rate to IVIG therapy, duration of fever, or the incidence of CAL. This review reiterates the recommendation that exposing children to high-dose aspirin therapy in the acute phase of KD is unnecessary because available data show no appreciable benefit to IVIG therapy response, CAL formation, or fever duration.

Our recent study investigated 609 KD patients from 2 medical centers in Taiwan. The patients were divided into Group 1, receiving high-dose aspirin (N = 274), and Group 2, without high-dose aspirin (N = 335). There were no significant differences between Groups 1 and 2 in terms of gender ($p = 0.51$), IVIG resistance rate (34/274 vs. 26/335, $p = 0.06$), CAL formation rate (57/274 vs. 74/335, $p = 0.64$), and total hospital stay (6.3 ± 0.2 vs. 6.7 ± 0.2 days, $p = 0.13$). There were also no significant differences between total white blood cell counts, hemoglobin levels, platelet counts, and CRP levels before (within 1 day) and after (within 3 days) IVIG treatment of the 2 groups ($p > 0.1$). These results provide evidence that high-dose aspirin in the acute phase of KD does not affect the treatment results (CAL and IVIG resistance rate) or inflammatory condition. High-dose aspirin treatment in the acute phase of KD appears unnecessary, and further randomized controlled trials are needed.

However, Reye syndrome is a risk in children who receive salicylates while they are experiencing active infection with varicella or influenza and has been reported in patients receiving high-dose aspirin for a prolonged period after KD.(52) Taken together, it seems unnecessary to expose children to high-dose aspirin in acute KD, especially those with G6PD deficiency. However, as reported in the literature, due to the anti-platelet effect, low-dose aspirin has been prescribed for at least 6–8 weeks to prevent thrombocytosis in KD patients.(15) If patients are allergic or intolerant to a particular drug, clinicians must avoid using it and look for alternatives. Aspirin is used in most patients, often in conjunction with dipyridamole. Dipyridamole has been widely used to treat patients with a coronary aneurysm resulting from KD.(43, 53) The relationship between aspirin therapy and hemolytic disorder in G6PD-deficient patients is unclear. There are also no literature regarding usage of low-dose aspirin and the outcome of KD. G6PD deficiency, an X-linked disorder, is the most common enzymatic disorder of red blood cells in humans. The clinical expression of G6PD deficiency encompasses a spectrum of hemolytic syndromes. While affected patients are usually asymptomatic, some have episodic anemia, while a few have chronic hemolysis. With the most prevalent G6PD variants (G6PD A- and G6PD Mediterranean), severe hemolysis is induced by the sudden destruction of older, more deficient erythrocytes after exposure to drugs with a high redox potential or to fava beans, selected infections, or metabolic abnormalities. The likelihood of developing hemolysis and the severity of disease are determined by the magnitude of the enzyme deficiency, which in turn is determined by the biochemical characteristics of the G6PD variant. The World Health Organization has classified the different G6PD variants according to the magnitude of the enzyme deficiency and the severity of hemolysis.(54) Class I variants have severe enzyme deficiency (less than 10% of normal) and are associated with chronic hemolytic anemia. Class II variants also have severe enzyme deficiency, but are usually only intermittently associated with hemolysis. Class III variants have moderate enzyme deficiency (10–60% of normal), with intermittent hemolysis usually associated with infection or drugs. Class IV variants have no enzyme deficiency or hemolysis. Class V variants have increased enzyme activity, and classes IV and V are of no clinical significance. The incidence of hemolysis development in a patient with G6PD deficiency after taking aspirin is dosage-related.(55) G6PD deficiency is commonly considered a contraindication to aspirin intake. However, just few studies(56) have suggested that aspirin can be safely administered in therapeutic doses to G6PD-deficient subjects without nonspherocytic hemolytic anemia. Anti-platelet therapy is most commonly used to prevent thrombotic events for adults with atherosclerotic vascular disease, children with certain types of congenital heart disease, stroke, and KD.(57) Unfortunately, very little data on the efficacy and safety of anti-platelet therapy for pediatric patients, or even G6PD patients, are available. No prospective data exist to guide clinicians in selecting an optimal regimen. Therapeutic regimens used in patients with KD depend on the severity of CAL and include anti-platelet therapy with aspirin, with or without dipyridamole or clopidogrel; anticoagulant therapy with warfarin or low-molecular-weight heparin; or a combination of anticoagulant and anti-platelet therapy.(15)

A few articles have reported G6PD-deficient patients with sustained KD.(58) However, the question of whether aspirin is suitable for KD patients with G6PD deficiency remains

4.2. Intravenous immunoglobulin (IVIG or IVGG) responsiveness

The efficacy of IVIG administered in the acute phase of KD for reducing the incidence of coronary artery abnormalities is well established.(59) The mechanism of IVIG action is still under investigation. IVIG appears to have a generalized anti-inflammatory effect. Possible mechanisms of action include modulation of cytokine production, neutralization of bacterial super-antigens or other etiologic agents, augmentation of regulatory T cell activity (TGF-β),(23, 26) suppression of antibody synthesis and inflammatory markers (CD40-CD40L, nitric oxide, and iNOS expression),(60-62) provision of anti-idiotypic antibodies, Fc-gamma receptor,(63) and balancing Th1/Th2 responses.(28-30)

KD patients should be treated with a single 12-hour infusion of 2 g/kg IVIG together with aspirin in the acute phase with fever or inflammation progression without fever.(3, 4, 15) This therapy should be administered within 10 days of illness onset, and if this is not possible, within 7 days of illness onset. Treatment of KD before day 5 of illness appears no more likely to prevent cardiac sequelae than treatment on days 5–9. However, it may be associated with an increased need for repeat IVIG treatment.(64, 65) In the presence of 4 of 5 classic criteria for KD, US and Japanese experts agree that only 4 days of fever are necessary before initiating treatment with IVIG.(15, 66)

The efficacy of treating patients using IVIG after 10 days of illness is unknown; therefore, early diagnosis and treatment is desired. IVIG should be administered to children presenting after day 10 of illness (i.e., children with delayed diagnosis or incomplete KD) if they have either persistent fever without explanation or aneurysms and ongoing systemic inflammation, as manifested by elevated ESR or CRP.(4, 67-69) Burns *et al.* also suggested that any child with KD who has evidence of persisting inflammation, including fever or high concentrations of inflammatory markers with or without coronary artery abnormalities, should be treated even if the diagnosis is made after 10 days of illness.(4)

4.3. IVIG resistance (or IVIG unresponsiveness, initial IVIG treatment failure)

The incidence of IVIG resistance varies from 9.4% to 23% between centers (but it can be as high as 38%, as reported in one US cohort).(70) Recent studies have identified demographic and laboratory characteristics as predictors of IVIG resistance, including age, illness day, platelet count, ESR, hemoglobin concentration, CRP, eosinophils, lactate dehydrogenase, albumin, and ALT.(5, 29, 71-73) As IVIG-resistant patients are at a higher risk for CAL formation, it is important to identify those who may benefit from more aggressive therapy. As shown in Figure 1 (modified from Newburger *et al.*(15)), there are no definite treatment principles available for the management of KD patients with initial IVIG resistance or unresponsiveness to other adjuvant therapies. A second dose of IVIG (1 or 2 g/kg),(15, 74) methylprednisolone (MP) pulse therapy,(75) tumor necrosis factor (TNF)-α blockade,(76) cytotoxic agents (cyclophosphamide, cyclosporine A [CyA], or MTX(77)), plasmapheresis,(78) and plasma exchange(79) have been reported to benefit KD patients with initial IVIG treatment failure. These other treatment modalities will be discussed.

4.4. Methylprednisolone pulse therapy

At present, the usefulness of steroids in the initial treatment of KD is not well established.(15) Newburger *et al.* reported that, compared to conventional IVIG therapy for routine primary treatment of KD in children, a single-pulse dose of intravenous MP (IVMP) does not improve treatment outcome.(22) However, IVMP therapy appears to benefit IVIG-resistant KD patients.(80) Miura *et al.* revealed the effectiveness of IVMP therapy for KD patients that were previously unresponsive to initial IVIG treatment. IVMP suppresses cytokine levels faster, and subsequently, the outcomes are similar to those of IVIG-responsive patients who receive a second dose of IVIG.(81) Furukawa *et al.* reported similar findings.(82) IVMP appears to have the same effect on IVIG-resistant KD patients compared to an additional IVIG treatment.(83) The cost-benefit differences between IVMP and additional IVIG should be carefully considered, taking into account different medical conditions or health insurance policies among countries. The first dose of IVIG is well established, while IVMP or additional IVIG for IVIG-resistant KD patients requires further investigation. Ogata *et al.*(83) reported that IVMP was useful for reducing fever duration and medical costs for KD patients with initial IVIG resistance. IVMP (N = 13) and additional IVIG treatment (N = 14) were not significantly different in terms of preventing the development of coronary artery aneurysm. IVMP (30 mg/kg MP per day for 3 days) or a second dose of IVIG (2 g/kg) was prescribed to KD patients with fever and marked inflammation (i.e., non-exudative conjunctival injection, strawberry tongue, fissured lips, and erythematous change at the BCG inoculation site) 48 hours after initial IVIG treatment.(22, 82-84)

The safety of IVMP therapy in patients with KD is uncertain. Miura *et al.*(85) reported that IVMP (N = 11) incurred a higher incidence of sinus bradycardia and hyperglycemia when compared with the additional IVIG group (N = 11). Hypertension between IVMP and IVIG groups did not differ significantly. All of the adverse effects were transient. There were no convulsions, gastrointestinal symptoms, infections, malignant arrhythmias, or sudden death in any subject.(85) Taken together, IVMP is safe for KD patients as additional or adjuvant therapy of initial IVIG treatment.(22, 86, 87) After additional IVIG therapy, IVMP is considered for KD patients with persistently poor responses to the second IVIG treatment.(74, 88) Kobayashi *et al.*(89) reported that the addition of prednisolone (2 mg/[kg·day] administered over 15 days) to the standard regimen of IVIG improves coronary artery outcomes in patients with severe KD in Japan.

4.5. Tumor necrosis factor-α blockade

TNF-α levels are elevated in children with KD,(90) and the TNF-α (–308) genetic polymorphism is associated with KD susceptibility, suggesting a role for TNF-α receptor blocking in the treatment of KD, especially for those patients/cases refractory to IVIG. The early administration of TNF-α receptor antagonists in KD may provide effective adjunctive therapy. Infliximab, which binds the pro-inflammatory cytokine TNF-α, has been evaluated in several studies and shown to have a significant effect in KD patients with IVIG resistance.(91-93) Recently, etanercept, a more suitable TNF-α receptor blocker for children with refractory juvenile idiopathic arthritis,(94, 95) was reported to benefit the treatment of IVIG-resistant KD as an adjuvant therapy to initial IVIG.(96, 97) A TNF-α receptor blocker may be administered after initial IVIG treatment failure or after a second dose of IVIG therapy.

4.6. Statins

Chronic vascular inflammation and endothelial dysfunction persists in KD patients with CAL, even long after the acute stage.(98, 99) There is currently no specific treatment for ongoing vascular inflammation and endothelial dysfunction. Low-dose aspirin can be prescribed until CAL normalizes, but it does not have an effect on inflammation or endothelial dysfunction. Lipid abnormalities in the acute phase of KD, with decreased triglycerides and high-density lipoprotein cholesterol (HDL-C) levels have been reported in previous studies.(100, 101)

Statins, hydroxymethylglutaryl coenzyme A reductase inhibitors, have been shown to reduce cholesterol levels as well as improve surrogate markers of atherosclerosis and cardiovascular disease.(102) Huang *et al.*(103) reported that short-term (3 months) statin treatment (simvastatin, 10 mg/day as a single dose at bedtime) in KD patients complicated with CAL (N = 11) can significantly reduce total cholesterol and low-density lipoprotein cholesterol levels and increase HDL-C levels. Chronic vascular inflammation is also significantly improved, as is endothelial dysfunction, with no adverse effects. However, long-term and randomized control trials are needed before further conclusions can be drawn.

Recently, Blankier *et al.*(104) also reported that atorvastatin is able to inhibit critical steps (T cell activation and proliferation, production of the pro-inflammatory cytokine TNF-α, and upregulation of matrix metalloproteinase-9 and an elastolytic protease) known to be important in the development of coronary aneurysms in an animal model of KD (murine model with injection of *Lactobacillus casei* cell wall extract), suggesting that statins may have therapeutic benefits in KD patients. Taken together, statins may be beneficial as an adjuvant therapy in KD patients with CAL. However, the association between dyslipidemia and atherosclerosis in KD patients is not certain.

4.7. Other treatments

Acute KD can lead to the development of large coronary artery aneurysms that may persist for years. Abciximab, a platelet glycoprotein IIb/IIIa receptor inhibitor, is associated with resolution of thrombi and vascular remodeling in adults with acute coronary syndromes. Williams *et al.*(105) reported that KD patients who were treated with abciximab demonstrated greater regression in aneurysm diameter at early follow-up than patients who received standard therapy alone. McCandless *et al.*(106) also reported that abciximab treatment might be associated with vascular remodeling in patients with aneurysms. Abciximab appears to benefit KD patients, especially those who develop aneurysms.

There are still no well-defined treatments for refractory KD. Suzuki *et al.(107)* reported that CyA treatment is considered safe and well tolerated and may serve as a promising option for patients with refractory KD. Hyperkalemia developed in 9 of 28 (32%) patients 3–7 days after commencing CyA treatment. Adverse effects such as arrhythmias should be monitored with CyA. Kuijpers *et al.*(108) described a case of mortality, and a review of the literature showed that immunosuppressive medication such as CyA may not influence coronary inflammation and proliferation. Further trials are needed to clarify the optimal dose, safety, and timing of CyA treatment.

Specific changes in inflammatory markers (such as white blood cell count, neutrophil count, CRP, IL-6, soluble IL-2 receptor [sIL-2R](109), Th17/regulatory T-cell imbalance(110), and IL-1 pathway(111)) have been reported to disrupt immunological functions and result in KD with IVIG resistance and CAL formation. This indicates the possible treatment role of plasma exchange (PE) for KD with IVIG resistance. Mori *et al.*(79) studied 46 children who had not responded to the second IVIG treatment and subsequently received PE, and compared them with 59 children that received a third dose of IVIG therapy. No complications occurred with PE therapy. CAL developed in 8 of the 46 children (17.3%) who received PE and in 24 of the 59 (40.7%) who received a third course of IVIG ($p < 0.001$). PE is considered safe and effective in the prevention of CAL in KD that is refractory to IVIG therapy. PE can be performed at an early stage, as soon as fractional increases in inflammatory markers are found after the first or second dosage of IVIG therapy.(79)

5. Genetic association study in Kawasaki disease

The higher incidence of KD in Asia, in conjunction with a higher incidence of the disease in Asian descendants compared with other ethnic populations in the United States and Europe, suggests that genetic predisposition might play an important role in the susceptibility to this disease.(3, 4, 9, 15) There is also evidence that the incidence of KD is higher among siblings than in the general population.(112) A growing number of research reports provide evidence that genetic polymorphisms contribute to the susceptibility to KD. For example, single-nucleotide polymorphisms (SNPs) in the monocyte chemoattractant protein 1 (*MCP-1*),(113) *IL-10*,(114-116) *CD40L*,(117) *CD40*,(62) *IL-4*,(32) *CASP3*,(24) *IL-18*,(118) *IL-1B*,(119) *HLA-E*,(120) C-C chemokine receptor 5 (*CCR5*),(121-124) and *ITPKC*(20, 125, 126) and TGF-β receptors(23) have been reported to be associated with the development of KD. Although genetic association studies have been widely performed in KD, several studies have produced inconsistent results. Some genes were proposed in one population; however, the findings could not be replicated in another population. In addition, the genes that are responsible for KD susceptibility may not be involved in CAL formation. Thus, studies addressing this question are plagued with inconsistencies. Three possibilities may explain these inconsistencies. First, some studies were performed in a small sample size that may not have been able to provide sufficient power to detect minor genetic effects. Second, it is becoming clear that there are different genetic backgrounds within populations that, due to variations in allele frequencies or heterogeneity of the phenotypes, may also influence the results. Third, the incidence of KD in Asia is much higher than that in other regions. Thus, the environmental factors or infectious agents between countries should also be considered carefully.

6. Genetic polymorphisms of the ITPKC signaling pathway in Kawasaki disease

A major advancement in the genetic study of KD was made by the discovery of *ITPKC* in the RIKEN SNP center Japan. In 2008, Onouchi and colleagues first identified a functional polymorphism of *ITPKC* (rs28493229) that significantly associated with the susceptibility of KD and CAL in both Japanese and US children.(125) By using cell-based functional studies, Onouchi *et al.* further provided evidence to indicate that the risk C allele of *ITPKC* can reduce the splicing efficiency of the *ITPKC* mRNA that, in turn, contributes to the hyperactivation of Ca^{2+}-dependent NFAT pathways in T cells. Thus, in the model of Onouchi *et al.*, ITPKC is a negative regulator of T cells, and it may function as a calcium channel modulator that is involved in controlling immune systems. Interestingly, replication studies in the Taiwanese populations are strikingly controversial. The first replication study was by Chi *et al.* A total of 385 KD patients and 1158 normal subjects were genotyped.(127) However, no significant association was observed. Lin *et al.* took similar approaches in another independent medical center in Taipei. Their results indicated that the C allele of rs28493229 is associated with KD susceptibility.(128) Recently, data by meta-analysis support the correlation between rs28493229 of *ITPKC* and susceptibility of KD in the Taiwanese population.(23) Due to the increase in genetic diversity between cities in the south or north of Taiwan, we attribute the controversial results in the Taiwanese population to population migration.

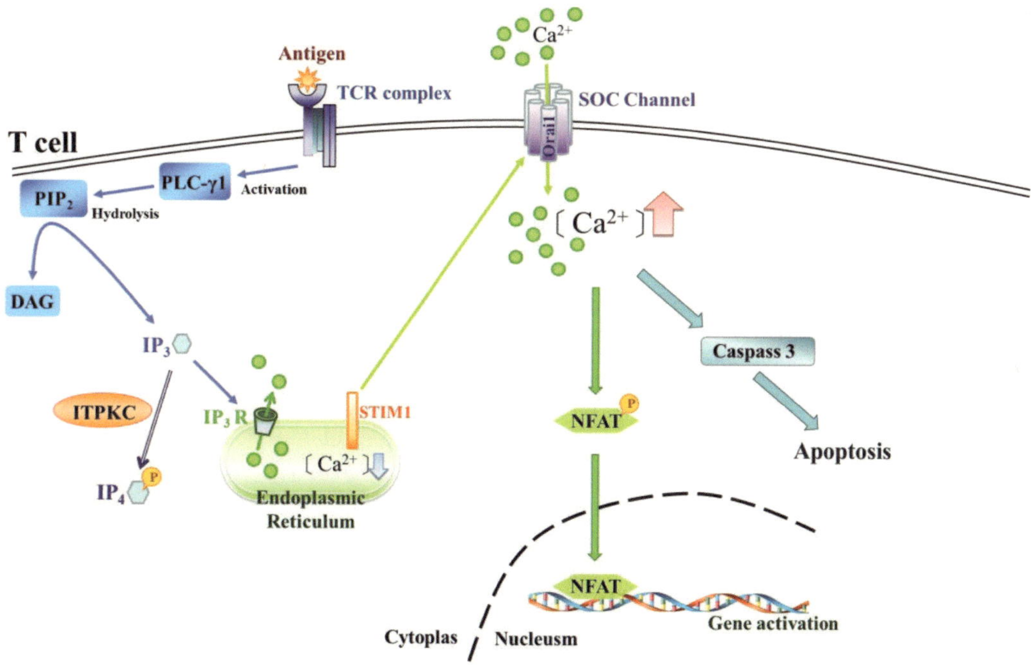

Figure 10. Model depicting the cellular pathways of ITPKC/calcium signaling in T cells.

In the non-excitable cells such as T and B cells, calcium entry is mainly through store-operated calcium channels (SOC). The activation of SOC can be controlled by the expression level of IP3, which is the substrate of ITPKC protein. As ITPKC is involved in the Ca^{2+}-dependent NFAT signaling in T cells, genetic association studies between calcium pathways and susceptibility of KD were performed. The calcium-dependent downstream gene CASP3 is a good example. Onouchi *et al.* reported that a G-to-A substitution in the

5′-untranslated region of *CASP3* (rs72689236) is associated with susceptibility to KD in Japanese and in Americans of European descent.(129) In the sample year, *CASP3* (rs72689236) was replicated in the KD children in the Taiwanese population.(24) Kuo *et al.* confirmed that the A allele of rs72689236 is very likely a risk allele in the development of aneurysms in patients with KD. Another 2 important molecules in the SOC are *ORAI1* (also known as *CRACM1*) and *STIM1*. Feske *et al.* identified *ORAI1* in 2006. Modified linkage analysis completed on data generated by SNP arrays and RNA interference screening led to an important finding. A single missense mutation in *ORAI1* was found in patients with severe combined immune deficiency syndrome. In 2011, genetic polymorphisms of *ORAI1* were reported to associate with the risk and recurrence of calcium nephrolithiasis(130) and HLA-B27-positive AS(131). In the KD study, no significant association between OARI1 genotypes *ORAI1* and KD clinical parameters (such as CAL formation or IVIG treatment responses) was found. However, a novel genetic polymorphism in the STIM1 gene was detected that associated with CAL formation in KD patients (data not shown). As STIM1 is a key initiator of SOC, DNA sequencing for the STIM1 gene family in a larger population may be helpful to identify novel polymorphisms. Future studies are needed to address the mechanism by which calcium signaling contributes to the development of KD. (Figure 10)

7. Genetic polymorphisms of the TGF-β signaling pathway in Kawasaki disease

TGF-β is an important molecule that is involved in the regulation of cytokine expression and immune response. It has been shown that TGF-β-mediated signaling pathways are mainly via transcription factors, Smads, which include at least 3 common proteins: Smad2, Smad3, and Smad4. The binding of TGF-β to its receptor results in the phosphorylation of Smad2 or Smad3, which heterodimerizes with Smad4. The formation of the Smad complex further translocates to the nulclus to regulate activation of the target genes. In the cardiovascular system, which is an important target of KD, TGF-β signaling is involved in the pathogenesis of multiple cardiovascular diseases via aberrant vascular remodeling. Low expression levels of endogenous TGF-β activity in the blood may contribute to the development of atherosclerotic cardiovascular disease. In 2011, a large genetic study revealed a significant association between the polymorphisms in TGF-β pathways and KD susceptibility or CAL formation in the European and US populations. In this study, Shimizu *et al.*(23) were the first to identify 16 SNPs in 6 genes (*TGFB2, TGFBR2, SMAD1, ENG, ACVRL1,* and *SMAD3*) associated with the susceptibility to KD. The significance of genetic variation in 3 genes (*SMAD3, TGFB2,* and *TGFBR2*) could be replicated in the multiethnic TDT analysis from the independent United States/United Kingdom/Australia subjects.

Kuo *et al.*(26) performed a replication study of 12 polymorphisms in 950 Taiwanese children. It was confirmed that genetic polymorphisms of *SMAD* as well as *TGFB2* contribute to the susceptibility of KD. These observations, in combination with those of the recent study, support the importance of TGF-β pathways for the susceptibility or severity of KD.

8. Genome-wide association study (GWAS) in Kawasaki disease

In 2009, Burgner *et al.* were the first to perform a genome-wide association study (GWAS) on 119 Caucasian KD cases and 135 matched controls. Forty SNPs and 6 haplotypes were confirmed in an independent cohort of KD families.(132) This insightful work led to the identification of an SNP within the N-acetylated alpha-linked acidic dipeptidase-like 2 gene (*NAALADL2*; rs17531088), which was significantly associated with the susceptibility to KD. Although the function of *NAALADL2* remains unclear, mutations in the gene may be involved in the development of Cornelia de Lange syndrome. In 2010, Kim *et al.* conducted another GWAS in a Korean population.(133) In total, 786 subjects (186 KD patients and 600 controls) were recruited. A locus in the 1p31 region was identified as a susceptibility locus for KD. Furthermore, the PELI1 gene locus in the 2p13.3 region was confirmed to associate with the development of CAL in KD patients. In 2012, two independent research groups by Lee *et al.*(62) and Onouchi *et al.*(134) published GWAS data from Taiwanese and Japanese populations, respectively. The results suggested that *BLK* (encoding B-lymphoid tyrosine kinase) and *CD40* are novel susceptibility genes for KD. Consistent with this findings, Kuo *et al.* conducted a case–control genetic association study and identified another polymorphism in the CD40 gene that associated with susceptibility to KD. Hence, the results from independent groups support a significant role of immune-related genes such as *CD40* for KD and CAL formation.(135)

9. Conclusion

Several major advances have been made in understanding the genetic effects of the susceptibility and clinical status of KD over the past decade. Very recently, genome-wide association led 2 groups (Lee *et al.* and Onouchi *et al.*) to identify the same novel susceptibility loci as being important for KD in the Asian population. Although the exact functional role of these genes in KD is still unclear, at present, these loci could provide a new direction for future studies. We can expect to see more insightful research beginning to elucidate the genes responsible for KD susceptibility.

Author details

Ho-Chang Kuo
Department of Pediatrics, Kaohsiung Chang Gung Memorial Hospital, Taiwan
Chang Gung University College of Medicine, Kaohsiung, Taiwan

Wei-Chiao Chang
Department of Medical Genetics, College of Medicine, Kaohsiung Medical University, Taiwan
Cancer Center, Kaohsiung Medical University Hospital, Taiwan
School of Pharmacy, College of Pharmacy, Taipei Medical University,Taipei, Taiwan

Acknowledgement

We thank Siou-Jin Chiu (Kaohsiung Medical University) for help with the figure and the National Science Council (Taiwan) for their research support.

10. References

[1] Kawasaki T, Kosaki F, Okawa S, Shigematsu I, Yanagawa H. A new infantile acute febrile mucocutaneous lymph node syndrome (MLNS) prevailing in Japan. Pediatrics. 1974 Sep;54(3):271-6.

[2] Kawasaki T. [Acute febrile mucocutaneous syndrome with lymphoid involvement with specific desquamation of the fingers and toes in children]. Arerugi. 1967 Mar;16(3):178-222.

[3] Wang CL, Wu YT, Liu CA, Kuo HC, Yang KD. Kawasaki disease: infection, immunity and genetics. Pediatr Infect Dis J. 2005 Nov;24(11):998-1004.

[4] Burns JC, Glode MP. Kawasaki syndrome. Lancet. 2004 Aug 7-13;364(9433):533-44.

[5] Kuo HC, Liang CD, Wang CL, Yu HR, Hwang KP, Yang KD. Serum albumin level predicts initial intravenous immunoglobulin treatment failure in Kawasaki disease. Acta Paediatr. 2010 Oct;99(10):1578-83.

[6] Hinks A, Ke X, Barton A, Eyre S, Bowes J, Worthington J, *et al.* Association of the IL2RA/CD25 gene with juvenile idiopathic arthritis. Arthritis Rheum. 2009 Jan;60(1):251-7.

[7] Dendrou CA, Plagnol V, Fung E, Yang JH, Downes K, Cooper JD, *et al.* Cell-specific protein phenotypes for the autoimmune locus IL2RA using a genotype-selectable human bioresource. Nat Genet. 2009 Sep;41(9):1011-5.

[8] Qu HQ, Verlaan DJ, Ge B, Lu Y, Lam KC, Grabs R, *et al.* A cis-acting regulatory variant in the IL2RA locus. J Immunol. 2009 Oct 15;183(8):5158-62.

[9] Huang WC, Huang LM, Chang IS, Chang LY, Chiang BL, Chen PJ, *et al.* Epidemiologic features of Kawasaki disease in Taiwan, 2003-2006. Pediatrics. 2009 Mar;123(3):e401-5.

[10] Park YW, Han JW, Park IS, Kim CH, Cha SH, Ma JS, *et al.* Kawasaki disease in Korea, 2003-2005. Pediatr Infect Dis J. 2007 Sep;26(9):821-3.

[11] Nakamura Y, Yashiro M, Uehara R, Sadakane A, Chihara I, Aoyama Y, *et al.* Epidemiologic features of Kawasaki disease in Japan: results of the 2007-2008 nationwide survey. J Epidemiol. 2010;20(4):302-7.

[12] Nakamura Y, Yashiro M, Uehara R, Oki I, Kayaba K, Yanagawa H. Increasing incidence of Kawasaki disease in Japan: nationwide survey. Pediatr Int. 2008 Jun;50(3):287-90.

[13] Nakamura Y, Yashiro M, Uehara R, Sadakane A, Tsuboi S, Aoyama Y, *et al.* Epidemiologic Features of Kawasaki Disease in Japan: Results of the 2009-2010 Nationwide Survey. J Epidemiol. 2012 Mar 10. doi:10.2188/ jea.JE20110126.

[14] Townley RG, Barlan IB, Patino C, Vichyanond P, Minervini MC, Simasathien T, *et al.* The effect of BCG vaccine at birth on the development of atopy or allergic disease in young children. Ann Allergy Asthma Immunol. 2004 Mar;92(3):350-5.

[15] Newburger JW, Takahashi M, Gerber MA, Gewitz MH, Tani LY, Burns JC, *et al.* Diagnosis, treatment, and long-term management of Kawasaki disease: a statement for health professionals from the Committee on Rheumatic Fever, Endocarditis and Kawasaki Disease, Council on Cardiovascular Disease in the Young, American Heart Association. Circulation. 2004 Oct 26;110(17):2747-71.

[16] Akagi T, Rose V, Benson LN, Newman A, Freedom RM. Outcome of coronary artery aneurysms after Kawasaki disease. J Pediatr. 1992 Nov;121(5 Pt 1):689-94.

[17] Shulman ST, De Inocencio J, Hirsch R. Kawasaki disease. Pediatr Clin North Am. 1995 Oct;42(5):1205-22.

[18] Yu HR, Kuo HC, Sheen JM, Wang L, Lin IC, Wang CL, *et al.* A unique plasma proteomic profiling with imbalanced fibrinogen cascade in patients with Kawasaki disease. Pediatr Allergy Immunol. 2009 Nov;20(7):699-707.

[19] Wu MT, Hsieh KS, Lin CC, Yang CF, Pan HB. Images in cardiovascular medicine. Evaluation of coronary artery aneurysms in Kawasaki disease by multislice computed tomographic coronary angiography. Circulation. 2004 Oct 5;110(14):e339.

[20] Kuo HC, Yang KD, Juo SH, Liang CD, Chen WC, Wang YS, *et al.* ITPKC single nucleotide polymorphism associated with the Kawasaki disease in a Taiwanese population. PLoS ONE. 2011;6(4):e17370.

[21] Kuo HC, Yang KD, Chang WC, Ger LP, Hsieh KS. Kawasaki disease: an update on diagnosis and treatment. Pediatr Neonatol. 2012 Feb;53(1):4-11.

[22] Newburger JW, Sleeper LA, McCrindle BW, Minich LL, Gersony W, Vetter VL, *et al.* Randomized trial of pulsed corticosteroid therapy for primary treatment of Kawasaki disease. N Engl J Med. 2007 Feb 15;356(7):663-75.

[23] Shimizu C, Jain S, Davila S, Hibberd ML, Lin KO, Molkara D, *et al.* Transforming growth factor-beta signaling pathway in patients with Kawasaki disease. Circ Cardiovasc Genet. 2011 Feb;4(1):16-25.

[24] Kuo HC, Yu HR, Juo SH, Yang KD, Wang YS, Liang CD, *et al.* CASP3 gene single-nucleotide polymorphism (rs72689236) and Kawasaki disease in Taiwanese children. J Hum Genet. 2011 Feb;56(2):161-5.

[25] Kuo HC, Lin YJ, Juo SH, Hsu YW, Chen WC, Yang KD, *et al.* Lack of association between ORAI1/CRACM1 gene polymorphisms and Kawasaki disease in the Taiwanese children. J Clin Immunol. 2011 Aug;31(4):650-5.

[26] Kuo HC, Onouchi Y, Hsu YW, Chen WC, Huang JD, Huang YH, *et al.* Polymorphisms of transforming growth factor-beta signaling pathway and Kawasaki disease in the Taiwanese population. J Hum Genet. 2011 Dec;56(12):840-5.

[27] Sabharwal T, Manlhiot C, Benseler SM, Tyrrell PN, Chahal N, Yeung RS, *et al.* Comparison of factors associated with coronary artery dilation only versus coronary artery aneurysms in patients with Kawasaki disease. Am J Cardiol. 2009 Dec 15;104(12):1743-7.

[28] Kuo HC, Wang CL, Liang CD, Yu HR, Chen HH, Wang L, *et al.* Persistent monocytosis after intravenous immunoglobulin therapy correlated with the development of coronary artery lesions in patients with Kawasaki disease. J Microbiol Immunol Infect. 2007 Oct;40(5):395-400.

[29] Kuo HC, Yang KD, Liang CD, Bong CN, Yu HR, Wang L, *et al.* The relationship of eosinophilia to intravenous immunoglobulin treatment failure in Kawasaki disease. Pediatr Allergy Immunol. 2007 Jun;18(4):354-9.

[30] Liang CD, Kuo HC, Yang KD, Wang CL, Ko SF. Coronary artery fistula associated with Kawasaki disease. Am Heart J. 2009 Mar;157(3):584-8.

[31] Brosius CL, Newburger JW, Burns JC, Hojnowski-Diaz P, Zierler S, Leung DY. Increased prevalence of atopic dermatitis in Kawasaki disease. Pediatr Infect Dis J. 1988 Dec;7(12):863-6.

[32] Burns JC, Shimizu C, Shike H, Newburger JW, Sundel RP, Baker AL, *et al.* Family-based association analysis implicates IL-4 in susceptibility to Kawasaki disease. Genes Immun. 2005 Aug;6(5):438-44.

[33] Liew WK, Lim CW, Tan TH, Wong KY, Tai BC, Quek SC, *et al.* The effect of Kawasaki disease on childhood allergies - a sibling control study. Pediatr Allergy Immunol. Aug;22(5):488-93.

[34] Webster RJ, Carter KW, Warrington NM, Loh AM, Zaloumis S, Kuijpers TW, *et al.* Hospitalisation with infection, asthma and allergy in Kawasaki disease patients and their families: genealogical analysis using linked population data. PLoS ONE.6(11):e28004.

[35] Kuo HC, Wang CL, Liang CD, Yu HR, Huang CF, Wang L, *et al.* Association of lower eosinophil-related T helper 2 (Th2) cytokines with coronary artery lesions in Kawasaki disease. Pediatr Allergy Immunol. 2009 May;20(3):266-72.

[36] Hirao J, Hibi S, Andoh T, Ichimura T. High levels of circulating interleukin-4 and interleukin-10 in Kawasaki disease. Int Arch Allergy Immunol. 1997 Feb;112(2):152-6.

[37] Abe J, Ebata R, Jibiki T, Yasukawa K, Saito H, Terai M. Elevated granulocyte colony-stimulating factor levels predict treatment failure in patients with Kawasaki disease. J Allergy Clin Immunol. 2008 Nov;122(5):1008-13 e8.

[38] Matsubara T, Katayama K, Matsuoka T, Fujiwara M, Koga M, Furukawa S. Decreased interferon-gamma (IFN-gamma)-producing T cells in patients with acute Kawasaki disease. Clin Exp Immunol. 1999 Jun;116(3):554-7.

[39] Kuo HC, Wang CL, Wang L, Yu HR, Yang KD. Patient characteristics and intravenous immunoglobulin product may affect eosinophils in Kawasaki disease. Pediatr Allergy Immunol. 2008 Mar;19(2):184-5.

[40] Lin LY, Yang TH, Lin YJ, Yu HR, Yang KD, Huang YC, *et al.* Comparison of the Laboratory Data Between Kawasaki Disease and Enterovirus After Intravenous Immunoglobulin Treatment. Pediatr Cardiol. 2012 Mar 25. doi: 10.1007/s00246-012-0293-9.

[41] Liu YC, Hou CP, Kuo CM, Liang CD, Kuo HC. [Atypical kawasaki disease: literature review and clinical nursing]. Hu Li Za Zhi. 2010 Dec;57(6):104-10.

[42] Azab SS, Salama SA, Abdel-Naim AB, Khalifa AE, El-Demerdash E, Al-Hendy A. 2-Methoxyestradiol and multidrug resistance: can 2-methoxyestradiol chemosensitize resistant breast cancer cells? Breast Cancer Res Treat. 2009 Jan;113(1):9-19.

[43] Kobayashi T, Sone K. Effect of dipyridamole on the blood flow in coronary aneurysms resulting from Kawasaki disease. Pediatr Cardiol. 1994 Nov-Dec;15(6):263-7.

[44] Uehara R, Igarashi H, Yashiro M, Nakamura Y, Yanagawa H. Kawasaki disease patients with redness or crust formation at the Bacille Calmette-Guerin inoculation site. Pediatr Infect Dis J. 2010 May;29(5):430-3.

[45] Sonobe T, Kiyosawa N, Tsuchiya K, Aso S, Imada Y, Imai Y, *et al.* Prevalence of coronary artery abnormality in incomplete Kawasaki disease. Pediatr Int. 2007 Aug;49(4):421-6.

[46] Chang FY, Hwang B, Chen SJ, Lee PC, Meng CC, Lu JH. Characteristics of Kawasaki disease in infants younger than six months of age. Pediatr Infect Dis J. 2006 Mar;25(3):241-4.

[47] Burns JC, Wiggins JW, Jr., Toews WH, Newburger JW, Leung DY, Wilson H, *et al.* Clinical spectrum of Kawasaki disease in infants younger than 6 months of age. J Pediatr. 1986 Nov;109(5):759-63.
[48] Heuclin T, Dubos F, Hue V, Godart F, Francart C, Vincent P, *et al.* Increased detection rate of Kawasaki disease using new diagnostic algorithm, including early use of echocardiography. J Pediatr. 2009 Nov;155(5):695-9 e1.
[49] Kuo HC, Liang CD, Yu HR, Wang CL, Lin IC, Liu CA, *et al.* CTLA-4, position 49 A/G polymorphism associated with coronary artery lesions in Kawasaki disease. J Clin Immunol. 2011 Apr;31(2):240-4.
[50] Hsieh KS, Weng KP, Lin CC, Huang TC, Lee CL, Huang SM. Treatment of acute Kawasaki disease: aspirin's role in the febrile stage revisited. Pediatrics. 2004 Dec;114(6):e689-93.
[51] Kuo HC, Wang CL, Liang CD, et al. Revisit high dose aspirin in acute stage of Kawasaki disease. Taipei, Taiwan: 6th Asian Society for Pediatric Research; 2010.
[52] Lee JH, Hung HY, Huang FY. Kawasaki disease with Reye syndrome: report of one case. Zhonghua Min Guo Xiao Er Ke Yi Xue Hui Za Zhi. 1992 Jan-Feb;33(1):67-71.
[53] Tizard EJ, Suzuki A, Levin M, Dillon MJ. Clinical aspects of 100 patients with Kawasaki disease. Arch Dis Child. 1991 Feb;66(2):185-8.
[54] Beutler E. The molecular biology of G6PD variants and other red cell enzyme defects. Annu Rev Med. 1992;43:47-59.
[55] Youngster I, Arcavi L, Schechmaster R, Akayzen Y, Popliski H, Shimonov J, *et al.* Medications and glucose-6-phosphate dehydrogenase deficiency: an evidence-based review. Drug Saf. 2010 Sep 1;33(9):713-26.
[56] Beutler E. G6PD deficiency. Blood. 1994 Dec 1;84(11):3613-36.
[57] Li JS, Newburger JW. Antiplatelet therapy in pediatric cardiovascular patients. Pediatr Cardiol. 2010 May;31(4):454-61.
[58] Cattaneo G, Galvagno G, Mussa F. [Kawasaki disease in a subject with G6PD deficiency]. Pediatr Med Chir. 1989 Mar-Apr;11(2):219-21.
[59] Newburger JW, Takahashi M, Beiser AS, Burns JC, Bastian J, Chung KJ, *et al.* A single intravenous infusion of gamma globulin as compared with four infusions in the treatment of acute Kawasaki syndrome. N Engl J Med. 1991 Jun 6;324(23):1633-9.
[60] Wang CL, Wu YT, Liu CA, Lin MW, Lee CJ, Huang LT, *et al.* Expression of CD40 ligand on CD4+ T-cells and platelets correlated to the coronary artery lesion and disease progress in Kawasaki disease. Pediatrics. 2003 Feb;111(2):E140-7.
[61] Wang CL, Wu YT, Lee CJ, Liu HC, Huang LT, Yang KD. Decreased nitric oxide production after intravenous immunoglobulin treatment in patients with Kawasaki disease. J Pediatr. 2002 Oct;141(4):560-5.
[62] Lee YC, Kuo HC, Chang JS, Chang LY, Huang LM, Chen MR, *et al.* Two new susceptibility loci for Kawasaki disease identified through genome-wide association analysis. Nat Genet. 2012 Mar 25;44(5):522-5.
[63] Khor CC, Davila S, Breunis WB, Lee YC, Shimizu C, Wright VJ, *et al.* Genome-wide association study identifies FCGR2A as a susceptibility locus for Kawasaki disease. Nat Genet. 2011 Dec;43(12):1241-6.

[64] Muta H, Ishii M, Egami K, Furui J, Sugahara Y, Akagi T, *et al.* Early intravenous gamma-globulin treatment for Kawasaki disease: the nationwide surveys in Japan. J Pediatr. 2004 Apr;144(4):496-9.
[65] Fong NC, Hui YW, Li CK, Chiu MC. Evaluation of the efficacy of treatment of Kawasaki disease before day 5 of illness. Pediatr Cardiol. 2004 Jan-Feb;25(1):31-4.
[66] Dajani AS, Taubert KA, Gerber MA, Shulman ST, Ferrieri P, Freed M, *et al.* Diagnosis and therapy of Kawasaki disease in children. Circulation. 1993 May;87(5):1776-80.
[67] Minich LL, Sleeper LA, Atz AM, McCrindle BW, Lu M, Colan SD, *et al.* Delayed diagnosis of Kawasaki disease: what are the risk factors? Pediatrics. 2007 Dec;120(6):e1434-40.
[68] Anderson MS, Todd JK, Glode MP. Delayed diagnosis of Kawasaki syndrome: an analysis of the problem. Pediatrics. 2005 Apr;115(4):e428-33.
[69] Marasini M, Pongiglione G, Gazzolo D, Campelli A, Ribaldone D, Caponnetto S. Late intravenous gamma globulin treatment in infants and children with Kawasaki disease and coronary artery abnormalities. Am J Cardiol. 1991 Sep 15;68(8):796-7.
[70] Tremoulet AH, Best BM, Song S, Wang S, Corinaldesi E, Eichenfield JR, *et al.* Resistance to intravenous immunoglobulin in children with Kawasaki disease. J Pediatr. 2008 Jul;153(1):117-21.
[71] Rigante D, Valentini P, Rizzo D, Leo A, De Rosa G, Onesimo R, *et al.* Responsiveness to intravenous immunoglobulins and occurrence of coronary artery abnormalities in a single-center cohort of Italian patients with Kawasaki syndrome. Rheumatol Int. 2010 Apr;30(6):841-6.
[72] Sleeper LA, Minich LL, McCrindle BM, Li JS, Mason W, Colan SD, *et al.* Evaluation of kawasaki disease risk-scoring systems for intravenous immunoglobulin resistance. J Pediatr. 2011 May;158(5):831-5 e3.
[73] Egami K, Muta H, Ishii M, Suda K, Sugahara Y, Iemura M, *et al.* Prediction of resistance to intravenous immunoglobulin treatment in patients with Kawasaki disease. J Pediatr. 2006 Aug;149(2):237-40.
[74] Kuo HC, Wu CC, Yang TH, Yu HR, Liang CD, Chen YJ, *et al.* Non-Langerhans cell histiocytosis in a child with Kawasaki disease. BMJ Case Rep. 2009;2009.
[75] Wright DA, Newburger JW, Baker A, Sundel RP. Treatment of immune globulin-resistant Kawasaki disease with pulsed doses of corticosteroids. J Pediatr. 1996 Jan;128(1):146-9.
[76] Son MB, Gauvreau K, Burns JC, Corinaldesi E, Tremoulet AH, Watson VE, *et al.* Infliximab for intravenous immunoglobulin resistance in Kawasaki disease: a retrospective study. J Pediatr. 2011 Apr;158(4):644-9 e1.
[77] Ahn SY, Kim DS. Treatment of intravenous immunoglobulin-resistant Kawasaki disease with methotrexate. Scand J Rheumatol. 2005 Mar-Apr;34(2):136-9.
[78] Pinna GS, Kafetzis DA, Tselkas OI, Skevaki CL. Kawasaki disease: an overview. Curr Opin Infect Dis. 2008 Jun;21(3):263-70.
[79] Mori M, Imagawa T, Katakura S, Miyamae T, Okuyama K, Ito S, *et al.* Efficacy of plasma exchange therapy for Kawasaki disease intractable to intravenous gamma-globulin. Mod Rheumatol. 2004;14(1):43-7.

[80] Chen HH, Liu PM, Bong CN, Wu YT, Yang KD, Wang CL. Methylprednisolone pulse therapy for massive lymphadenopathy in a child with intravenous immunoglobulin-resistant Kawasaki disease. J Microbiol Immunol Infect. 2005 Apr;38(2):149-52.
[81] Miura M, Kohno K, Ohki H, Yoshiba S, Sugaya A, Satoh M. Effects of methylprednisolone pulse on cytokine levels in Kawasaki disease patients unresponsive to intravenous immunoglobulin. Eur J Pediatr. 2008 Oct;167(10):1119-23.
[82] Furukawa T, Kishiro M, Akimoto K, Nagata S, Shimizu T, Yamashiro Y. Effects of steroid pulse therapy on immunoglobulin-resistant Kawasaki disease. Arch Dis Child. 2008 Feb;93(2):142-6.
[83] Ogata S, Bando Y, Kimura S, Ando H, Nakahata Y, Ogihara Y, *et al.* The strategy of immune globulin resistant Kawasaki disease: a comparative study of additional immune globulin and steroid pulse therapy. J Cardiol. 2009 Feb;53(1):15-9.
[84] Han RK, Silverman ED, Newman A, McCrindle BW. Management and outcome of persistent or recurrent fever after initial intravenous gamma globulin therapy in acute Kawasaki disease. Arch Pediatr Adolesc Med. 2000 Jul;154(7):694-9.
[85] Miura M, Ohki H, Yoshiba S, Ueda H, Sugaya A, Satoh M, *et al.* Adverse effects of methylprednisolone pulse therapy in refractory Kawasaki disease. Arch Dis Child. 2005 Oct;90(10):1096-7.
[86] Okada K, Hara J, Maki I, Miki K, Matsuzaki K, Matsuoka T, *et al.* Pulse methylprednisolone with gammaglobulin as an initial treatment for acute Kawasaki disease. Eur J Pediatr. 2009 Feb;168(2):181-5.
[87] Lang BA, Yeung RS, Oen KG, Malleson PN, Huber AM, Riley M, *et al.* Corticosteroid treatment of refractory Kawasaki disease. J Rheumatol. 2006 Apr;33(4):803-9.
[88] Ogata S, Ogihara Y, Nomoto K, Akiyama K, Nakahata Y, Sato K, *et al.* Clinical score and transcript abundance patterns identify Kawasaki disease patients who may benefit from addition of methylprednisolone. Pediatr Res. 2009 Nov;66(5):577-84.
[89] Kobayashi T, Saji T, Otani T, Takeuchi K, Nakamura T, Arakawa H, *et al.* Efficacy of immunoglobulin plus prednisolone for prevention of coronary artery abnormalities in severe Kawasaki disease (RAISE study): a randomised, open-label, blinded-endpoints trial. Lancet. 2012 Mar 7. doi: 10.1016/S0140-6736(11)61930-2.
[90] Maury CP, Salo E, Pelkonen P. Elevated circulating tumor necrosis factor-alpha in patients with Kawasaki disease. J Lab Clin Med. 1989 May;113(5):651-4.
[91] Burns JC, Best BM, Mejias A, Mahony L, Fixler DE, Jafri HS, *et al.* Infliximab treatment of intravenous immunoglobulin-resistant Kawasaki disease. J Pediatr. 2008 Dec;153(6):833-8.
[92] Oishi T, Fujieda M, Shiraishi T, Ono M, Inoue K, Takahashi A, *et al.* Infliximab treatment for refractory Kawasaki disease with coronary artery aneurysm. Circ J. 2008 May;72(5):850-2.
[93] Zulian F, Zanon G, Martini G, Mescoli G, Milanesi O. Efficacy of infliximab in long-lasting refractory Kawasaki disease. Clin Exp Rheumatol. 2006 Jul-Aug;24(4):453.
[94] Kuo HC, Yu HR, Wu CC, Chang LS, Yang KD. Etanercept treatment for children with refractory juvenile idiopathic arthritis. J Microbiol Immunol Infect. 2011 Jan 13;44(1):52-6.

[95] Guo MM, Yang KD, Yu HR, Kuo HC. Hypersensitive joint reaction after etanercept treatment in a patient with juvenile rheumatoid arthritis. J Rheumatol. 2011 Mar;38(3):577-9.

[96] Choueiter NF, Olson AK, Shen DD, Portman MA. Prospective open-label trial of etanercept as adjunctive therapy for kawasaki disease. J Pediatr. 2010 Dec;157(6):960-6 e1.

[97] Portman MA, Olson A, Soriano B, Dahdah N, Williams R, Kirkpatrick E. Etanercept as adjunctive treatment for acute Kawasaki disease: study design and rationale. Am Heart J. 2011 Mar;161(3):494-9.

[98] Senzaki H, Chen CH, Ishido H, Masutani S, Matsunaga T, Taketazu M, *et al.* Arterial hemodynamics in patients after Kawasaki disease. Circulation. 2005 Apr 26;111(16):2119-25.

[99] Borzutzky A, Gutierrez M, Talesnik E, Godoy I, Kraus J, Hoyos R, *et al.* High sensitivity C-reactive protein and endothelial function in Chilean patients with history of Kawasaki disease. Clin Rheumatol. 2008 Jul;27(7):845-50.

[100] Weng KP, Hsieh KS, Huang SH, Lin CC, Huang DC. Serum HDL level at acute stage of Kawasaki disease. Zhonghua Min Guo Xiao Er Ke Yi Xue Hui Za Zhi. 1998 Jan-Feb;39(1):28-32.

[101] Newburger JW, Burns JC, Beiser AS, Loscalzo J. Altered lipid profile after Kawasaki syndrome. Circulation. 1991 Aug;84(2):625-31.

[102] Rodenburg J, Vissers MN, Wiegman A, Trip MD, Bakker HD, Kastelein JJ. Familial hypercholesterolemia in children. Curr Opin Lipidol. 2004 Aug;15(4):405-11.

[103] Huang SM, Weng KP, Chang JS, Lee WY, Huang SH, Hsieh KS. Effects of statin therapy in children complicated with coronary arterial abnormality late after Kawasaki disease: a pilot study. Circ J. 2008 Oct;72(10):1583-7.

[104] Blankier S, McCrindle BW, Ito S, Yeung RS. The role of atorvastatin in regulating the immune response leading to vascular damage in a model of Kawasaki disease. Clin Exp Immunol. 2011 May;164(2):193-201.

[105] Williams RV, Wilke VM, Tani LY, Minich LL. Does Abciximab enhance regression of coronary aneurysms resulting from Kawasaki disease? Pediatrics. 2002 Jan;109(1):E4.

[106] McCandless RT, Minich LL, Tani LY, Williams RV. Does abciximab promote coronary artery remodeling in patients with Kawasaki disease? Am J Cardiol. 2010 Jun 1;105(11):1625-8.

[107] Suzuki H, Terai M, Hamada H, Honda T, Suenaga T, Takeuchi T, *et al.* Cyclosporin A treatment for Kawasaki disease refractory to initial and additional intravenous immunoglobulin. Pediatr Infect Dis J. 2011 Oct;30(10):871-6.

[108] Kuijpers TW, Biezeveld M, Achterhuis A, Kuipers I, Lam J, Hack CE, *et al.* Longstanding obliterative panarteritis in Kawasaki disease: lack of cyclosporin A effect. Pediatrics. 2003 Oct;112(4):986-92.

[109] Suzuki H, Suenaga T, Takeuchi T, Shibuta S, Yoshikawa N. Marker of T-cell activation is elevated in refractory Kawasaki disease. Pediatr Int. 2010 Oct;52(5):785-9.

[110] Jia S, Li C, Wang G, Yang J, Zu Y. The T helper type 17/regulatory T cell imbalance in patients with acute Kawasaki disease. Clin Exp Immunol. 2010 Oct;162(1):131-7.

[111] Fury W, Tremoulet AH, Watson VE, Best BM, Shimizu C, Hamilton J, *et al.* Transcript abundance patterns in Kawasaki disease patients with intravenous immunoglobulin resistance. Hum Immunol. 2010 Sep;71(9):865-73.

[112] Dergun M, Kao A, Hauger SB, Newburger JW, Burns JC. Familial occurrence of Kawasaki syndrome in North America. Arch Pediatr Adolesc Med. 2005 Sep;159(9):876-81.

[113] Jibiki T, Terai M, Shima M, Ogawa A, Hamada H, Kanazawa M, *et al.* Monocyte chemoattractant protein 1 gene regulatory region polymorphism and serum levels of monocyte chemoattractant protein 1 in Japanese patients with Kawasaki disease. Arthritis Rheum. 2001 Sep;44(9):2211-2.

[114] Weng KP, Hsieh KS, Hwang YT, Huang SH, Lai TJ, Yuh YS, *et al.* IL-10 polymorphisms are associated with coronary artery lesions in acute stage of Kawasaki disease. Circ J. 2010 May;74(5):983-9.

[115] Hsieh KS, Lai TJ, Hwang YT, Lin MW, Weng KP, Chiu YT, *et al.* IL-10 promoter genetic polymorphisms and risk of Kawasaki disease in Taiwan. Dis Markers. 2011;30(1):51-9.

[116] Hsueh KC, Lin YJ, Chang JS, Wan L, Tsai YH, Tsai CH, *et al.* Association of interleukin-10 A-592C polymorphism in Taiwanese children with Kawasaki disease. J Korean Med Sci. 2009 Jun;24(3):438-42.

[117] Onouchi Y, Onoue S, Tamari M, Wakui K, Fukushima Y, Yashiro M, *et al.* CD40 ligand gene and Kawasaki disease. Eur J Hum Genet. 2004 Dec;12(12):1062-8.

[118] Chen SY, Wan L, Huang YC, Sheu JJ, Lan YC, Lai CH, *et al.* Interleukin-18 gene 105A/C genetic polymorphism is associated with the susceptibility of Kawasaki disease. J Clin Lab Anal. 2009;23(2):71-6.

[119] Weng KP, Ho TY, Chiao YH, Cheng JT, Hsieh KS, Huang SH, *et al.* Cytokine genetic polymorphisms and susceptibility to Kawasaki disease in Taiwanese children. Circ J. 2010 Nov;74(12):2726-33.

[120] Lin YJ, Wan L, Wu JY, Sheu JJ, Lin CW, Lan YC, *et al.* HLA-E gene polymorphism associated with susceptibility to Kawasaki disease and formation of coronary artery aneurysms. Arthritis Rheum. 2009 Feb;60(2):604-10.

[121] Burns JC, Shimizu C, Gonzalez E, Kulkarni H, Patel S, Shike H, *et al.* Genetic variations in the receptor-ligand pair CCR5 and CCL3L1 are important determinants of susceptibility to Kawasaki disease. J Infect Dis. 2005 Jul 15;192(2):344-9.

[122] Breunis WB, Biezeveld MH, Geissler J, Kuipers IM, Lam J, Ottenkamp J, *et al.* Polymorphisms in chemokine receptor genes and susceptibility to Kawasaki disease. Clin Exp Immunol. 2007 Oct;150(1):83-90.

[123] Jhang WK, Kang MJ, Jin HS, Yu J, Kim BJ, Kim BS, *et al.* The CCR5 (-2135C/T) polymorphism may be associated with the development of Kawasaki disease in Korean children. J Clin Immunol. 2009 Jan;29(1):22-8.

[124] Mamtani M, Matsubara T, Shimizu C, Furukawa S, Akagi T, Onouchi Y, *et al.* Association of CCR2-CCR5 haplotypes and CCL3L1 copy number with Kawasaki Disease, coronary artery lesions, and IVIG responses in Japanese children. PLoS ONE. 2010;5(7):e11458.

[125] Onouchi Y, Gunji T, Burns JC, Shimizu C, Newburger JW, Yashiro M, *et al.* ITPKC functional polymorphism associated with Kawasaki disease susceptibility and formation of coronary artery aneurysms. Nat Genet. 2008 Jan;40(1):35-42.

[126] Onouchi Y, Suzuki Y, Suzuki H, Terai M, Yasukawa K, Hamada H, *et al.* ITPKC and CASP3 polymorphisms and risks for IVIG unresponsiveness and coronary artery lesion formation in Kawasaki disease. Pharmacogenomics J. 2011 Oct 11. doi: 10.1038/tpj.2011.45.

[127] Chi H, Huang FY, Chen MR, Chiu NC, Lee HC, Lin SP, *et al.* ITPKC gene SNP rs28493229 and Kawasaki disease in Taiwanese children. Hum Mol Genet. 2010 Mar 15;19(6):1147-51.

[128] Lin MT, Wang JK, Yeh JI, Sun LC, Chen PL, Wu JF, *et al.* Clinical Implication of the C Allele of the ITPKC Gene SNP rs28493229 in Kawasaki Disease: Association With Disease Susceptibility and BCG Scar Reactivation. Pediatr Infect Dis J. 2011 Feb;30(2):148-52.

[129] Onouchi Y, Ozaki K, Buns JC, Shimizu C, Hamada H, Honda T, *et al.* Common variants in CASP3 confer susceptibility to Kawasaki disease. Hum Mol Genet. 2010 Jul 15;19(14):2898-906.

[130] Chou YH, Juo SH, Chiu YC, Liu ME, Chen WC, Chang CC, *et al.* A polymorphism of the ORAI1 gene is associated with the risk and recurrence of calcium nephrolithiasis. J Urol. 2011 May;185(5):1742-6.

[131] Wei JC, Yen JH, Juo SH, Chen WC, Wang YS, Chiu YC, *et al.* Association of ORAI1 haplotypes with the risk of HLA-B27 positive ankylosing spondylitis. PLoS ONE. 2011;6(6):e20426.

[132] Burgner D, Davila S, Breunis WB, Ng SB, Li Y, Bonnard C, *et al.* A genome-wide association study identifies novel and functionally related susceptibility Loci for Kawasaki disease. PLoS Genet. 2009 Jan;5(1):e1000319.

[133] Kim JJ, Hong YM, Sohn S, Jang GY, Ha KS, Yun SW, *et al.* A genome-wide association analysis reveals 1p31 and 2p13.3 as susceptibility loci for Kawasaki disease. Hum Genet. 2011 May;129(5):487-95.

[134] Onouchi Y, Ozaki K, Burns JC, Shimizu C, Terai M, Hamada H, *et al.* A genome-wide association study identifies three new risk loci for Kawasaki disease. Nat Genet. 2012 Mar 25. doi: 10.1038/ng.2220. Nat Genet. 2012 Mar 25;44(5):517-21.

[135] Kuo HC CM, Hsu YW, Lin YC, Huang YH, Yu HR, Hou MF, Liang CD, Yang KD, Chang WC and Wang CL. CD40 gene polymorphisms associated with susceptibility of Kawasaki disease and coronary artery lesions in the Taiwanese population. The scientific world journal. 2012:(In press). ScientificWorldJournal. 2012;2012:520865.

Chapter 11

Vascular Smooth Muscle Cells and the Comparative Pathology of Atherosclerosis

Hafidh I. Al-Sadi

Additional information is available at the end of the chapter

http://dx.doi.org/10.5772/48378

1. Introduction

Atherosclerosis (ATH) is a condition characterized by thickening of the arterial intima as a result of the accumulation of fatty materials (mainly cholesterol and cholesterol esters). Lesions of the disease (atheromas or atherosclerotic plaques) have three distinct components, (1) the atheroma which is the nodular accumulation of a soft, flaky yellowish material at the center of the large plaques, composed of macrophages nearest the lumen of the artery, (2) underlying accumulations of cholesterol crystals, and (3) calcification of the older or more advanced lesions. In human beings, ATH is major cause of adult mortality in the developed world [1]. In animals, ATH occurs rarely, and only infrequently leads to a clinical disease such as infarction of the heart and brain. Several studies have indicated that the pig, rabbit, and chicken are susceptible to experimental induction of the disease through feeding of a high- cholesterol diet and that the dog, cat, cow, goat, and rat are resistant. Naturally – occurring disease has been reported in aged pigs and birds and in dogs suffering from hypothyroidism that accompanies hypercholesterolemia [2].

The smooth muscle cells (SMCs) constitute the predominant cellular element of the vascular media, cause vasoconstriction or dilation in response to physiological or pharmacological stimuli, synthesize extracellular matrix (collagen, elastin and proteoglycans); elaborate growth factors and cytokines; and migrate to the intima and proliferate after vascular injury. These activities of SMCs are important in both normal vascular repair and pathological processes such as ATH. Physiological regulation of the migratory and proliferative activities of the SMCs is regulated by growth promoters and inhibitors. Among the promoters are the platelet – derived growth factor derived from platelets (and endothelial cells and macrophages), basic fibroblast growth factor, and interleukin 1. Inhibitors include heparin sulfates, nitric oxide, interferon γ, and transforming growth factor β.

Injury of the wall of blood vessels stimulates SMCs growth by disrupting the physiological balance between inhibition and stimulation. Repair of the injured vascular wall constitutes a physiologic healing response with the formation of a neointima, in which SMCs: (1) migrate from the media to the intima, (2) proliferate as intimal SMCs, and (3) elaborate extracellular matrix. During the healing process, the SMCs in the intima lose the capacity to contract and gain the capacity to divide. Within the intima, the SMCs may return to the nonproliferative state when either the overlying endothelial layer is re – established after acute injury or the chronic inflammation ceases. However, intimal thickening occurs when the healing response is exaggerated and this can cause stenosis or occlusion of small and medium sized blood vessels.

2. Functional roles of VSMC in atherosclerosis

Functional roles of VSMC in atherosclerosis include: (1) phenotypic switching (from quiescent "contractile "phenotype to active "synthetic "state; (2) extracellular matrix (ECM) deposition; (3) proliferation; (4) migration; (5) inflammatory gene expression, (6) oxidant stress; and (7) monocyte retention (monocyte – VSMC binding) (Figure 1).

2.1. Phenotypic switching

Several studies demonstrated that the vascular smooth muscle cells have phenotypes that differ in the media and atherosclerotic lesions, and that phenotypic switching of SMCs plays a central role in atherosclerosis according to Ross's hypothesis [3, 4]. It has been proposed that before migration from the media into the intima, a transition of the SMCs phenotype is required [5]. In the media, the SMCs have a contractile phenotype that enables them to regulate the vascular tone. When these cells proliferate they acquire a synthetic phenotype. During the proliferative state the SMC requires extensive changes in gene expression and protein synthesis [6].

A major challenge in understanding differentiation of the SMCs is their ability to appear as a wide range of different phenotypes at different stages of development, and even in adult organism the cells are not terminally differentiated and are capable of major changes in their phenotype in response to changes in their local environment [7, 8, 9]. During early stages of vasculogensis SMCs are highly migratory and undergo rapid cell proliferation. Many studies indicated that there is a remarkable amount of movement of SMCs and SMC progenitor cells as part of the complex morphogenic events that leads to the formation of the cardiovascular system [10, 11]. During vascular development, SMCs also show high rate of synthesis of extracellular matrix components including collagen, elastin, proteoglycans, cadherins, and integrins that share the formation of blood vessel mass. In this stage of development, SMCs form abundant gap junctions with endothelial cells, and the process of investment of endothelial tube with SMCs or pericytes is necessary for vascular maturation and vessel remodeling [12]. In comparison, the SMCs in adult blood vessels show very low rate of proliferation turnover, are largely nonmigratory, show a very low rate of synthesis of extracellular matrix components, and are committed only to carry their contractile function [9]. Mature fully differentiated SMCs express a repertoire of appropriate receptors, ion channels, signal transudation molecules, calcium regulatory protein, and contractile protein

[7]. Following vascular injury the "contractile" SMCs are capable of undergoing transient modification of this phenotype to a highly "synthetic ' phenotype, and they play an important role in the repair of vascular injury. However, that repair of vascular injury is carried out principally (or exclusively) by reversible phenotype modulation of preexisting SMCs is a matter of controversy. Many studies proposed two alternative mechanisms although in reality none is mutually exclusive. In the first one there is evidence that the circulating bone marrow derived SMC progenitor cells play a major role in the repair of vascular injury [13, 14, 15]. The second proposed mechanism is supported by the evidence that SMC populations within blood vessels are extremely heterogeneous with resident stable populations of preexisting SMCs that are phenotypically distinct from the classical definition of a contractile SMCs [16, 17] and that these cells accomplish the injury repair [9].

Some recent studies have provided evidence that circulating cells, presumably derived from bone marrow, can contribute to neointima formation and repair of vascular injury [13, 15, 18]. However, these studies involved very extensive damage to medial SMCs (almost complete destruction of the media and SMCs death), and / or immunologic injury due to genetic mismatch of host and donor tissues following tissue transplantation combined with lack of adequate immunosuppression therapy. However, there is no single study in the severe mechanical injury models have provided strong evidence that bone marrow cells within lesions express definitive SMC markers such as smooth muscle (SM) myosin heavy chain (MHC) and smoothelin. In addition no studies have been made on the possibility of fusion of circulating progenitor cells with resident SMCs [9].

A number of studies are available demonstrating that there are heterogeneities between SMCs within the blood vessel with retention of a resident stable population of cells that have a "synthetic phenotype" [17, 19]. In the study of [19] a panel of antibodies specific for different markers of SMC differentiation including SMC α – actin, SM MHC, calponin, desmin, and meta – vinculin were used to perform immunofluorescence labeling studies on cryosections of adult and fetal bovine main pulmonary arteries. The authors performed also Western analyses of these marker genes in the three different layers of the adult bovine pulmonary artery. These authors reported the presence of four distinct populations or clusters of MSCs based on morphology, cell orientation, pattern of elastic lamellae, and immunostaining patterns and proposed that these populations may represent unique lineages performing different functions within the arterial media, and respond in different ways to stimuli. Although strong evidence exists for the presence of heterogenous populations of SMCs in vivo no studies were done to demonstrate that these represent distinct stable SMC lineages that play a preferential role in carrying out repair of vascular injury in vivo. The studies of Clowes and co-workers [20, 21, 22] would seem to refute such a possibility in that they showed SMC growth fractions (the fraction of medial SMC at time O that leave Go and reenter the cell cycle) of up to 60% following balloon injury of the rat carotid artery, indicating that the majority of SMCs within the media possess the ability to reenter the cell cycle and contribute to repair of vascular injury in adult animals. Thus the preexisting "subpopulation "of SMC capable of phenotypic switching is much greater than frequencies reported by [19] and represent a large fraction of SMCs in the vessel wall. [23] using the generation of complex SMC ancestor tables for the entire SMC population within the thoracic and abdominal aorta based on pulse – chase labeling with [^{3}H] thymidine in hypercholesteremic swine models of atherosclerosis provided evidence that supports the previous findings. They found that the intimal lesions were polyclonal and derived from multiple histologically discrete medial SMCs that initiated DNA replication and subsequently underwent several rounds of DNA replication. These findings are in discrepancy with a model in which only a small fraction of medial SMCs contribute to lesion formation.

Results of another study [24] reported that distinct populations of rat cultured SMCs (adult and embryonic) when implanted into a rat carotid artery in vivo retained some phenotypic differences. These findings suggest that there is considerable stability in the phenotype of these cells. However, it is possible that the stable epigenetic reprogramming of these cells was a function of their extensive growth in culture. Additionally, since large number of cells were transplanted, it is possible that the transplanted cells could created their own "microenvironmental domain or milieu "and that autocrine and paracrine effects contributed to the retention of phenotypic differences. [25] suggested the existence of a subpopulation of terminally differentiated SMC that is incapable of cell cycle reentery. Evidence for this suggestion was based on studies showing failure of a subpopulation of SMC derived from dog aorta to proliferate in culture. However, this finding may repersent the lack of appropriate culture reagents and / or conditions necessary to support growth of these cells.

In summary, it seems that the principal source of SMCs responsible for repair of vascular injury under "normal "circumstances are the preexisting SMCs that undergo transient and reversible phenotypic modulation. However, circulating bone marrow cells, cells derived from the adventitia, and / or preexisting subpopulations of phenotypically modified SMC can participate to some extent as well. The role of each of these different populations differs according to the nature of the vascular injury or the disease state.

2.2. ECM deposition

ECM comprises > 50 % of the atherosclerotic lesion and it consists of a mixture of vastly different macromolecules including collagen, elastin, glycoproteins, fibronectin, laminin, vitronectin, and thrombospondin. These matrix proteins are produced largely by the activated VSMCs [26, 27] and confers tensile strength and viscoelasticity to the arterial wall. Each of these ECM components possesses unique structural properties that determine its own roles during the development of atherosclerotic plaques. In addition to its role in supporting the plaque, ECM participates in many key events such as cell migration and proliferation, lipoprotein retention and thrombosis.

Matrix metalloproteinases (MMPs) are endopeptidases produced by SMCs and macrophages and they contribute significantly to the degradation and remodeling of the plaque extracellular matrix [29]. MMps that are induced to be expressed by environmental factors present within the lesion can actively modify the matrix in which SMCs reside and actively contribute to further phenotypic switching of the SMC. SMC phenotype is prone to

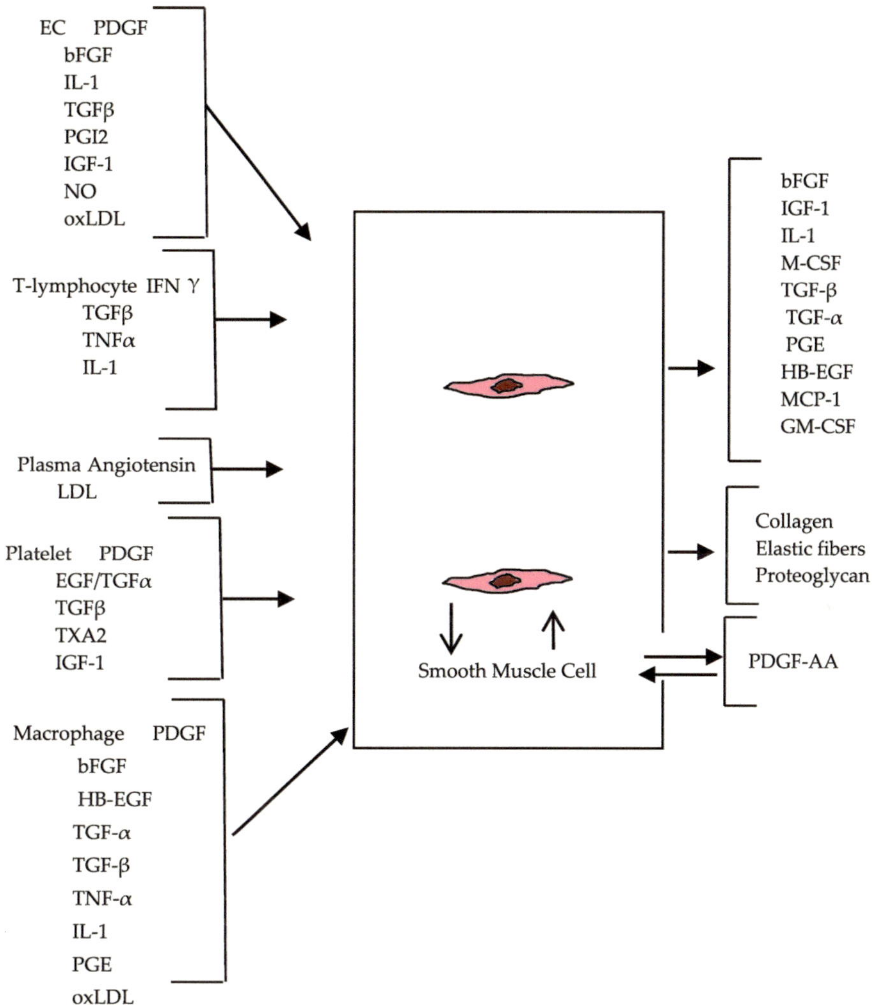

Figure 1. Redrawn and modified from Ross [28].

modification by a number of factors including PDGF [30], TGF β [31], intric oxide [32], and reactive oxygen species [33]. These factors have been demonstrated to modify MMP production in cultured SMCs, although information concerning the mechanisms and factors that control expression of MMPs in vivo are still lacking. A major determinant of plaque stability is the existence of a balance between production of matrix degrading MMPs, the inhibitors MMPs or tissue inhibitor of metalloproteinase (TIMPs), and matrix production by SMCs [34, 35]. In normal arteries of human and laboratory animals, MMP – 2 (72 – Kda gelatinase) and TIMp – 1 and TIMP – 2 are constitutively expressed at levels creating a stable balance between endogenous matrix proliferation and matrix degradation m and a normal MMP – to – TIMP ratio [29]. In the developing atherosclerotic lesion, this ratio is tipped towards MMPs as described in part by an increase in MMP – 3 (stromelysin) and MMP – 9 (92 – Dka gelatinase), and increased MMP – to – TIMP ratio [29]. It would seem that MMP overexpression is important for the migration and formation of a SMC – rich fibrous plaque and subsequent plaque stabilization. Galis *et al.* [36] presented evidence that MMP – 3 and MMP – 9 could be also expressed by SMCs in the shoulder region of human atherosclerotic plaques. [37] found a correlation between decrease of SM – 1 and SM – 2 marker expression and increase in PDGF and MIM – 3/ 9 expression in the rabbit neointima. Knoockout gene studies for MMP – 9 have demonstrated decreased intimal SMC hyperplasia and reduced late lumen loss in the mouse carotid artery flow cessation model [38] and in the carotid wire injury model [39]. During the natural progression of atherosclerosis in APOE – 1- mice, differential expression of MMP – 9 increases with time in lesional areas versus nonlesion areas [40]. Lemaitere *et al.* [35] showed that in TIMP – 1 / - mice crossed to APOE – 1- mice there was increased aortic medial ruptures compared with control mice after 10 weeks of Western diet feeding, indicating imbalance between MMP expression and TIMP expression can lead to lesion instability.

2.3. SMC proliferation

Immediately following activation by the injury, growth factor, or cytokine, the SMC undergoes phenotypic change that leads to a migratory and secretory cell that migrates into the neointima [41]. Stimulation by growth factor or cytokine causes the SMCs to proliferate and secrete matrix proteins and enzymes. Whereas the complex atherosclerotic lesions contain a mixed type of cells including SMCs, lipid laeden macrophages, and lymphocytes, the vascular SMCs are the dominant cellular component of the de novo and in-stent restenotic lesions. Results of both experimental and randomized clinical trials have shown that significant retardation of in-sent stenosis could be achieved through the delivery of inhibitors of inflammation and of SMCs proliferation such as sirolimus and paclitaxel to the site of intervention on drug – eluting stents [42, 43].

Proliferative phenotype of vascular SMCs in both physiological and pathological events involve critical changes in the gene expression patterns of the proliferating cells. Growth-promoting factors such as angiotensin 11 or oxidized low density lipoprotein stimulate the vascular SMCs and lead to increased expression of genes coding for iron transporters, extracellular matrix components, cell – cell adhesion molecules, cytoskeletal proteins, transcription factors and cell cycle regulatory proteins [44, 45]. Comparison between the healthy vascular SMCs and vascular SMCs isolated from various disease conditions, such as primary atherosclerosis and in – stent stenosis reveals similar changes in gene expression patterns [46].

Changes in gene expression programs are influenced largely by transcriptional events, but the contribution of posttranscriptional events (such as mRNA processing, transport, turnover, and translation) is becoming increasingly recognized [47]. One of the most important of these events, stability of regulated mRNA critically contributes to the implementation of gene expression patterns during the cellular response to mitogens, immunological triggers, stress, and differentiation agents [48, 49]. A growing list of proteins central to the execution of such responses (P21, Hsp70, MnsoD, catalase, Cdc25, cyclin A, cyclin B1, c-Fos, c-Jun, c-Myc, Egr-1, etc) are encoded by labile mRNA, which have tightly regulated half – lives [47].

Among the well – studied determinants of transcript stability are U – rich or A + U rich elements (collectively termed AREs) generally present in the 3' -untranslated regions of labile mRNA [50]. Several RNA – binding proteins have been found to bind to AREs and leads to transcript decay and they include BRF1, AUF1, and KSRP [51, 52, 53]. HuR, a member of the Hu / ELAV protein family binds to AREs and instead promote transcript stabilization. HuR stabilizes many mRNAs that encode growth factors, cell division proteins, and cytokines. An important role of HuR during the response to immune factors and proliferate signals has been described in many cell systems [54, 55, 56, 57, 58].

HuR was found to regulate cell proliferation in a mouse model of skeletal muscle development and regeneration, purportedly through its effect on the expression of proteins governing cell growth and differentiation [59, 60]. Pullmann, Jr. *et al.* [47] found that treatment of hVSMCs with platelet - derived growth factor increased HuR levels in the cytoplasm, thereby influencing the expression of metabolic, proliferative, and structural genes. In addition, knockdown of HuR expression by using RNA interference was reported to cause a reduction of hVSMC proliferation, both basally and following platelet – derived growth factor treatment. These authors postulated that HuR contributes to regulating hVSMC growth and homeostasis in pathologies associated with VSMC proliferation [47]. In other studies, HuR was found to increase the rate of proliferation of human diploid fibroblasts, shortened the cell division time of colon cancer cells, and accelerated the development of tumors in nude mice [55, 56, 61]. The stimulatory effect of HuR on cell proliferation is proposed to be though increasing the stability of several mRNAs, including those that encode cyclin A, cyclin B1, c- Fos, c – Myc, and cyclin D1, thereby enhancing the expression of the corresponding proteins,. which promote progression through the cell division cycle [62]. Pullmann, Jr. et al. [47] have identified the cDk2 mRNA as a novel target of HuR. This is an interesting finding in view of the regulatory influence of cdk2 and cyclin A on rat carotid artery VSMC proliferation. It has been proposed that HuR may be a contributing factor to smooth muscle cell and neointima proliferation and consequently to atherosclerosis [47].

In summary, the migratory and proliferative activities of VSMCs are regulated by a balance between growth promoters such as platelet derived growth factors (PGF), endthelin – 1 (ET – 1), thrombin, fibroblast growth factor (FGF), interleukin -1 (IL - 1) and inhibtors such as, heparin sulfates, intric oxide (NO), transforming growth factor (TGF) – beta. A role has been suggested for matrix metallo – proteinases (MMPs) could catalyze and remove the basement membrane around VSMC and facilitate contacts with the interstitial matrix. VSMCs are stimulated to proliferate and migrate by some kinds of cytokines, growth factors, and angiotensin II (Ang – II). Apoptosis, proliferation and migration of VSMCs are essential features of the pathogenesis of atherosclerosis and plaque rupture. Rupture of the plaque is associated with increased number of fibrous cap macrophages, increased VSMC apoptosis and reduced fibrous cap VSMCs. Within the plaques, the VSMCs are the only cells capable of synthesizing structurally important collagen isoforms, and the apoptosis of VSMC might promote plaque rupture [63].

2.4. SMC migration

Agents that induce VSMC migration include growth factors (Angiotensin II, PDGF, bFGF, HB – EGF, IGF – 1, VEGF, Thrombin), cytokines / chemokines (IL – 1 β,. IL – 6, TL -6, TFβ 1, TNF – alpha, MCP – 1), extracellular matrix components (collagen 1, IV , collagen VIII, fibronectin, hyaluronan, laminin, osteopontin thrombospondin, vitronectin), bioactive lipids (LPA, Hydroxyeicosatrienoic acids [12 or 15 (s) – HETE], diabetogenic agents [high glucose (25 mmol / L)], advanced glycation end products (AGEs), RAGE ligands S 100 B, and other molecules (ATP, UTP, norepinephrine, histamine, and serotonin) [64]. Angiotensin II – induces VSMC migration through P38 – MAPK activated c – Src through two distinct but redundant pathways, one via Syk, and other via EGFR transactivation through ERK 1/2 and partially through p38 MAPK [64]. Receptor for advanced glycation end products (RAGE) ligands induce inflammatory genes and VSMC migration via Src kinase [64]. Additionally, RAGE ligand (S100B) activates Src and MAP kinases in VSMCs [64]. In diabetic db / db mice, it has been found that there was enhancement of RAGE expression, Src activation, and migration in VSMCs [64].

2.5. Inflammatory gene expression

The atherogenic cytokines that are released from SMCs, endothelial cells (EC), macrophages, T lymphocytes, and B lymphocytes include IFN (released from SMC, macrophages, T lymphocytes, and causes increased SMC migration and proliferation, increased ECM remodeling, and increased adhesion molecule expression), IL – 1 (released from SMCs, EC, macrophages, and T and B lymphocytes and increases SMC migration and proliferation, monocyte accumulation, and adhesion molecule expression) , IL – 18 (released from SMC, EC, and macrophages, and causes increased adhesion molecule expression, increased SMC accumulation, and ECM remodeling), MCP – 1 (produced by the SMC, EC, macrophages, and T lymphocytes, and increases recruitment of monocytes , SMC migration and ECM synthesis and remodeling), PDGF – BB (produced by the SMC, EC, and macrophages, and increases SMC migration and increases SMC migration anf proliferation), and TGFβ (elaborated by SMC, EC, macrophages, and proliferation and increases SMC migration and proliferation, and ECM synthesis) [64]. Cytokines that are involved in atherogenesis include TNF – α, IL – 1, IL – 6, IL – 8, IL – 12, IL – 15. IL – 18, IL 32, and MCP – 1 produced by macrophages under the influence of resisting leptin, and adiponectin elaborated from adipose tissue. Cytokines released from macrophages stimulate SMC to produce the cytokines TNF – α, IL – 1, IL – 6, and IFN – (?). Those same cytokines also activate the EC to prduce VCAM – 1, ICAM – 1, E – selectin, P – selectin, IL – 1, IL – 6, IL – 8, IL – 18, IL PB, and MCP – 1. Other cytokines also produced by Tho (IL – 2, IL – 3), Th 1 (IL – 2, INF, IL – 17) Th 2 (IL – 4, IL – 5 < IL – 6, IL – 10, IL – 13), Treg (IL – 10, TGFβ), NKT cells (IFN), and macrophages (IL – 10, TGFβ, IL – lra, IL – 18BP) [64].

2.6. Oxidative stress

An important feature of the pathological porcess in atherogenesis is an increased generation of reactive oxygen species (ROS). All components of the atherosclerotic lesion has been shown to increase production of ROS mainly superoxide anion ($O2^-$) [26, 65]. The ROS are produced by the VSMCs, endothelial cells, fibroblasts, and infiltrating leukocytes [66]. These ROS affect gene transcription, damages DNA, and increases production of inflammatory transcription factors [67]. Oxidation of LDL and scavenging of endothelium – derived NO are the best studied effects [26].

The mechanism of oxidative modification of LDL is unknown but there is always oxidized LDL in atherosclerotic lesions. Experimental studies indicated that the level of oxidized LDL, as measured by autoantibody titers, is reflective of the atherosclerotic burden [68]. Many of the atherogenic processes are induced by the oxidized LDL and these include transcription of proatherogenic genes, production of matrix metalloproteinases and tissue factor, antagonism of endothelial cell production of NO, and promotion of VSMC apoptosis [69]. The increased production of superoxide anion rapidly reacts with NO to produce peroxynitrite, a potent oxidant [70]. It has been stated that scavenging of NO increases inflammation, platelet activation, and vasoconstriction [26].

Studies of antioxidant vitamins, including the Guppo Italiano per 10 Studio della Sopravvivenza nell' Infarto myocradico (GISSI) Prevention Trial, the Heart Outcomes prevention Evalution Study (HOPE), and the Heart protection Study (HPS), did not show any reduction in clinical events with antioxidant vitamin E therapy [71, 72]. However, in these trials there were limitations that precluded the adequate test of the hypothesis ([73]. One of these limitations is that the rate constant for reaction of vitamin E or C with superoxide anion is much slower than for superoxide anion with NO or endogenous antioxidant enzymes [74]. Oral intake only slightly increases the levels of vitamin in plasma and tissue, and this may not affects the events in the vascular wall, where it is not concentrated. Vitamin E contributes little in the form of antioxidant protection in the cytoplasm, nucleus, or interstitial space since it is concentrated in the lipid bilayers. Additionally, a delay in treatment may have abolished its effect on the development of lesions and with little effect on plaque rupture and clinical events. However, some other therapeutic studies in the same trials (e.g. HOPE and HPS) indicated significant benefit [67, 75]. Of equal importance is that conventional antipletlet therapy has antioxidant effects as it has the ability to limit ROS production by activated platelets. The fact that oxidative stress plays important role in the pathogenesis of atherosclerosis makes clear that the limitations of current therapies should not conclude therapeutic interest in this area but stimulates studies into new ways of treatment.

2.7. Monocyte – VSMC binding

Monocytes in circulation adhere and migrate across the endothelium in response to atherogenic stimuli. Initially, these processes may be reversible, whereas subsequent accumulation and retention of monocytes – macrophages into the intima become a central pathogenic process in atherogenesis [76]. The mechanisms by which the monocytes are retained in the subendothelium and the role of VSMCs in this process are not known. Adhesive reactions between marginated monocytes and VSMCs have been proposed to contribute to monocyte- macrophage retention in the intima. Evidence for the possible interaction between VSMCs and the monocytes comes from the fact that VSMC express adhesion molecules within the atherosclerotic lesions but not in normal vessels [77]. In addition, a highly significant association was demonstrated between VSMC vascular cell adhesion molecule (VCAM) – 1 expression and the content of the intimal macrophage [78, 79]. Furthermore, a significant focal expression of intercellular adhesion molecule (ICAM) – 1 on VSMC in regions prone to atherosclerosis was found preceding mononuclear cell infiltration in man, which indicates a causative role in lesion development [80]. It has been shown that cell – to – cell interactions between monocytes and VSMC enhanced the procoagulant activity of monocytes and increased the production in both cell types of atherosclerosis – related materials such as metalloproteinase – 1 [81]. These findings indicated that the VSMCs and monocytes – macrophages are not merely neighbors in the vessel wall but that VSMC – monocyte interactions constitute additional signals in the pathogenesis of atherosclerosis. However, the cellular, molecular, and signal transudation mechanisms need to be elucidated.

A well-known fact is that angiotensin II (ANG II) and platelet – derived growth factor (PDGF) – BB have a significant role in vascular remodeling and atherosclerosis [82]. They are capable of inducing VSMC migration, hypertrophy, and proliferation [82, 83]. Cai *et al.* [84] investigated the effects of angiotensin II (ANG II) and PDGF – BB on VSMC – monocyte interactions. They found that treatment of human aortic VSMC (HVMC) with ANG II or PDGF – BB significantly increased binding to human monocytic THP-1 cells and to peripheral blood monocytes. This was inhibited by antibodies to monocyte β1 – and β2 – integrins. Attenuation of the binding was also achieved through blocking of VSMC arachidonic acid (AA) metabolism by inhibitors of 12 / 15 – lipoxygenase (12 / 15 – LO) or cyclooxygenase -2 (COX - 2). On the other hand, enhancement of binding was obtained by overexpression of 12 / 15 – LO or COX – 2. Binding was also enhanced by direct treatment of HVSMC with AA or its metabolites. Additionally, VSMC derived from 12 / 15 – LO knockout mice showed reduced binding to mouse monocytic cells in comparison with genetic control mice. Using specific signal transudation inhibitors, Cai *et al.* [84] showed the involvement of Src, phosphoinositide 3 – kinase, and MAPKs in ANG 11- or PDGF – BB induced binding. These authors also found that after coculture with HVSMC, THP – 1 cells surface expression of the scavenger receptor CD36 was increased. In conclusion, results of the work of Cai *et al.* [84] indicated that the growth factors may play additional roles in atherosclerosis by increasing monocyte binding to VSMC via AA metabolism and key signaling pathways. This process can lead to monocyte subendothelial retention, CD36 expression, and foam cell formation.

3. VSMCs in the pathogenesis of atherosclerosis

3.1. Humans

Three major types of cells that are commonly seen in the atherosclerotic lesions are the SMCs (which dominates the fibrous cap), macrophages (inflammatory cells) that infiltrate around the necrotic core, and the lymphocytes (intracellular and intercellular lipid) which have been mainly ascribed to the fibrous cap [85, 86]. A very complex interplay exists between these cells in the various developmental stages of the atherosclerotic lesion. This complex interrelation is further complicated by a number of risk factors that contribute to the clinical manifestation of the disease, and they include the abnormal vasomotor function, the thrombogenicity of the blood vessel wall, the state of activation of the coagulation cascade, the fibrinolytic system, SMC migration and proliferation, and inflammation process [76, 86, 85]. However, the exact cause of development of the atherosclerosis is not completely understood. Similarly much more information is needed concerning regulation of the key role of SMCs in vascular injury repair and in the development and / or progression of atherosclerosis. Finally, additional work is needed to clarify the specific contributions of the SMC versus other cell types within the lesion, such as macrophages and endothelial cells, to the end-stage clinical sequelae of atherosclerosis including plaque rupture, thrombosis, infarction, vasospasm, myocardial ischemia, and death.

3.2. Birds

Various avian species such as pigeons [87, 88], turkeys [89], and chickens [90] have been shown to be the convenient experimental animals for induction of atherosclerosis. Shih *et al.* [91] mentioned that the Japanese quail is an ideal laboratory animal for long – term experiment because of its small size, short life cycle, and low feed consumption. Atherosclerosis was conventionally induced in various experimental animals through feeding of cholesterol and fat or oil [88, 89, 90]. Oku *et al.* [92] demonstrated that dietary feeding of 2 % cholesterol and 15 % corn oil for 3 months can induce typical atherosclerotic lesions more frequently in the ascending aorta and its large branches than in abdominal aorta in Japanese quails. This finding is comparable to that reported by Toda *et al.* [93] in the chickens. Atherosclerosis was also induced in quails by feeding them with 2 % cholesterol and 0.5 % cholic acid for 15 weeks [94]. Morrissey and Danalbson [95] demonstrated that atherosclerosis could be induced by feeding quails with 1 % cholesterol and 10 % fat for 10 weeks. Wexler [96] reported that both the male and female Japanese quails developed spontaneous atherosclerosis at 2 years of age. Jarrold *et al.* [97] and Velleman *et al.* [98] have demonstrated that the cholesterol – induced atherosclerosis (CIA) line of Japanese quails is a valid animal model to study ECM remodeling induced by hypercholesterolemia. Jarrold *et al.* [97] showed that the proteoglycan decorin was localized in the foam cell regions and collagen type I was found to surround the foam cells where decorin accumulated. Velleman *et al.* [99] showed that remodeling of the collagen component of the dorsal aorta extra – cellular matrix during the progression of atherosclerosis in Japanese quails selected for CIA. Toda *et al.* [93] reported that fibroblasts rather than smooth muscle cells are the main cellular component in the development of atherosclerosis in Japanese quails as in chicken. Oku *et al.* [92] suggested that phenotype transformation of intimal cell migrating from the tunica media to play an important role in the initiation and the development of atherosclerosis. Casale *et al.* [100] studied the cellular events of quail atherosclerosis using monoclonal antibodies to alpha-actin and chicken macrophages and effectively identified the presence of SMC and macrophages, respectively, as constituents of the atherosclerotic lesions. The presence of macrophage, as well as SMC proliferation, was observed in early lesions. Although it was not possible to acertain the first cell type to be involved in the initial stages of atherosclerosis, results suggested early intervention of macrophages and SMC. Bavelaar *et al.* [101] studied the possibility that feeding of α-linolenic acid instead of linoleic acid or saturated fatty acids would diminish the degree of atherosclerosis in cholesterol-fed quails. The authors concluded that a differential effect on the development of atherosclerosis of α-linolenic acid, linoleic acid and saturated fatty acids could not be demonstrated.

Fabricant and Fabricant [102] investigated the roles of both Marek's disease herpesvirus (MDV) and dietary cholesterol in atherosclerosis in chickens. The birds were examind 7 months after MDV infection with and without cholesterol feeding for gross and microscopic arterial lesions. Typical lesions of atherosclerosis were observed only in infected normocholesterolemic or hypercholesterolemic birds. They were not detected in uninfected birds even if the birds were hypercholesterolemic. Furthermore, immunization with turkey herpesvirus vaccine or SB – 1 vaccine prevented atherosclerotic lesions. Hajjar *et al.* [103] demonstrated histologically that infection of normocholesterolemic, specific-pathogen-free chickens with Marek's disease herpesvirus (MDV) lead to chronic atherosclerosis like that in humans (Figures 2 and 3).

Spontaneous atherosclerosis is of common occurrence in captive parrots [104]. It occurs in all parrot species but with the highest occurrence in African Grey parrots and Amazons. Old birds are more commonly affected, and the disease has been seen in both males and females. Sudden death is the most common signs, but clinical symptoms include dyspnea, lethargy and nervous sign, such as paresis and collapses. Atherosclerosis is mostly an unexpected finding at necropsy because clinical signs of the condition are seldomly seen and the difficulty associated with diagnosis. In parrots, age and species are determinants of atherosclerosis. Risk factors have been suggested to include an elevated plasma cholesterol level, diet composition, social stress, and inactivity [104]. Frick *et al.* [105] studied the incidence of spontaneous atherosclerosis in 62 African grey parrots (*Psittacus erithacus*) and 35 Amazon parrots (*Amazona spp.*). Incidence of atherosclerosis was 91.9% in African grey parrots and 91.4% in Amazon parrots. According to the missing lymphocytes and macrophages and the absence of invasion and proliferation of SMCs, the authors concluded that the "response-to-injury hypothesis" is inapplicable in parrots.

3.3. Pigs

In recent years an increase occurred in use of pigs as a promising species for morphological, biochemical, and metabolic studies of cardiovascular diseases particularly atherosclerosis [106, 107, 108]. Anatomy and physiology of the cardiovascular system of pigs resemble that of man. Among the similarities are size and distribution of arteries, location of thickened intima in the normal state, blood pressure, heart rate, plasma lipoprotein patterns, and responses to diets rich in fat and cholesterol [108]. Spontaneous atherosclerosis is known to develop in pigs with increased age, and have lipoprotein profiles and metabolism similar to

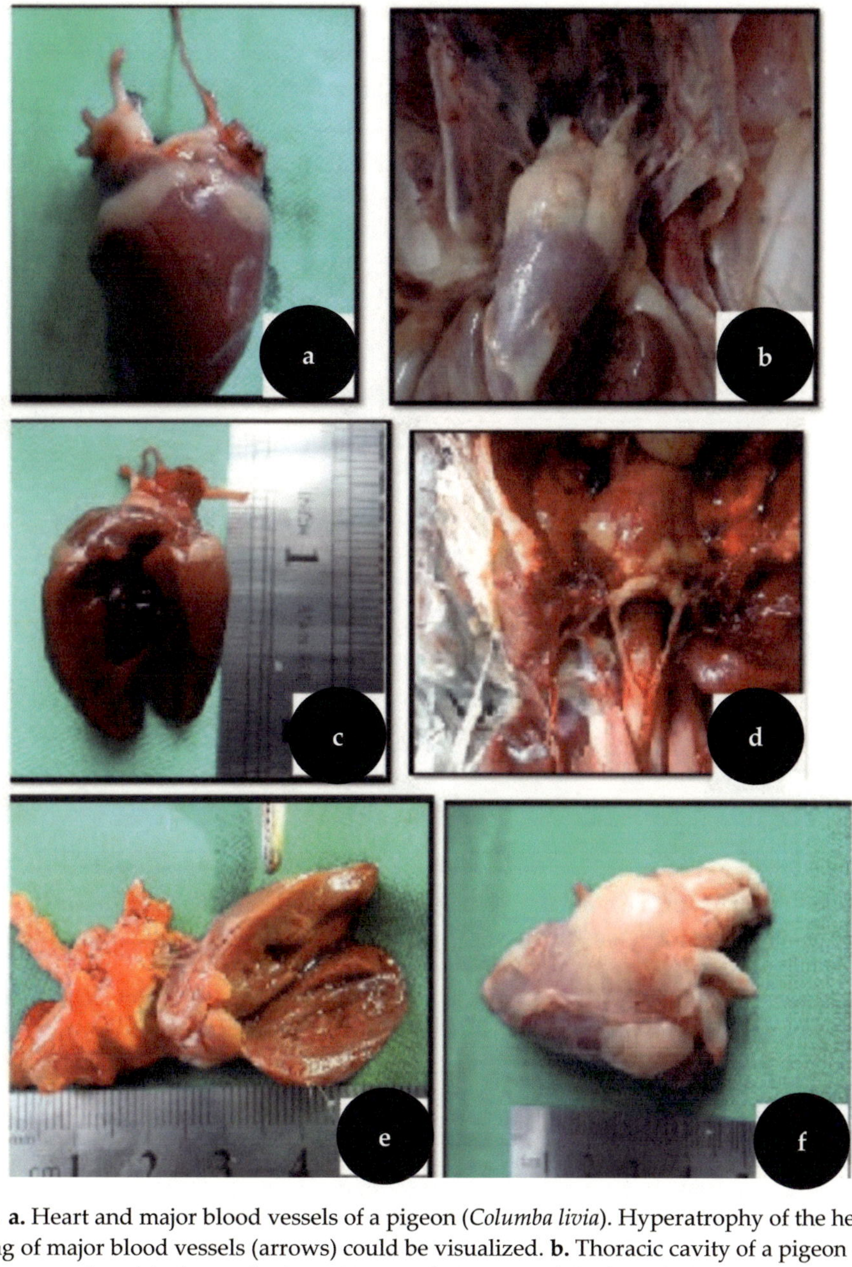

Figure 2. **a.** Heart and major blood vessels of a pigeon (*Columba livia*). Hyperatrophy of the heart and thickening of major blood vessels (arrows) could be visualized. **b.** Thoracic cavity of a pigeon (*Columba livia*). Hyperatrophy of the heart, fat deposition on the heart, and thickened blood vessels originating from the heart (arrows) could be visualized. **c.** Same heart shown in figure 2 following opening the chambers and shows narrowing of chambers as result of myocardial hypertrophy (arrows). **d.** A view of thoracic cavity of a chicken .A rounded hyperatrophied heart (H) with thickened major blood vessels (arrows) could be seen. **e.** Heart and major blood vessels of a chicken following opening of heart to show narrowing of ventricles as result of myocardial hyperatrophy. **f.** Heart and major blood vessels of a New Zealand White rabbit. Note fat deposition on pericardium and thickening of the walls of major blood vessels.

humans [106, 107]. Use of high fat, high cholesterol diets in pigs leads to elevation of total and LDL plasma cholesterol [106]. Several studies have shown the ability to induce aortic lesions in pigs very similar to those seen in human atherosclerosis disease [106].

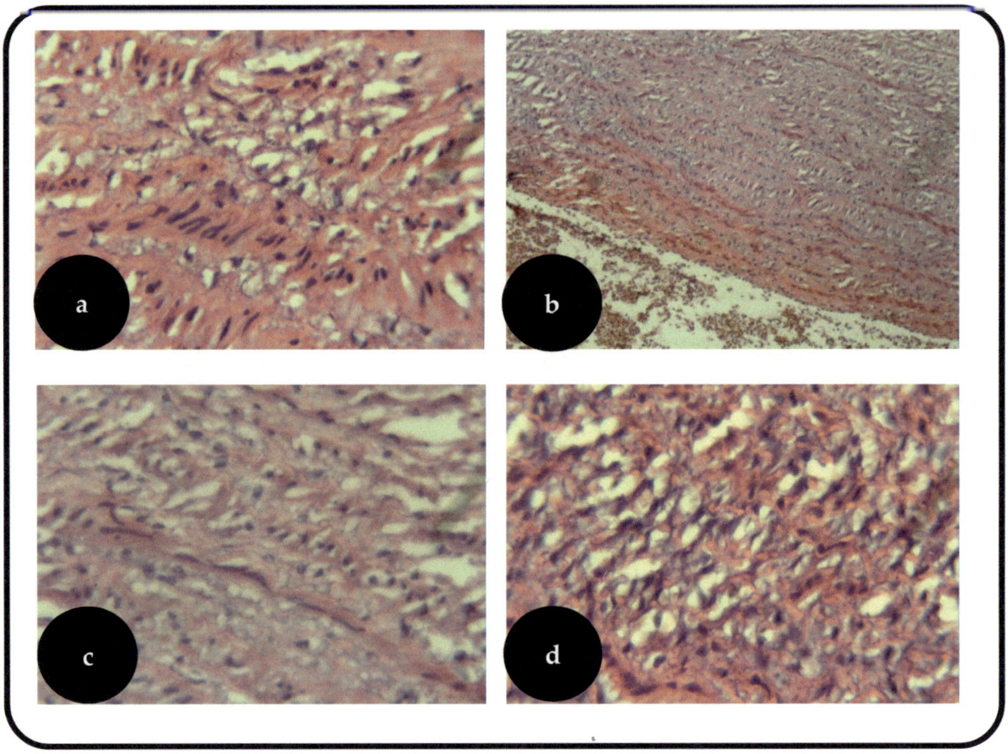

Figure 3. **a.** Microphotography of arterial wall in a pigeon (*Columba livia*) showing lipid accumulation in tunica media . H&E. **b.** Microphotography of arterial wall in a pigeon (*Columba livia*) showing lipid accumulation in tunica media and destruction of some of the elastic lamellae. H&E.
c. Microphotography of arterial wall in a pigeon (*Columba livia*) showing destruction of the elastic lamellae . H&E. **d.** Microphotography of arterial wall in a chicken showing lipid accumulation in the intima. H&E.

3.4. Rabbits

Among animals that have been used as models for experimental atherosclerosis, the rabbit is the only one that has the tendency to exhibit hypercholesterolemia within a few days of administration of a high cholesterol diet [110, 111]. Furthermore, the late stages lesions of human atherosclerosis are similar to those caused in rabbits when a diet low in cholesterol is administered for extended periods [110]. More advanced lesions in the thoracic and abdominal aorta, could be induced in rabbits through a high cholesterol diet can be combined with a single or double balloon injury. Balloon injury enhances the formation of atheromatic lesions and leads to the production of plaques with a lipid core covered by a fibrous cap with a high amount of SMCs. Such lesions are more similar to human atherosclerotic lesions than those produced by feeding rabbits with a high cholesterol diet alone [37, 112].

Hypercholesterolemia that is induced in rabbits through diet is caused by the accumulation of exogenous cholesterol. Rabbits cannot increase the excretion of sterols and this explains their high susceptibility to inducement of atherosclerosis [113]. Consequently, increased quantities of lipoproteins rich in cholesterol esters enter the blood circulation. LDL and β – VLDL are the main transporters of cholesterol in plasma. They remain for prolonged time in blood circulation [113].

In rabbits, morphological features of the lesion could be modified by the percentage of cholesterol added to the diet and the duration of the diet [111, 113]. Diets with a percentage of cholesterol of more than 2 % and given for short duration, cause hypercholesterolemia, and atherosclerotic lesions rich in foam cells originate from macrophages. In contrast of this, a diet with a low cholesterol content, and long duration causes atherosclerosis lesions, which are rich in SMCs and contain cholesterol deposits leading to atherosclerosis lesions more similar to those of humans [113]. More advanced lesions were found to be formed not with continuous but with intermittent atherogenic diets [113]. Additionally, increasing the percentage of cholesterol in the diet to more than 0.15 %, cholesterol esters were detected in the lesion [113]. Spagnoli *et al.* [114] found that the formation of advanced lesion depends on the age of the animal. Thus, old rabbits (3 – 4.5 years old) exhibit fibrotic plaques while young rabbits (4 months old) do not have such advanced lesions (Figure 2-f).

Author details

Hafidh I. Al-Sadi
Comparative Pathology, Department of Pathology, College of Veterinary Medicine, University of Mosul, Mosul, Iraq

Acknowledgement

I apologize for not being able to cite many excellent original papers on this subject. The field has been in existence for decades so the depth and breadth of the material is very large. I am indebted to Azhar Abd Al-Jabbar Albaker for typing the manuscript.

4. References

[1] Caplice NM, Bunch TJ, Stalboerger PG, Wang S, Simper D, Miller DV, Russell SJ, Litzow M R,, Edwards WD. Smooth Muscle Cells in Human Coronary Atherosclerosis can Originate From Cells Administered at Marrow Transplantation. PNAS 2003; 100 (8) 4754-4750.

[2] McGavin, D. and Zachary, J.F. (2007). Pathological basis of veterinary diseases. 4th ed. Mosby, Philadelphia.

[3] Ross, R. and Glomest, J.A. (1976a). The pathogenesis of atherosclerosis. N. Engl. J. Med., 295: 269-377.

[4] Ross, R. and Glomset, J.A. (1976b). The pathogenesis of atherosclerosis. N. Engl. J. Med., 295: 420-425.

[5] Owens, G.K. (1995). Regulation of differentiation of vascular smooth muscle cells. Physiol. Rev., 75: 487-517.

[6] Schwartz, S.M.; Heimark, R.L. and Mjesky, M.W. (1990). Developmental mechanisms underlying pathology of arteries. Physiol. Rev., 70-1177-1209.

[7] Owens, G.K.; Kumar, M.S. and Wamhoff, B.R. (2003). Molecular regulation of vascular smooth muscle cell differentiation in development and disease. Physiol. Rev., 84 (3): 767-801.

[8] Isogai, S.; Horiguchi, M. and Weinstein, B.M. (2001). The vascular anatomy of the developing zebrafish: an atlas of embryonic and early larval development. Dev. Biol., 230: 278-301.

[9] Czirok, A.; Rupp, P.A.; Rongish, B.J.; and Little, C.D. (2002). Multi- field 3D scanning – light microscopy of early embryogenesis. J. Microsc., 206: 209-217.

[10] Hungerford, J.E. and Little, C.D. (1999). Developmental biology of the vascular smooth muscle cell: building a multilayered vessel wall. J. Vasc. Res., 36: 2-27.

[11] Han, C.I.; Campbell, G. and Campbell, J.H. (2001). Circulation bone marrow cells can contribute to neointimal formation. J. Vasc. Res., 38: 113-119.

[12] Sata, M.; Saiura,. A.; Kunisato, A.,; Tojo, A.; Okada, S. (2002). Hematopoietic stem cells differentiate into vascular cells in pathogenesis of atherosclerosis. Nature Medicine, 8 (4): 403 – 409.

[13] Shimizu, K.; Sugyama, S.; Aikawa, M.; Fukumoto, Y.; Rabkin, E.; Libby, P. and Mitchell, R.N. (2001). Host bone – marrow cells are a source of donor intimal smooth – muscle – like cells in murine aortic transplant arteriopathy. Nature Med., 7: 738-741.

[14] Frid, M.G.; Kale, V.A. and Stenmark, K.R. (2002). Mature vascular endothelium can give rise of smooth muscle cells via endothelial - mesenchymal trasndifferentiation: in vitro analysis. Cir. Res., 90: 1189-1196.

[15] Hao, H.; Ropraz, P.; Verin, V.; Camenzind, E.; Geinoz, A.; Pepper, M.S.; Gabbiani, G. and Bochaton – Piallat, M.L. (2002). Heterogeneity of smooth muscle cell populations cultured from pig coronary artery. Arterioscler. Thromb. Vasc. Biol., 22: 1093-1099.

[16] Schhulick, A.H.; Taylor, A.J.; Zuo, W.Qiu, C.B.; Dong, G.; Woodward, R.N.; Agah, R.; Roberts, A.B.; Virmani, R. and Dichek, D.A. (1998). Overexpression of transforming growth factor beta 1 in arterial endothelium causes hyperplasia, apoptosis, and cartilaginous metaplasia. Proc. Natl. Acad. Sci., USA, 95: 6983-6988.

[17] Frid, M.; Moiseeva, E.P. and Stenmark, K.R. (1994). Multiple phenotypically distinct smooth muscle cell population exist in the adult and developing bovine pulmonary arterial media in vivo. Cir. Res., 75: 669-681.

[18] Clowes, A.W.; Clowes, M.M.; FIngerle, J. and Reidy, M.A. (1989). Kinetics of cellular proliferation after arterial injury. V. role of acute distension in the induction of smooth muscle proliferation. Lab. Invest., 60: 360-364.

[19] Clowes, A.W.; Clowes, M.M. and Reidy, M.A. (1986). Kinetics of cellular proliferation after arterial injury. I. smooth muscle growth in the absence of endothelium. Lab. Invest., 49: 327-333.

[20] Clowes, A.W.; Reidy, M.A. and Clowes, M.M. (1983). Mechanisms of stenosis after arterial injury. Lab. Invest., 49: 208-215.

[21] Thomas, W.A.; Florenting, R.A.; Reiner, J.M.; Lee, W.M. and Lee, R.T. (1976). Alterations in population dynamics of arterial smooth muscle cells during atherogenesis. IV. evi-

dence for a polyclonal origin of hypercholesterolemic diet- induced atherosclerotic lesions in young swine. Exp. Mol. Pathol., 24: 244-260.
[22] Bochaton _ Piallat, M.L.; Clowes, A.W.; Clowes, M.M.; Fishcer, J.W.; Redard, M.; Gabbiani, F. and Gabbiani, G. (2001). Cultured arterial smooth muscle cells maintain distinct phenotypes when implanted into carotid artery. Arterioscler. Thromb. Vasc. Biol., 21: 949-954.
[23] Seidel, C.L.; Helgason, T.; Allen, J.C. and Wilson, C. (1997). Migratory abilities of different vascular cells from the tunica media of canine vessels. Am. J. Physiol. Cell Physiol., 272: C847-C852.
[24] Faxon, D. P.; Fuster, V.; Libby, P.; Beckman, J. A.; Hiatt, W. R.; Thompson, R. W.; Topper, J. N.; Annex, B. H.; Rundback, J. H.; Fabunmi, R. P.; Robertsn, R. M. and Loscalzo, J. (2004) Atherosclerotic vascular disease conference writing group III: pathophysiology. Circulation, 109: 2617 – 2625.
[25] Raines, E.W. (2000). The extracellular matrix can regulate vascular cell migration, proliferation, and survival: relationships to vascular disease. Int. J. Exp. Pathol., 81: 173-182.
[26] Ross, R. (1993). The pathogenesis of atherosclerosis: a perspective for the 1990s. N. Engl. J. Med., 362: 801-809.
[27] Galis, Z.S. and Khatri, J.J. (2002). Matrix metalloproteinases in vascular remodeling and atherogenesis: the good; the bad, and the ugly. Circ. Res., 90: 251-262.
[28] Cho, A.; Graves, J. and Reidy, M.A. (2000). Mitogen – activated protein kinase mediated matrix metalloproteinase – 9 expression in vascular smooth muscle cells. Arterioscler. Thromb. Vasc. Biol., 20: 2527-2532.
[29] Ma, C. and Chegini, N. (1999). Regulation of matrix metalloproteinases (MMPs) and their tissue inhibitors in human myometrial smooth muscle cells by TGF-β1. Mol. Hum. Reprod., 5: 950-954.
[30] Gurjar, M.V.; Deleon, J.; Sharma, R.V. and Bhalla, R.C. (2010a) Mechanism of inhibition of matrix metalloproteinase-9 induction by NO in vascular smooth muscle cells. J. Appl. Physiol., 91: 1380-1386.
[31] Gurjar, M.V.; Deleaon, J.; Sharma, R.V. and Bhalla, R.C. (2010b). Role of reactive oxygen species in IL-1 beta- stimulate sustained ERK activation and MMD-9 induction. Am. J. Physiol. Heart. Circ. Physiol., 281: H2568-H2574.
[32] Fabunmi, R.P.; Sukhova, G.K.; Sugiyama, S. and Libby, P. (1998). Expression of tissue inhibitor of metalloproteinase -3 in human atheroma and regulation in lesion – associated cells: a potential protective mechanism in plaque stability. Cir. Res., 83: 270-278.
[33] Lemaitre, V.; Soloway, P.D. and D'Armiento, J. (2003). Increased medial degradation with pseudo-aneurysm formation in apoliportein E-knockout mice deficient in tissue inhibitor of metalloproteinases-1. Circulation, 107: 333-338.
[34] Galis, Z.S.; Sukhova, G.K.; Lark, M.W.; and Libby, P. (1994). Increase expression of matrix metalloproteinases and matrix degrading activity in vulnerable regions in human atherosclerotic plaques. J. Clin. Invest., 94: 2493-2503.
[35] Aikawa, M.; Rabkin, E.; Voglic, S.J.; Shing, H.; Nagai, R.; Schoen, F.J.; And Libby, P. (1998a). Lipid lowering promotes accumulation of mature smooth muscle cells expressing smooth muscle myosin heavy chain isoforms in rabbit atheroma. Circ. Res., 83: 1015-1026.

[36] Galis, Z.S.; Johnson, C.; Godin, D.; Magid, R.; Shipley, J.M.; Senior, R.M. and Ivan, E. (2002). Targeted disruption of the matrix metalloproteinase – 9 gene impairs smooth muscle cell migration and geometrical arterial remolding. Circ. Res., 91: 852-859.

[37] Cho, A. and Reidy, M.A. (2002). Matrix metalloproteinase - 9 is necessary for the regulation of smooth muscle cell replication and migration after arterial injury. Circ. Res., 91: 845-851.

[38] Jormsjo, S,; Wuttge, D.M.; Sirsji, A.; Whatling, C.; Hamsten, A.; Stemme, S. and Eriksson, P. (2002). Differential expression of cysteine and aspartic proteases during progression of atherosclerosis in apoliportein E-deficient mice. Am. J. Pathol., 161: 939-945.

[39] Rivard, A. and Andres, V. (2000). Vascular smooth muscle cells proliferation in the pathogenesis of atherosclerotic cardiovascular diseases. Histol. Histopathol., 15 (2): 557 – 571.

[40] Morice, M.C.; Serruys, P.W.; Sousa, J.E. et al., (2002). A randomized comparison of sirlimus – eluting stent with a standard stent for coronary revascularization. N. Engl. J. Med., 346: 1773-1780.

[41] Grube, E.; Silber, S.M.; Hauptmann, K.E. (2001). Taxus 1: prospective randomized, double – blind comparison of NIRxTM stent coated with paclitaxel in a polymer carrier in de-novo coronary lesions compared with uncoated controls. Circulation, 104 (Suppl. II): 463 (Abstract).

[42] Sukhanov, S.; Song, Y.H. and Delafontaine, P. (2003) Biochem. Biophys. Res. Comm., 306:443-449.

[43] Campos, A.H.; Zhao, Y.; Pollman, M.J. and Gibbons, G.H.(2003). DNA Microarray Profiling to Identify Angiotensin-Responsive Genes in Vascular Smooth Muscle Cells - Potential Mediators of Vascular Disease. Circ. Res., 92:111-118.

[44] Zhang, G.J; Goddard, M.; Shanaha, C.; Shapiro, L.and Bennett, M. (2002). Arterioscler. Thromb. Vasc. Biol., 22:2030-2036.

[45] Pullmann, R. Jr.; Juhaszova, M.; de silanes, I L.; Kawai, T.; Mazan – Mamczarz, K.; Halushka, M. K. and Gorospe, M. (2005). Enhanced proliferation of cultured human vascular smooth muscle cells linked to increased function of RNA – binding protein HR. J. Biol. Chem., 280 (24): 22819- 22826.

[46] Fan, J.; Yang, X.; Wang, W.; Wood, W, W. H. III; Becker, K.G.; and Gorospe, M. (2002). Global analysis of stress-regulated mRNA turnover by using cDNA arrays Proc. Natl. Acad. Sci. USA, 99:10611-10616.

[47] Wilusz, C.J.; Wormington, M. and Peltz, S.W. (2001). The cap-to-tail guide to mRNA turnover. Nat. Rev. Mol. Cell Biol., 2:237-246.

[48] Xu, N.; Chen, C.Y.; and Shyu, A.B. (1997). Modulation of the fate of cytoplasmic mRNA by AU-rich elements: key sequence features controlling mRNA deadenylation and decay. Mol. Cell Biol., 17:4611-4621.

[49] Gherzi, R.; Lee, K.Y.; Briata, P.; Wegmuller, D.; Moroni, C.; Karin, m.; and Chen, C.Y. (2004). A KH domain RNA binding protein, KSRP, promotes ARE-directed mRNA turnover by recruiting the degradation machinery. Mol. Cell, 14:571-583.

[50] Zhang W.; Wagner, B.J.; Ehrenman, K.; Schaefer, A. W.; DeMaria, C.T.; Carter, D,; DeHaven, K.; Long, L.; and Brewer, G. (1993). Purification, characterization, and cDNA cloning of an AU-rich element RNA-binding protein, AUF1. Mol. Cell Biol., 13:7652-7665.

[51] Stoecklin, G.; Colombi, M.; Raineri, I.; Leuenberger, S.; Mallaun, M.; Schmidlin, M.; Gross, B.; Lu, M.; Kitamura, T.; and Moroni, C. (2002). Functional cloning of BRF1, a regulator of ARE-dependent mRNA turnover. EMBO J.21:4709-4718.
[52] Wang, W.; Caldwell, M. C.; Lin, S.; Jurneaux, H.; and Gorospe, M. (2000). HuR regulates cyclin A and cyclin B1 mRNA stability during cell proliferation. EMBO J., 19:2340-2350.
[53] Lopez de Silanes, I.; Fan, J.; Yang, X.; Potapona, O.; Zonderman, A.B.; Pizer, E.S.; and Gorospe, M. (2003). Role of the RNA-binding protein HuR in colon carcinogenesis. Oncogene, 22:7146-7154.
[54] Wang, W.; Yang, X.; Cristofalo, V. J. Holbrook, N.J.; and Borospe, M. (2001). Loss of HuR is linked to reduced expression of proliferative genes during replicative senescence. Mol. Cell Biol., 21:5889-5898.
[55] Brennan, C.M. and Steitz, J.A. (2001). HuR and mRNA stability. Cell Mol. Life Sci., 58:266-277.
[56] Atasoy, U.; Curry, S.L.; Lopez de Silanes, I.; Shyu, A.B.; Casolaro, V.; Gorospe, M.; and Stellato, C. (2003). Regulation of eotaxin gene expression by TNF-alpha and IL-4 through mRNA stabilization: involvement of the RNA-binding protein HuR. J. Immunol., 171:4369-3478.
[57] Fegueroa, A.; Cuadrado, A.; Fan, J.; Atasoy, U,; Muscat, G.E.; Munoz-Canoves, P.; Gorospe, M.; and Munoz, A. (2003). Role of HuR in skeletal myogenesis through coordinate regulation of muscle differentiation genes. Mol. Cell Biol., 23:4991-5004.
[58] Van der Giessin, K.; Di-Marco, S.; Clair, E.; and Gallouzi, I.E. (2003). RNAi-mediated HuR depletion leads to the inhibition of muscle cell differentiation. J. Biol. Chem. 278:47119-47128.
[59] Wang, L.; Zheng, J.; Du, Y.; Huang, Y.; Li, J.; Liu, B.; Liu, C-J.: Zhu, Y.; Gao, Y., Xu, Q.; Kong, W. and Wang, X. (2010). Cartilage oligometric matrix protein maintains the contractile phenotype of vascular smooth muscle cells by interacting with α 7 β1 integrin. Circul. Res., 106: 514 – 524.
[60] Lopez de Silanes, I.; Zhang, M.; Lal, A.; Yang, X.; and Gorospe, M. (2004). Identification of a target RNA motif for RNA-binding protein HuR. Proc. Natl. Acad. Sci. USA, 101:2978-2992.
[61] Rudijanto, A. (2007). The role of vascular smooth muscle cells on the pathogenesis of atherosclerosis. Acta Med. Indones., 39 (2): 86-93.
[62] Natarajan, R. Role of vascular smooth muscle cells in the pathology of atherosclerosis. Department of diabetes Beckman Research Institute of City of Hope, Duarte, CA 91010 (internet).
[63] Maytin, M.; Leopold, J.; Loscalzo, J. (1999). Oxidative stress in the vasculature. Curr. Atheroscler. Rep., 1: 156-164.
[64] Zalba, G.; Beamount, J.; San Jose, G.; et al., (2000). Vascular oxidant stress: molecular mechanisms and pathophysiological implications. J. Physiol. Biochem., 56: 57-64.
[65] Griendling, K.K.; Harruison, D.G. (2001). Out, damned dot: studies of the NADPH oxidase in atherosclerosis. J. Clin. Invest., 108: 1423-1424.
[66] Tsimikas, S.; Palinski, W.; Witztum, J.L. (2001). Circulating auto – antibodies to oxidized LDL correlate with arterial accumulation and depletion of oxidized LDL in LDL receptor deficient mice. Arterioscler. Thromb, Vasc. Biol., 21: 95-1000.

[67] Kita, T.; Kume, N.; Minami, M. et al., (2001). Role of oxidized: LDL in atherosclerosis. Ann. NY. Acad. Sci., 947: 199-205.

[68] Beckman, J.S.; Beckman, T.W.; Chen, J.; et al., (1990). Apparent hydroxyl radical production by peroxynitrite: implications for endothelial injury from nitric oxide and superoxide. Proc. Natl. Acad. Sci., USA, 87: 1620-1624.

[69] de Gaetano, G. (2001). Low- dose aspirin and vitamin E in people at cardiovascular risk: a randomized trial in general practice. Collaborating Group of the Primary Prevention Project. Lancet, 357: 89-95.

[70] Yusuf, S.; Dagenais, G.; Pogue, J.; et al., (2000). Vitamin E supplementation and cardiovascular events in high – risk patients. The Heart Outcomes Prevention Evaluation Study Investigators. N. Engl. J. Med., 342: 154-160.

[71] Landmesser, U.; Harrison, D.G. (2001). Oxidant stress as a marker for cardiovascular events: ox marks the spot. Circulation, 104: 2638-2640.

[72] Jackson, T.S.; Xu, A.; Vita, J.A. et al., (1998). Ascorbate prevents the interaction of superoxide and nitric oxide only at very high physiological concentrations. Cir. Res., 83: 916-929.

[73] Libby, P. and Aikawa, M. (2002). Vitamin C, collagen, and cracks in the plaque. Circulation, 105: 139-1398.

[74] Libby, P. (2002). Inflammation in atherosclerosis. Nature, 420: 868-874.

[75] Braun, M.; Pietsch, P.; Schror, K.; Baumann, G. and Felix, S.B. (1999). Cellular adhesion molecules on vascular smooth muscle cells. Cardiovasc. Res., 41: 395-401.

[76] Huo, Y. and Ley, K. (2001). Adhesion molecules and atherogenesis. Acta Physiol. Scand., 173: 35-43.

[77] O'Brien, K.D.; Allen, M.D.; McDonald, T.O.; Chait, A.; Harlan, J.M.; Fishbein, D; McCarty, J.; Ferguson, M.; Hudkins, K. and Benjamin, C.D. (1993). Vascular cell adhesion molecule – 1 is expressed in human coronary atherosclerotic plaques: implication for the mode of progression of advanced coronary atherosclerosis. J. Clin. Invest., 92: 945-951.

[78] Endres, M.; Laufs, U.; Merz, H. and Kaps, M. (1997). Focal expression of intercellular adhesion molecule – 1 in the human carotid bifurcation. Stroke, 28: 77-82.

[79] Zhu, Y.; Hojo, Ikeda, U.; Takahasi, M.; and Shimada, K.J. (2002). Interaction between monocytes and vascular smooth muscle cells enhances matrix metalloproteinase – 1 production. J. Cardiovasc. Pharmacol., 36: 152-161.

[80] Weiss, D.; Sorescu, D. and Taylor, W.R. (2001). Angiotensin II and atherosclerosis. Am. J. Cardiol., 87: 25c-32c.

[81] Bornfeldt, K.E.; Raines, E.W.; Gravesm L.M.; Skinner, M.P.; Krebs, E.G.; and Ross, R. (1995). Platelet- derived growth factor, distinct signal transudation pathways associated with migration versus proliferation. Ann. NY. Acad. Sci., 766: 416-430.

[82] Cai, Q.; Lanting, L. and Natarajan, R. (2004). Growth factors induce monocyte binding to vascular smooth muscle cells: implications for monocyte retention in atherosclerosis. Am. J. Physiol. Cell Physiol., 286: C707-C714.

[83] Margariti, A.: Zeng, L. and Xu; Q. (2006). Stem cells, vascular smooth muscle cells and atherosclerosis. Histol. Histopathol., 21: 979 -985.

[84] Hansson, G. K. (2001). Immune mechanisms in atherosclerosis. Arterioscler. thromb.. Vasc. – Biol., 21: 1876 – 1890.

[85] Al-Sadi, H.I.; and Abdullah, A.K. (2011). Spontaneous atherosclerosis in free-living pigeons in Mosul area, Iraq.Pak.Vet. J.31(2):166-168.
[86] Clarkson, T. B.; Prichard, R. W.; Netsky, M. G.; and Lofland, H. B. (1959). Atherosclerosis in pigeons: its spontaneous occurrence and resemblance to human atherosclerosis. Arch. Pathol., 68: 143 – 147.
[87] Simpson, C. F. and Harms, R. H. (1968). Aortic atherosclerosis of turkeys induced by feeding of cholesterol. J. Atheroscler. Res., 10: 63 – 75.
[88] Moss, N. S. and Benditt, E. P. (1970). Spontaneous and experimentally induced arterial lesions. I. an ultrastructural survey of the normal chicken aorta. Lab. Invest., 22: 166 – 183.
[89] Shih, J. C. H.; Pullman, E. P.; and Kao, K. J. (1983). Genetic selection, general characterization, and histology of atherosclerosis susceptible and resistant Japanese quail. Atherosclerosis, 49: 41 – 53.
[90] Oku, H.; Toda, T.; Hamada, Y.; Kiyuna, M.; Chinen, I.; Toyomoto, M.; and Shinjo, A. (1990). Morphological and biochemical evaluation of the induction of atherosclerosis in Japanese quails. Acta Med. Nagasaki, 35: 81 – 87.
[91] Toda, T.; Nihimori, I.; and Kummerow, F. A. (1983). Animal model of atherosclerosis, experimental atherosclerosis in the chicken animal model. J. Jpn. Atheroscler. Soc., 11: 755 – 761.
[92] Day, C. G.; Stafford, W. W.; and Schurr, P. E. (1977). Utility of a selected line (SEA) of the Japanese quail (Coturnix coturnix Japonica) for the discovery of new anti – atherosclerosis drug. Lab. Anim. Sci., 27: 817 – 821.
[93] Morrissey, R.B. and Danalbson, W.E. (1977). Rapid accumulation of cholesterol in serum, liver, and aorta of Japanese quail. Poult. Sci., 56: 2003-2008.
[94] Wexler, B.C. (1977). Spontaneous atherosclerosis in the Japanese quail. Artery, 3: 507-516.
[95] Jarrold, B.B.; Bacon, W.L. and Velleman, S.G (1999). Expression and localization of the production of the proteoglycan decorin during the progression of cholesterol induced atherosclerosis in Japanese quail: Implication for interaction with collagen type I and lipoproteins. Atherosclerosis, 146: 299-308.
[96] Velleman, S.G.; Bacon, W.; Whitmoyer, R. and Hosso, S.J. (1998). Changes in distribution of glycosaminoglycans during the progression of cholesterol induced atherosclerosis in Japanese quail. Atherosclerosis, 137: 63-70.
[97] Velleman, S.G.; McCormick, R.J.; E:y, D.; Jarrold, B.B.; Patterson, R.A.; Scott, C.B.; Daneshvar, H. and Bacon, W. (2001). Collagen characteristics and organization during the progression of cholesterol – induced atherosclerosis in Japanese quail. Exp. Biol. Med., 226(4): 328-333.
[98] Casale et al. (1992) studied the cellular events of quail atherosclerosis using monoclonal antibodies to alpha – actin and chicken macrophages and effectively identified the presence of SMC and macrophages, respectively, as constituents of the atherosclerotic lesions. The presence of macrophage, as well as SMC proliferation, was observed in early lesions. Although it was not possible to acertain the first cell type to be involved in the initial stages of atherosclerosis results suggested early intervention of macrophages and SMC.

[99] Bavelaar, F.J. and Beynen, A.C. (2004). The relation between diet, plasma cholesterol and atherosclerosis in pigeons, quails and chickens, Int. J. poult. Sci., 3(11):671-684.
[100] Fabricant, C. G. and Fabricant, J. (1999). Atherosclerosis induced by infection with Marek's disease herpesvirus in chickens. Am. Heart J., 138 (5 pt2): 466-468.
[101] Hajjar et al. (1986) demonstrated histologically that infection of normocholesterolemic, specific – pathogen – free chickens with Marek's disease herpesvirus (MDV) lead to chronic atherosclerosis like that in humans.
[102] Bavelaar et al. (2004) studied the possibility that feeding of α – linolenic acid instead of linoleic acid or saturated fatty acids would diminish the degree of atherosclerosis in cholesterol – fed quails. The authors concluded that a differential effect on the development of atherosclerosis of α – linolenic acid, linoleic acid and saturated fatty acids could not be demonstrated.
[103] Frick, C.; Schmidt, V.; Cramer, K.; Krautwald – Junghanns M. E.; Dorrestein, G. M. (2009). Characterization of atherosclerosis by histochemical and immunohistochemical methods in African grey parrots (psittacus erithacus) and amazon parrots (Amazon spp.). Avian Dis., 53 (3): 466 – 472.
[104] Jensen, T.W.; Mazurm Pettigew, J.E.; Perez-Mendoza, V.G.; Zachary, J. and Schook, L.B. (2010). A cloned pig model for examining atherosclerosis induced by high fat, high cholesterol diets. Animal Biotechnol., 21: 179-187.
[105] Hughes, G.C. et al., (2003). Translational physiology: Porcine models of human coronary artery disease: implication for preclinical trails of therapeutic angiogenesis. J. APpl. Physiol., 94(5): 1689-1701.
[106] Lee, K.T. (1987). Experimental atherosclerosis in pigs. Yonsei Med. J., 28(1):1-5.
[107] Bell, F.P. and Gerrity, R.G. (1992). Evidence for an altered lipid metabolic state in circulation blood monocytes under condition of hyperlipemia in swine and its implications an arterial lipid – metabolism. Arterioscler. Throm., 12(2): 155-162.
[108] Yanni, A.E. (2004). The laboratory rabbit: an animal model of atherosclerosis research. Lab. Anim., 38: 246-256.
[109] Bocan, T.M.; Muller, S.B.; Mazur, M.J.; Uhlendorf, P.D.; Brown, E.Q.; and Kieft, K.A. (1993). The relationship between the degree of dietary – induced hypercholesterolemia in the rabbit and atherosclerotic lesion formation. Atherosclerosis, 102: 9-22.
[110] Aikawa, M.; Rabkin, E.; Voglic, S.J.; Shing, H.; Nagai, R.; Schoen, F.J.; And Libby, P. (1998b). Lipid lowering by diet reduces matrix metalloproteinase activity and increase collagen content of rabbit atheroma: a potential mechanism of lesion stabilization. Circulation, 97: 2433-2444.
[111] Kolodgie, F.D.; Katoces, A.S.; Largis, E.E.; Wrenn, S.M.; Cornhill, J.F.; Herderick, E.E.; Lee, S.J.; Virman, R. (1996). Hypercholesterolemia in the rabbit induced by feeding graded amounts of low level cholesterol. Arteriosclerosis Thrombosis and Vascular Biology, 16: 1454-1464.
[112] Spagnoli, L. G.; Bonanno, E.; Sangiorgi, G. and Mauriello, A. (2007). Role of inflammation in atherosclerosis. J. Nucl. Meo., 48 (11): 1800-1815.

Chapter 12

Choroidal Vessel Wall: Hypercholesterolaemia-Induced Dysfunction and Potential Role of Statins

J.M. Ramírez, J.J. Salazar, R. de Hoz,
B. Rojas, B.I. Gallego, A.I. Ramírez and A. Triviño

Additional information is available at the end of the chapter

http://dx.doi.org/10.5772/47794

1. Introduction

The choroid, the most important vascular tissue of the eye, is made up of vascular layers of descending thicknesses, from the large-sized choroidal vessels close to the sclera, to the choriocapillaris underlying the retina. Some 85% of the blood flow of the eye circulates through the choroidal vessels. The most important functions of the choroid include the regulation of eye temperature and the nutrient supply to the outer retinal layers, in which ischaemia would seriously compromise visual function. The integrity of the cells of the vascular endothelium and of the smooth-muscle cells therefore proves essential to maintain the choroidal flow [1].

The rise of the plasma-cholesterol levels is known to be accompanied by an overexpression of monocytic chemotactic protein-1 (MCP-1) by the macrophages and vascular smooth-muscle cells, which play a pivotal role in the development of fatty streaks [2]. In this way, an endothelium altered by hypercholesterolaemia leads to a pre-thrombosis state that begins with the aggregation of platelets. This is contributed to by aggregates that accumulate in the intima of the arteriolar wall as a result of the interaction of low-density lipoprotein (LDL) with the extracellular matrix [3,4]. Endothelial dysfunction precedes atherosclerosis, which is characterized by the expression of two adhesion molecules, the vascular adhesion molecule-1 (VCAM-1) and the inter-cellular adhesion molecule-1 (ICAM-1). Both participate in the adhesion and extravasation of the monocytes into the subendothelial space, where they are transformed into macrophages. These subendothelial macrophages participate in transforming the LDL into highly oxidized LDL (oxLDL), which after being taken up by the macrophages contributes to the formation of foam cells [5] such as those seen in the large choroidal vessels and in the suprachoroid of hypercholesterolaemic rabbits [6]. The saturation of the foam cells leads to their death and the release of toxic products such as esterified and oxidized cholesterol, a scenario that inflicts greater endothelial damage and encourages the progression of the atherosclerotic lesion.

The change in cell activation is measured by factors such as the nuclear factor kappa-beta (NF-kB) detected in macrophages, vascular endothelial cells, and smooth-muscle cells of atherosclerotic lesions [7]. The proliferation of smooth-muscle cells towards the intima, associated with an increase in apoptosis and a decrease in inflammatory cells, could indicate an attempt to limit the lesion. These processes described in hyperlipaemic aortas are also found in the walls of the large choroidal vessels, where, in addition to hypertrophy of the smooth-muscle cells, a great quantity of lipid inclusions appear and, in the case of choroidal arterioles, an increase in collagen fibre of the intima and the adventitia [8].

Several mechanisms contribute to impaired vascular tone in atherosclerosis, such as the activation of the smooth-muscle cells, by e.g. the synthesis of the extracellular matrix induced by the macrophage-derived TGF-beta, the intimal thickening and the changes in the vascular endothelium, including the release of mediators that promote vascular constriction [9]. Clinical studies have suggested that hyperlipidaemia alone can prompt structural changes in the choroidal vascular and retinal system which over time could provoke retinal dysfunction [10]. Chronic ischaemia is also present in causes of blindness as prevalent as aged-related macular degeneration, glaucoma or diabetic retinopathy [11-13].

In summary, this chapter illustrates the structural and ultrastructural changes that occur experimentally in the choroids of animals subjected to a hyperlipaemic diet; the changes that appear in this tissue when the plasma-cholesterol levels are normalized after the hyperlipaemic diet is replaced by a standard one, and the effects in vascular tissue, fundamentally the mitochondria and the caveolar system of the endothelial and smooth-muscle cells, caused by the treatment with statins at such low rates that they do not alter the high cholesterol levels (pleitropic effects) but are effective for controlling chronic ischaemia.

2. Anatomo-physiology of the choroid

2.1. Anatomy of the choroid

The *choroid* constitutes the most posterior region of the *tunica intermedia* of the eyeball or uvea. This is formed by a pigmented vascular tissue underlying the sclera and overlying the retina. Histologically, the choroid has three layers, which from the sclera towards the retina are: the suprachoroid, the vascular layer, and Bruch's membrane (Figure 1A).

2.1.1. Suprachoroid

The suprachoroid is the outermost layer of the choroid, a transition zone between the innermost part of the sclera and the large-sized-vessel layer (Fig. 1A). It is composed of tightly packed collagen fibres, melanocytes [14], fibroblasts [15], elastic fibres [16], smooth-muscle cells [16-18], nerve plexus, and intrinsic choroidal neurons (ICNs) [19-21]. In primates, birds, and rabbits, this layer contains large, endothelium-lined spaces, which empty into veins.

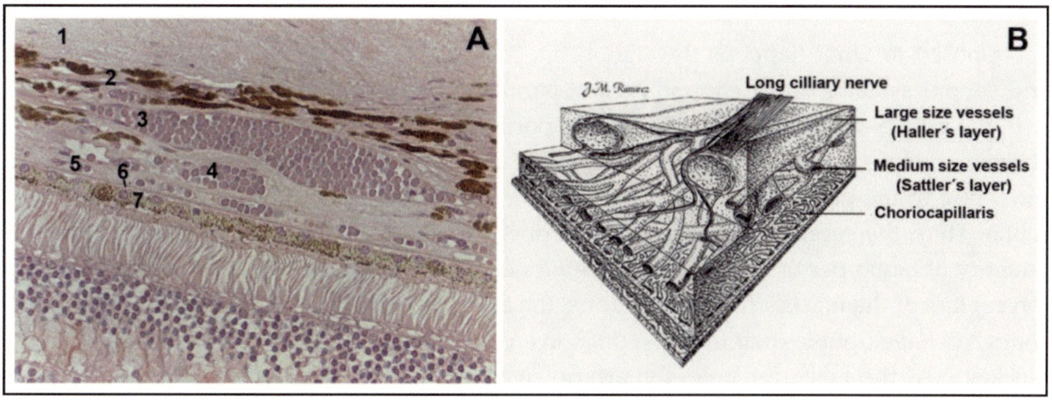

Figure 1. Choroidal vascular layers. A: Histological section (hematoxylin/eosin). B: Tridimensional scheme (Modified with permission from Ramírez et al. [19]). 1: Sclera; 2: Suprachoroid; 3: Large-sized-vessel layer (Haller´s Layer); 4: Medium-sized-vessel layer (Sattler´Layer); 5: Choriocapillaris; 6: Bruch´s membrane; 7: retinal pigment epithelium.

2.1.2. Choroidal-vessel layers

Most of the choroid consists of blood vessels that decrease in diameter from outer to inner as the vessels branch. It is made up of arteries, veins, arterioles, and a vascular stroma (Figure 1A,B). The latter contains collagen and elastic fibres, fibroblasts, non-vascular smooth cells, numerous melanocytes, mast cells, macrophages, and lymphocytes. The arterial vessels are branches of the posterior ciliary arteries (PCAs) (Figure 2) and some recurring branches of the major arterial circle of the iris [22]. The vessels become smaller in a branching hierarchy towards the capillary bed, enabling the identification of the three vessel layers of decreasing calibre [23]: an outer layer of large-sized vessels (Haller's layer), an intermediate layer of medium-sized vessels (Sattler's layer), and inner layer of interconnected capillaries (the choriocapillaris) (Figure 1A,B).

Large- and medium-sized-vessel layers

Before penetrating the sclera, the PCAs are subdivided into various branches that surround the optic disc: two long posterior ciliary arteries (LPCA) [22,24] and 15-20 short posterior ciliary arteries (SPCA) [25] (Figure 2). Depending on their scleral penetration, the SPCA can be further subdivided into paraoptic SPCA (closer to the optic disc) and distal SPCA [26,27]. Overall, 2 paraoptic SPCA pass through the sclera and surround the optic disc, thus forming the Zinn-Haller arterial circle, which provides blood flow to the circumpapillary choroid and the prelaminary and laminary regions of the optic nerve head [27,28] (Figures 2,3). The rest of the SPCA, both paraoptic as well as distal, once in the vascular choroid, divide sectorially, forming triangular areas towards the 4 regions of the eyeball. The macular region is irrigated by a dense network of distal branches of the SPCA [29,30].

The LPCA penetrate the eyeball at some 4 mm from the optic disc in the horizontal meridians following a rectilinear path in the most superficial part of the layer of large vessels [24,27] (Figure 2). These arteries help constitute the major arterial circle of the iris [25,31,32] and sends out recurrent branches toward the peripheral choroid [32,33].

As opposed to what happens throughout most of the human vascular system, the arterial and venous systems of the choroid are not parallel, as most of the veins are located primarily in the most external choroid [33]. The proportion of arteries to veins varies throughout the choroid, as does capillary density. In the latter case, density progressively diminishes from the centre to the periphery, coinciding with the fall in the number of photoreceptors of the retina. Thus, the submacular choroid is the one with the greatest capillary density (greatest quantity of blood per unit of area), this being necessary to nourish the great number of photoreceptors of the macula [23]. In this zone, the arteriole/venule ratio is 3:1, so that the foveal cones can have double their blood supply, even at the cost of greater risk of oedema by vein blockage. At the posterior pole, the arteriole:veinule ratio is 1:2 to 1:4 [34,35]. As a result, these zones of the choriocapillaris could be more susceptible to local blockages of arteries, with the consequent risk of ischaemic damage.

The voriticose veins drain the entire choroid, the ciliary body, and the iris. These veins are located at the equator (1 per each quadrant) and form a bottle shape receptacle before penetrating the sclera, which in turn is constituted by the meeting of 2 to 4 ampuliform dilations [27,36]. The choroidal veins and venules are larger than the arteries and maintain a rectilinear path, joining at many acute angles before ending in the vorticose veins [15,23,33].

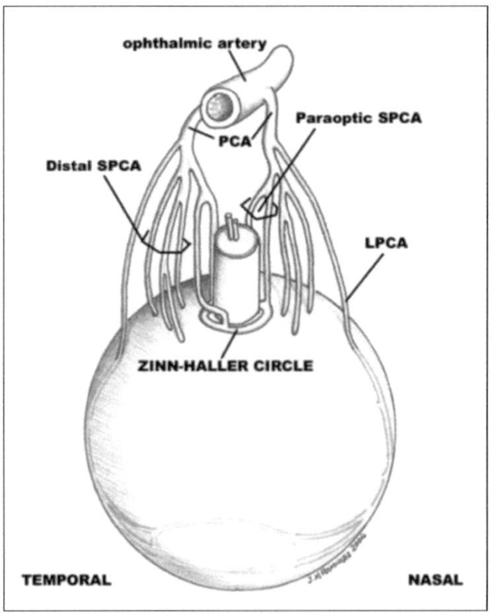

Figure 2. Scheme that represents the entry of the ciliary arteries in the eyeball. LPCA: Long posterior ciliary arteries; SPCA: Short posterior ciliary arteries; PCA: Posterior ciliary arteries. (Modified from Ramírez et al [72]).

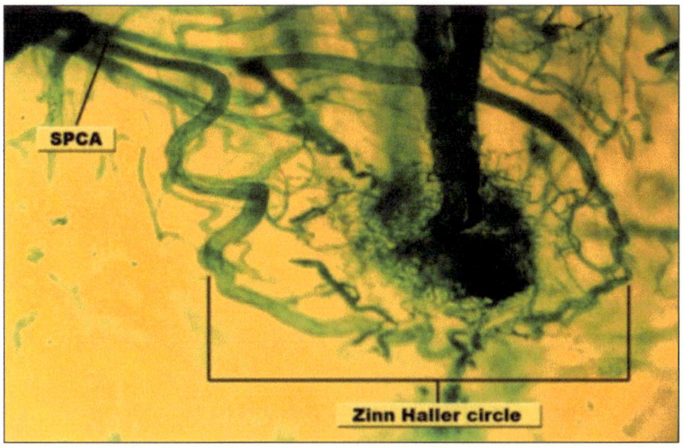

Figure 3. Diafanization and repletion with colored polymers of two short posterior ciliary arteries (SPCA) forming the Zinn-Haller arterial circle. (Modified from Ramírez et al [72]).

The choriocapillaris

The choriocapillaris is made up of capillaries situated between the medium-sized vessel layer (Sattler's layer) and Bruch's membrane (Figures 1,4). The position of these capillaries, flattened and elliptical in transverse section, create a large surface for metabolic exchange with the retina. With the same purpose, the vessels of the choriocapillaris are composed of tubes of endothelial cells (EC) and pericytes that do not completely surround it, appearing only towards the scleral side. Meanwhile, the ratio of pericytes:endothelial cells in the human retina is 1:1, and in the choriocapillaris it is 1:6 [23]. Given the contractile character of these cells, which enables the blood supply to be regulated in other tissues, their lower number in the choriocapillaris could suggest that the regulation of the choroidal blood flow by contraction is practically nil [23].

The EC presents fenestrations in the side oriented towards Bruch's membrane and retinal pigment epithelium (RPE) [37-39] (Figure 5). These fenestrations are of great physiological importance, as they permit the passage of nutrients towards the retina. The choroid capillaries are permeable to small molecules such as glucose (20-fold more than in cardiac muscle and 80-fold more than in skeletal muscle) [40] and amino acids [41], as well as large molecules such as γ-immunoglobulin and vitamin A [40,42,43]. Endothelial cells send out prolongations towards Bruch's membrane [44] that could physically stabilize the angioarchitecture of the inner choroid [45]. In addition, these prolongations could be involved in processes of phagocytosis for the elimination of waste products of Bruch's membrane [46], which would give them an important function in the metabolic exchange between the choroid and the RPE [47].

Among the capillaries of this layer are collagen fibres that form intercapillary septa [48]. These septa are reinforced and intermixed with fibres that come from the collagen area of Bruch's membrane [23]. In this way, the capillaries are immersed in a network of relatively rigid collagen that prevents collapse [15].

2.1.3. Bruch's membrane

The innermost layer of the choroid is Bruch's membrane. This is an acellular structure, easily distinguishable from the retina and the choroid, and contains elements of both tissues. According to studies made with the electron microscope [15] [49][49][49] it can be divided into five layers, which, from the choroid side towards the retina are: The basal membrane of the choriocapillaris, the outer collagen layer, the elastic-fibre layer, the inner collagen layer, and the basal membrane of the RPE (Figure 5).

The basal membranes, which on one side separate Bruch's membrane from the choriocapillaris, and on the other from the RPE, are not joined by the hemidesmosomes to their adjacent layers [50]. The basal membrane of the choriocapillaris is discontinuous [15], being present on the side of the endothelium of the capillaries but absent in the intercapillary spaces [50].

The two collagen layers, with a thickness of 1 μm in young individuals, surround the layer of elastic fibres [15]. Some collagen fibres are arranged parallel to the plane of the tissue, particularly at the level of the inner collagen. Others pass from one side to the other of the elastic-fibre layer, interconnecting both collagen layers [15,33]. Through the external collagen, fibres penetrate the interruptions of the basal membrane to join the collagen fibres of the intercapillary septa (as described above). This arrangement can help to join Bruch's membrane to the choriocapillaris.

The elastic fibre layer lies between the inner collagen layer on the inside and the outer collagen layer on the outside. It is made-up of inter-woven bands of elastic fibres 2-4 μm thick. Between these appear irregular spaces through which the collagen fibres pass [15].

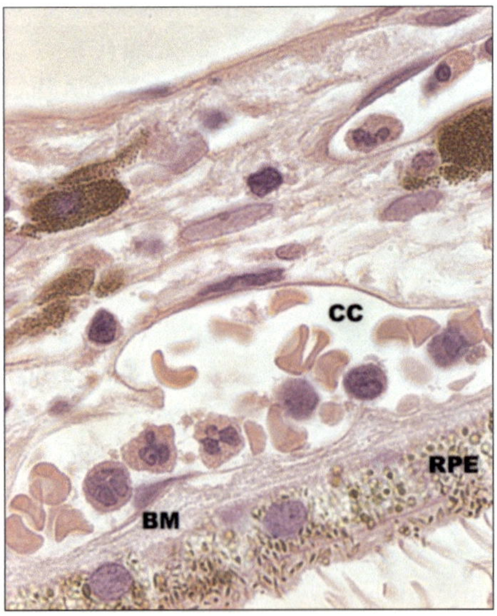

Figure 4. Histologic section of the choriocapillaris (CC), Bruch's membrane (BM) and retinal pigment epithelium (RPE). Hematoxylin/eosin. (Modified from Ramírez et al [72]).

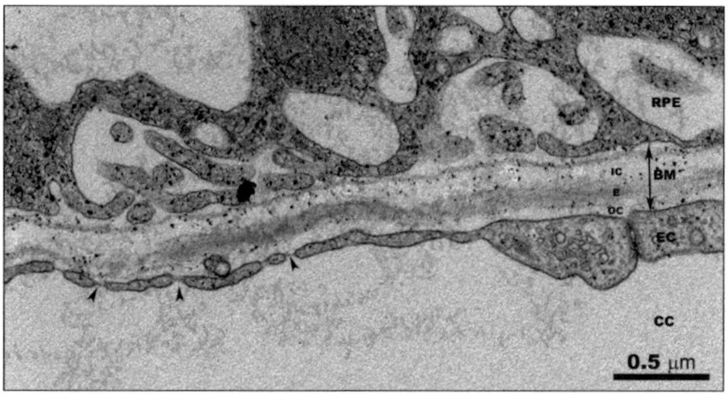

Figure 5. Electron micrograph of Bruch´s membrane (BM) and choriocapillaris (CC). Bruch´s membrane is composed of several layers: the basal membrane of the retinal pigment epithelium (RPE), the inner collagenous layer (IC), the elastic layer (E), the outer collagenous layer (OC) and the basal membrane of the choriocapillary endothelium. Endothelial cells (EC) of the choriocapillaris present fenestrations (arrow-head).

2.1.4. Choroidal innervation.

The abundant choroidal innervation is made up of fibre bundles from the sympathetic and parasympathetic (autonomic nervous system) and sensitive (central nervous system). The sympathetic fibres are from the superior cervical ganglion, the parasympathetic from the ciliary and pterygopalatine ganglion, and the sensitive fibres from the trigeminal ganglion. These nerve fibres reach the choroid in three ways: i) going around the SPCA, these fibres stem from the plexuses of the internal carotid and continue with the ophthalmic artery until reaching the SPCA in the choroid; ii) through the short posterior ciliary nerves that stem from the ciliary ganglion; or iii) through the long posterior ciliary nerves that come from the nasociliary nerve [19,20] (Figure 6). In addition to the nerve fibres, the choroid has intrinsic choroidal neurons (ICNs). Both nervous structures are more abundant in the suprachoroid than in the rest of the vascular layers, with no innervation in the choriocapillaris [20].

Nerve Fibres

The long ciliary nerves (nasal and temporal) enter the choroid together with the LPCA, in the horizontal meridian of the eye, near the optic disc. They head towards the anterior part of the eyeball, with few branches in the part towards the suprachoroid or the choroidal vessel layers [20] (Figure 6). The short ciliary nerves enter the choroid together with the SPCA around the optic nerve and proceed towards the ciliary body. In this trajectory, there are many branches for the suprachoroid and the choroidal vessel layers [20] (Figure 6).

The short and long ciliary nerves send out branches to the large-sized-vessel layer of the choroid that are arranged in a paravascular form (Figure 7). Immunohistochemical techniques have demonstrated that these paravascular axons carry information which is sympathetic [NPY(+) and TH(+)] [21]; parasympathetic [(VIP(+)], and sensitive [SP(+) and CGRP(+)] [51]. In addition, in the vascular walls, sympathetic perivascular ends [NPY(+) and TH(+)] as well as sensitive ones [SP(+) and CGRP(+)] have been described. These perivascular fibres can present small end dilations on reaching the vascular wall. Some axons penetrate to a deeper level of the medium-sized-vessel layer to form a polygonal plexus. This axons are sympathetic [NPY(+) and TH(+)] and sensitive [SP(+) and CGRP(+)]. Innervation has not been demonstrated in the choriocapillaris [21,51,52].

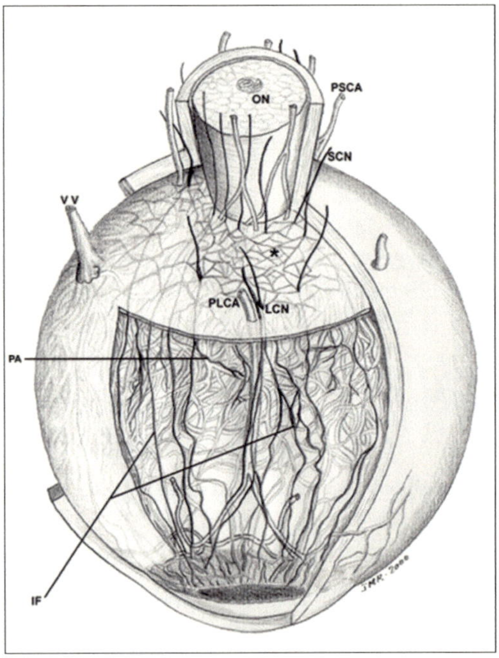

Figure 6. Tridimensional scheme showing the topographic distribution of the short ciliary nerves (SCN) and long ciliary nerves (LCN) in human choroid. The short ciliary nerves form a plexus of nerve fibers in the suprachoroid. In the vascular layers, the scheme shows branches from this plexus adapting to the vessel contours - paravascular fibers (PF) and intervascular fibers (IF). ON: optic nerve; PSCA: short posterior ciliary arteries; PLCA: Long posterior ciliary arteries; VV: vorticose vein; asterisk: posterior pole. (From Triviño et al. [20]

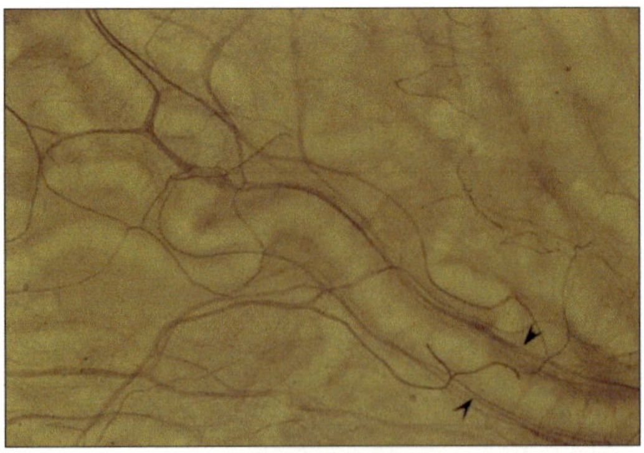

Figure 7. Human choroidal wholemount. Nerve fibers adapted to the choroidal vascular morphology (paravascular fibers) lying parallel to the large arteries (arrow-head). Anti NF-200 PAP. Bar: 250 μm. (From Triviño et al. [20])

Intrinsic choroidal neurons (ICNs)

Besides nerve fibres, the human choroid possesses abundant ICNs (from 1300 to 1500), which are located mainly in the suprachoroid (Figure 8). The greatest concentration of ICNs is found in the central and temporal region adjacent to the macula. This location could be related to the existence of the macula. In humans, ICNs could be responsible for regulating the rapid vasoregulatory reflexes, which are so important in the fovea [20].

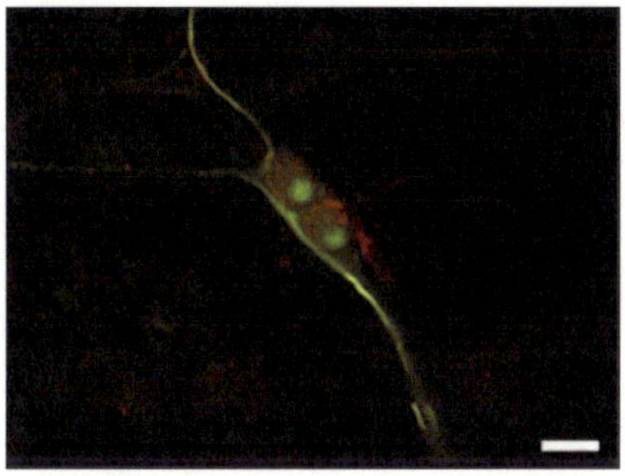

Figure 8. Intrinsic choroidal neurons of the suprachoroid forming a microganglion. Double immunofluorescence for NF 200 (green) and TH (red). Bar: 20 µm. (From Triviño et al. [21])

The ICNs can present immunoreactivity to VIP (parasympathetic) and to nitric oxide synthase (NOS), indicating the use of nitric oxide (NO) as a neurotransmitter. Both the NO and the VIP would be involved in the vasodilation through the regulation of the vascular smooth-muscle fibre. Recently, ICNS have been reported to be immunoreactive to NPY and TH (sympathetic) (Figure 8) and to SP and CGRP (sensitive) [52]. All these are located preferentially in the central region of the temporal area (submacular choroid). Sympathetic ICNs could exert a protective mechanism to prevent over-perfusion and breakdown of barriers [21]. Sensitive ICNs could be involved in regulating the ocular blood supply, inflammatory processes, and vascular architecture [52].

2.1.5. Ultrastructural anatomy

Choroidal arteries and veins

The choroidal arteries have the structure of small arteries. They are constituted by a tunica intima, which consists of a monolayer of EC, a basal membrane often containing a cluster of fine osmophilic particles and of interspersed elastic fibres [15]. The intima is surrounded by a thick tunica media made up by one or more layers of smooth-muscle cells (Figure 12A). In larger arteries, there are two to three layers of smooth-muscle cells, the innermost arranged circumferentially and the outermost obliquely or longitudinally. These cells are separated from each other by basal membranes [53]. Over the tunica media lies an adventitia of collagen in a circular arrangement but without an outer elastic membrane. In the adventitia, there are often melanocytes and fibroblasts [54]. The choroidal arterioles are constituted by: an intima, formed only by an endothelium and a basal membrane without inner elastic; a tunica media made up of a discontinuous muscle layer; and an adventitia formed by a fine layer of circular collagen [15].

Choroidal veins have a structure similar to the rest of the veins of the organism, are composed of an endothelia, a basal membrane, one or two layers of muscle cells, and an adventitious layer of collagen [23]. Clusters of unmyelinated axons and synaptic terminals reach the outer muscle layers of the veins and arteries. Terminal boutons make contact with the smooth-muscle cells [53].

The EC, both of the arteries as well as of the choroid veins, present evaginations that protrude towards the inside of the vessel and towards the basal membrane. These later pass through the membrane and make contact with the closest muscle cells. In the cytoplasm of the EC the normal organelles can be found: mitochondria, Golgi apparatus, rough endoplasmic reticulum, caveolar system, free polyribosomes, cytoplasmic vesicles and Weibel-Palade bodies, which are membrane-bound granules containing an electrodense material [54,55]. The processes of the EC overlap and are joined together by different types of cell bonds: zonula adherens, tight junctions, and long gap junctions [38] (Figure 13A).

The vascular smooth-muscle cells (VSMC) contain an elongated, highly heterochromatic nucleus, a cytoplasm with elongated mitochondria having smooth and rough-surfaces, endoplasmic reticulum, and numerous contractile filaments oriented parallel to one other [54]. Among the bundles of filaments appear dense osmiophilic structures that resemble Z bands, which probably represent areas of attachment of the filaments. Other filaments are oriented towards thickenings of the cell membrane; the filaments join these thickenings in the same way as hemidesmosomes [15]. Alternating with the dense structures, rows of caveoles appear along the surface of the cell membrane [55]. The muscle cells are joined together by macular adherent junctions, punctiform junctions, and gap junctions [54] (Figure 13A).

Choriocapillaris

The capillaries of the choriocapillaris differ from those of the retina by the great diameter of its lumen, in such a way that red blood cells can pass through at least 2 or 3 at a time [23].The EC of these capillaries have their nuclei located on the side of the vessels opposite Bruch's membrane and protrude slightly into the lumen. Its cytoplasm, contains scattered mitochondria, small portions of Golgi apparatus, short cisterns of rough endoplasmic reticulum, free polyribosomes, cytoplasmic vesicles, caveolar system, and occasional Weibel-Palade bodies [54].

The EC are joined together by thin discontinuous bonds (zonula occludens), desmosomes, and communicating junctions (gaps) [37,38,56]. In addition, they have the peculiarity of presenting fenestrations on the side oriented towards Bruch's membrane and the RPE [37-39]. These fenestrations have a mean pore size of 60-80 nm [15,23] and are crossed by a diaphragm with a central density [37] (Figure 5). Endothelial-cell processes protrude into Bruch's membrane, typically at sites of focally thickened, nonfenestrated regions of the EC [46]. The endothelium is enveloped by a thin basal lamina which also surrounds the pericytes [14,57,58]. Pericytes appear only on the scleral face of the capillaries, incompletely wrapping them [23]. In their cytoplasm, pericytes bear small bundles of microfilaments, preferentially located along the plasma membrane [54,59].

The stroma (extravascular tissue) contains collagen and elastic fibres, fibroblasts, non-vascular smooth-muscle cells and numerous very large melanocytes closely apposed to the blood vessels. As in other types of connective tissue, numerous mast cells, macrophages and lymphocytes appear ([60] (Figures 11A,12A).

2.2. Choroidal physiology

The choroid is one of the most vascularized tissues of the human body and therefore it has traditionally been thought that its main function was to nourish the external layers of the retina. However, today it is known that it has many other functions, including: light absorption, elimination of the aqueous humour from the unveoscleral pathway, adjustment of the position of the retina by changes in the choroid thickness, help in the control intraocular pressure, and thermoregulation. The latter two functions of the choroidal blood flow have a fundamental role.

2.2.1. The role of the choroidal blood flow

The choroidal blood flow represents some 85% the total blood in the eye [34]. This high flow rate considerably surpasses that of other richly vascularized tissues, being 10-fold higher than in the grey matter of the brain, and 4-fold that of the kidney. This high choroidal blood-flow rate (800-2000 ml/min/100 gr of tissue) is probably due to the great calibre of the vascular lumen of the choriocapillaris, and therefore to its low flow resistance [24,61]. Furthermore, the blood flow is not uniform, as it presents regional differences throughout the choroid, so that on the periphery the flow is 6- to 7-fold lower than in the central regions (foveal and peripapillary) [62].

The exact function of the high choroidal blood flow is still not known with exactitude. Physiologically, one of the main functions of the choroid is to nourish (supply O_2 and glucose) to the most external layers of the retina (fundamentally to the photoreceptors and to the RPE). In many species, the retina depends entirely on choroidal circulation for its metabolic needs. The human choroid provides the metabolic requirements to the entire thickness of the retina only in the macular region. The high choroidal blood flow results in a steep gradient for the diffusion of O_2 towards the external layers of the retina and a low concentration of waste products, facilitating their elimination from the retina [1]. If we take into account that vein blood is approximately 95% of that found in the arterial blood, we see that the O_2-extraction rate from choroidal arterial blood is only 3-5% [63]. However, the high blood flow through its vessels permit the choroid to provide a high percentage of oxygen consumed by the retina.

Nevertheless, the choroid appears to be perfused at a proportion that exceeds its nutritive needs, suggesting therefore an additional role for the high rate of choroidal flow. Thus, it is thought that the choroidal blood flow could help maintain the regulation of the intraocular pressure (IOP) [43] and, on the other hand, offer thermoregulation by the following mechanisms: dissipating the heat generated during the visual transduction process [64], preventing overheating of the outer retina during exposure to bright light [65], and, finally, heating the intraocular structures that could be cooled by exposure to extreme outside conditions [66].

2.2.2. *Regulatory mechanisms of choroidal blood-flow*

The control of circulation in most tissue is quite complex, as there are many factors that influence vascular resistance, such as: local myogenic responses, substances derived from the endothelium, local metabolic factors, and the autonomous nervous system. The existence or not of choroidal self-regulating mechanisms has long been the object of debate. The pressure that promotes blood flow through the tissues is called perfusion pressure. Ocular perfusion pressure is the difference between the pressure of the arteries that reach the eye (aP) and that of the veins that leave the eye (vP) [1]. The perfusion pressure and vascular resistance (R) determine blood flow (F) according to the following formula: F=(aP-vP)/R. As in the eye, the vP is practically equal to the IOP both under normal conditions as well as in cases of increased IOP, so that the formula becomes: F=(aP-IOP)/R. According to this formula, a decrease in the perfusion pressure (by relieved arterial pressure or greater IOP) could result in a proportional restriction of blood flow without a compensatory reduction in vascular resistance [33].

In most tissues, blood flow tends to remain constant despite moderate variations in perfusion pressure, thanks to the mechanism known as self-regulation [24]. However, in the choroid, as opposed to the retinal vessels and the prelaminary region of the optic disc, increases (even moderate ones) in the IOP, cause concomitant reductions of the choroidal blood flow [63]. This circumstance is not controlled by any self-regulated mechanism [67,68]. Nevertheless, some authors have considered there to be a certain self-regulatory capacity of choroidal circulation under very specific conditions. Thus, it has been demonstrated that the choroidal circulation is sensitive to tensions of CO_2 as well as to acidic metabolic products which cause vasodilation. Studies on cats have shown a marked vasodilation in the choroid when the CO_2 concentration is raised [69]. It has been observed that the flow augments with the respiration of air containing 10% CO_2 and diminishes with respiration of air saturated with 100% O_2 [70]. Other studies reveal evidence of choroidal self-regulation when the average arterial pressure gradually weakens. This mechanism is IOP dependent, so that its effect is far more pronounced at low IOP levels (< 5 mmHg) and does not occur within normal or higher IOP ranges. Therefore, the choroid would be capable of regulating its circulation at extremely low perfusion pressures induced by a fall in average arterial pressure.

Neural control of choroidal blood-flow

In recent years, the importance of the neural control in choroidal blood-flow regulation has been stressed. Several mechanisms are involved in the neuroregulation of the uveal flow. In a direct way, this flow is regulated by perivascular innervation, which permits a balance between the vasoconstriction and vasodilation necessary for the maintenance of blood flow. Indirect regulation comes from the paravascular fibres, both by means of classical neurotransmitters as well as neuropeptides released by sympathetic, parasympathetic, and stroma-sensitive nerve endings, and diffusing factors such as NO [20].

Sympathetic stimulation causes sharp choroidal vasoconstriction and a fall in intraocular pressure due to a decline in the ocular blood volume (reductions in choroidal flow of up to 60%) [62]. This response comes fundamentally from the stimulation of α-adrenergic receptors located in the VSMC [1]. The sympathetic innervation places the choroid under vasoconstriction tone, suggesting that this could protect the retina and the optic nerve head from the hyperfusion and break of the ocular barriers that could appear under certain circumstances as for example arterial hypertension. The systemic hypotension would, by barroreflex, increase sympathetic activity and depress parasympathetic activity, which would in turn bolster vascular resistance in the choroid, with the consequent restriction in choroidal blood flow. This situation has not been confirmed in experimental studies [71], indicating that there could be a local mechanism in the choroid capable of ignoring neurogenic vasoconstriction, or the choroidal nerves do not become activated during the discharge of the barroreceptors. Other authors [62,63] have demonstrated that, although during the direct stimulation of the sympathetic activity, it has little effect on the diminished choroidal flow during the hypotension induced by a haemorrhage. The existence of sympathetic neurons in the choroid (ICNs) immunopositive for NPY and TH [21], could partly explain the self-regulatory capacity of the choroid.

The role of the parasympathetic innervation it is not as clearly defined as the sympathetic one. However, it has been observed that the choroid responds to cholinergic parasympathetic stimulation (which arrives through the short ciliary nerves) by vasodilation. This vasodilation would explain the increase light-induced choroidal blood flow [72].

Recently, it has been postulated that the sensitive peripheral nerves play an important role in choroidal blood-flow regulation. Thus, SP could have a viscero-motor function, regulating the choroidal flow during ocular irritation. In addition, a role in vasodilation has been attributed to the CGRP as a cholinergic co-mediator with the SP [52].

Neural control of choroidal blood flow in ocular diseases

The importance exerted by nerve control on the regulation of choroidal blood flow appears to imply that damage in the choroidal innervation could be involved in the vascular alterations that occur in some ocular diseases. Experimental studies have demonstrated that sympathetic innervation is critical in the regulation of choroidal vascularization [73], and that the chronic loss of sympathetic activity can contribute to the anomalous vascular proliferation noted in diseases such as age-related macular degeneration (ARMD) [74] and diabetic retinopathy [75,76]. Furthermore, the loss of this innervation can cause oedema in the retina [71], a circumstance that could be important in illnesses such as diabetes or hypertension, in which automatic control is altered [56].

The axonal damage in the sympathetic nervous system is a notable fact of diabetic neuropathy. In addition, a dysfunction of the sympathetic nerves of the eye in diabetic patients has been suggested, postulating that the episodes of hyperglucaemia could determine an increase in the choroidal flow and in the pressure of the vessels in the submacular choroid, as well as changes in the RPE. In this way, the extravasation of fluid from the submacular choroid would be exacerbated. The excess of intraretinal liquid caused by diabetic macular oedema would not come from the retinal vessels alone but also from the choroid, reaching the retina through lesions of the RPE near the choriocapillaris affected [77].

In ARMD, haemodynamic anomalies have been reported as potential causal agents as part of the pathological process. By laser Doppler flowmetry, it has been observed that the choroidal blood flow diminishes with age [78], and it is lower in the non-exudative stages of macular degeneration than in control. This effect is due to a descent in the volume of blood flow [79].

Recently, it has been postulated that the sensitive nerves could be involved in regulating choroidal flow through different inflammatory mechanisms that measure vasodilation and plasma extravasation. Also, its role has been examined in processes of maintenance and vascular renovation with substantial implications in visual function [23]. Thus, it has been suggested that changes in choroidal thickness would play a main role in the regulation of this ocular refractive state, particularly in recuperation from myopia [22]. Additionally, the diabetes-like conditions induced by streptozotocin reduce the content of CGRP in the sensory nerves and exogenous CGRP-mediated vasodilation. CGRP is likely an important regulator of vascular tone, and compromising its function could contribute to nerve ischaemia and diabetic neuropathy [80]. It has been postulated that the dysfunction of SP(+) and/or CGRP(+) ICNs could be involved in the physiopathology of ocular diseases associated with peripheral innervation damage. The majority distribution of sympathetic and sensitive ICNs in the submacular region suggests the possibility that vascular pathologies of certain ocular diseases such diabetic macular oedema or age-related macular degeneration are related to the possible dysfunction of these cells [20,21,52].

It appears that the peripheral innervation, both sensitive as well as sympathetic, would have a broad and significant role in regulating the choroidal vascular architecture. Moreover, given the special susceptibility to damage presented by peripheral innervation under a great variety of conditions (age, arterial hypertension, ocular hypertension, diabetes), sympathetic and sensitive choroidal nerve dysfunction could intervene in the aetiology of ocular diseases that appear in association with these conditions [21,52].

3. Hypercholesterolaemia as a risk factor for smooth-muscle and endothelial cell dysfunction

Atherosclerosis results from a local imbalance between the production of reactive oxygen species (ROS) and antioxidant enzymes. The EC, macrophages, and VSMC are targets for ROS-dependent atherogenic signalling [81]. Our data suggest that changes in arterial ROS production may occur very early in hypercholesterolaemia. Atherosclerosis is initiated by lipids deposited in the subendothelial layer of the artery wall. These lipids and modified LDL, oxidized low-density lipoprotein (Ox-LDL), induce the expression of adhesion molecules and chemotactic molecules including monocyte chemoattractant protein-1 (MCP-

1/CCL2). After activating EC in this way, monocytes/ macrophages enter the intima and differentiate [82]. Lipid-laden macrophages, known as foam cells, promote the progression of the atherosclerotic plaque [83]. Specifically, pro-inflammatory cytokines and growth factors secreted by foam cells induce local inflammatory responses, ROS in the lesion, and accelerate the migration of VSMC from the media to the intima [84]. Intimal VSMC also take up modified lipoproteins, contributing to foam-cell formation, and synthesise extracellular matrix proteins (collagen, elastin, and proteoglycans) that lead to the development of a fibrous cap [85].

In VSMC, ROS mediates various functions, including growth, migration, matrix regulation, inflammation, and contraction [86], which are critical factors in the progression and complication of atherosclerosis. VSMC regulate plaque stability by modulating inflammation and apoptosis and by producing or degrading matrix proteins [85,87]. Under atherogenic conditions, VSMC in the media of the artery undergo phenotypic changes from the normal contractile state to the active synthetic state, which produces collagen, elastin, and proteoglycans. Activated VSMC migrate to the intima and proliferate [88]. In addition, ROS produced via NADPH oxidase induce VSMC to produce and secrete matrix metalloproteinases involved in the degradation and reorganization of the extracellular matrix [89]. In VSMC, ROS also mediate inflammation (e.g., MCP-1 expression via TNF-α) [90] and apoptosis via p53 and Bax/Bad [91].

The junction integrity of EC in blood vessels regulates leukocyte transmigration. In this process, ICAM-1 and VCAM-1 clustering prompts actin remodelling and junction disruption via ROS production [82].

It has been reported that Ox-LDL affects endothelium-dependent responses [92]. Some studies demonstrate that Ox-LDL decreases the endothelial release of NO [93] and that endothelium-dependent relaxations are improved by lipid-lowering therapy in patients with hypercholesterolaemia [94]. Another study suggests a mechanism for the cholesterol-induced impairment of NO production through endothelial nitric oxide synthase (eNOS) regulation by inhibitory interaction with caveolin [95].

4. Animal models of hypercholesterolaemia

Animal models provide a controlled environment in which to study disease mechanisms and to devise technologies for diagnosis and therapeutic intervention for human atherosclerosis. Different species have been used for experimental purposes (cat, pig, dog, rabbit, rat, mouse, zebra fish). The larger animal models more closely resemble human situations of atherosclerosis and transplant atherosclerosis and can also be easily used in (molecular) imaging studies of cardiovascular disease, in which disease development and efficacy of (novel) therapies can be monitored objectively and non-invasively. Imaging might also enable early disease diagnosis or prognosis [96]. On the other hand, the benefits of genetically modified inbred mice remain useful, especially in quantitative trait locus (QTL)-analysis studies (a genetic approach to examine correlations between genotypes and phenotypes and to identify (new) genes underlying polygenic traits [96].

4.1. Mice

Wild-type mice are quite resistant to atherosclerosis as a result of high levels of anti-atherosclerotic HDL and low levels of pro-atherogenic LDL and very-low-density-lipoproteins (VLDL). All of the current mouse models of atherosclerosis are therefore based on perturbations of lipoprotein metabolism through dietary or genetic manipulations [97].

ApoE-knockout mice

In apoliprotein-deficient mice (apoE$^{-/-}$) the homozygous delection of the apoE gene results in a pronounced rise in the plasma levels of LDL and VLDL attributable to the failure of LDL-receptor (LDLr-) and LDL-related proteins (LRP-) mediated clearance of these lipoproteins. As a consequence, apoE$^{-/-}$ mice develop spontaneous atherosclerosis. Of the genetically engineered models, the apoE-deficient model is the only one that develops extensive atherosclerotic lesions on a low-fat cholesterol-free chow diet (<40g/kg). The development of atherosclerosis lesion can be strongly accelerated by a high-fat, high-cholesterol (HFC) diet [98]. ApoE-knockout mice have played a pivotal role in understanding the inflammatory background of atherosclerosis, a disease previously thought to be mainly degenerative. The apoE-deficient mouse model of atherosclerosis can be used to: i) identify atherosclerosis-susceptibility-modifying genes; ii) define the role of various cell types in atherogenesis; iii) characterize environmental factors affecting atherogenesis; and iv) to assess therapies [99]. Because of the rapid development of atherosclerosis and the resemblance of lesion to human counterparts, the apoE$^{-/-}$ model have been widely used. However, some drawbacks are associated with the complete absence of apoE proteins: i) the model is dominated by high levels of plasma cholesterol; ii) most plasma levels are confined to VLDL and not to LDL particles, as in humans; and iii) apoE protein has additional antiatherogenic properties besides regulating the clearance of lipoproteins such as antioxidant, antiproliferative (smooth-muscle cells, lymphocytes), anti-inflammatory, antiplatelet, and also has NO-generating properties or immunomodulatory effects [100-102]. The study of the above processes and the effects of drugs thereupon is restricted in this model.

LDLreceptor-deficient mice (LDLr$^{-/-}$ mice)

In humans, mutations in the gen for the LDLr cause familial hypercholesterolaemia. Mice lacking the gene for LDL receptor (LDLr$^{-/-}$ mice), develops atherosclerosis, especially when fed a lipid-rich diet [103]. The morphology of the lesions in LDLr$^{-/-}$ mice is comparable to that in apoE$^{-/-}$, while the main plasma lipoprotein in LDLr$^{-/-}$ mice are LDL and high-density-lipoprotein (HDL) [104].

*ApoE*3Leiden (E3L) transgenic mouse*

ApoE*3Leiden (E3L) transgenic mice are being generated by introducing a human *ApoE*3-Leiden* construct into C57B1/6 mice. E3L mice develop atherosclerosis on being fed cholesterol. Because they are highly responsive to diets containing fat, sugar, and cholesterol, plasma lipid levels can easily be adjusted to a desired concentration by titrating the amount of cholesterol and sugar in the diet. E3L mice have a hyperlipidaemic phenotype with a prominent increase in VLDL- and LDL-sized lipoproteins fractions [105] and are more sensitive to lipid-lowering drugs than are $apoE^{-/-}$ and $LDLr^{-/-}$ mice [97].

4.2. Minipigs

Because of their well-known physiological and anatomical similarities to humans, swine are considered to be increasingly attractive toxicological and pharmacological models. Pigs develop plasma cholesterol levels and atherosclerotic lesions similar to those of humans, but their maintenance is more difficult and expensive than that of smaller animals [96]. The minipig, smaller than the domestic swine, has served as a model of hypercholesterolaemia for more than two decades now. In 1986, Jacobsson reported that the Göttingen strain had more susceptibility to alimentary hypercholesterolaemia and experimental atherosclerosis than did domestic swine of the Swedish Landrace [106]. Clawn[107], Yucatan, Sinclair, and Handford are among other general minipigs used for experimental use [107-109]. Down-sized Rapacz pigs are minipigs with familial hypercholesterolaemia caused by a mutation in the low-density lipoprotein receptor. It is a model of advanced atherosclerosis with human like vulnerable plaque morphology that has been used to test an imaging modality aimed at vulnerable plaque detection [110]. The Microminipig (MMP) is the smallest of the minipigs used for experimental atherosclerosis [111]. One of its advantages is that in 3 months an atherosclerosis very similar in location, pathophysiology and pathology to that in humans can be induced [112]. The easy handling and mild character of the MMP make it possible to draw blood and conduct CT scanning under non-anaesthesized conditions [113].

4.3. Zebra fish

Cholesterol-fed zebra fish represent a novel animal model in which to study the early events involved in vascular lipid accumulation and lipoprotein oxidation [114,115]. Feeding zebra fish a high-cholesterol diet results in hypercholesterolaemia, vascular lipid accumulation, myeloid cell recruitment, and other pathological processes characteristic of early atherogenesis in mammals [113]. The advantages of the zebra-fish model include the optical transparency of the larvae, which enables imaging studies.

4.4. Rabbits

Investigation has continued on hypercholesterolaemic rabbits since 1913, when Anitschkow demonstrated that, in rabbits fed a hypercholesterolaemic diet underwent atherosclerotic changes at the level of the arterial intima similar to those in atherosclerotic humans. The atheromatose lesions in this animal are similar to those in humans also in sequence, as confirmed en aortic atherosclerosis [116], making this animal a universal model for studying the anti-atherogenic activity of many drugs [117-120]. For the characteristics detailed below, the New Zealand rabbit is an excellent model to reproduce human atheromatosis because: i) it is possible to induce hypercholesterolaemia in a few days after administration of a high-cholesterol diet [121]; ii) it is sensitive to the induction of atheromatose lesions [116]; iii) hypercholesterolaemia results from excess LDL [122]; iv) excess cholesterol is eliminated from the tissues to be incorporated in high-density lipoproteins (HDL) [4]; vi) it is capable of forming cholesterol-HDL complexes associated with apoE which are transported by the blood to the liver [4]; vii) the lipoprotein profile is similar in size to that of humans in the highest range, with HDl being practically the same [123]; viii) it presents postprandial hyperlipaemia for the existence of chilomicron remnants [124]; ix) the hyperlipaemic diet increases apoE [125]; and x) the sustained alteration of lipids after feeding with a cholesterol-rich diet is reversible when the diet is replaced by a normal one [121].

Studies on hypercholesterolaemic rabbits have improved our knowledge of human atherosclerosis by delving into different aspects of the disease such as lipoproteins, mitogenes, growth factors, adhesion molecules, endothelial function, and different types of receptors. At the vascular level, the importance of endothelial integrity and cell adhesion has been investigated [126]. It has been demonstrated that the high levels of lysosomal iron start the oxidation of the LDL, spurring the formation of lesions [127]. In addition, the expression of VCAM-1 preceding the infiltration of the subendothelial space by macrophages has been studied [128], as have the proteins, including MCP-1. In hypercholesterolaemic rabbits, this protein is over-expressed when the serum-cholesterol levels rise in macrophages and smooth-muscle cells, contributing to the development of fatty streaks [2].

In hypercholesterolaemic rabbits, the expression de Fas-L in cells of the arterial wall help us to understand the progression of the atherosclerotic lesion, as this expression indicates an increase in cell injury, as well as a greater accumulation in the intima of smooth-muscle cells [8]. Also, a hyperlipaemic diet causes a selective alteration of the functioning of certain regulatory proteins that are involved in gene expression, as occurs with the nuclear B factor, which stimulates the proliferation of macrophages and smooth-muscle cells [129].

In this model, a study was also made of the pre-thrombosis state triggered by the platelet aggregation in an altered endothelium and the possibilities of its inhibition [130], as well as the interactions of the LDL with the extracellular matrix to form aggregates that accumulate in the intima of the artery wall [9].

The consequences of hypercholesterolaemia in ischaemic cardiopathy and cerebrovascular pathology are well known. The same does not occur with the functional repercussions of the hypercholesterolaemia at the ocular level, partly because the underlying structural changes are not well known. The hypercholesterolaemic rabbit constitutes a useful model to explore the repercussions of excess lipids at the ocular level. This is because rabbits are susceptible to both systemic as well as the ocular alterations. One of the broadest contributions made to the implications of experimental hypercholesterolaemia at the ocular level was that of Françoise and Neetens [131]. These authors, apart from analysing the changes in the liver, spleen, adrenaline glands, heart, aorta, and supra-aortic trunk, described the most significant ocular findings, such as the accumulation of lipids in the choroid, retinal disorganization, and lipid keratopathy. With respect to the retinal macroglia, the synthesis of the ApoE by the Müller cells, its subsequent secretion *in vitro,* and its being taken up by the axons and transported by the optic nerve enabled the detection of ApoE in the latter geniculate body and in the superior colliculus [132].

Studies with electron microscopy on hypercholesterolaemic rabbits have revealed hypercellularity and optically empty spaces in the corneal stroma. These optically empty spaces, with an elongated or needle shape, were previously occupied by crystals of cholesterol monohydrate or crystals of cholesterol esters [133]. In other studies, the analysis in the form adopted for the crystallizations of the different types of lipids revealed that the needles corresponded to esterified cholesterol, and the short, thin ones to triglycerides [4]. Both crystallizations appear to be associated with other components such as collagen.

In addition, the formation of foam cells as a consequence of phagocytes from the macrophage-oxidized LDL has also been detected, with the retention of cholesterol in the vascular wall and the activation of ACAT (acetyl-cholesterol-acyl-transferase) [5], this point being key to the role of macrophages in the progression or regression of the lesions [4].

Watanabe

The Watanabe heritable hyperlipidaemic (WHHL) rabbit is an animal model for hypercholesterolaemia due to genetic defects in LDL receptors [134] and a lipoprotein metabolism very similar to that of humans [135]. These features make WHHL rabbits a true model of human familial hypercholesterolaemia. The first paper on the WHHL rabbit was published in 1980 [136]. The original WHHL rabbits had a very low incidence of coronary atherosclerosis and did not develop myocardial infarction. Several years of selective breeding led to the development of coronary atherosclerosis-prone WHHL rabbits, which showed metabolic syndrome-like features, and myocardial infarction-prone WHHLMI rabbits. WHHL rabbits have been used in studies of several compounds with hypocholesterolaemic and/or anti-atherosclerotic effects with special relevance for statins [135]. Recently, WHHLMI rabbits have been used in studies of the imaging of atherosclerotic lesions by MRI [137], PET [138] and intravascular ultrasound [139].

5. Statins

Hypercholesterolaemia is a known risk factor for cardiovascular disease, and statin therapy has led to a significant reduction in morbidity and mortality from adverse cardiac events, stroke, and peripheral arterial disease [140,141]. Statins block the enzyme necessary for the production of L-mevalonate, an intermediary product in cholesterol synthesis. One of the main actions of statins is to lower circulating cholesterol levels. Cholesterol is produced from acetoacetyl coenzyme A in a process consisting of 28 steps. Statins block the second step, the conversion of 3-hydroxy-3-methylglutaryl coenzyme A (HMG-CoA) into L-mevalonate. This is also the rate-limiting step of cholesterol synthesis, and is catalysed by HMGCoA reductase. Lower levels of cholesterol prompt the cell to up-regulate the LDL-receptor. However, statin treatment also increases LDL-receptor degradation, so that the surface expression of the receptor remains unchanged. The receptor cycling is possibly increased, thus boosting the import of LDL-bound cholesterol into the cell and lowering the levels of circulating cholesterol as well [142].

Statin structure can be divided into three basic parts: an analogue of HMG-CoA, a hydrophobic ring structure that aids in binding to HMG-CoA reductase, and side groups on the rings. These side groups determine statin solubility and, as a result, many of the pharmacokinetic properties of statins. Atorvastatin, fluvastatin, lovastatin, pitavastatin, cerivastatin, and simvastatin are considered lipophilic, whereas pravastatin and rosuvastatin are considered hydrophilic as a result of polar side groups [140]. Although all statins can enter hepatic cells through either active or passive transport, hydrophilic statins, such as pravastatin and rosuvastatin are less likely to enter non-hepatic cells, while lipophilic statins, e.g. atorvastatin and simvastatin are more likely to enter hepatic and non-hepatic cells through passive diffusion. This difference in tissue permeability and metabolism may account for some of the differential pleiotropic effects among the statins [143,144]. Not all statins cross the blood-brain barrier (BBB). Short-term statin treatment does not alter cholesterol levels in the brain. The more lipophobic statins such as pravastatin cannot cross the BBB. However, the lipophilic statins lovastatin and simvastatin are capable of such a crossing, although they reach only a relatively low concentration [145].

Recent compelling evidence suggests that the beneficial effects of statins may be due not only to their cholesterol-lowering effects, but also to their cholesterol-independent or pleiotropic effects [140,142,143]. These cholesterol-independent effects include improving endothelial function, attenuating vascular and myocardial remodelling, inhibiting vascular inflammation and oxidation, and stabilizing atherosclerotic plaques [144]. The mechanism underlying some of these pleitropic effects is to inhibit the conversion of HMG-CoA to L-mevalonic acid. This inhibition prevents the synthesis of important isoprenoid intermediates of the cholesterol biosynthetic pathway, such as farnesylpyrophosphate (FPP) and geranylgeranylpyrophosphate (GGPP) [146]. Both intermediates have indirect but important roles in vascular structure and function. These two isoprenoids are both used to provide proteins with lipophilic attachments to the cell membrane. Two protein families that require these lipophilic attachments for appropriate localization within the membrane and proper functional activity are the Ras and Rho families of small G proteins. Without modification of Ras by farnesylation and Rho by geranylgeranylation, neither protein can function properly or localize to its appropriate place within the cell membrane. Furthermore, both Ras and Rho are vital components of second messenger systems known to affect vascular inflammation, hypertrophy, and hyperplasia. These second messenger systems are also intimately involved in promoting vascular remodelling in disease states such as atherosclerosis and diabetes [140].

Hypercholesterolaemia interferes with endothelial function, resulting in impaired synthesis, release, and activity of endothelial NO [147]. The mechanism by which LDL-cholesterol (LDL-C) causes endothelial dysfunction and decreases NO bioactivity involves downregulation of endothelial NOS expression, diminished receptor-mediated NO release, and less NO bioavailability owing to greater ROS production [146]. LDL-C apheresis can improve endothelium-dependent vasodilatation, which indicates that statins could restore endothelial function, in part by lowering serum LDL-C levels. Cholesterol lowering alters atherosclerotic-plaque biology, thereby decreasing vascular inflammation and leukocyte activation. Thus, statins can improve endothelial function by lowering serum cholesterol levels. However, in some studies, statins improve endothelial function before significant change in serum-cholesterol levels. Statins accelerate endothelial NO production by stimulating and upregulating endothelial NO synthase (eNOS), especially in the presence of hypoxia and oxidized LDL [147].

Statins affect eNOS expression and activity mainly through three mechanisms. First, statins increase eNOS expression by prolonging eNOS mRNA half-life rather than by inducing eNOS gene transcription. Second, statins reduce caveolin-1 abundance, an integral membrane protein that binds to eNOS in caveolae, thereby directly inhibiting NO production. Third, statins can activate the phosphatidylinositol 3-kinase (PI3K)/protein kinase Akt pathway. Akt is a serine/threonine kinase that regulates various cell functions, such as survival, growth and proliferation. Because Akt, in turn, phosphorylates and activates eNOS, statins can also increase eNOS activity through the PI3K/Akt pathway [146].

Statins may also improve endothelial function through their antioxidant effects. For example, statins enhance endothelium-dependent relaxation by inhibiting production of reactive oxygen species, such as superoxide and hydroxyl radicals, from aortas of cholesterol-fed rabbits [148]. Although lipid lowering by itself can reduce vascular oxidative stress, some of these antioxidant effects of statins appear to be cholesterol independent. Indeed, they can attenuate angiotensin II-induced production of the highly oxidative (free radical) species in VSMC and downregulate angiotensin-1 receptor expression. Because NO is scavenged by reactive oxygen species, these findings indicate that the antioxidant properties of statins may also contribute to their ability to improve endothelial function [147].

Vascular smooth-muscle cell migration and proliferation are two other major components of both atherogenesis and neointimal hyperplasia, and both of these components are affected by statin administration [140]. The small G proteins Ras and Rho, which play a large role in VSMC migration and proliferation, are plausible targets for the direct antiproliferative vascular effects of statins [143]. Statin administration is able to attenuate these effects of both small G-protein families [140]. Simvastatin inhibits both VSMC proliferation and migration in a series of experiments with human saphenous vein grafts [149].

In summary, statins could be beneficial not only for their lipid-lowering properties but for their other effects referred to as pleiotropic -that is, their capacity to restore endothelial function, to stabilize atherosclerotic plaque, or to alleviate oxidative stress and vascular inflammation [150-153]. Such effects have been demonstrated in rabbits fed cholesterol at a dose insufficient to reduce plasma-cholesterol levels [55,148].

6. Choroidal changes induced by hypercholesterolaemia

Few experimental studies examine the effects of hypercholesterolaemia on the posterior segment of the eye [154-159]. Hypercholesterolaemic rabbits constitute a useful model to delve into the repercussions of excess lipids at the ocular level. Rabbits fed a 0.5% cholesterol-enriched diet for 8 months showed a statistical increase in total serum cholesterol and in the sclera-choroid complex thickness in comparison to control [6,157,158,160,161] (Figure 9).

The increased choroidal thickness is due mainly to clusters of lipid-charged macrophages (foam cells) which are surrounded by collagen fibres in the suprachoroid. In some instances, the foam cells and fibroblasts have ultrastructural features of necrosis, with a rupture of the cytomembranes. In addition, the clusters of foam cells are encircled by cholesterol clefts [6,160] (Figure 10A). As a result of the lipid accumulation in the suprachoroid and in the large- and medium-size-vessel layers, the lumens of the choroidal vessels are mechanically compressed and the choroidal blood flow constricted (Figures 11B,12B). This situation also affects the choriocapillaris, which is compressed against Bruch's membrane, the capillary lumens being reduced to the point of collapse in some instances [6,160] (Figure 11B). However, the lumen reduction in the choroidal vessels of hypercholesterolaemic rabbit is due not only to the compression caused by the lipid build-up in the suprachoroid and vessel layers but to the changes of the VSMC and EC. The VSMC in the large- and medium-sized-vessel layer are hypertrophic with a compact appearance (Figure 12B), while EC in these layers and in the choriocapillaris show rarefactions in the cytoplasm that contribute to vascular lumen reduction (Figure 12B). The cytoplasm organelles show similar changes in VSMC and EC. The endoplasmic reticulum and Golgi cisterns are dilated, the ribosomes are disassembled, the mitochondria appear swollen, the caveolar system is smaller than in normal choroid (Figure 13A,E,F), and the cytoplasm contains droplets of lipids and dense bodies. In VSMC the myofilaments are disorganized, with areas of focal necrosis affecting the dense plaques (Figure 13F). In the areas where the tissue was highly disorganized, other necrotic features (swelling, vacuolization, cytomembrane necrosis, and the disappearance of specific ultrastructural characteristics) are visible in both cell types [6,160] (Figure 13E,F).

The endothelial alterations described above could be explained, at least partly, by the Posieulle's law, which states that tangential tension is directly proportional to the blood-flow viscosity and inversely proportional to the third power of the inside radius. Accordingly, reduction of the choroidal vascular lumens in hypercholesterolaemic rabbits could increase the local tangential force exerted by the choroidal blood flow on the vessel walls. With time, these changes in blood flow could lead to a reorganization of the endothelial phenotype, resulting in endothelial dysfunction [162]. Endothelial changes and basal-membrane thickening of ocular vessels in hypercholesterolaemic rabbits may precede the atherosclerotic process. In the hypercholesterolaemic rabbits, the basal membrane of EC and VSMC contained electrodense and electrolucent particles and were thicker than in the normal choroid.

In the atherosclerotic process, the adhesion molecules VCAM-1 and ICAM-1 favour monocyte binding and migration to the subendothelial space, where they transform into macrophages [3,4]. In iris arterioles of hypercholesterolaemic rabbits, the internalization of lipids by macrophages results in the formation of foam cells [163]. Foam cells are also found in the suprachoroid and large- and medium-sized-vessel layers of hypercholesterolaemic rabbits [6,160]. Before and after dying, foam cells may release products such as cholesterol (oxidized and esterified) that increase endothelial damage, thus encouraging the atherosclerotic lesion. The foam cells underneath the basal membrane of the large-sized-vessel layer [6,160], where the vessels have a diameter of up to 90 μm [15], could represent the microatheromas described in brain circulation [164]. In the medium-sized-vessel layer, arterioles with diameters ranging from 20 to 40 micrometers [15], combined with the increased collagen detected in the adventitia and the lipid deposits (electron-dense vesicles) inside muscle cells [6,160], could produce arteriolar hyalinosis [164].

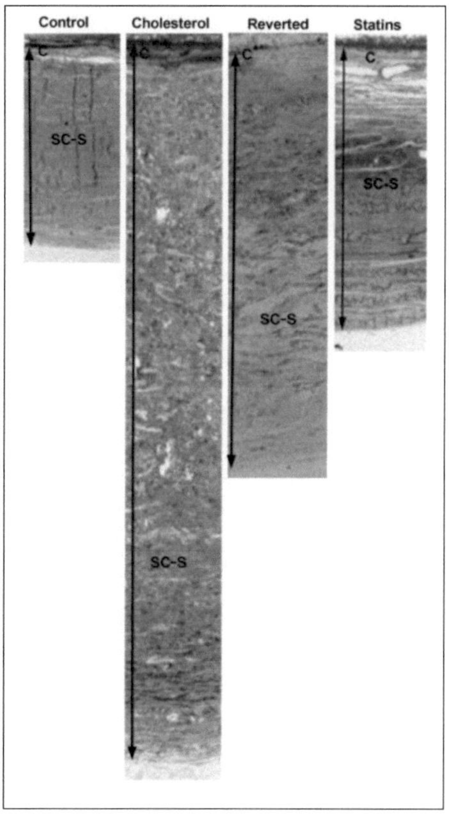

Figure 9. Thickening of the sclera-choroid complex (double arrow) in the study groups. Semithin sections (toluidine blue).(C: vascular layers of the choroid; SC-S: suprachoroidea-sclera). [Control: standard diet; Hypercholesterolemic: 0.5% cholesterol-enriched diet for 8 months; Reverted: 0.5% cholesterol-enriched diet for 8 months plus standard diet for another 6 months; Statins: Hypercholesterolemic diet + fluvastatin sodium or pravastatin sodium for 8 months]. Scale bar, 100 μm. (Modified from Salazar et al [6] and Rojas et al. [55]).

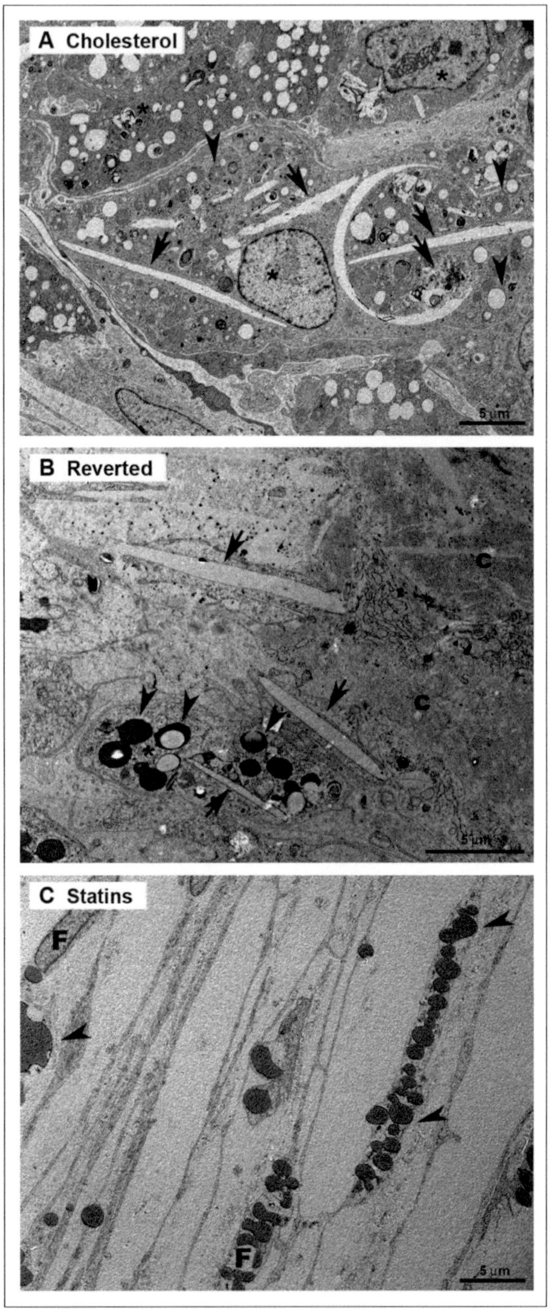

Figure 10. Electron micrographs of the suprachoroidea. A: Hypercholesterolemic rabbit. The image shows a large amount of lipid containing foam cells as well as collagen fibers and cholesterol clefts between them. B: Reverted animals. The collagen fibres predominate over the foam cells and cholesterol clefts. C: Statin-treated animals. An increase in collagen is observed. The lipids are located mainly inside the fibroblast. [Foam cells (asterisk); lipids (arrowhead); cholesterol clefts (arrow); collagen (C); fibroblast (F)]. Scale bars: 5 µm. (Image B from Salazar et al. [6]).

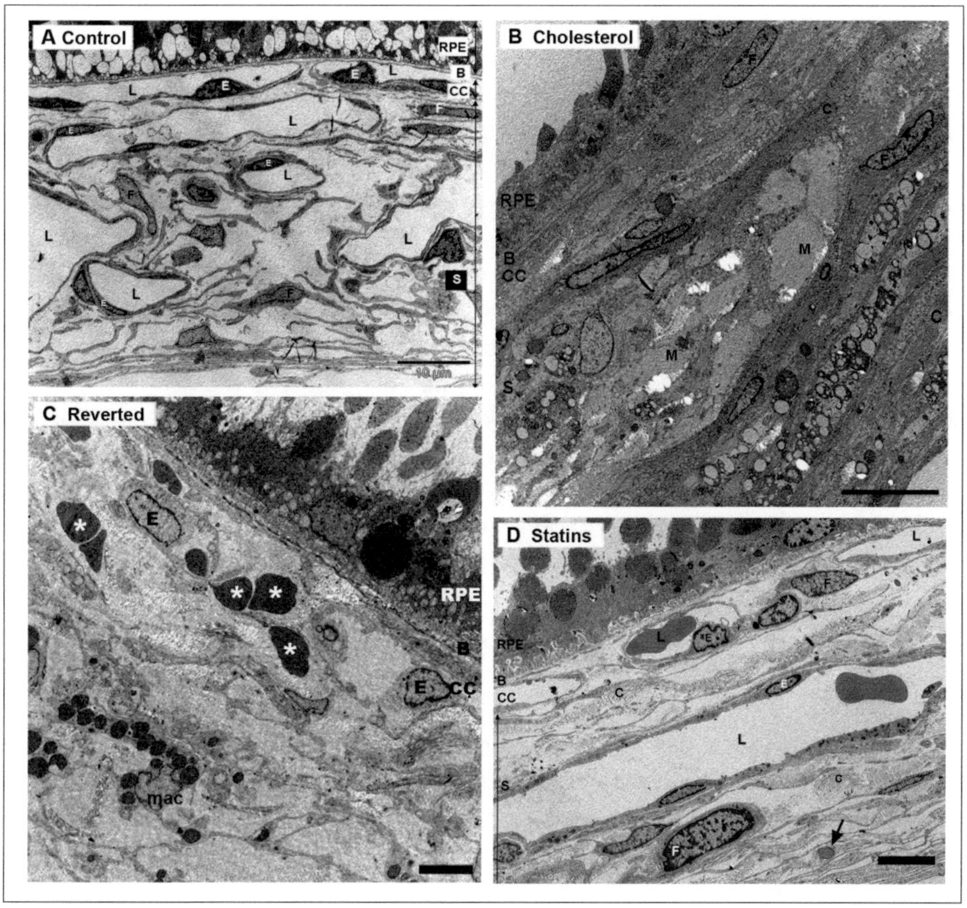

Figure 11. Electron micrographs of large- and medium sized-vessel layers and choriocapillaris. A: Spongy structure of the choroid of a control animal. B: In the hypercholesterolemic group the vascular lumens were reduced to the point of collapse in some instances. C: In the reverted animals, choroidal changes make difficult to differentiate the choroidal layers. D: The vascular lumens of statin-treated animals were opened in part due to the reduction of suprachoroidal lipids. [Bruch's membrane (B); collagen (C); choriocapillaris (CC); endothelial cell (E); fibroblast (F); vascular lumen (L); lipids (arrow); vascular smooth-muscle cells (M); retinal-pigment epithelium (RPE); large- and medium-sized vessel layers (S); Red blood cells (*) outside the vascular lumens; macrophage (mac)]. [Bars: A,B, 10 μm; C,D 5 μm]. (Modified from Salazar et al [6] and Rojas et al. [55]).

Studies examining the aortic preparations of hypercholesterolaemic rabbits indicate that, at sites where the luminal side was covered by fatty streaks, the number of caveolae in the EC was significantly lower [165]. Caveolae are distinctive cholesterol-enriched invaginations of the plasma membrane of the most mammalian cells. Caveolin is an essential component of caveolas, and their expression in cells results in the assembly of caveolae at the cell surface. There is an emerging role of caveolae, in organizing and modulating the basic functions of smooth muscles [166] and vascular functions dependent on NO and Ca^{2+} signalling [166,167]. Caveolin proteins tightly bind cholesterol and contribute to the regulation of cholesterol fluxes and distribution within cells [168]. It has been reported that in the early stages of hypercholesterolaemia, caveolin-1 synthesis increases to remove the excess cholesterol from the cells. However, after long periods of cholesterol feeding, caveolin-1 synthesis was repressed [169]. In [95] reported the cholesterol-induced impairment of NO production through the modulation of caveolin abundance in EC as a mechanism that may be involved in the pathogenesis of endothelial dysfunction. In hypertensive rats, it has been postulated that a decreased number of caveolae could be the reason for the impaired relaxation by NO donors. Additionally, in rabbits, after a long period (12 weeks) of ingesting a high-cholesterol diet, both the level of caveolin-1 and the activity of NOS declined [169]. EC and VSMC in hypercholesterolaemic rabbits (8 months of hypercholesterolaemic diet) exhibit less caveolae than did control animals (Figure 13A,F), a situation that could contribute to the deposit of circulating molecules, such as LDL, which are detected as electrolucent or electrodense particles of varying size. A decrease in caveola abundance has been postulated to represent a novel mechanism of endothelial dysfunction in atheromatous rabbit aorta [165,170]. This idea is supported by the fact that a pharmacological disruption of endothelial caveolae also results in the attenuation of acetylcholine-induced relaxation [165,170].

7. Choroidal changes after diet-induced normalization of plasma lipid levels

It has been established that the atherosclerotic lesions can undergo regression in experimental animals such as rabbits, dogs, and non-human primates [171]; and the lack of progression or even regression can occur in humans, especially with the introduction of new therapeutical options [172].

Knowledge of the changes arising after the normalization of the cholesterol values in a vascular tissue such as the choroid can be useful to evaluate the effects of the different drugs for hypocholesterolaemia. Animal models offer a useful tool for studying lesion regression after cholesterol-serum values normalize. When excessive cholesterol is withdrawn from the diet of rabbits, these animals recover some of the biochemical and histological parameters that are altered in cholesterol-fed animals [6,173]. It has been reported that the serum concentration of total cholesterol, triglycerides, phospholipids, VLDL, HDL, LDL, and intermediate-density lipoprotein (IDL) increased in rabbits fed a 0.5% cholesterol-enriched diet for 8 months. When the same animals were then fed a standard diet for another 6 months, (reverted animals), lipid values returned to normal. Notably, the normalization of serum values was not followed by a complete recovery of the histology of the retina [158], choroid, or thoracic aorta [6]. In reverted rabbits, the sclera-choroid complex thinned compared with that of hypercholesterolaemic animals, due to having far fewer clusters of foam cells encircled by cholesterol clefts in the suprachoroid (Figure 9). However, these tissues had not yet reached the normal thickness values of control due to the increase in collagen fibres [6] (Figures 10B,11C). This marked reduction in the suprachoroidal build-up of lipids implies less compression of the vessels. Such findings have also been reported in the iris and cornea of hypercholesterolaemic reverted rabbits [174]. However, although in most cases fewer lipids were found in the suprachoroidal spaces in reverted rabbits, there were still numerous ultrastructural changes in comparison with hypercholesterolaemic and normal rabbits. These changes could contribute to chronic ischaemia, for several reasons: i) The large- and medium-sized-vessel layers still exhibited alterations in VSMC and EC that could impair vascular function. Specifically, the cytoplasm organelles (endoplasmic reticulum, Golgi cisterns, ribosomes, mitochondria, and caveolar system) of VSMC and EC in the reverted rabbits showed changes similar to hypercholesterolaemic animals. Also, in zones where the tissue was highly disorganized, necrotic features (swelling, vacuolization, cytomembrane necrosis, and disappearance of specific ultrastructural characteristics) were noted [6] (Figures 11C,12C). ii) The EC of the choriocapillaris extended cytoplasmic projections that reduced the vascular lumen (Figure 12C). iii) Finally, the intervascular spaces of the choroid contained fewer lipids than in hypercholesterolaemic animals but more collagen fibres than in control (Figures 11A,C,12A,C).

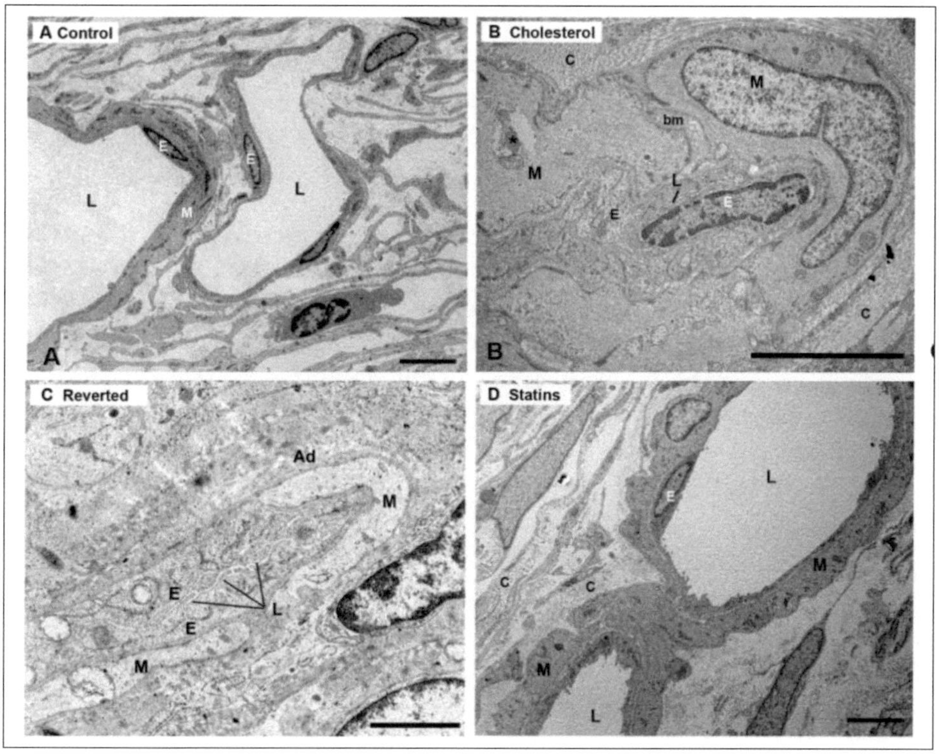

Figure 12. Electron micrographs of large- and medium sized-vessel layers of the choroid. A: Control animals. B: In the hypercholesterolemic animals the vascular lumens are reduced. The muscle cells are hypertrophic and contain drops of lipids (*) in the cytoplasm. Additionally, the endothelial cells are hypertrophic and/or necrotic. The basal membrane is thick. C: Reverted animals maintained the reduction of the vascular lumen and the thickening of the adventitia. D: In animals treated with statins vascular lumens, vascular smooth-muscle cells, endothelial cells and basal membrane were more similar to the control group. [collagen (C) ; endothelial cell (E); vascular lumen (L); vascular smooth-muscle cells (M); basal membrane (bm); adventitia (Ad)]. [Bars: A,B, D 5 μm; C 2 μm]. (Modified from Salazar et al [6] and Rojas et al. [55]).

This increase in collagen fibres was also found in the intima, among VSMC and in the adventitia (Figure 12C) of the large- and medium-sized-vessel layer and in the basal membrane of the EC of the choriocapillaris [6]. It is well known that the reabsorption of both esterified and free cholesterol is followed by intense sclerogenic activity [175], due to the capacity of cholesterols and their esters to induce inflammation. Some inflammatory cytokines (TNFα and IL-1) control the remodelling activity of macrophages and smooth-muscle cells, which have the capacity to produce several matrix metalloproteinases (MMPs) [176,177]. It has been observed that such MMPs are synthesised mainly where the concentration of foam cells is greatest and that there is a relationship between lipid uptake and MMP activity [177]. The observations cited above underline the strong connection between inflammation, tissue remodelling, and lipid metabolism.

In summary, in hypercholesterolaemic rabbits the replacement of a hyperlipidaemic diet by a standard one normalized serum-lipid levels and eliminated most lipid build-up in the posterior segment of the eye. However, this lipid reduction was not followed by a reversal of the changes in choroidal vessels, where persisting ultrastructural changes in VSMC, EC, and extracellular matrix were compatible with a chronic ischaemia.

8. Choroidal changes after low-dose statin treatment

As mentioned above, it is well known that statins can exert cholesterol-independent or pleitropic effects which involve the restoration of endothelial function, stabilization of the atheromatous plaque, reduced oxidative stress, lower vascular inflammation, immunomodulation, and inhibition of VSMC proliferation [146,150,153,178]. Fluvastatin sodium (a lipophilic synthetic statin metabolised by the liver) and pravastatin sodium (a hydrophilic statin of fungal origin) can induce the regression of atherosclerosis besides reducing plasma cholesterol [148,179].

It has been reported that 2 mg/kg/day of fluvastatin sodium and pravastatin sodium exerts an anti-atherosclerotic effect that appeared not to be mediated by the lipid-lowering properties of the drugs [150,180,181]. A 12.5-50 mg/Kg/day dosage of fluvastatin effectively reduces plasma lipids in Watanabe heritable hyperlipidaemic rabbits [182]. On the contrary, 2 mg/Kg/day of fluvastatin did not lower plasma-cholesterol levels but did contribute to the impaired endothelium-dependent relaxation response in aortic rings [148] and femoral arteries of cholesterol-fed rabbits [150].

According with reference [148], hypercholesterolaemic rabbits treated with fluvastatin sodium and pravastatin sodium at a non-lipid-lowering dosage (2 mg/kg/day) showed a very small descent in plasma cholesterol and triglycerides levels [55]. In comparison with hypercholesterolaemic rabbits, low-dose statin-treated animals had far fewer lipids in the suprachoroid (the foam cells were decreased or absent), triggering a significant thinning of the sclera-suprachoroid complex ($p< 0.000$) (Figure 9). This reduction in the build-up of lipids relieved the mechanical compression of the vascular layers of the choroid and lead to normalization of the spongy texture of the choroid [55] (Figure 11D). However, statin-treated animals bore some alterations not present in control animals, such as clumps of electrodense particles, some foam cells, and fibroblast-containing lipids in the suprachoroid (Figure 10C) or a significant thickening of the vascular layers because the vascular lumens were open [55] (Figures 11D,12D). The lack of lumen closure in low-dose statin-treated rabbits (unlike that observed in hypercholesterolaemic rabbits) suggests that the choroidal blood flow was not compromised and that endothelial dysfunction would not result in part because of the lesser stress on EC by means of a weakening of the local tangential forces exerted on the vessel walls when the inner vascular radius decreased [55]. In addition, in low-dose statin-treated animals, signs of necrosis were rarely found (Figure 13B,C), the EC looked similar to those of control (Figure 13A-D), the cytoplasmic organelles were normal (Figure 13B,D), and the caveolar system was more numerous than in hypercholesterolaemic rabbits [55] (Figure 13B,D-F). This increment of caveolae detected in the low-dose statin-treated animals could account for the better preservation of endothelial function and could contribute to the normalization of vascular lumens observed in these animals. It has been proposed that caveolae could be specialized plasmalemmal regions involved in the integration of extracellular contractile signals and intracellular effectors in VSMC [183]. Three findings in low-dose statin-treated rabbits led to speculation that treatment improved VSMC function: first, the increased amount of caveolae observed and their location close to the underlying network of peripheral sarcoplasmic reticulum; second, VSMC were not hypertrophic and exhibited the elongated shape observed during relaxation [55]. This contrasted with the compact, round, and short shape exhibited by VSMC in hypercholesterolaemic rabbits, which was compatible with a contractile state; and third, the absence of areas of focal necrosis related to the dense plaques in low-dose statin animals.

A noteworthy finding in the cytoplasm of the EC and VSMC in the low-dose statin-treated group was that the cytoplasmic organelles were normal. Particularly the mitochondria maintained a normal electrodensity of the matrix, and the cristae and membranes were well preserved (Figure 13D). This is of outstanding importance, taking into account that these organelles represent a fundamental structure in the aerobic metabolism [184] and that they are extremely sensitive to hypoxia. The importance of mitochondria is well known in oxidative phosphorylation to generate ATP. Mitochondrial phosphorylation is impaired in ischaemic situations and, as a consequence of oxidative stress, DNA, proteins, and lipids, is damaged by a surge in free radicals [185]. From the mitochondrial integrity found in low-dose statin-treated rabbits, good functioning of EC and VSMC could be deduced. The normal ultrastructure of EC and VSMC suggests that the endothelium-depending relaxation is maintained, thus reducing the ischaemia. According to ultrastructural examinations in the choroidal vascular tissue, the effects of low-dose statins are probably mediated by an overregulation of the endothelial NO and a downregulation of endothelin-1, as reported in studies on the direct consequences of the endothelial dysfunction in extraocular tissues [180]. This hypothesis should be tested by functional studies on ocular vascular reactivity. Another remarkable observation in low-dose statin-treated rabbits was that the basal membranes resembled control. However, it should be mentioned that an increase in collagen fibres in the vascular adventitia and in the intervascular spaces, and even fibroblast containing clumps of lipids could still be detected (Figures 11D,12D). These features can be explained firstly by the low lipid-decreasing capability of statins at the dosage of 2 mg/kg/day used in the study and, secondly, to macrophage activity and their ability to produce MMPs [176,177] in the areas of higher concentrations of foam cells, where esterified and free-cholesterol reabsorption is followed by sclerogenic activity [175].

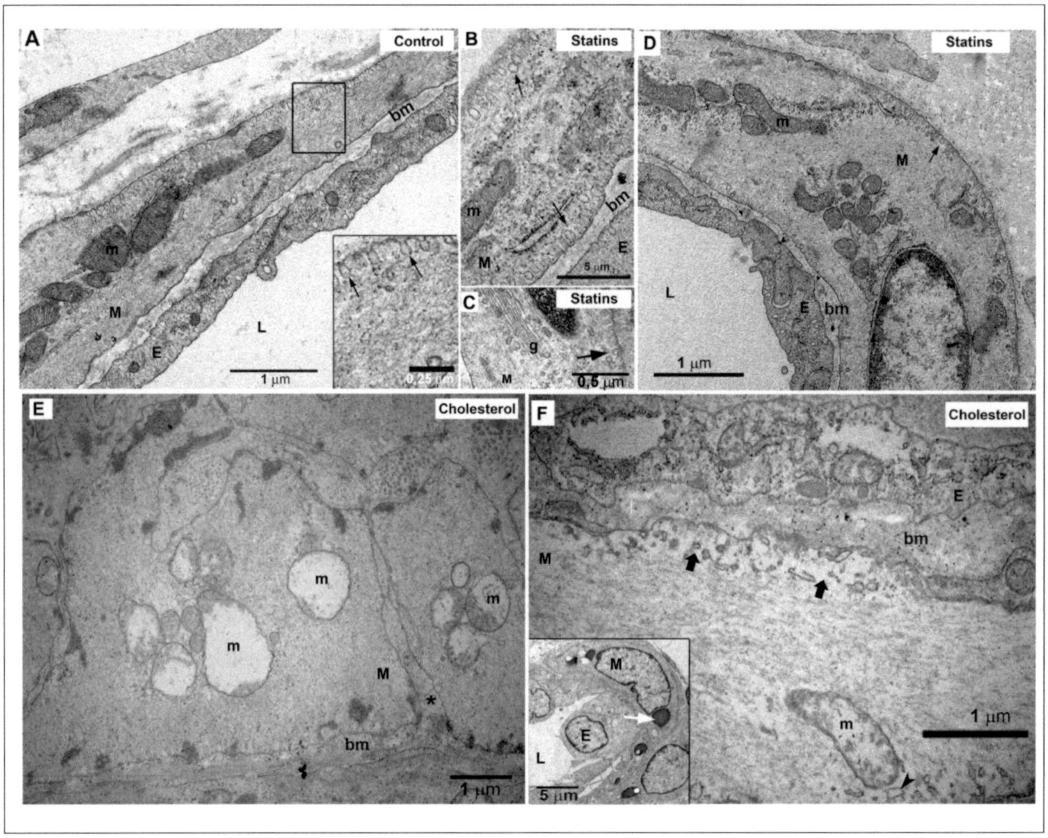

Figure 13. Electron micrographs of endothelial and vascular smooth-vessel cells (VSMC). A: Control animal. The insert shows the normal ultrastructure of the myofilaments and caveolar system. B, C, D: Statin-treated rabbits. B,C: Intact caveolar system and cytoplasmic organelles. D: Elongated VSMC with well-preserved mitochondria, caveolae, and peripheral sarcoplasmic reticulum. The nucleus and perinuclear cisterns are normal. E, F: Hypercholesterolemic rabbits. E: Rounded and hypertrophic VSMC. Swollen mitochondria with loss of the cristae. The caveolae are decreased and the intercellular space increased (asterisk) in some instances. F: Area of focal necrosis below the plasma membrane (large black arrow). Dilated endoplasmic reticulum (arrowhead). The myofilaments are disorganized. Drops of lipids (white arrow) in VSMC (insert). [basal membrane (bm); endothelial cell (E); Golgi (g); vascular lumen (L); vascular smooth muscle cells (M); mitochondria (m); Caveolae (small black arrow)]. [Scale bars: A,D,E,F, 1 µm; A insert, 0.25 µm; B,F insert 5 µm; C, 0.5 µm.]. (From Rojas et al. [55])

In hypercholesterolaemic New Zealand rabbits, treatment with fluvastatin sodium and pravastatin sodium at a dose (2mg/Kg/day) insufficient to normalize plasma-lipid levels prevents the progression of atherosclerosis in the different vascular layers of the choroid; the most striking effects being seen in EC and VSMC.

9. Conclusions and perspectives

Clinical studies have suggested that hyperlipidaemia alone can prompt structural changes in the choroidal vascular and retinal system that, over time, could provoke retinal dysfunction. This situation of chronic ischaemia is also present in causes of blindness as prevalent as aged-related macular degeneration, glaucoma or diabetic retinopathy [11-13]. This situation is extremely important, as today we know from epidemiological studies the relation between vascular retinal lesions and the incidence of lesions in non-ocular tissues that appear to be linked to common factors having microvascular effects. Statins are cholesterol-lowering medications. The treatment of hypercholesterolaemic rabbits with a non-lipid-lowering dose of statins dramatically reduces the ultrastructural choroidal damage induced by a sustained cholesterol-enriched diet [55]. This implies better choroidal flow and consequently oxygenation of the external retina, reducing chronic ischaemia [157]. It is well known that statins can exert cholesterol-independent or pleitropic effects [11] which involve the restoration of endothelial function, stabilization of atheromatous plaque, reduced oxidative stress, or lower eye inflammation [150,153,178]. These effects are potentially decisive in maintaining normal endothelial and smooth-muscle function and in improving endothelium-dependent relaxation. It bears noting that the most striking effects of low-dose statins in hypercholesterolaemic rabbits are the normalization of the ultrastructure of EC and VSMC, specifically cytoplasm organelles and caveolar system. Such effects suggest a possible role of statins in those ocular diseases having endothelial dysfunction in their physiopathology such as AMD, glaucoma or diabetic retinopathy, among others.

Author's details

J.M. Ramírez *
Instituto de Investigaciones Oftalmológicas Ramón Castroviejo,
Universidad Complutense de Madrid, Madrid, Spain

Departamento de Oftalmología, Facultad de Medicina,
Universidad Complutense de Madrid, Madrid, Spain

J.J. Salazar
Instituto de Investigaciones Oftalmológicas Ramón Castroviejo,
Universidad Complutense de Madrid, Madrid, Spain

Escuela Universitaria de Óptica, Universidad Complutense de Madrid, Madrid, Spain

*Corresponding Author

R. de Hoz
Instituto de Investigaciones Oftalmológicas Ramón Castroviejo,
Universidad Complutense de Madrid, Madrid, Spain

Escuela Universitaria de Óptica, Universidad Complutense de Madrid, Madrid, Spain

B. Rojas
Instituto de Investigaciones Oftalmológicas Ramón Castroviejo,
Universidad Complutense de Madrid, Madrid, Spain

Departamento de Oftalmología, Facultad de Medicina,
Universidad Complutense de Madrid, Madrid, Spain

B.I. Gallego
Instituto de Investigaciones Oftalmológicas Ramón Castroviejo,
Universidad Complutense de Madrid, Madrid, Spain

Escuela Universitaria de Óptica, Universidad Complutense de Madrid, Madrid, Spain

A.I Ramírez
Instituto de Investigaciones Oftalmológicas Ramón Castroviejo,
Universidad Complutense de Madrid, Madrid, Spain

Escuela Universitaria de Óptica, Universidad Complutense de Madrid, Madrid, Spain

A. Triviño
Instituto de Investigaciones Oftalmológicas Ramón Castroviejo,
Universidad Complutense de Madrid, Madrid, Spain

Departamento de Oftalmología, Facultad de Medicina,
Universidad Complutense de Madrid, Madrid, Spain

Acknowledgement

The authors would like to thank David Nesbitt for correcting the English version of this work. This work was supported by RETICs Patología Ocular del Envejecimiento, Calidad Visual y Calidad de Vida (Grant ISCIII RD07/0062/0000, Spanish Ministry of Science and Innovation); Fundación Mutua Madrileña (Grant 4131173); BSCH-UCM GR35/10-A Programa de Grupos de Investigación Santander-UCM. Beatriz Gallego is currently supported by a predoctoral fellowship from the Universidad Complutense de Madrid.

10. References

[1] Bill A, Sperber GO. Control of retinal and choroidal blood flow. Eye (London, England) 1990;4(Pt 2) 319-325.
[2] Chen Y, Chang Y, Jyh Jiang M. Monocyte chemotactic protein-1 gene and protein expression in atherogenesis of hypercholesterolemic rabbits. Atherosclerosis 1999;143(1) 115-123.

[3] Ross R. Atherosclerosis — An Inflammatory Disease. New England Journal of Medicine 1999;340(2)15-126.

[4] Crispin S. Ocular lipid deposition and hyperlipoproteinaemia. Progress in Retinal and Eye Research 2002;21(2) 169-224.

[5] Rong JX, Shen L, Chang YH, Richters A, Hodis HN, Sevanian A. Cholesterol Oxidation Products Induce Vascular Foam Cell Lesion Formation in Hypercholesterolemic New Zealand White Rabbits. Arteriosclerosis, Thrombosis, and Vascular Biology 1999;19(9) 2179-2188.

[6] Salazar JJ, Ramírez AI, de Hoz R, Rojas B, Ruiz E, Tejerina T, Triviño A, Ramírez JM. Alterations in the choroid in hypercholesterolemic rabbits: reversibility after normalization of cholesterol levels. Experimental Eye Research 2007 ;84(3) 412-422.

[7] Martínez-González J, Llorente-Cortés V, Badimon L. Biología celular y molecular de las lesiones ateroscleróticas. Revista Española de Cardiología 2001;54 218-231.

[8] Schneider DB, Vassalli G, Wen S, Driscoll RM, Sassani AB, DeYoung MB, Linnemann R, Virmani R, Dichek DA. Expression of Fas Ligand in Arteries of Hypercholesterolemic Rabbits Accelerates Atherosclerotic Lesion Formation. Arteriosclerosis, Thrombosis, and Vascular Biology 2000;20(2) 298-308.

[9] Öörni K, Pentikäinen MO, Ala-Korpela M, Kovanen PT. Aggregation, fusion, and vesicle formation of modified low density lipoprotein particles: molecular mechanisms and effects on matrix interactions. Journal of Lipid Research 2000;41(11) 1703-1714.

[10] Klein R, Sharrett AR, Klein BEK, Chambless LE, Cooper LS, Hubbard LD, Evans G. Are Retinal Arteriolar Abnormalities Related to Atherosclerosis?: The Atherosclerosis Risk in Communities Study. Arteriosclerosis, Thrombosis, and Vascular Biology 2000;20(6) 1644-1650.

[11] Wong TY, Klein R, Couper DJ, Cooper LS, Shahar E, Hubbard LD, Wofford MR, Sharrett AR. Retinal microvascular abnormalities and incident stroke: the Atherosclerosis Risk in Communities Study. Lancet 2001;358(9288) 1134-1140.

[12] Klein R, Klein BEK, Tomany SC, Wong TY. The relation of retinal microvascular characteristics to age-related eye disease: the Beaver Dam eye study. American Journal of Ophthalmology 2004;137(3) 435-444.

[13] Wong TY, McIntosh R. Systemic associations of retinal microvascular signs: a review of recent population-based studies. Ophthalmic and Physiological Optics 2005;25(3) 195-204.

[14] Matsusaka T. Cytoarchitecture of choroidal melanocytes. Experimental Eye Research 1982;35(5) 461-469.

[15] Hogan MJ, Alvarado JA, Weddell JE. Histology of the human eye: an atlas and textbook. Toronto: W.B. Saunders Company Ed; 1971.

[16] Tamm ER, Flügel-Koch C, Mayer B, Lütjen-Drecoll E. Nerve cells in the human ciliary muscle: ultrastructural and immunocytochemical characterization. Investigative Ophthalmology & Visual Science 1995;36(2) 414-426.

[17] Poukens V, Glasgow BJ, Demer JL. Nonvascular contractile cells in sclera and choroid of humans and monkeys. Investigative Ophthalmology & Visual Science 1998;39(10) 1765-1774.

[18] Flügel-Koch C, May CA, Lütjen-Drecoll E. Presence of a contractile cell network in the human choroid. Ophthalmologica. 1996;210(5) 296-302.

[19] Ramírez JM, Triviño A, De Hoz R, Ramírez AI, Salazar JJ, García-Sánchez J. Immunohistochemical study of rabbit choroidal innervation. Vision Research 1999;39(7) 1249-1262.

[20] Triviño A, De Hoz R, Salazar JJ, Ramírez AI, Rojas B, Ramírez JM. Distribution and organization of the nerve fiber and ganglion cells of the human choroid. Anatomy and Embryology 2002;205(5-6) 417-430.

[21] Triviño A, de Hoz R, Rojas B, Salazar JJ, Ramírez AI, Ramírez JM. NPY and TH innervation in human choroidal whole-mounts. Histology and Histopathology 2005;20(2) 393-402.

[22] Triviño A, Ramírez JM. Anatomofisiología de la coroides. In: Gómez-Ulla F, Marín F, Ramírez JM, Triviño A. (ed) La circulación coroidea. Barcelona: EDIKA-MED. S.A; 1989. p7-29.

[23] Bron AJ, Tripathi RC, Tripathi BJ. The choroid and uveal vessels. In: Bron AJ, Tripathi RC, Tripathi BJ. (ed) Wolff's Anatomy of the Eye and Orbit (Eighth edition). London: Chapman & Hall Medical; 1997. p371-410.

[24] Riva CE, Alm A, Pournaras CJ. Ocular circulation. In: Levin LA, Nilsson SFE, Ver Hoeve J, Wu SM, Kaufman PL, Alm A. (ed) Adler's Physiology of the Eye. Edinburgh: Elsevier Saunders; 2011. p243-273.

[25] Hayreh SS. The long posterior ciliary arteries. An experimental study. Albrecht von Graefes Archiv fur Klinische und Experimentelle Ophthalmologie 1974;192(3) 197-213.

[26] Ducournau DH. A new technique for the anatomical study of the choroidal blood vessels. Ophthalmologica. 1982;184(4) 190-197.

[27] Triviño A, Ramírez JM, García-Sánchez J. Study of the choroidal circulation in the human eye: experimental model. In: Flower RW. (ed) II Internacional Symposium on the Choroid. Maryland (USA); 1989. p32-42.

[28] Risco JM, Grimson BS, Johnson PT. Angioarchitecture of the Ciliary Artery Circulation of the Posterior Pole. Archives of Ophthalmology 1981;99(5) 864-868.

[29] Hayreh SS. Submacular choroidal vascular pattern. Experimental fluorescein fundus angiographic studies. Albrecht von Graefes Archiv fur Klinische und Experimentelle Ophthalmologie. 1974;192(3) 181-196.

[30] Shimizu K, Ujiie K. Morphology of the submacular choroid: vascular structure. Ophthalmologica. 1981;183(1) 5-10.

[31] Weiter JJ, Ernest JT. Anatomy of the choroidal vasculature. American Journal of Ophthalmology 1974;78(4) 583-590.

[32] Ramírez JM, Triviño A, De Hoz R, González C, Borrego R, Salazar JJ, García-Sanchez J. Estudio de la vascularización ciliar en el conejo albino. Archivos de la Sociedad Española de Oftalmología 1990;5 823-30.

[33] Buggage RR, Torcynski E, Grossniklaus HE. The uveal tract. In: Duane TD, Jaeger EA. (ed) Biomedical Foundations of Ophathalmology. CD-Rom. Philadelphia: Harper & Row Publishers; 2004.

[34] Olver JM, Sharma A. Anatomy and physiology of the uveal tract. In: Easty JM, Sparrow JM. (ed) Oxford Textbook of Ophthalmology. New York: Oxford Medical Publications; 1999. p501-508.
[35] Fryczkowski AW, Sherman MD, Walker J. Observations on the lobular organization of the human choriocapillaris. International Ophthalmology 1991;15(2) 109-120.
[36] Triviño A, Ramírez JM, García-Sánchez J. Estudio comparativo entre la vascularización coroidea del hombre y el animal de experimentación. Archivos de la Sociedad Española de Oftalmología 1986;51 305-312.
[37] Spitznas M. The fine structure of the chorioretinal border tissues of the adult human eye. Advances in Ophthalmology 1974; 2878-174.
[38] Spitznas M, Reale E. Fracture faces of fenestrations and junctions of endothelial cells in human choroidal vessels. Investigative Ophthalmology 1975; 14(2) 98-107.
[39] Melamed S, Ben-Sira I, Ben-Shaul Y. Ultrastructure of fenestrations in endothelial choriocapillaries of the rabbit--a freeze-fracturing study. British Journal of Ophthalmology 1980;64(7) 537-543.
[40] Bill A, Tornquist P, Alm A. Permeability of the intraocular blood vessels. Transactions of the Ophthalmological Societies of the United Kingdom 1980;100(3) 332-336.
[41] Törnquist P. Capillary permeability in cat choroid, studied with the single injection technique (II). Acta Physiologica Scandinavica 1979;106(4) 425-430.
[42] Bill A. Blood circulation and fluid dynamics in the eye. Physiological Reviews 1975;55(3) 383-417.
[43] Bill A, Sperber G, Ujiie K. Physiology of the choroidal vascular bed. International Ophthalmology 1983;6(2) 101-107.
[44] Yamamoto T, Fukuda S, Obata H, Yamashita H. Electron microscopic observation of pseudopodia from choriocapillary endothelium. Japanese Journal of Ophthalmology 1994;38(2) 129-138.
[45] Guymer R, Luthert P, Bird A. Changes in Bruch's membrane and related structures with age. Progress in Retinal and Eye Research 1999;18(1) 59-90.
[46] Guymer RH, Bird AC, Hageman GS. Cytoarchitecture of Choroidal Capillary Endothelial Cells. Investigative Ophthalmology Visual Science 2004;45(6) 1660-1666.
[47] Manche EE, Korte GE. Ultrastructural evidence of remodelling in the microvasculature of the normal rabbit and human eye. Acta Anatomica 1990;138(2) 89-96.
[48] Torczynski E, Tso MO. The architecture of the choriocapillaris at the posterior pole. American Journal of Ophthalmology 1976;81(4) 428-440.
[49] Hogan MJ, Feeney L. Electron microscopy of the human choroid. III. The blood vessels. American Journal of Ophthalmology 1961;51 1084-1097.
[50] Oyster CW. The human eye. Structure and function. Sunderland (Massachusetts): Sinauer Associates; 1999.
[51] Triviño A, de Hoz R, Rojas B, Salazar JJ, Ramírez AI, Gallego B, Ramírez JM. The human choroid posseses substance P and calcitonine gene-related peptide intrinsic neurons. Acta Ophthalmologica 2009;87 (s244).

[52] de Hoz R, Ramírez AI, Salazar JJ, Rojas B, Ramírez JM, Triviño A. Substance P and calcitonin gene-related peptide intrinsic choroidal neurons in human choroidal whole-mounts. Histology and Histopathology 2008;23(10) 1249-1258.

[53] Ruskell GL. Facial parasympathetic innervation of the choroidal blood vessels in monkeys. Experimental Eye Research 1971;12(2) 166-172.

[54] De Stefano ME, Mugnaini E. Fine structure of the choroidal coat of the avian eye. Vascularization, supporting tissue and innervation. Anatomy and Embryology 1997;195(5) 393-418.

[55] Rojas B, Ramírez AI, Salazar JJ, de Hoz R, Redondo A, Raposo R, Mendez T, Tejerina T, Triviño A, Ramírez JM. Low-dosage statins reduce choroidal damage in hypercholesterolemic rabbits. Acta Ophthalmologica 2011;89(7) 660-669.

[56] Pournaras CJ, Rungger-Brändle E, Riva CE, Hardarson SH, Stefansson E. Regulation of retinal blood flow in health and disease. Progress in Retinal and Eye Research 2008;27(3) 284-330.

[57] Feeney L, Hogan MJ. Electron microscopy of the human choroid. I. Cells and supporting structure. American Journal of Ophthalmology 1961;51 1057-1072.

[58] Cavallotti C, Corrado BG, Feher J. The human choriocapillaris: evidence for an intrinsic regulation of the endothelium? Journal of Anatomy 2005;206(3) 243-247.

[59] Le Beux YJ, Willemot J. Actin- and myosin-like filaments in rat brain pericytes. Anatomical Record 1978;190(4) 811-826.

[60] Nickla DL, Wallman J. The multifunctional choroid. Progress in Retinal and Eye Research 2010;29(2) 144-168.

[61] Alm A, Bill A, Young FA. The effects of pilocarpine and neostigmine on the blood flow through the anterior uvea in monkeys. A study with radioactively labelled microspheres. Experimental Eye Research 1973;15(1) 31-36.

[62] Alm A, Bill A. Ocular and optic nerve blood flow at normal and increased intraocular pressures in monkeys (Macaca irus): a study with radioactively labelled microspheres including flow determinations in brain and some other tissues. Experimental Eye Research 1973;15(1) 15-29.

[63] Alm A, Bill A. Blood flow and oxygen extraction in the cat uvea at normal and high intraocular pressures. Acta Physiologica Scandinavica 1970;80(1) 19-28.

[64] Parver LM, Auker CR, Carpenter DO. The stabilizing effect of the choroidal circulation on the temperature environment of the macula. Retina 1982;2(2) 117-120.

[65] Parver LM, Auker CR, Carpenter DO. Choroidal blood flow. III. Reflexive control in human eyes. Archives of Ophthalmology 1983;101(10) 1604-1606.

[66] Parver LM, Auker CR, Carpenter DO, Doyle T. Choroidal blood flow II. Reflexive control in the monkey. Archives of Ophthalmology 1982;100(8) 1327-1330.

[67] Yu DY, Alder VA, Cringle SJ, Brown MJ. Choroidal blood flow measured in the dog eye in vivo and in vitro by local hydrogen clearance polarography: validation of a technique and response to raised intraocular pressure. Experimental Eye Research 1988;46(3) 289-303.

[68] Friedman E. Choroidal blood flow. Pressure-flow relationships. Archives of Ophthalmology 1970;83(1) 95-99.

[69] Dollery CT, Bulpitt CJ, Kohner EM. Oxygen supply to the retina from the retinal and choroidal circulations at normal and increased arterial oxygen tensions. Investigative Ophthalmology 1969;8(6) 588-594.
[70] Flower RW, Fryczkowski AW, McLeod DS. Variability in choriocapillaris blood flow distribution. Investigative Ophthalmology & Visual Science 1995;36(7) 1247-1258.
[71] Kiel JW, Shepherd AP. Autoregulation of choroidal blood flow in the rabbit. Investigative Ophthalmology & Visual Science 1992;33(8) 2399-2410.
[72] Ramírez JM, Ramírez AI, Salazar JJ, de Hoz R, Rojas B, Triviño A. Anatomofisiología de la úvea posterior: coroides. In: Monés J, Gómez-Ulla F. (ed) Degeneración macular asociada a la edad. Barcelona: Prous Science; 2005. p1-28.
[73] Steinle JJ, Pierce JD, Clancy RL, G. Smith P. Increased Ocular Blood Vessel Numbers and Sizes Following Chronic Sympathectomy in Rat. Experimental Eye Research 2002;74(6) 761-768.
[74] Schmidt RE, Beaudet LN, Plurad SB, Dorsey DA. Axonal cytoskeletal pathology in aged and diabetic human sympathetic autonomic ganglia. Brain Research 1997;769(2) 375-383.
[75] Ishikawa S, Bensaoula T, Uga S, Mukuno K. Electron-microscopic study of iris nerves and muscles in diabetes. Ophthalmologica. 1985;191(3) 172-183.
[76] Fulk GW, Bower A, McBride K, Boatright R. Sympathetic denervation of the iris dilator in noninsulin-dependent diabetes. Optometry and Vision Science 1991;68(12) 954-956.
[77] Ernest JT. Regulatory mechanism of the choroidal vasculature in health and disease. In: Tso MOM. (ed) Retinal diseases: biomedical foundations and clinical management Philadelphia: JB Lippincott.; 1988. p125-130.
[78] Grunwald JE, Hariprasad SM, DuPont J. Effect of aging on foveolar choroidal circulation. Archives of Ophthalmology 1998;116(2) 150-154.
[79] Grunwald JE, Metelitsina TI, Dupont JC, Ying GS, Maguire MG. Reduced foveolar choroidal blood flow in eyes with increasing AMD severity. Investigative Ophthalmology & Visual Science 2005;46(3) 1033-1038.
[80] Yorek MA, Coppey LJ, Gellett JS, Davidson EP. Sensory nerve innervation of epineurial arterioles of the sciatic nerve containing calcitonin gene-related peptide: effect of streptozotocin-induced diabetes. Experimental Diabesity Research 2004;5(3) 187-193.
[81] Park JG, Oh GT. The role of peroxidases in the pathogenesis of atherosclerosis. BMB Reports 2011;44(8) 497-505.
[82] Wittchen ES. Endothelial signaling in paracellular and transcellular leukocyte transmigration. Frontiers in Bioscience 2009;14 2522-2545.
[83] Shibata N, Glass CK. Regulation of macrophage function in inflammation and atherosclerosis. Journal of Lipid Research 2009;50 SS 277-81.
[84] Madamanchi NR, Vendrov A, Runge MS. Oxidative stress and vascular disease. Arteriosclerosis, Thrombosis, and Vascular Biology 2005;25(1) 29-38.
[85] Glass CK, Witztum JL. Atherosclerosis. The road ahead. Cell 2001;104(4) 503-516.
[86] Taniyama Y, Griendling KK. Reactive oxygen species in the vasculature: molecular and cellular mechanisms. Hypertension 2003;42(6) 1075-1081.

[87] Libby P, Okamoto Y, Rocha VZ, Folco E. Inflammation in atherosclerosis: transition from theory to practice. Circulation Journal 2010;74(2) 213-220.

[88] Rudijanto A. The role of vascular smooth muscle cells on the pathogenesis of atherosclerosis. Acta Medica Indonesiana 2007;39(2) 86-93.

[89] Grote K, Flach I, Luchtefeld M, Akin E, Holland SM, Drexler H, Schieffer B. Mechanical stretch enhances mRNA expression and proenzyme release of matrix metalloproteinase-2 (MMP-2) via NAD(P)H oxidase-derived reactive oxygen species. Circulation Research 2003;92(11) 80-6.

[90] De Keulenaer GW, Ushio-Fukai M, Yin Q, Chung AB, Lyons PR, Ishizaka N, Rengarajan K, Taylor WR, Alexander RW, Griendling KK. Convergence of redox-sensitive and mitogen-activated protein kinase signaling pathways in tumor necrosis factor-alpha-mediated monocyte chemoattractant protein-1 induction in vascular smooth muscle cells. Arteriosclerosis, Thrombosis, and Vascular Biology 2000;20(2) 385-391.

[91] von Harsdorf R, Li PF, Dietz R. Signaling pathways in reactive oxygen species-induced cardiomyocyte apoptosis. Circulation 1999;99(22) 2934-2941.

[92] Zhu P, Dettmann ES, Resink TJ, Luscher TF, Flammer J, Haefliger IO. Effect of Ox-LDL on endothelium-dependent response in pig ciliary artery: prevention by an ET(A) antagonist. Investigative Ophthalmology & Visual Science 1999;40(5) 1015-1020.

[93] Tanner FC, Noll G, Boulanger CM, Luscher TF. Oxidized low density lipoproteins inhibit relaxations of porcine coronary arteries. Role of scavenger receptor and endothelium-derived nitric oxide. Circulation 1991;83(6) 2012-2020.

[94] Anderson TJ, Meredith IT, Yeung AC, Frei B, Selwyn AP, Ganz P. The effect of cholesterol-lowering and antioxidant therapy on endothelium-dependent coronary vasomotion. New England Journal of Medicine 1995;332(8) 488-493.

[95] Feron O, Dessy C, Moniotte S, Desager JP, Balligand JL. Hypercholesterolemia decreases nitric oxide production by promoting the interaction of caveolin and endothelial nitric oxide synthase. Journal of Clinical Investigation 1999;103(6) 897-905.

[96] Donners MMPC, Heeneman S, Daemen MJAP. Models of atherosclerosis and transplant arteriosclerosis: the quest for the best. Drug Discovery Today: Disease Models 2004;1(3) 257-263.

[97] Zadelaar S, Kleemann R, Verschuren L, de Vries-Van der Weij J, van der Hoorn J, Princen HM, Kooistra T. Mouse Models for Atherosclerosis and Pharmaceutical Modifiers. Arteriosclerosis, Thrombosis, and Vascular Biology 2007;27(8) 1706-1721.

[98] Nakashima Y, Plump AS, Raines EW, Breslow JL, Ross R. ApoE-deficient mice develop lesions of all phases of atherosclerosis throughout the arterial tree. Arteriosclerosis and Thrombosis 1994;14(1) 133-140.

[99] Jawien J. The role of an experimental model of atherosclerosis: apoE-knockout mice in developing new drugs against atherogenesis. Current Pharmaceutical Biotechnology 2012; [Epub ahead of print] PMID: 22280417

[100] Davignon J. Apolipoprotein E and atherosclerosis: beyond lipid effect. Arteriosclerosis, Thrombosis, and Vascular Biology 2005;25(2) 267-269.

[101] Ali K, Middleton M, Pure E, Rader DJ. Apolipoprotein E suppresses the type I inflammatory response in vivo. Circulation Research 2005;97(9) 922-927.

[102] Grainger DJ, Reckless J, McKilligin E. Apolipoprotein E modulates clearance of apoptotic bodies in vitro and in vivo, resulting in a systemic proinflammatory state in apolipoprotein E-deficient mice. Journal of Immunology 2004;173(10) 6366-6375.

[103] Knowles JW, Maeda N. Genetic modifiers of atherosclerosis in mice. Arteriosclerosis, Thrombosis, and Vascular Biology 2000;20(11) 2336-2345.

[104] Ishibashi S, Goldstein JL, Brown MS, Herz J, Burns DK. Massive xanthomatosis and atherosclerosis in cholesterol-fed low density lipoprotein receptor-negative mice. Journal of Clinical Investigation 1994;93(5) 1885-1893.

[105] van Vlijmen BJ, van den Maagdenberg AM, Gijbels MJ, van der Boom H, HogenEsch H, Frants RR, Hofker MH, Havekes LM. Diet-induced hyperlipoproteinemia and atherosclerosis in apolipoprotein E3-Leiden transgenic mice. Journal of Clinical Investigation 1994;93(4) 1403-1410.

[106] Jacobsson L. Comparison of experimental hypercholesterolemia and atherosclerosis in Gottingen mini-pigs and Swedish domestic swine. Atherosclerosis 1986;59(2) 205-213.

[107] Turk JR, Henderson KK, Vanvickle GD, Watkins J, Laughlin MH. Arterial endothelial function in a porcine model of early stage atherosclerotic vascular disease. International Journal of Experimental Pathology 2005;86(5) 335-345.

[108] Kamimura R, Miura N, Suzuki S. The hemodynamic effects of acute myocardial ischemia and reperfusion in Clawn miniature pigs. Experimental Animals 2003;52(4) 335-338.

[109] Liang Y, Zhu H, Friedman MH. The correspondence between coronary arterial wall strain and histology in a porcine model of atherosclerosis. Physics in Medicine and Biology 2009;54(18) 5625-5641.

[110] Thim T. Human-like atherosclerosis in minipigs: a new model for detection and treatment of vulnerable plaques. Danish Medical Bulletin 2010;57(7) B4161.

[111] Miyoshi N, Horiuchi M, Inokuchi Y, Miyamoto Y, Miura N, Tokunaga S, Fujiki M, Izumi Y, Miyajima H, Nagata R, Misumi K, Takeuchi T, Tanimoto A, et al. Novel microminipig model of atherosclerosis by high fat and high cholesterol diet, established in Japan. In Vivo 2010;24(5) 671-680.

[112] Kawaguchi H, Miyoshi N, Miura N, Fujiki M, Horiuchi M, Izumi Y, Miyajima H, Nagata R, Misumi K, Takeuchi T, Tanimoto A, Yoshida H. Microminipig, a non-rodent experimental animal optimized for life science research:novel atherosclerosis model induced by high fat and cholesterol diet. Journal of Pharmacological Sciences 2011;115(2) 115-121.

[113] Stoletov K, Fang L, Choi SH, Hartvigsen K, Hansen LF, Hall C, Pattison J, Juliano J, Miller ER, Almazan F, Crosier P, Witztum JL, Klemke RL, et al. Vascular lipid accumulation, lipoprotein oxidation, and macrophage lipid uptake in hypercholesterolemic zebrafish. Circulation Research 2009;104(8) 952-960.

[114] Fang L, Harkewicz R, Hartvigsen K, Wiesner P, Choi SH, Almazan F, Pattison J, Deer E, Sayaphupha T, Dennis EA, Witztum JL, Tsimikas S, Miller YI. Oxidized cholesteryl esters and phospholipids in zebrafish larvae fed a high cholesterol diet: macrophage binding and activation. Journal of Biological Chemistry 2010;285(42) 32343-32351.

[115] Fang L, Green SR, Baek JS, Lee SH, Ellett F, Deer E, Lieschke GJ, Witztum JL, Tsimikas S, Miller YI. In vivo visualization and attenuation of oxidized lipid accumulation in hypercholesterolemic zebrafish. Journal of Clinical Investigation 2011;121(12) 4861-4869.

[116] Yanni AE. The laboratory rabbit: an animal model of atherosclerosis research. Laboratory Animals 2004;38(3) 246-256.

[117] Daugherty A, Zweifel BS, Schonfeld G. Probucol attenuates the development of aortic atherosclerosis in cholesterol-fed rabbits. British Journal of Pharmacology 1989;98(2) 612-618.

[118] Del Rio M, Chulia T, Merchan-Perez A, Remezal M, Valor S, Gonzalez J, Gutierrez JA, Contreras JA, Lasuncion MA, Tejerina T. Effects of indapamide on atherosclerosis development in cholesterol-fed rabbits. Journal of Cardiovascular Pharmacology 1995;25(6) 973-978.

[119] Huff MW, Carroll KK. Effects of dietary protein on turnover, oxidation, and absorption of cholesterol, and on steroid excretion in rabbits. Journal of Lipid Research 1980;21(5) 546-548.

[120] Zauberman H, Livni N. Experimental vascular occlusion in hypercholesterolemic rabbits. Investigative Ophthalmology & Visual Science 1981;21(2) 248-255.

[121] Finking G, Hanke H. Nikolaj Nikolajewitsch Anitschkow (1885-1964) established the cholesterol-fed rabbit as a model for atherosclerosis research. Atherosclerosis 1997;135(1) 1-7.

[122] Redgrave TG, Dunne KB, Roberts DCK, West CE. Chylomicron metabolism in rabbits fed diets with or without added cholesterol. Atherosclerosis 1976;24(3) 501-508.

[123] Chapman MJ. Animal lipoproteins: chemistry, structure, and comparative aspects. Journal of lipid research 1980;21(7) 789-853.

[124] Roth RI, Gaubatz JW, Gotto AM,Jr, Patsch JR. Effect of cholesterol feeding on the distribution of plasma lipoproteins and on the metabolism of apolipoprotein E in the rabbit. Journal of Lipid Research 1983;24(1) 1-11.

[125] Reddy C, Stock EL, Mendelsohn AD, Nguyen HS, Roth SI, Ghosh S. Pathogenesis of experimental lipid keratopathy: corneal and plasma lipids. Investigative Ophthalmology & Visual Science 1987;28(9) 1492-1496.

[126] Holm P, Andersen HL, Arroe G, Stender S. Gender gap in aortic cholesterol accumulation in cholesterol-clamped rabbits: role of the endothelium and mononuclear-endothelial cell interaction. Circulation 1998;98(24) 2731-2737.

[127] Ponraj D, Makjanic J, Thong PS, Tan BK, Watt F. The onset of atherosclerotic lesion formation in hypercholesterolemic rabbits is delayed by iron depletion. FEBS Letters 1999;459(2) 218-222.

[128] Hanyu M, Kume N, Ikeda T, Minami M, Kita T, Komeda M. VCAM-1 expression precedes macrophage infiltration into subendothelium of vein grafts interposed into carotid arteries in hypercholesterolemic rabbits--a potential role in vein graft atherosclerosis. Atherosclerosis 2001;158(2) 313-319.

[129] Kálmán J, Kudchodkar BJ, Krishnamoorthy R, Dory L, Lacko AG, Agarwal N. High cholesterol diet down regulates the activity of activator protein-1 but not nuclear factor-kappa B in rabbit brain. Life Sciences 2001;68(13) 1495-1503.

[130] de la Peña NC, Sosa-Melgarejo JA, Ramos RR, Méndez JD. Inhibition of platelet aggregation by putrescine, spermidine, and spermine in hypercholesterolemic rabbits. Archives of Medical Research 2000;31(6) 546-550.

[131] Francois J, Neetens A. Vascular manifestations of experimental hypercholesteraemia in rabbits. Angiologica 1966;3(1) 1-20.

[132] Amaratunga A, Abraham CR, Edwards RB, Sandell JH, Schreiber BM, Fine RE. Apolipoprotein E is synthesized in the retina by Muller glial cells, secreted into the vitreous, and rapidly transported into the optic nerve by retinal ganglion cells. Journal of Biological Chemistry 1996;271(10) 5628-5632.

[133] Sebesteny A, Sheraidah GA, Trevan DJ, Alexander RA, Ahmed AI. Lipid keratopathy and atheromatosis in an SPF laboratory rabbit colony attributable to diet. Laboratory Animals 1985;19(3) 180-188.

[134] Yamamoto T, Bishop RW, Brown MS, Goldstein JL, Russell DW. Deletion in cysteine-rich region of LDL receptor impedes transport to cell surface in WHHL rabbit. Science 1986;232(4755) 1230-1237.

[135] Shiomi M, Ito T. The Watanabe heritable hyperlipidemic (WHHL) rabbit, its characteristics and history of development: A tribute to the late Dr. Yoshio Watanabe. Atherosclerosis 2009;207(1) 1-7.

[136] Watanabe Y. Serial inbreeding of rabbits with hereditary hyperlipidemia (WHHL-rabbit). Atherosclerosis 1980;36(2) 261-268.

[137] Steen H, Lima JA, Chatterjee S, Kolmakova A, Gao F, Rodriguez ER, Stuber M. High-resolution three-dimensional aortic magnetic resonance angiography and quantitative vessel wall characterization of different atherosclerotic stages in a rabbit model. Investigative Radiology 2007;42(9) 614-621.

[138] Ogawa M, Ishino S, Mukai T, Asano D, Teramoto N, Watabe H, Kudomi N, Shiomi M, Magata Y, Iida H, Saji H. (18)F-FDG accumulation in atherosclerotic plaques: immunohistochemical and PET imaging study. Journal of Nuclear Medicine 2004;45(7) 1245-1250.

[139] Iwata A, Miura S, Imaizumi S, Zhang B, Saku K. Measurement of atherosclerotic plaque volume in hyperlipidemic rabbit aorta by intravascular ultrasound. Journal of Cardiology 2007;50(4) 229-234.

[140] Sadowitz B, Maier KG, Gahtan V. Basic science review: Statin therapy--Part I: The pleiotropic effects of statins in cardiovascular disease. Vascular and Endovascular Surgery 2010;44(4) 241-251.

[141] Sadowitz B, Seymour K, Costanza MJ, Gahtan V. Statin therapy--Part II: Clinical considerations for cardiovascular disease. Vascular and Endovascular Surgery 2010;44(6) 421-433.

[142] van der Most PJ, Dolga AM, Nijholt IM, Luiten PGM, Eisel ULM. Statins: Mechanisms of neuroprotection. Progress in Neurobiology 2009;88(1) 64-75.

[143] Zhou Q, Liao JK. Statins and cardiovascular diseases: from cholesterol lowering to pleiotropy. Current Pharmaceutical Design 2009;15(5) 467-478.

[144] Zhou Q, Liao JK. Pleiotropic effects of statins. - Basic research and clinical perspectives -. Circulation Journal 2010;74(5) 818-826.

[145] Botti RE, Triscari J, Pan HY, Zayat J. Concentrations of pravastatin and lovastatin in cerebrospinal fluid in healthy subjects. Clinical Neuropharmacology 1991;14(3) 256-261.

[146] Wang CY, Liu PY, Liao JK. Pleiotropic effects of statin therapy: molecular mechanisms and clinical results. Trends in Molecular Medicine 2008;14(1) 37-44.

[147] Athyros VG, Kakafika AI, Tziomalos K, Karagiannis A, Mikhailidis DP. Pleiotropic effects of statins--clinical evidence. Current Pharmaceutical Design 2009;15(5) 479-489.

[148] Rikitake Y, Kawashima S, Takeshita S, Yamashita T, Azumi H, Yasuhara M, Nishi H, Inoue N, Yokoyama M. Anti-oxidative properties of fluvastatin, an HMG-CoA reductase inhibitor, contribute to prevention of atherosclerosis in cholesterol-fed rabbits. Atherosclerosis 2001;154(1) 87-96.

[149] Porter KE, Naik J, Turner NA, Dickinson T, Thompson MM, London NJ. Simvastatin inhibits human saphenous vein neointima formation via inhibition of smooth muscle cell proliferation and migration. Journal of Vascular Surgery 2002;36(1) 150-157.

[150] Mitani H, Egashira K, Kimura M. HMG-CoA reductase inhibitor, fluvastatin, has cholesterol-lowering independent "direct" effects on atherosclerotic vessels in high cholesterol diet-fed rabbits. Pharmacological Research 2003;48(5) 417-427.

[151] Hall NF, Gale CR, Syddall H, Phillips DI, Martyn CN. Risk of macular degeneration in users of statins: cross sectional study. British Medical Journal 2001;323(7309) 375-376.

[152] McCarty CA, Mukesh BN, Guymer RH, Baird PN, Taylor HR. Cholesterol-lowering medications reduce the risk of age-related maculopathy progression. Medical Journal of Australia 2001;175(6) 340.

[153] Yamada K, Sakurai E, Itaya M, Yamasaki S, Ogura Y. Inhibition of Laser-Induced Choroidal Neovascularization by Atorvastatin by Downregulation of Monocyte Chemotactic Protein-1 Synthesis in Mice. Investigative Ophthalmology Visual Science 2007;48(4) 1839-1843.

[154] Yamakawa K, Bhutto IA, Lu Z, Watanabe Y, Amemiya T. Retinal vascular changes in rats with inherited hypercholesterolemia--corrosion cast demonstration. Current Eye Research 2001;22(4) 258-265.

[155] Ong JM, Zorapapel NC, Rich KA, Wagstaff RE, Lambert RW, Rosenberg SE, Moghaddas F, Pirouzmanesh A, Aoki AM, Kenney MC. Effects of cholesterol and apolipoprotein E on retinal abnormalities in ApoE-deficient mice. Investigative Ophthalmology & Visual Science 2001;42(8) 1891-1900.

[156] Kouchi M, Ueda Y, Horie H, Tanaka K. Ocular lesions in Watanabe heritable hyperlipidemic rabbits. Veterinary Ophthalmology 2006;9(3) 145-148.

[157] Triviño A, Ramírez AI, Salazar JJ, de Hoz R, Rojas B, Padilla E, Tejerina T, Ramírez JM. A cholesterol-enriched diet induces ultrastructural changes in retinal and macroglial rabbit cells. Experimental Eye Research 2006;83(2) 357-366.

[158] Ramírez AI, Salazar JJ, de Hoz R, Rojas B, Ruiz E, Tejerina T, Ramírez JM, Triviño A. Macroglial and retinal changes in hypercholesterolemic rabbits after normalization of cholesterol levels. Experimental Eye Research 2006;83(6) 1423-1438.

[159] Shibata M, Sugiyama T, Hoshiga M, Hotchi J, Okuno T, Oku H, Hanafusa T, Ikeda T. Changes in optic nerve head blood flow, visual function, and retinal histology in hypercholesterolemic rabbits. Experimental Eye Research 2011;93(6) 818-824.

[160] Triviño A, Rojas B, Ramírez AI, Salazar JJ, de Hoz R, Ramajo M, Redondo S, Navarro-Dorado J, Tejerina T, Ramírez JM. Low-dosage statins reduce choroidal damage in hypercholesterolemic rabbits. Acta Ophthalmologica 2011;89(7) 660-669.

[161] Torres RJ, Muccioli C, Maia M, Noronha L, Luchini A, Alessi A, Olandoski M, Farah ME, Precoma DB. Sclerochorioretinal abnormalities in hypercholesterolemic rabbits treated with rosiglitazone. Ophthalmic Surgery, Lasers & Imaging 2010;41(5) 562-571.

[162] Topper JN, Gimbrone MA,Jr. Blood flow and vascular gene expression: fluid shear stress as a modulator of endothelial phenotype. Molecular Medicine Today 1999;5(1) 40-46.

[163] Wu CC, Chang SW, Chen MS, Lee YT. Early change of vascular permeability in hypercholesterolemic rabbits. Arteriosclerosis, Thrombosis, and Vascular Biology 1995;15(4) 529-533.

[164] Sánchez-Pérez RM, Molto JM, Medrano V, Beltrán I, Diaz-Marín C. Atherosclerosis and brain circulation. Revista de Neurologia 1999;28(11) 1109-1115.

[165] Darblade B, Caillaud D, Poirot M, Fouque M, Thiers J, Rami J, Bayard F, Arnal J. Alteration of plasmalemmal caveolae mimics endothelial dysfunction observed in atheromatous rabbit aorta. Cardiovascular Research 2001;50(3) 566-576.

[166] Hardin CD, Vallejo J. Caveolins in vascular smooth muscle: form organizing function. Cardiovascular Research 2006;69(4) 808-815.

[167] Drab M, Verkade P, Elger M, Kasper M, Lohn M, Lauterbach B, Menne J, Lindschau C, Mende F, Luft FC, Schedl A, Haller H, Kurzchalia TV. Loss of caveolae, vascular dysfunction, and pulmonary defects in caveolin-1 gene-disrupted mice. Science 2001;293(5539) 2449-2452.

[168] Bosch M, Mari M, Gross SP, Fernandez-Checa JC, Pol A. Mitochondrial cholesterol: a connection between caveolin, metabolism, and disease. Traffic 2011;12(11) 1483-1489.

[169] Lin WW, Lin YC, Chang TY, Tsai SH, Ho HC, Chen YT, Yang VC. Caveolin-1 expression is associated with plaque formation in hypercholesterolemic rabbits. Journal of Histochemistry and Cytochemistry 2006;54(8) 897-904.

[170] Xu Y, Buikema H, van Gilst WH, Henning RH. Caveolae and endothelial dysfunction: filling the caves in cardiovascular disease. European Journal of Pharmacology 2008;585(2-3) 256-260.

[171] Malinow MR. Experimental models of atherosclerosis regression. Atherosclerosis 1983;48(2) 105-118.

[172] Lusis AJ. Atherosclerosis. Nature 2000;407(6801) 233-241.

[173] Saso Y, Kitamura K, Yasoshima A, Iwasaki HO, Takashima K, Doi K, Morita T. Rapid induction of atherosclerosis in rabbits. Histology and Histopathology 1992;7(3) 315-320.

[174] Rodger FC. A new preparation for the study of experimental atherosclerosis progressive-regressive changes in albino rabbit iris. Experimental Eye Research 1972;14(1) 1-6.

[175] Abdulla YH, Adams CW, Morgan RS. Connective-tissue reactions to implantation of purified sterol, sterol esters, phosphoglycerides, glycerides and free fatty acids. Journal of Pathology and Bacteriology 1967;94(1) 63-71.

[176] Henney AM, Wakeley PR, Davies MJ, Foster K, Hembry R, Murphy G, Humphries S. Localization of stromelysin gene expression in atherosclerotic plaques by in situ hybrid-

ization. Proceedings of the National Academy of Sciences of the United States of America 1991;88(18) 8154-8158.

[177] Galis ZS, Sukhova GK, Lark MW, Libby P. Increased expression of matrix metalloproteinases and matrix degrading activity in vulnerable regions of human atherosclerotic plaques. Journal of Clinical Investigation 1994;94(6) 2493-2503.

[178] McGwin G,Jr, Xie A, Owsley C. The use of cholesterol-lowering medications and age-related macular degeneration. Ophthalmology 2005;112(3) 488-494.

[179] Jukema JW, Bruschke AVG, van Boven AJ, Reiber JHC, Bal ET, Zwinderman AH, Jansen H, Boerma GJM, van Rappard FM, Lie KI. Effects of Lipid Lowering by Pravastatin on Progression and Regression of Coronary Artery Disease in Symptomatic Men With Normal to Moderately Elevated Serum Cholesterol Levels : The Regression Growth Evaluation Statin Study (REGRESS). Circulation 1995;91(10) 2528-2540.

[180] Corsini A, Pazzucconi F, Pfister P, Paoletti R, Sirtori CR. Inhibitor of proliferation of arterial smooth-muscle cells by fluvastatin. Lancet 1996;348(9041) 1584.

[181] Lee TM, Lin MS, Chou TF, Tsai CH, Chang NC. Effect of pravastatin on left ventricular mass by activation of myocardial K ATP channels in hypercholesterolemic rabbits. Atherosclerosis 2004;176(2) 273-278.

[182] Kurokawa J, Hayashi K, Toyota Y, Shingu T, Shiomi M, Kajiyama G. High dose of fluvastatin sodium (XU62-320), a new inhibitor of 3-hydroxy-3-methylglutaryl coenzyme A reductase, lowers plasma cholesterol levels in homozygous Watanabe-heritable hyperlipidemic rabbits. Biochimica et Biophysica Acta 1995;1259(1) 99-104.

[183] Taggart MJ. Smooth muscle excitation-contraction coupling: a role for caveolae and caveolins? News in Physiological Sciences 2001;16 61-65.

[184] Poche R, de Mello Mattos CM, Rembarz HW, Stoepel K. The mitochondrial-myofibril ratio in rat myocardial cells in hypertensive cardiac hypertrophy. Virchows Archives A: Pathology.1968;344(1) 100-110.

[185] Simon DK, Johns DR. Mitochondrial disorders: clinical and genetic features. Annual Review of Medicine 1999;50 111-127.

Section 3

Factors Influencing Structure and Function of Smooth Muscle Cells and Tissues

Chapter 13

Different Modulators of Airways and Distal Lung Parenchyma Contractile Responses in the Physiopathology of Asthma

Carla Máximo Prado, Edna Aparecida Leick,
Fernanda Degobbi Tenório Quirino dos Santos Lopes,
Milton A. Martins and Iolanda de Fátima Lopes Calvo Tibério

Additional information is available at the end of the chapter

http://dx.doi.org/10.5772/48347

1. Introduction

Asthma outcomes from an allergen-driven Th2 (T helper 2) response in which airway hyperresponsiveness (AHR) is associated with chronic airway inflammation and airway remodeling have crucial clinical importance (1-3).

Recent investigations have emphasized the importance of lung tissue alterations in the pathophysiology of this syndrome. Additionally, current investigations have shown that patients who died of asthma presented important alterations in the lung parenchyma (4-7) that could also be observed in animal models of chronic allergic inflammation (8-11). In this regard, the importance of the mechanical properties of the lung parenchyma has been characterized as one of the major determinants of physiological function (8, 12-15).

Asthma physiopathology is highly complex and involves a diverse immune response and the release of different types of mediators. The bronchial and tissue inflammation is caused by eosinophils, mast cells and T lymphocytes (16), and the persistence of inflammation induces changes in the structural components of the airway and alveolar walls (5, 8, 17).

The airway smooth muscle (ASM) has been considered the main effector of the AHR in asthma (17-19) and is also believed to contribute to airway remodeling and inflammation due to its increased sensitivity to different bronchoconstrictor stimuli.

The continuous bronchial inflammation process associated with the release of various mediators is thought to be responsible for asthma symptoms directly and indirectly by inducing the constriction of the ASM, enhancing airway responsiveness to different stimuli, and inducing changes in the structural components of the airway wall, leading to airway remodeling.

Inhaled corticosteroids, which are the gold-standard treatment for asthmatic patients, are more involved in counteracting the airway inflammation than in acting in the ASM. Although some studies have shown the potential of corticosteroids in causing bronchodilation, their role in airway smooth muscle relaxation is controversial. In its formulation (hydrofluoroalkane-HFA), this inhaled corticosteroid is delivered to the distal airways more effectively (68.3%) than chlorofluorocarbon formulations (19.7%) (20, 21). Although eosinophilic infiltration could be adequately controlled in the distal airways, whether both distal lung parenchyma eosinophilic infiltration and extracellular matrix remodeling may be sufficiently modulated by this new treatment is not clear (8, 20).

We discuss in this chapter the role of different mediators and modulators in the contractile responses of the airways and lung distal parenchyma. These studies contribute to the understanding of the mechanisms involved in asthma physiopathology and in smooth muscle contraction and also open opportunities to develop new therapeutic tools to treat asthma. In this regard, we will address the importance of the modulation of iNOS, arginase and Rho kinase pathways, the impact of inducing oral tolerance and the effects of exercise. In addition, aspects of neuroimmunomodulation, including stress effects, will be discussed.

2. Airway and lung parenchyma hyperresponsiveness and smooth muscle alterations in asthma

AHR is the hallmark of asthma, and it is characterized by an increase in the airway response to bronchoconstrictor stimuli. There are two components of AHR. AHR has a variable component that mainly reflects the current airway inflammation (22, 23) and an irreversible component that probably reflects pulmonary remodeling (24).

As described above, the ASM is the major effector of the AHR in asthma (17-19). There are two phenotypes of ASM cells in asthmatics: the contractile, which is responsive to contractile agonists and has an increased expression of contractile proteins, and the synthetic-proliferative, which lacks the responsiveness to contractile stimuli and has a reduced expression of contractile proteins (17). Both phenotypes can coexist or not in the airways of the same person (25-29). Depending on the triggers, it can also induce the proliferation of the synthetic-proliferative cells or induce the maturation of these cells into contractile cells (17, 19).

In patients with asthma, the ASM was thought to generate more force and consequently a greater extent of contraction in response to different stimuli (30). Cultures of ASM cells isolated from lung tissue (trachea, bronchi) were used to study the contractile responses and the mitogenic and synthetic responses, which revealed that these cells are active players in inflammation (25, 31, 32).

In addition, ASM can contribute to lung inflammation. Many studies showed that there was an increased number of mast cells in the asthmatic ASM layer (33-38). Brightling et al. (32) evaluated patients with asthma and eosinophilic bronchitis and observed that both groups showed an increase in eosinophils but that the patients with eosinophilic bronchitis were not hyperresponsive to bronchoconstrictor stimuli. The analysis of the ASM layers in these patients showed that only the asthmatics showed a higher number of mast cells and a worsening of respiratory function, suggesting that the mast cells present in the ASM of asthmatics are responsible for the enhancement of airway narrowing.

The ASM cells release chemotactic agents for mast cells, such as CCL11 (25), CXCL10 (34) and CX3CL1 (35). Because the mast cells are in the airways, they adhere to the ASM cells and produce, together with the eosinophils, contractile mediators, such as prostaglandins (PGF2α, PGD2, and thromboxane TXA2) (39).

Clinically, the AHR symptoms are described as cough, tightness of the chest and wheezing after exercise or exposure to cold air or other environmental irritants (40). Some studies suggest that monitoring of the AHR in asthmatic patients can serve as a guide to asthma therapy (24).

In clinical and experimental studies, AHR is evaluated by the aerosol administration of bronchoconstrictor agonists, such as histamine, methacholine or carbachol. This methodology considers that the ASM in asthmatics exposed to exogenous bronchoconstrictor stimuli showed an increased tonus and a concomitant bronchoconstriction. The hyperresponsiveness occurs due to an increase in both the sensitivity and/or reactivity of the airways (Figure 1). The increase in sensitivity is a reduction in the minimal dose that is necessary to induce bronchoconstriction, whereas the increase in reactivity is described by an increase in the intensity of the bronchoconstriction.

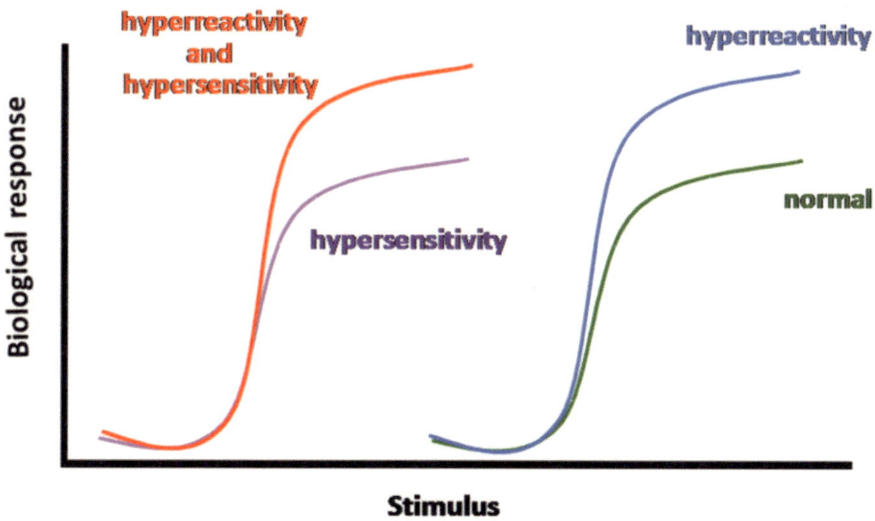

Figure 1. Airway hyperresponsiveness.

Considering that lung parenchyma strips have long been used to study the behavior of the peripheral lung, they are commonly used to evaluate the mechanics and pharmacological properties of the lung periphery (41). Dolhnikoff et al. (15) concluded that human lung tissue strips respond to an acetylcholine (ACh) challenge with changes in their dynamic mechanical behavior. In addition, Lanças et al. (10) have recently shown that the lung tissue is involved in the late asthmatic response in guinea pigs with chronic allergic lung inflammation, which is correlated to lung tissue eosinophilic recruitment and extracellular matrix remodeling.

Although the *in vivo* apparatus of oscillatory mechanics permits the evaluation of large and small airways, the oscillatory mechanics *in vitro* provide a tool for the specific evaluation of the lung periphery with minimal interference with the compartment represented by the small airways (10, 15). In addition, this *in vitro* methodology permits the specific analysis of the effects of several mediator/modulators in the lung periphery while avoiding other compensatory mechanisms that could be activated in *in vivo* studies. Lung parenchyma strips exclusively represent the distal units of the lung tissue and offer a better assessment of pure tissue properties. Thus, studies using this technique have been performed to evaluate the mechanical and pharmacological properties of the lung periphery (10, 42, 43).

Several authors have discussed the importance of these structures in the mechanical behavior of lung tissue, including the consequences of stiffening the extracellular matrix network and of elastin and collagen digestion in these responses (44, 45). In the subpleural region, there was a small number of bronchial and blood vessels (less than 30%). Romero et al. (46) concluded that pneumoconstriction significantly modifies the intrinsic mechanical properties of the connective matrix via a mechanism differing from that of passive stretching. In fact, the contractile cells could be accepted as being able to modulate the mechanical properties of the connective matrix.

3. Mediators involved in airways and distal lung parenchyma contractile responses

A large quantity of extracellular agonists (inflammatory mediators or neurotransmitters) released in an inflammatory milieu can stimulate the contraction of ASM in asthma. Mediators that are found in high concentrations in asthma, including leukotrienes (produced by inflammatory cells) (47), prostaglandins such as PGF2α, PGD2, and thromboxane TXA2 (produced by mast cells and/or eosinophils) (39) and endothelin (produced by epithelial or endothelial cells) (48, 49), are direct contractile agonists of ASM. Neurotransmitters, such as ACh or neurokinins, are highly present in asthma and are also potent contractile messengers of ASM (50, 51).

To increase the release of the contractile mediators, there is also a lower release of relaxant mediators, such as vasoactive intestinal peptide (VIP), PGE 2, adrenaline and NO (35, 52). These mediators are involved in the mechanisms responsible for many of the structural and functional lung alterations observed in asthmatic patients and in animal models of chronic pulmonary allergic inflammation (53-55).

3.1. Excitatory non-adrenergic non-cholinergic mediators: Neurokinins and Substance P

Neurokinins and substance P are involved in the excitatory NANC responses and modulate several histopathological alterations observed in asthmatics, such as airway smooth muscle contraction, peribronchial edema formation and airway mucous secretion. In this regard, substance P (SP) and neurokinin A (NKA) play significant roles in priming and recruiting eosinophils and lymphocytes in models of allergic lung inflammation (56-58).

Asthmatic patients are hyperresponsive to the SP and NK1 expression that is augmented in their bronchi (59). Tibério et al. (60) showed that capsaicin infusion induced an increase in the respiratory system resistance that was attenuated mainly by a NK2 receptor antagonist. The NK receptors are also involved in eosinophil recruitment, which contributes to the hyperresponsiveness. Using a model of experimental asthma in guinea pigs, Tibério et al. (57) evaluated the airway inflammation induced by repeated exposure to ovalbumin and the effects of neurokinin depletion on these responses. These authors showed that neurokinin depletion reduced the peribronchial edema, CD4 lymphocytes and the hyperresponsiveness to the antigen challenge. In addition, Prado et al. (61) showed that the bronchodilation observed after 14 days of capsaicin infusion could be related to the increase in NO produced by nNOS, which counteracts the bronchoconstriction.

Emphasizing that SP has a preferential affinity for NK1 receptors and that neurokinin A has a preferential affinity for NK2 receptors is important (58). However, each neurokinin also exhibits activity at other NK receptors. In this regard, Regoli et al. (62) showed that NKA has 25% of the affinity of SP for the dog carotid artery, a preparation that contains only NK1 receptors. Tibério et al. (60) investigated the role of substance P (SP) and neurokinin A (NKA) and their receptor antagonists (RAs) SR140333 and SR48968 (respectively for the NK(1) and NK(2) receptors) in the pulmonary eosinophil influx induced by the stimulation of capsaicin (CAP)-sensitive nerve terminals. Both SP and NKA contribute to eosinophil lung recruitment in the distal airways and the alveolar wall, and these findings suggest that neurokinins may contribute to the development of eosinophilic inflammation in both allergic asthma and hypersensitive pneumonitis.

3.2. Cysteinyl leukotrienes

Cysteinyl leukotrienes (cysLTs) are synthesized *de novo* from arachidonic acid, and most of their actions are mediated by the CysLT1 receptor, a G protein-coupled receptor (63). CysLTs have many pulmonary actions, including human airway smooth muscle contraction, chemotaxis, mucous secretion, smooth muscle proliferation and increased vascular permeability (64-66).

The cysteinyl leukotrienes (LTC4, LTD4, LTE4) produced by inflammatory cells and endothelin, produced by epithelial or endothelial cells, are increased in asthma. They are also potent contractile agonists of ASM (48, 67).

Leukotriene antagonists have been shown to reduce sputum and mucosal eosinophils in subjects with asthma (68, 69). However, recent long-duration trials have evaluated the impact of CysLT receptor antagonists compared with glucocorticoids and showed that spirometry, symptoms, β2-agonist use and the quality of life were improved to a greater extent with glucocorticoids (70-72). Corroborating this idea, the blockade of leukotriene activity does not cause an improvement in airflow as intense as that obtained with glucocorticoids (70, 73).

Considering studies in animal models, Gardiner et al. (74) observed that the inhibition of leukotriene synthesis resulted in an attenuation of OVA-induced airway contraction in sensitized animals. Liu et al. (75) demonstrated that the CysLT1 receptor antagonists pranlukast and zafirlukast inhibited OVA-induced mucus secretion in the trachea of a sensitized guinea pig. Comparing the effects of montelukast and corticosteroid treatments in a guinea pig model, Leick-Maldonado et al. (76) showed that although montelukast, an antagonist of leukotriene, reduced some aspects of inflammation, this treatment was not able to attenuate the changes in lung mechanics.

3.3. Complex NOS-arginases

Nitric oxide derived either from constitutive isoforms (nNOS and eNOS) or from other NO-adduct molecules (nitrosothiols) modulates bronchomotor and vascular tone. In addition, NO derived from inducible isoenzyme (iNOS) is mainly involved in the immunomodulation (77-80).

Prado et al. (81) tested the differences between chronic and acute nitric oxide inhibition by *N*-nitro-L-arginine methyl ester (L-NAME) treatment in lung mechanics, inflammation, and airway remodeling in an experimental asthma model in guinea pigs. Both acute and chronic L-NAME treatment reduced the exhaled nitric oxide in sensitized animals. Chronic L-NAME treatment increased the baseline and maximal responses after an antigen challenge (ovalbumin) of the respiratory system resistance and reduced peribronchial edema and airway infiltration by mononuclear cells. Acute administration of L-NAME increased the maximal values of respiratory system elastance and reduced the mononuclear cells and eosinophils in the airway wall, supporting the hypothesis that, in this model, nitric oxide acts as a bronchodilator in the airways.

iNOS enzyme activation has been found in many types of inflammatory cells, such as eosinophils, neutrophils and macrophages, as well as in respiratory epithelial cells. In fact, NO produced from this isoenzyme is related to the amplification of the inflammatory and remodeling responses (54, 78, 79, 82). Considering these aspects, a specific inhibition of iNOS-derived NO has been considered to be a future therapeutic strategy for several diseases, such as asthma, sepsis and acute lung inflammation (82-85).

Considering the smooth muscle responses, NO mainly derived from cNOS relaxes the airway smooth muscle. Many studies have focused on the role of NO in the modulation of airway smooth muscle contraction in different models of experimental pulmonary allergic inflammation (78, 81, 85-87). NO that is mainly derived from the constitutive isoforms of NOS has been shown to attenuate the bronchoconstriction induced by allergens in sensitized experimental animals (54, 85, 88). In contrast, others have observed that nNOS-derived NO could contribute to airway constriction (61). We previously evaluated the effects of NO in respiratory system resistance using a guinea pig model of asthma and compared the cNOS and iNOS inhibition. We showed that chronic treatment with L-NAME, a false substrate that nonspecifically inhibits the production of NO, increased the respiratory system resistance in sensitized animals, whereas the iNOS-specific inhibition by 1400W reduced this response (54). Our results suggested a protective effect of NO derived from cNOS. In addition, we showed that iNOS contributes to the airway hyperresponsiveness in this model. Interestingly, in naïve animals, we observed that both L-NAME and 1400W treatments increased the resistance of the respiratory system. Because the role of iNOS is more pronounced in inflammatory situations, few studies have evaluated the effects of iNOS inhibition in physiologic situations. We have previously shown that there is a basal expression of iNOS in resident cells around the airways in guinea pigs not exposed to an inflammatory stimulus (54, 78). In addition, Guo and colleagues (89) showed that iNOS is continuously produced by the airway epithelium in normal humans. These data suggested that NO produced by iNOS under physiological conditions can also contribute to the control of the airway smooth muscle responses.

Analyzing the nitrergic nerve density, there appears to be a progressive reduction throughout the bronchial tree (90). In fact, Prado et al. (54) demonstrated that the inhibition of NO by chronic L-NAME treatment amplified the elastance responses. Considering that the respiratory system elastance responses are related to alterations in the distal airways and lung tissue, the authors suggested that NO could also be involved in the modulation of lung tissue constriction. Dupuy et al. (90) proposed that inhaled NO only affects the distal airways at high doses, suggesting that, although less intensive, NO can also modulate the responses of the distal airways and/or lung tissue.

Angeli et al. (11) evaluated the effects of chronic L-NAME treatment, a false substrate for all nitric oxide enzymes, on the modulation of lung tissue mechanics, eosinophilic inflammation and extracellular matrix tissue remodeling in guinea pigs with chronic lung inflammation. The authors suggested that nitric oxide plays an important role in lung tissue constriction and elastic fiber deposition within the alveolar septa in this animal model of chronic pulmonary inflammation. The activation of the pulmonary oxidative stress pathway, mainly via 8-iso-PGF2α, may contribute to these responses.

Starling et al. (9) demonstrated that iNOS activation contributes to lung parenchyma inflammatory and remodeling alterations in guinea pigs with chronic pulmonary allergic inflammation. 1400W, an iNOS-specific inhibitor, diminished the lung tissue elastance and resistance as well as the eosinophilic infiltration, collagen and elastic fiber content and volume proportion of actin in lung tissue. To our knowledge, this study has provided the first evidence of the effects of iNOS inhibition on the distal lung parenchyma.

In addition, the authors showed that specifically blocking iNOS reduced 8-isoprostane expression in the alveolar septa, which had previously been increased by repeated ovalbumin exposures (9). These findings suggest that the effects of iNOS-derived NO in the lung parenchyma depend, at least partially, on the activation of the oxidative stress pathway. The inhibition of NO production derived from iNOS activation also reduced the actin content (9). These results suggest an iNOS-derived effect on the myofibroblasts, which were believed to be the major cells responsible for the production of the extracellular matrix and the contraction of the parenchyma (91).

Another pathway to be discussed involves the arginases. These enzymes convert L-arginine into L-ornithine and urea and are the key enzymes of the urea cycle in the liver (arginase 1) but are also expressed in cells and tissues that lack a complete urea cycle, e.g., arginase 2 expression in the lung (88). Arginases are involved in cell growth and tissue repair via the increased production of L-ornithine, a precursor of polyamines and proline (88).

Que et al. (92) demonstrated the expression of arginase in the bronchial epithelium and in peribronchial connective tissue fibroblasts. In addition, Meurs et al. (87) showed that arginase appears to modulate the tone of the airway smooth muscle and potentiates methacholine-induced airway constriction. Arginase accomplishes these actions by forcing the common substrate L-arginine away from epithelial cNOS to diminish the agonist-induced production of NO. Arginases and NOS compete for the bioavailability of the same substrate, L-arginine, and are involved indirectly in the regulation of NO synthesis (53, 88). Corroborating this idea, Morris et al. (93) showed that there is a reduction in the levels of plasma arginine in asthmatic patients compared with patients without asthma but with increased serum arginase activity. Together, these results suggest that increased arginase activity in asthma may be a contributing factor to the decrease in the circulating levels of L-arginine and the consequent NO deficiency. Thus, blocking NO production could be a tool to study the indirect involvement of arginase in various pathophysiological processes (82, 87).

Several powerful drugs have been used to investigate the role of arginases in the pathophysiology of asthma, including nor-NOHA (Nω-hydroxy-nor-L-arginine), which is one of the most potent inhibitors of arginase (88). Meurs et al. (87), studying in vitro tracheal ring-sensitized guinea pigs, demonstrated that treatment with nor-NOHA reduced the hyperresponsiveness to methacholine, and this effect was reversed by treatment with L-NAME.

We demonstrated that chronic distal lung inflammation was associated with an increase in arginase content and iNOS-positive cells (data not published). These results were associated with constriction of the distal lung parenchyma. The increased iNOS expression leads to activation of the oxidative stress pathway and formation of PGF2α, which had a procontractile effect. In addition, we showed that the mechanism involved in the activation of arginase and the iNOS pathways may be related to the modulation of NF-kB expression. Finally, we demonstrated that the association of iNOS and arginase 2 inhibitions potentiated the reduction of PGF2α and NF-kB expression in the distal lung of guinea pigs with chronic pulmonary inflammation (data not published).

Airway inflammation is accompanied by a marked upregulation of iNOS expression, particularly in the airway epithelium (94), which has been associated with the activation of nuclear factor-kB (NF-kB), a transcription factor that is implicated in the induction of multiple genes expressed during the inflammatory response (95). Ckless et al. (96) showed that the activation of NF-kB may induce an increase in NOS and arginases. Furthermore, NF-κB activity can be affected by reactive oxygen species (ROS) and by reactive nitrogen species (RNS) (97).

Several mechanisms reported in the literature have tried to explain how NO could interfere with airway tone. The ability of NO to control airway tone could be related to both GMPc-dependent and GMPc-independent mechanisms (98-100).

Although the mechanisms involving the effects of NO in airway constriction have been extensively described, the exact mechanism involved in the effect of NOS inhibition on reducing lung parenchyma constriction is not completely understood. Another pathway discussed by some authors is related to the fact that the release of NO by NOS activation also contributes to oxidative stress, amplifying the deleterious and harmful effects on the lungs (9, 77).

The potent oxidant peroxynitrite is formed by the interaction of NO and superoxide by a rapid iso-stoichiometric reaction (77). Haddad et al. (101) suggested that peroxynitrite may contribute to the injury of pulmonary surfactant. Bhandari et al. (102) demonstrated that increased peroxynitrite formation was associated with a dose-dependent increase in the apoptotic cell death of type II pneumocytes. However, in strip preparations perfused with Krebs solution, the importance of reducing pulmonary surfactant was poorly associated with the pulmonary mechanical responses.

In contrast, peroxynitrite formation leads to lipid peroxidation and the generation of iso-prostanes (8-iso-$PGF_{2\alpha}$). Jourdan et al. (103) showed that L-NAME treatment greatly inhibits 8-iso-$PGF_{2\alpha}$. Therefore, isoprostanes appear to induce airway and vascular smooth muscle contractions by acting through tyrosine kinase, Rho and Rho kinase, leading to the decreased activity of myosin light chain phosphatase. The net response is associated with an increased level of phosphorylated myosin light chain and contraction (104).

3.4. Rho kinase pathway

The protein Rho, a member of the Ras superfamily of small monomeric GTPases, controls a variety of downstream effector proteins, including Rho kinase. Rho exhibits GDP- and GTP-binding and GTPase activity and is able to alternate between a GDP-bound inactive state and a GTP-bound active state. This alternation allows Rho to function as a molecular switch to control downstream signal transduction, influencing the level of smooth muscle tone and changes in the actin cytoskeleton, which contributes to cell adhesion, motility, migration, and contraction (105). Effects on the airway smooth muscle responses may be one of the most important factors that need to be considered for the development of new therapies for asthmatics (106).

The influence of Rho kinase on airway hyperresponsiveness is considered to be at least partly related to agonist-mediated Ca^{2+} sensitization. Ca^{2+} sensitization, which is also observed in the airways, is the increase in smooth muscle tension and/or phosphorylation of the 20-kDa regulatory light chain of myosin (MLC_{20}) at a constant Ca^{2+} concentration (107). In a variety of smooth muscles, this Ca^{2+} sensitization is mediated by a small G protein, RhoAp21, and its target protein, the Rho kinase (108), which is especially important during the sustained phase of contraction in smooth muscle (107).

Several studies have shown that the use of Rho kinase inhibitors might be beneficial for the treatment of airway diseases. Y-27632((+)-(R)-*trans*-4-(1-aminoethyl)-*N*-(4-pyrydil) cyclohexanecarboxamide, monohydrate) is one of the drugs that arose as a possible treatment for asthma. Y-27632 is a highly selective inhibitor of the Rho kinase pathway, capable of reversing G-protein sensitization and consequently relaxing the airway smooth muscle (108).

The effects of the acute inhibition of Rho kinase in sensitized animals have been analyzed by several authors. Schaafsma et al. (109) showed that the inhalation of Y-27632 at 30 min prevents the development of airway hyperresponsiveness both after the early and late airway reaction. Y-27632 reduces also reduces the cholinergic nerve-mediated contractions in the tracheal preparations of guinea pigs and mice in a dose-dependent manner (110). Witzenrath et al. (111) verified that the use of Y-27632 attenuated the methacholine-provoked airway response in the sensitized lungs.

Some studies suggested that the RhoA/ROCK system plays a role in eosinophil recruitment and Th-1 and Th-2 cytokine secretion (105, 112). In this regard, Henry et al. (112) demonstrated that pretreatment with Y-27632 reduced the number of eosinophils recovered from the bronchoalveolar lavage (BAL) fluid of OVA-sensitized mice.

Taki et al. (105) showed that another Rho kinase inhibitor, fasudil, reduced the presence of eosinophils in the BAL fluid, airways and blood vessels. In the BAL fluid, this Rho kinase inhibitor also diminished the augmented levels of IL-5, IL-13 and eotaxin. Aihara et al. (113) showed that Y-27632 suppressed the release of Th-1 cytokines and partially suppressed the release of Th-2 cytokines in healthy persons but reduced the release of IL-2 and IL-5 and weakly reduced the release of IL-4 and IFN-gamma in asthmatic patients.

Recently, we showed the chronic inhibition of Rho kinase reduced the airway and distal lung mechanical responses to an antigenic challenge with an associated reduction in NO_{EX}, eosinophilic infiltration, IL-2-, IL-4-, IL-5- and IL-13-positive cells, extracellular matrix remodeling and NF-κB-positive cells in the airways and distal lung. In addition, there was a significant reduction in the activation of the oxidative stress pathway, which was correlated with the attenuation of the maximal mechanical responses after antigen challenge (data not published).

These data suggest that treatment with an inhaled Rho kinase inhibitor contributes to the attenuation of the distal lung functional and structural changes induced by chronic allergic inflammation, both in the airways and distal lung. Taken together, this evidence suggests that Rho kinase inhibitors may be potential pharmacological tools to control distal lung asthmatic functional and histopathological alterations.

4. Modulators involved in airways and distal lung parenchyma contractile responses

4.1. Modulation of the lung contractile responses by physical exercises

The role of physical exercise in asthma is somewhat controversial. Exercise can induce bronchoconstriction in humans (114). Recently, however, various studies have shown that physical training, particularly at a moderate intensity, can improve lung function and is related to a reduction in asthma symptoms and AHR. Fanelli et al. (115) associated physical training improvements in the physiological variables at peak and submaximal exercise, and these authors also showed that trained patients have a reduction in the daily doses of inhaled steroids.

Studying adults, Mendes et al. (116) showed that 3 months after supervised training, patients presented a reduction in inflammation and asthma exacerbation and an increase in asthma symptom-free days. Although the authors did not directly measure the AHR, the reduction in symptoms and exacerbations indirectly reflects a reduction in the airway responsiveness. These authors clearly suggest that aerobic training might be useful as an adjuvant therapy in asthmatic patients under optimized medical care. In addition, physical training reduced the anxiety and depression levels with a significant correlation between improvements in the aerobic capacity and days without asthma symptoms (117).

Considering the experimental studies, Silva et al. (118) showed that aerobic training in mice with allergic chronic inflammation reduced both tissue elastance and resistance. These effects of aerobic training on lung mechanics could be at least partly mediated by the epithelium (119).

Based on these data, although AHR was frequently found among competitive athletes (120, 121), physical training may be beneficial to asthmatics, particularly when performed with supervision and at a moderate intensity.

4.2. Modulation of the lung contractile responses by stress

The stress response, which can be defined as the psychological reaction of the body to a variety of emotional or physical stimuli that threaten homeostasis (122), results in the activation of the hypothalamic-pituitary-adrenal (HPA) axis and the sympathetic and adrenomedullary systems. Although acute stress was shown to have anti-inflammatory effects, some studies have demonstrated that stressful situations and emotional states are triggers of asthmatic symptoms (123-125) and can influence the course and treatment of atopic diseases (126, 127). Chronic stress may induce a down-regulation of the expression and/or function of glucocorticoid receptors, leading to glucocorticoid resistance and contributing to the worsening of lung inflammation and pulmonary hyperreactivity.

Capelozzi et al. (128) showed that swimming-induced stress amplified mononuclear cell recruitment to the lungs in guinea pigs that performed 31 days of the stress protocol. These authors also showed that the amount of these cells was reduced when the animals were treated with fluoxetine. Recently, Leick et al. (129), studying the effects of stress induced by forced swimming in bronchoconstriction, observed that stress amplified the airway response to ovalbumin in guinea pigs. In addition, Marques et al. (130) showed that the malefic effects of stress in asthma are related not only to the airways but to the lung distal parenchyma. In sensitized animals, they showed that repeated stress increased the distal lung constriction associated with an augmentation of actin content, which is indirect evidence of the alveolar smooth muscle content. The authors also showed that iNOS inhibition attenuated the effects of stress in the lung parenchyma response in this animal model.

Considering humans, Ritz and Steptoe (125) observed a negative association between mood states and a reduction in the forced expiratory volume in the first second in asthmatic patients. Höglund et al. (131) studied 41 undergraduate students 22 with allergies, 16 asthmatics and 19 controls in a low-stress period and in a period associated with a large exam. The values of the forced expiratory volume in the first second of the control group differed significantly from that of the group of asthmatics only during the exam stress phase. These results collectively reinforced the idea that stress is an important modulator of the AHR present in asthma.

Collectively, these studies showed that chronic stress is harmful to asthmatic individuals and is involved in the AHR.

4.3. Oral tolerance

Immunotherapy has been considered a possible therapeutic strategy for asthma. Oral tolerance has been recognized as an alternative treatment to autoimmune and allergic diseases (132-134). Oral tolerance has classically been defined as the specific suppression of the cellular and/or humoral immune response to an antigen by the prior administration of the antigen by the oral route (135). There are two primary effector mechanisms of oral tolerance: the induction of regulatory T cells that mediate the active suppression and the induction of clonal anergy or deletion (135-137). In atopic patients, the oral, sublingual, or inhaled administration of antigens leads to a reduction in symptoms and local inflammation as well as a reduction in dyspnea and airway hyperresponsiveness. Some meta-analyses found that sublingual immunotherapy is beneficial for asthma treatment, although the magnitude of the effect is not very large (138-140).

Some authors (141-143) have previously evaluated the effects of oral tolerance in experimental models of airway disease. In an animal model, oral tolerance induced an attenuation of airway eosinophilic recruitment, bronchial hyperresponsiveness, and mucous secretion (143, 144). Russo et al. (141, 142) observed that animals submitted to an oral antigen administration protocol presented low levels of Th2 cytokines in the bronchoalveolar lavage fluid and a reduction in the production of ovalbumin-specific antibodies. The tolerance process is known to attenuate B-cell responses. Hasegawa et al. (145) demonstrated that B-cells have been implicated in myofibroblast activation mainly by secreting IL-6, IL-9, and fibroblast growth factor. Thus, considering that myofibroblasts are one of the contractile elements that modulate lung parenchyma responses is important (146, 147).

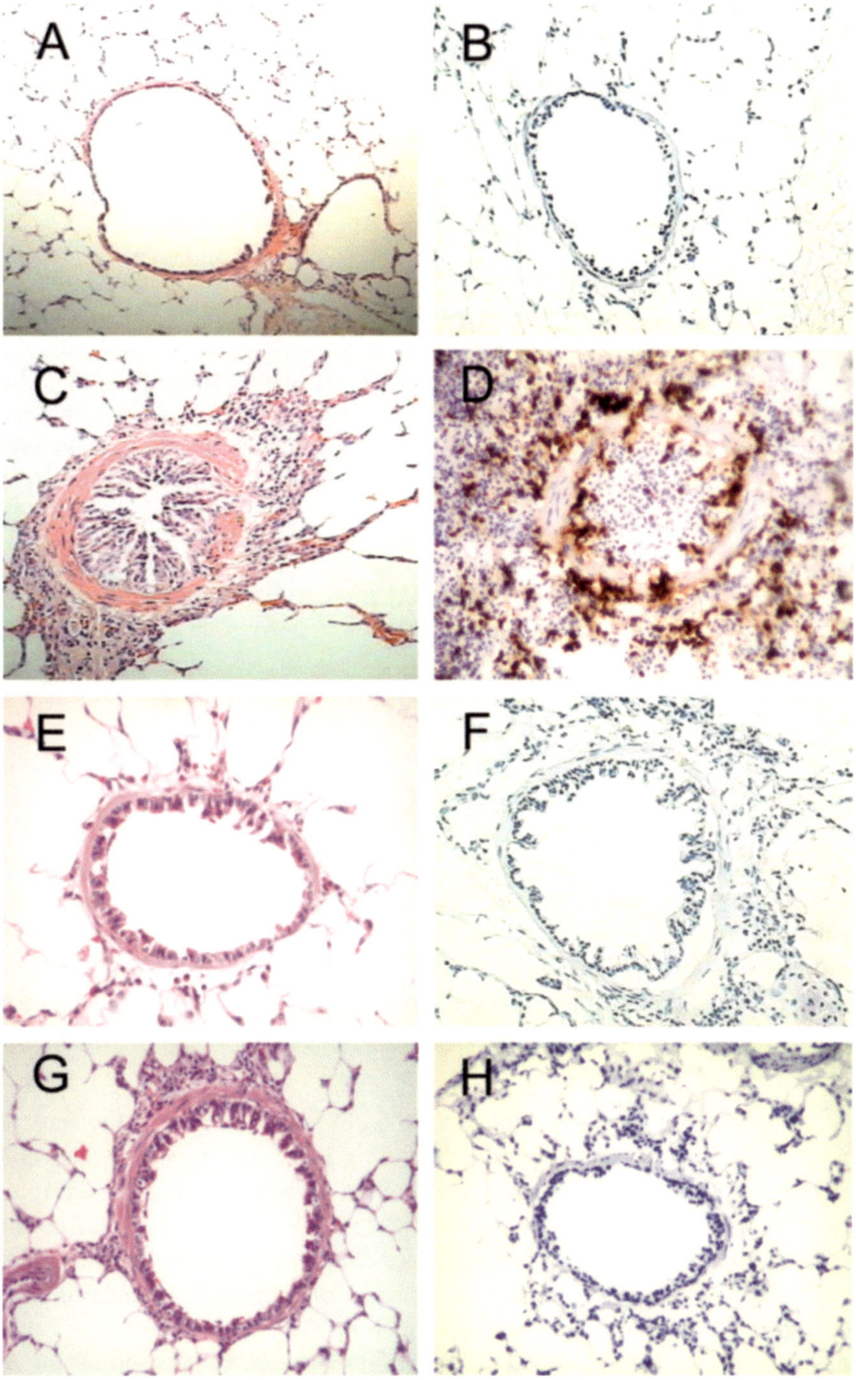

Figure 2. Photomicrographs of distal airways from the guinea pig (×200), stained with haematoxylin-eosin (left panels) and EPO+ eosinophils (right panels). Panels A and B: NS group. Panels C and D: OVA group. Panels E and F: OT1 group. Panels G and H: OT2 group. Reproduced with permission. *Published in Ruiz Schtüz et al. (143).*

Our group evaluated the airway responses in two different models of oral tolerance (ovalbumin-exposed and treat with oral tolerance beginning together with the 1st inhalation (OT1 group) and ovalbumin-exposed and treated with oral tolerance beginning after the 4th inhalation (OT2 group), and showed that both models counteract the bronchoconstriction induced by a specific antigen (ovalbumin) and by a nonspecific challenge using methacholine (143) (Figure 2). These data suggested that oral tolerance is an effective treatment to induce the relaxation of airway smooth muscle in asthma.

Although previous investigations showed that oral tolerance attenuated the airway responses, few studies have provided evidence of the effects of oral tolerance in lung periphery responses in an experimental model of chronic lung inflammation. In this regard, Nakashima et al. (43) showed that inducing oral tolerance attenuates peripheral lung tissue responsiveness, eosinophilic inflammation and extracellular matrix remodeling in an experimental model of chronic allergic pulmonary inflammation (Figure 3), suggesting that this approach could attenuate or prevent the distal lung functional and structural changes induced by chronic allergic inflammation.

5. Contribution of the airway and distal parenchyma structural changes to the pulmonary contractile responses.

The underlying persistent component of AHR, by contrast, is likely related to the structural (and/or physiological) airway changes often collectively referred to as airway remodeling. Structural changes in the airways and in the distal lung parenchyma, which were recently addressed, are involved in the remodeling process and include the epithelium basal membrane thickness, subepithelial fibrosis, mucous gland and goblet cell hypertrophy and hyperplasia, neoangiogenesis, increased ASM mass (hypertrophy of the smooth muscle cell and wall thickening), increased amount of actin and changes in the extracellular matrix (ECM), such as the deposition of fibronectin, laminin, and collagen fiber, alterations in the airway elastic fibers, and the increased expression of several metalloproteinases (MMP-1, MMP-2 and MMP-9) (45, 54). Such airway structural alterations or airway remodeling is associated with airway hyperresponsiveness to diverse triggers and with a decrease in the lung function of asthmatic patients.

In addition, an important structural change of the airways is related to the smooth muscle. One of the pathological consequences of remodeling is airway hyperresponsiveness. Myocyte hypertrophy and hyperplasia and myofibroblast hyperplasia are known to contribute to this hyperresponsiveness and the worsening of lung function in these patients (148). Throughout breathing, airway stiffening is a feasible contributor to airway hyperresponsiveness through the attenuation of the transmission of a potently bronchodilating cyclical stress to the ASM (37). ASM hyperplasia is characterized by a proliferation of cells, a reduction in the apoptosis of the ASM cells and migration of myofibroblasts within the ASM layer (19). Hence, alterations in the smooth muscle, either in the airways or in regions that are associated with perturbed alveolar attachments, may be factors that affect airway-parenchyma uncoupling and alterations in the mechanical properties of the distal lung that lead to constriction.

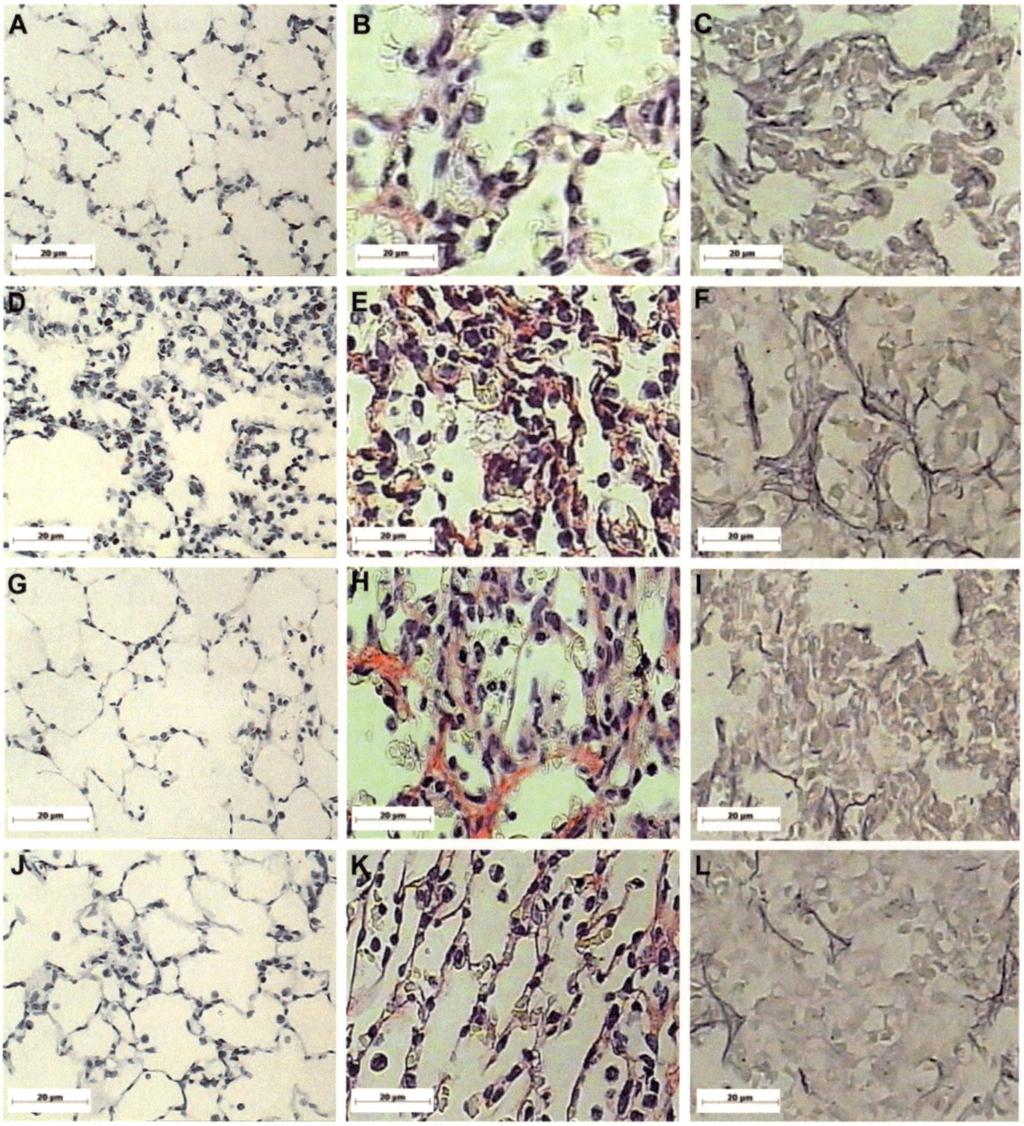

Figure 3. Photomicrographs of lung parenchymal strips eosinophilic infiltration (A, D, G, and J – x400), collagen density (B, E, H, and K – x1000) and elastic fibers (C, F, I, and L – x1000) in saline-exposed (NS group - panels A to C), ovalbumin-exposed (OVA group – panels D to F), ovalbumin-exposed and treat with oral tolerance beginning together with the 1st inhalation (OT1 group – panels G to I) and ovalbumin-exposed and treated with oral tolerance beginning after the 4th inhalation (OT2 group – panels J to L). Ovalbumin-exposed animals showed a significant increase in eosinophilic infiltration as well collagen and elastic density compared to saline-exposed ones. Both oral-induced tolerance protocols attenuated all these responses in ovalbumin-exposed animals. Reproduced with permission. *Published in Nakashima et al. (43).*

5.1. Mechanisms involved in lung remodeling

A chronic inflammatory process is almost invariably related to tissue damage and healing. The consequences of healing are repair and the replacement of injured cells by viable cells. Repair comprises regeneration (the replacement of damaged cells by cells of the same type) and replacement (by connective tissue). Chronic inflammatory processes have a wide variety of consequences leading from the complete or partial restoration of the affected structure to fibrotic processes. The mechanisms underlying remodeling move from the highly dynamic process of cell migration, differentiation, and maturation to changes in the connective tissue deposition and to the altered restitution of the structures (149).

The airway epithelium constitutes a continuous physical barrier, crucial to maintaining tissue homeostasis, which lines the airway lumen and separates the underlying tissue from environmental antigens (150, 151). Currently, the airway epithelium is acknowledged to also sense and react to antigens by regulating innate (through pattern-recognition receptors, including Toll-like receptors [(TLRs]) and adaptive immune mechanisms, driving both allergic sensitization and airway remodeling through the release of inflammatory cytokines and chemokines. In addition, direct physical interactions with immune cells protect the internal milieu of the lung (152) and therefore contribute to airway narrowing. Furthermore, the increased loss of epithelial barrier integrity is known to correlate with more severe airway hyperresponsiveness, which may lead to the augmented exposure of the ASM to inhaled contractile agonists (153). Therefore, epithelial cells participate in a wide range of repair mechanisms, including the epithelization of the nude luminal surface, the production of chemotactic factors, and the expression of some surface markers and a broad range of molecules that participate in the tissue repair, such as fibronectin, growth factors, cytokines and chemokines (149).

One of the mechanisms that may account for ASM hyperplasia is the migration of myofibroblasts within the ASM layer, which differentiate into ASM-like cells (154). Fibroblasts differentiate into the highly synthetic and contractile myofibroblast phenotype when exposed to substrates with an elastic modulus corresponding to pathologically stiff fibrotic tissue. Myofibroblasts, which are cells that display features intermediate between fibroblasts and smooth muscle cells, are involved in this process and are able to synthesize several extracellular matrix substances and contract the lung parenchyma (155).

Although the hypertrophy in ASM has been described in studies with tissue specimens from intermittent, mild, severe (156) and fatal (45) asthma, which have been characterized as having an increase in the ASM cell size, there are conflicting findings (157) that suggest that the ASM cell hypertrophy could be a hallmark of severe asthma because it can be used to differentiate between patients with severe asthma and patients with milder disease (156). In asthmatics, ASM cell proliferation occurs faster than in nonasthmatics (27), and it can be explained by alterations in the calcium homeostasis in these cells and a subsequent increase in mitochondrial biogenesis (158).

The main characteristics of myofibroblasts are the secretion of extracellular matrix components, the development of adhesion structures with the substrate by the incorporation of de novo expressed α-smooth muscle actin (α-SMA), and the formation of contractile bundles composed of actin and myosin, which help the myofibroblasts to develop a high contractile activity. These cytoskeletal features enable the myofibroblast to not only remodel and contract the extracellular matrix but also adapt its activity to changes in the mechanical microenvironment. In addition, immunohistochemistry and electron microscopy studies demonstrated that airway myofibroblasts and the smooth muscle bundles lie in close physical proximity in asthma (159, 160). The myofibroblasts have an intermediate phenotype between that of a fibroblast and that of a smooth muscle cell, which raises the possibility that these cells contribute to the increased smooth muscle mass because of their plasticity.

The arrangement and modification of the ECM involve dynamic processes of the production and degradation of matrix proteins, which are related to the ASM and parenchyma remodeling that are present and enhanced in asthma (161). The deposition of ECM proteins is increased by airway resident cells, such as epithelial cells, fibroblasts, myofibroblasts, and ASM cells. Some authors studying asthmatic bronchial samples demonstrated an increased deposition of ECM proteins in the bronchial wall, such as collagens I, III, and V, fibronectin, tenascin, hyaluronan, versican, laminin, lumican, and biglycan (162, 163), and a decreased deposition of collagen IV and elastin (164). Enhancing the ECM may be due to a reduced production of matrix metalloproteinases (MMPs), which degrade ECM proteins, and/or the enhanced production of tissue inhibitors of MMPs (TIMPs). Moreover, fibronectin and collagens III and V have been shown to enhance ASM migration (165) in the ASM cell contact with membranes coated with ECM components.

Notably, the epithelium in asthmatic children (aged 5-15 years) is stressed or injured without significant submucosal eosinophilic inflammation. This observation emphasizes the concept that the early pathological changes in asthma are linked to changes in the local tissue microenvironment related to epithelial stress and injury. The lamina reticularis from asthmatic biopsy sections was thicker than normal, with an increased deposition of collagen III. This alteration in the epithelial phenotype is associated with an enhanced collagen deposition in the lamina reticularis, suggesting that the epithelial mesenchymal trophic unit is active early in the natural history of asthma and may contribute to the pathogenesis of asthma (166).

ASM cells and the lung parenchyma have a crucial importance in the pathophysiology of asthma, leading to pulmonary remodeling, which remains unresponsive to conventional treatments, such as bronchodilators and anti-inflammatory drugs (167). Therefore, the development of new therapeutic tools targeting pulmonary remodeling is desirable.

6. Conclusions

ASM cells have a critical role in AHR in asthma, considering that these cells are part of the inflammatory process, have altered contractile, proliferative and secretory functions and contribute to airway remodeling.

Considering that many patients with AHR respond fairly well to conventional therapies, such as anti-inflammatory and bronchodilator drugs, and that ASM remodeling is insensitive to these treatments, further studies are necessary to evaluate ways to prevent or reverse ASM remodeling.

Author details

Carla Máximo Prado[1,2], Edna Aparecida Leick[1],
Fernanda Degobbi Tenório Quirino dos Santos Lopes[1],
Milton A. Martins[1] and Iolanda de Fátima Lopes Calvo Tibério[1]
[1]*School of Medicine, University of São Paulo, São Paulo, Brazil,*
[2]*Department of Biological Science, Universidade Federal de São Paulo, Diadema, Brazil*

7. References

[1] Brown RH, Pearse DB., Pyrgos G, Liu MC, Togias A, Permutt S. The Structural Basis of Airways Hyperresponsiveness in Asthma. Journal of Applied Physiology 2006;101 30-39.

[2] Yamauchi K. Airway Remodeling in Asthma and its Influence on Clinical Pathophysiology. Tohoku Journal Experimental Medicine 2006;209 75-87.

[3] Southam DS, Ellis R, Wattie J, Inman MD. Components of Airway Hyperresponsiveness and Their Associations with Inflammation and Remodeling In Mice. Journal of Allergy and Clinical Immunology 2007;119 848-854.

[4] Kraft M, Djukanovic R, Wilson S, Holgate ST, Martin RJ. Alveolar Tissue Inflammation in Asthma. American Journal of Respiratory and Critical Care Medicine 1996;154 1505.

[5] Kraft M. Part III: Location of Asthma Inflammation and the Distal Airways: Clinical Implications. Current Medical Research and Opinion 2007;3 S21-S27.

[6] Martin RJ. Therapeutic Significance of Distal Airway Inflammation in Asthma. Journal of Allergy and Clinical Immunology 2002;109(2 Suppl) S447-S460.

[7] Martin RJ. Exploring the Distal Lung: New Direction in Asthma. Israel Medical Association Journal 2008;10(12) 846-849.

[8] Xisto DG, Farias LL, Ferreira HC, Pincanc¸o MR, Amitrano D, Lapa e Silva JR, Negri EM, Carnielli D, F Silva LF, Capelozzi VL, Faffe DS, Zin WA, Rocco PM. Lung Parenchyma Remodeling in a Murine Model of Chronic Allergic Inflammation. American Journal of Respiratory and Critical Care Medicine 2005;171 829-837.

[9] Starling CM, Prado CM, Leick-Maldonado EA, Lanças T, Reis FG, Aristóteles LR, Dolhnikoff M, Martins MA, Tibério IF: Inducible Nitric Oxide Synthase Inhibition Attenuates Lung Tissue Responsiveness and Remodeling in a Model of Chronic Pulmonary Inflammation in Guinea Pigs. Respiratory Physiology and Neurobiology 2009:165 185-194.

[10] Lanças T, Kasahara DI, Prado CM, Tibério FLCI, Martins MA and Dolhnikoff M: Comparison of Early and Late Responses to Antigen of Sensitized Guinea Pig Parenchymal Lung Strips. Journal of Applied Physiology 2006;100(5) 1610-1616.

[11] Angeli P, Prado CM, Xisto DG, Silva PL, Passaro CP, Nakazato HD, Leick-Maldonado EA, Martins MA, Rocco PR, Tiberio IF: Effects of Chronic L-NAME Treatment Lung Tissue Mechanics, Eosinophilic and Extracellular Matrix Responses Induced By Chronic Pulmonary Inflammation. American Journal Physiology andd Lung Cell Molecular Physiology 2008;294 L1197-L1205.

[12] Fredberg JJ, Stamenovic D. On the Imperfect Elasticity of Lung Tissue. Journal of Applied Physiology 1989;67 2408-2419.

[13] Lauzon AM, Bates HT: Estimation of Time-Varying Respiratory Mechanical Parameters by Recursive Least Squares. Journal of Applied Physiology 1991;71 1159-1165.

[14] Dolhnikoff M, Mauad T, Ludwig MS: Extracellular Matrix and Oscillatory Mechanics of Rat Lung Parenchyma in Bleomycin- Induced Fibrosis. American Journal of Respiratory and Critical Care Medicine 1999;160 1750-1757.

[15] Dolhnikoff M, Morin J, Ludwig MS. Human Lung Parenchyma Responds to Contractile Stimulation. American Journal of Respiratory and Critical Care Medicine 1998;158 1607-1612.

[16] Busse WW, Lemanske RF Jr. Asthma. New England Journal of Medicine. 2001;344(5) 350-362.

[17] Zuyderduyn S, Sukkar MB, Fust A, Dhaliwal S, Burgess JK. Treating Asthma Means Treating Airway Smooth Muscle Cells. European Respiratory Journal 2008;32(2) 265-74.

[18] Black JL, Roth M. Intrinsic Asthma: Is it Intrinsic to the Smooth Muscle? Clinical and Experimental Allergy, 2009;9(7) 962-965.

[19] Ozier A, Allard B, Bara I, Girodet PO, Trian T, Marthan R, Berger P. The Pivotal Role of Airway Smooth Muscle in Asthma Pathophysiology. Journal of Allergy (Cairo) 2011;742710.

[20] Bergeron C, Hauber HP, Gotfried M, Newman K, Dhanda R, Servi RJ, Ludwig M, Hamid Q. Evidence of Remodeling in Peripheral Airways of Patients with Mild to Moderate Asthma: Effect of Hydrofluoroalkane-Flunisolide. Journal of Allergy and Clinical Immunology 2005;116 983-989.

[21] Micheletto C, Guerriero M, Tognella S, Dal Negro RW. Effects of HFA- and CFC-Beclomethasone Dipropionate on the Bronchial Response to Methacholine (Mch) in Mild Asthma. Respiratory Medicine 2005;99 850-855.

[22] de Monchy JG, Kauffman HF, Venge P, et al. Bronchoalveolar Eosinophilia During Allergen-Induced Late Asthmatic Reactions. American Review and Respiratory Diseases 1985;131 373-376.

[23] Hargreave FE. Late-Phase Asthmatic Responses and Airway Inflammation. Journal of Allergy and Clinical Immunology 1989;83 525-527.

[24] Cockcroft DW, Davis BE. Mechanisms of Airway Hyperresponsiveness. Journal of Allergy and Clinical Immunology 2006;118 551-559.

[25] Chan V, Burgess JK, Ratoff JC, O'connor BJ, Greenough A, Lee TH, Hirst SJ. Extracellular Matrix Regulates Enhanced Eotaxin Expression in Asthmatic Airway Smooth Muscle Cells. American Journal and Respiratory Critical Care Medicine 2006;174(4) 379-385.

[26] Johnson PR, Black JL, Carlin S, Ge Q, Underwood PA. The Production of Extracellular Matrix Proteins by Human Passively Sensitized Airway Smooth-Muscle Cells in Culture: The Effect of Beclomethasone. American Journal of Respiratory and Critical Care Medicine 2000;162: 2145-2151.

[27] Johnson PR, Roth M, Tamm M, Hughes M, Ge Q, King G, Burgess JK, Black JL. Airway Smooth Muscle Cell Proliferation is Increased in Asthma. American Journal of Respiratory and Critical Care Medicine 2001;164(3) 474-477.

[28] Burgess JK, Johnson PR, Ge Q, et al. Expression of Connective Tissue Growth Factor in Asthmatic Airway Smooth Muscle Cells. American Journal of Respiratory and Critical Care Medicine 2003;167 71-77.

[29] Sukkar MB, Stanley AJ, Blake AE, et al. ''Proliferative'' and ''Synthetic'' Airway Smooth Muscle Cells Are Overlapping Populations. Immunology & Cell Biology 2004;82 471-478.

[30] Jiang H, Rao K, Halayko AJ, Kepron W, Stephens NL. Bronchial Smooth Muscle Mechanics of a Canine Model of Allergic Airway Hyperresponsiveness. Journal of Applied Physiology 1992;72 39-45.

[31] Ammit AJ, Bekir SS, Johnson PR, Hughes JM, Armour CL, Black JL. Mast Cell Numbers Are Increased in the Smooth Muscle of Human Sensitized Isolated Bronchi. American Journal of Respiratory and Critical Care Medicine 1997;155 1123-1129.

[32] Brightling CE, Bradding P, Symon FA, Holgate ST, Wardlaw AJ, Pavord ID. Mast-Cell Infiltration of Airway Smooth Muscle in Asthma. New England Journal Medicine 2002;346 1699-1705.

[33] Begueret H, Berger P, Vernejoux JM, Dubuisson L, Marthan R, Tunon-De-Lara JM. Inflammation of Bronchial Smooth Muscle in Allergic Asthma. Thorax 2007;62(1) 8-15.

[34] Brightling CE, Ammit AJ, Kaur D et al. The CXCL10/ CXCR3 Axis Mediates Human Lung Mast Cell Migration to Asthmatic Airway Smooth Muscle. American Journal of Respiratory and Critical Care Medicine, 2005;171(10) 1103-1108.

[35] El-Shazly A, Berger P, Girodet PO et al. Fraktalkine Produced By Airway Smooth Muscle Cells Contributes to Mast Cell Recruitment in Asthma. Journal of Immunology, 2006;176(3) 1860-1868.

[36] Amin K, Janson C, Boman G, Venge P. The Extracellular Deposition of Mast Cell Products is Increased in Hypertrophic Airways Smooth Muscles in Allergic Asthma but not in Nonallergic Asthma. Allergy 2005;60(10) 1241-1247.

[37] Siddiqui S, Martin JG. Structural Aspects of Airway Remodeling in Asthma. Current Allergy and Asthma Report 2008;8(6) 540-547.

[38] Saha SK, Berry MA, Parker D et al. Increased Sputum and Bronchial Biopsy IL-13 Expression in Severe Asthma. Journal of Allergy and Clinical Immunology 2008;121(3) 685-691.

[39] Liu MC, Bleecker ER, Lichtenstein LM, et al. Evidence for Elevated Levels of Histamine, Prostaglandin D2, and Other Bronchoconstriction Prostaglandins in the Airways of Subjects with Mild Asthma. *American Review of Respiratory Disease* 1990;142(1) 126-132.

[40] Sterk PJ. The Place of Airway Hyperresponsiveness in the Asthma Phenotype. Clinical Experimental Allergy. 1995;25(2) 8-11; discussion 17-8.

[41] Rocco PR, Facchinetti LD, Ferreira HC, Negri EM, Capelozzi VL, Faffe DS, Zin WA. Time Course of Respiratory Mechanics and Pulmonary Structural Remodelling in Acute Lung Injury. Respiratory Physiology Neurobiology 2004;143(1) 49-61.

[42] Nagase T; Fukuchi Y; Dallaire MJ; Martin JG and Ludwig MS: In vitro Airway And Tissue Responses to Antigen in Sensitized Rats. American Journal of Respiratory and Critical Care Med 1995;153 81-86.

[43] Nakashima AS, Prado CM, Lanças T, Ruiz VC, Kasahara DI, Leick-Maldonado EA, Dolhnikoff M, Martins MA, Tibério IF: Oral Tolerance Attenuates Changes in In Vitro Lung Tissue Mechanics and Extracellular Matrix Remodeling Induced by Chronic Allergic Inflammation in Guinea Pigs. Journal of Applied Physiology 2008;104(6) 1778-1785.

[44] Romero PV, Rodriguez B, Lopez-Aguilar J, Manresa F. Parallel Airways Inhomogeneity and Lung Tissue Mechanics in Transition to Constricted State in Rabbits. Journal of Applied Physiology 1998;84 1040-1047.

[45] Mauad T, Silva LF, Santos MA, Grinberg L, Bernardi FD, Martins MA, Saldiva PH, Dolhnikoff M. Abnormal Alveolar Attachments with Decreased Elastic Fiber Content in Distal Lung in Fatal Asthma. American Journal and Respiratory and Critical Care Medicine 2004;170 857-862.

[46] Romero PV, Zin WA, Lopez-Aguilar J. Frequency Characteristics of Lung Tissue Strip during Passive Stretch and Induced Pneumoconstriction. Journal of Applied Physiology 2001;91 882-890.

[47] Barnes NC, Piper PJ, Costello JF. Comparative Effects of Inhaled Leukotriene C4, Leukotriene D4, and Histamine in Normal Human Subjects. *Thorax* 1984;39(7) 500-504.

[48] Perez-Zoghbi JF, Sanderson MJ. Endothelin-Induced Contraction of Bronchiole and Pulmonary Arteriole Smooth Muscle Cells is Regulated by Intracellular Ca2+ Oscillations and Ca2+ Sensitization. *American Journal of Physiology* 2007;293(4) L1000-L1011.

[49] Howarth PH, Redington AE, Springall DR, et al. Epithelially Derived Endothelin and Nitric Oxide in Asthma. Interarch of Allergy and Immunology 1995;107(1-3) 228-230.

[50] Gosens R, Zaagsma J, Grootte Bromhaar M, Nelemans A, Meurs H. Acetylcholine: A Novel Regulator of Airway Smooth Muscle Remodelling? European Journal of *Pharmacology* 2004;500(1-3) 193-201.

[51] Adcock IM, Peters M, Gelder C, Shirasaki H, Brown CR, Barnes PJ. Increased Tachykinin Receptor Gene Expression in Asthmatic Lung and its Modulation by Steroids. Journal Molecular and Endocrinology 1993;11(1) 1-7.

[52] Chambers LS, Black JL, Ge Q et al. PAR-2 Activation, PGE2, and COX-2 in Human Asthmatic and Nonasthmatic Airway Smooth Muscle Cells. American Journal of Physiology 2003;285(3) L619-L627.

[53] Meurs H, Hamer MAM, Pethe S, Goff SV, Boucher J, Zaagsma J: Modulation of Cholinergic Airway Reactivity and Nitric Oxide Production by Endogenous Arginase Activity. British Journal of Pharmacology 2000;130 1793-1798.

[54] Prado CM; Leick- Maldonado EA; Yano L Leme AS; Capelozzi VL; Martins MA Tibério IF: Effects of Nitric Oxide Synthases in Chronic Allergic Airway Inflammation and Remodeling. American Journal of Physiology and Lung Cellular Molecular Biology 2006;35(4) 457-465.

[55] Holgate S. Mechanisms of Allergy and Adult Asthma. Current Opinion in Allergy and Clinical Immunology 2001;1 47-50.

[56] Numao T, Agrawal DK. Neuropeptides Modulate Human Eosinophil Chemotaxis. Journal of Immunology 1992;149 3309-3315.

[57] Tibério IF, Turco GMG, Leick-Maldonado EA, Sakae RS, Paiva SO, do Patrocínio M, Warth TN, Lapa e Silva JR, Saldiva PH and Martins MA. Effects of Neurokinin Depletion on airway Inflammation Induced By Chronic Antigen Exposure. American Journal of Respiratory and Critical Care Medicine 1997;155 1739-1747.

[58] Piedimonte G. Neural Mechanisms of Airway Syncytial Virus-Induced Inflammation and Prevention of Airway Syncytial Virus Sequelae. American Journal of Respiratory and Critical Care Medicine 2001;163 S18-S21.

[59] O'Connor TM, O'Connell J, O'Brien DI, Goode T, Bredin CP, Shanahan F. The Role of Substance P in Inflammatory Disease. Journal of Cellular Physiology 2001, 167-180.

[60] Tibério IF, Leick-Maldonado EA, Miyahara L, Kasahara DI, Spilborghs GM, Martins MA, Saldiva PH. Effects of Neurokinins on Airway and Alveolar Eosinophil Recruitment. Experimental Lung Research 2003;29(3) 165-177.

[61] Prado CM, Leick-Maldonado EA, Miyamoto L, Yano LM, Kasahara DI, Martins MA, Tibério IF. Capsaicin-Sensitive Nerves and Neurokinins Modulate Non-Neuronal nNOS Expression in Lung. Respiratory Physiology and Neurobiology 2008;160(1) 37-44.

[62] Regoli D., Drapeau G, Dion S, Couture R. New Selective Agonists for Neurokinin Receptors: Pharmacological Tools for Receptor Characterization. Trends Pharmacology Science 1988;9 290-295.

[63] Holgate ST, Sampson AP. Antileukotriene Therapy: Future Directions. American Journal of Respiratory and Critical Care Medicine 2000;61 S147-S153.

[64] O'Byrne PM. Leukotriene Bronchoconstriction Induced by Allergen and Exercises. American Journal of Respiratory and Critical Care Medicine 2000;161 S68-S72.

[65] Drazen JM, Israel E, O'Byrne PM. Treatment of Asthma with Drugs Modifying the Leukotriene Pathway. New England Journal Medicine 1999;340 197-206.

[66] Drazen JM. Leukotrienes as Mediators of Airway Obstruction. American Journal of Respiratory and Critical Care Medicine 1998;158 S193-S200.

[67] Trakada G, Tsourapis S, Marangos M, Spiropoulos K. Arterial and Bronchoalveolar Lavage Fluid Endothelin-1 Concentration in Asthma. *Respiratory Medicine* 2000;94(10) 992-996.

[68] Pizzichini E, Leff JA, Reiss TF et al. Montelukast Reduces Airway Eosinophilic Inflammation in Asthma: A Randomized, Controlled Trial. European Respiratory Journal 1999;14 12-18.

[69] Nakamura Y, Hoshino M, Sim JJ, Ishii K, Hosaka K, Sakamoto T. Effect of the Leukotriene Receptor Antagonist Pranlukast on Cellular Infiltration in the Bronchial Mucosa of Patients with Asthma. Thorax 1998;53 835-841.

[70] Smith LJ. Comparative Efficacy of Inhaled Corticosteroids and Antileukotriene Drugs in Asthma. Biodrugs 2001;15 239-249.

[71] Malmstrom K, Rodriguez-Gomes G, Guerra J et al. Oral Montelukast, Inhaled Beclomethasone and Placebo for Chronic Asthma. Annals of Internal Medicine 1999;130 487-495.

[72] Bleecker ER, Welch MJ, Weinstein SF et al. Low-Dose Inhaled Fluticasone Propionate Versus Oral Zafirlukast in the Treatment of Persistent Asthma. Journal of Allergy and Clinnical Immunology 2000;105 1123-1129.

[73] Leff AR. Role of Leukotrienes in Bronchial Hyperresponsiveness and Cellular Responses In Airways. American Journal of Respiratory and Critical Care Medicine 2000;161 S125-S132.

[74] Gardiner PJ, Cuthbert NJ, Francis HP et al. Inhibition of Antigeninduced Contraction of Guinea-Pig Airways by a Leukotriene Synthesis Inhibitor, BAY x1005. European Journal Pharmacology 1994;258 95-102.

[75] Liu YC, Khawaja AM, Rogers DF. Effects of the Cysteinyl Leukotriene Receptor Antagonists Pranlukast and Zafirlukast on Tracheal Mucus Secretion in Ovalbumin-Sensitized Guinea-Pigs In Vitro. Brittish Journal of Pharmacology 1998;124 563-571.

[76] Leick-Maldonado EA, Kay FU, Leonhardt MC, Kasahara DI, Prado CM, Fernandes FT, Martins MA and Tibério IFLC: Comparison of Glucocorticoid and Cysteinyl Leukotriene Receptor Antagonist Treatments in an Experimental Model of Chronic Airway Inflammation in Guinea-Pigs. Clinical Experimental Allergy 2004,34 145-152.

[77] Ricciardolo FLM: Multiple Roles of Nitric Oxide in the Airways. Thorax 2003;58 175-182.

[78] Prado CM, Leick-Maldonado EA, Arata V, Kasahara DI, Martins MA, Tibério IF. Neurokinins and Inflammatory Cell iNOS Expression in Guinea Pigs with Chronic Allergic Airway Inflammation. American Journal of Physiology and Lung Cell Molecular Physiology 2005;288(4) L741-L748.

[79] Prado CM, Martins MA, Tiberio IF. Nitric Oxide in Asthma Physiopathology. International Scholarly Research Network. ISRN Allergy. 2011, 13 pages.

[80] Prado CM, Yano L, Rocha G, Starling CM, Capelozzi VL, Leick-Maldonado EA, Martins M de A, Tibério IF. Effects of Inducible Nitric Oxide Synthase Inhibition in Bronchial Vascular Remodeling-Induced by Chronic Allergic Pulmonary Inflammation. Experimental Lung Research 2011;137(5) 259-268.

[81] Prado CM, Leick-Maldonado EA, Kasahara DI, Capelozzi VL, Martins MA, Tiberio IF. Effects of Acute and Chronic Nitric Oxide Inhibition in an Experimental Model of Chronic Pulmonary Allergic Inflammation in Guinea Pigs. American Journal Physiology Lung Cellular Mollecular Biology 2005;289 L677-L683.

[82] Ricciardollo FL, Sterk PJ, Gaston B, Folkers G. Nitric Oxide in Health and Disease of the Respiratory System. Physiology Review 2004;84(3) 731-765.

[83] Garvey EP, Oplinger JA, Furfine ES, Kiff RJ, Laszlo F, Whittle BJR and Knowel RG: 1400W is a slow, Tight Binding and Highly Selective Inhibitor of Inducible Nitric Oxide Synthase *In Vitro* and *In Vivo*. Journal of Biochemistry 1997;272(8) 4959-63.

[84] De Boer J, Meurs H, Coers W, Koopal M, Bottone AE, Visser AC, Timens W, Zaagsma J: Deficiency of Nitric Oxide in Allergen-Induced Airway Hyperreactivity to Contractile Agonists after the Early Asthmatic Reaction: An Ex Vivo Study. Brittish Journal of Pharmacol 1996;119 1109-1116.

[85] Koarai A, Ichinose M, Sugiura H, Tomaki M, Watanabe M, Yamagata S, Komaki Y, Shirato K, Hattori T. iNOS Depletion Completely Diminishes Reactive Nitrogen-Species Formation after an Allergic Response. European Respiratory Journal 2002;20(3) 609-616.

[86] Eynott PR, Groneberg DA, Caramori G, Adcock IM, Donnely LE, Kharitonov S, Barnes PL, Chung KF. Role of Nitric Oxide in Allergic Inflammation and Bronchial Hyperresponsiveness. European Journal of Pharmacology 2002;452 123-133.

[87] Meurs H, McKay S, Maarsingh H, Hamer M , Macic L, Molendijk N, Zaagsma J: Increased Arginase Activity Underlies Allergen-Induced Deficiency of cNOS Derived Nitric Oxide and Airway Hyperresponsiveness. Brittish Journal of Pharmacology 2002;136 391-398.

[88] Meurs H, Maarsingh H, Zaagsma J. Arginase and Asthma: Novel Insights into Nitric Oxide Homeostasis and Airway Hyperresponsiveness. Trends Pharmacology in Science 2003;24 450-455.

[89] Guo FH, De Raeve HR, Rice TW, Stuehr DJ, Thunnissen FBJM,Erzurum SC. Continuous Nitric Oxide Synthesis by Inducible Nitric Oxide Synthase in Normal Human Airway Epithelium In Vivo. Procedings of the National Academy of Science USA 1995;92 7809-7813.

[90] Dupuy PM, Shore SA, Drazen JM, Frostell C, Hill Zapol WA. Bronchodilator Action of inhaled Nitric Oxide in Guinea Pigs. Journal of Clinical Investigation 1992;90 421-442.

[91] Hsu YC, Wang LF, Chien YW. Nitric Oxide In The Pathogenesis of Diffuse Pulmonary Fibrosis. Free Radical Biology Medicine 2007;42 599-607.

[92] Que LG, George SE, Gotoh T, Mori M, Huang YC. Effects of Arginase Isoforms on NO Production by nNOS. Nitric Oxide 2002;6(1) 1-8.

[93] Morris CR, Poljakovic M, Lavrisha L, Machado L, Kuypers FA, Morris SM: Decreased Arginine Bioavailability and Increased Serum Arginase Activity in Asthma. American Journal and Respiratory and Critical Care Medicine 2004;170 148-53.

[94] Hamid Q, Springall DR, Riveros-Moreno V, Chanez P, Howarth P, Redington A, Bousquet J, Godard P, Holgate S, Polak JM. Induction of Nitric Oxide Synthase in Asthma. Lancet 1993;342 1510-1513.

[95] Ckless K, Lampert, Reiss J, Kasahara D, Poynter ME, Irvin CG, Lundblad LKA, Norton R, Vliet A, Janssen-Heininger YMW. Inhibition of Arginase Activity Enhances Inflammation in Mice with Allergic Airway Disease, in Association with Increases in Protein S-Nitrosylation and Tyrosine Nitration1. The Journal of Immunology 2008;181 4255-4264.

[96] Ckless K, Van der Vliet A, Janssen-Heininger Y. Oxidative Nitrosative Stress and Post-Translational Protein Modifications: Implications to Lung Structure-Function Relations. Arginase Modulates NF-kappaB Activity Via a Nitric Oxide-Dependent Mechanism. American Journal and Respiratory Cellular Mollecular Biology 2007;36 645-53.

[97] Poynter ME, Cloots R, Van Woerkom T, Butnor KJ, Vacek P, Taatjes DJ, Irvin CG, Janssen-Heininger YM: NF-kB: Activation in airways modulates allergic inflammation but not hyperresponsiveness. Journal of Immunology 2004;173 7003–7009.

[98] Bannenberg G, Xue J, Engman L, Cotgreave I, Moldeus P, Ryrfeldt A. Characterization of Bronchodilator Effects and Fate of S-nitrosothiols in the Isolated Perfused and Ventilated Guinea Pig Lung. *Journal* of *Pharmacology* and Experimental Therapeutics 1995;272 1238-1245.

[99] Perkins WJ, Pabelick C, Warner DO, Jones KA. cGMP-Independent Mechanism of Airway Smooth Muscle Relaxation Induced by S-Nitrosoglutathione. American Journal of Physiology 1998;275 C468-C474.

[100] Janssen LJ, Premji M, Lu-Chao H, Cox G, Keshavjee S. NO but not NO Radical Relaxes Airway Smooth Muscle via cGMP-Independent Release of Internal Ca(2+). American Journal of Physiology Lung Cellular Molecular Physiology 2000;278 L899-L905.

[101] Haddad IY, Ischiropoulos H, Holm BA, Beckman JS, Baker JR, Matalon S. Mechanisms of Peroxynitrite-Induced Injury to Pulmonary Surfactants. American Journal Physiology 1993;265 L555-L564.

[102] Bhandari V, Johnson L, Smith-Kirwin S, Vigliotta G, Funanage V, Chander A. Hyperoxia and Nitric Oxide Reduce Surfactant Components (DSPC and Surfactant Proteins) and Increase Apoptosis in Adult and Fetal Rat Type II Pneumocytes. Lung 2002;180(6) 301-317.

[103] Jourdan KB, Mitchell JA, Evans TW. Release of Isoprostanes by Human Pulmonary Artery in Organ Culture: A Cyclo-Oxygenase and Nitric Oxide Dependent Pathway. Biochemical and Biophysical Research Communications 1997;233(3) 668-672.

[104] Janssen LJ. Isoprostanes: An Overview and Putative Roles in Pulmonary Pathophysiology. American Journal of Physiology and Lung Cellular Molecular Physiology 2001;280(6) L1067-L1082.

[105] Taki F, Kume H, Kobayashi T, Ohta H, Aratake H, Shimokata K. Effects of Rho-Kinase Inactivation on Eosinophilia and Hyper-Reactivity In Murine Airways by Allergen Challenges. Clinical and Experimental Allergy 2007;37 599-607.

[106] Gosens R, Schaafsma D, Grootte Bromhaar MM, Vrugt B, Zaagsma J, Meurs H, Nelemans SA. Growth Factor-Induced Contraction of Human Bronchial Smooth Muscle is Rho-Kinase-Dependent. European Journal of Pharmacology 2004;494 73-76.

[107] Heasman SJ, Ridley AJ. Multiple Roles for RhoA during T Cell Transendothelial Migration. Small Gtpases 2010;1(3) 174-179.

[108] Kume H. RhoA/Rho-kinase as a Therapeutic Target in Asthma. Current Medicinal Chemistry 2008;15(27) 2876-2885.

[109] Schaafsma D, Bos ST, Zuidhof AB, Zaagsma J, Meurs H. The Inhaled Rho Kinase Inhibitor Y-27632 Protects Against Allergen-Induced Acute Bronchoconstriction, Airway Hyperresponsiveness, and Inflammation. American Journal of Physiology and Lung Cellular Molecular Physiology 2008;295 L214-L219.

[110] Fernandes L, D'Aprile A, Self G, McGuire M, Sew T, Henry P, Goldie R A Rho-Kinase Inhibitor, Y-27632, Reduces Cholinergic Contraction but not Neurotransmitter Release. European Journal Pharmacology. 2006;550(1-3) 155-161.

[111] Witzenrath M, Ahrens B, Schmeck B, Kube SM, Hippenstiel S, Rosseau S, Hamelmann E, Suttorp N, Schütte H. Rho-Kinase and Contractile Apparatus Proteins in Murine Airway Hyperresponsiveness. Toxicology Pathology 2008;60(1) 9-15.

[112] Henry PJ, Mann TS, Goldie RG. A Rho Kinase Inhibitor, Y-27632 Inhibits Pulmonary Eosinophilia, Bronchoconstriction and Airways Hyperresponsiveness in Allergic Mice. Pulmonary Pharmacology Therapeutics 2005;18(1) 67-74.

[113] Aihara M, Dobashi K, Iizuka K, Nakazawa T, Mori M. Effect of Y-27632 on Release of Cytokines from Peripheral T Cells in Asthmatic Patients and Normal Subjects. International Immunopharmacology 2004;4 557-561.

[114] Spector S, Tan R. Exercise-Induced Bronchoconstriction Update: Therapeutic Management. Allergy Asthma Proceedings 2012;33(1) 7-12.

[115] Fanelli A, Cabral AL, Neder JA, Martins MA, Carvalho CR. Exercise Training on Disease Control and Quality of Life in Asthmatic Children. Medicine and Science Sports Exercise. 2007;39(9) 1474-1480.

[116] Mendes FA, Almeida FM, Cukier A, Stelmach R, Jacob-Filho W, Martins MA, Carvalho CR. Effects of Aerobic Training on Airway Inflammation in Asthmatic Patients. Medicine and Science Sports Exercise 2011;43(2) 197-203.

[117] Mendes FA, Gonçalves RC, Nunes MP, Saraiva-Romanholo BM, Cukier A, Stelmach R, Jacob-Filho W, Martins MA, Carvalho CR. Effects of Aerobic Training on Psychosocial Morbidity and Symptoms in Patients with Asthma: A Randomized Clinical Trial. Chest 2010;138(2) 331-337.

[118] Silva RA, Vieira RP, Duarte AC, Lopes FD, Perini A, Mauad T, Martins MA, Carvalho CR. Aerobic Training Reverses Airway Inflammation and Remodelling in an Asthma Murine Model. European Respiratory Journal 2010;35(5) 994-1002.

[119] Vieira RP, Toledo AC, Ferreira SC, Santos AB, Medeiros MC, Hage M, Mauad T, Martins Mde A, Dolhnikoff M, Carvalho CR. Airway Epithelium Mediates the Anti-Inflammatory Effects of Exercise on Asthma. Respiratory Physiology and Neurobiology 2011;175(3) 383-9.

[120] Stadelmann K, Stensrud T, Carlsen KH. Respiratory Symptoms and Bronchial Responsiveness in Competitive Swimmers. Medicine and Science Sports Exercise 2011;43(3) 375-381.

[121] Lund TK. Asthma in Elite Athletes: How do we Manage Asthma-Like Symptoms and Asthma in Elite Athletes? Clinnical Respiratory Journal 2009;3(2) 123.

[122] Forsythe P, Ebeling C, Gordon JR, Befus AD, Vliagoftis H: Opposing Effects of Short-and Asthma. Allergy Asthma Procedings 2000;21 241-246.

[123] Webster EL, Torpy DJ, Elenkov IJ, Chrousos GP: Corticotropin-Releasing Hormone and Inflammation. Annual NY Academic Science 1998;840 21-32.

[124] Ritz T, Steptoe A, DeWilde S, Costa M: Emotions and Stress Increase Respiratory Resistance in Asthma. Psychosomatic Medicine 2000;62 401-412.

[125] Ritz T, Steptoe A: Emotion and Pulmonary Function in Asthma: Reactivity in the Field and Relationship with Laboratory Induction of Emotion. Psychosomatic Medicine 2000;62 808-815.

[126] Marshall GD Jr, Agarwal SK: Stress, Immune Regulation, and Immunity: Applications for Long-Term Stress on Airway Inflammation. American Journal and Respiratoory Critical Care Med 2004;169 220-226.

[127] Miller GE, Cohen S, Ritchey AK: Chronic Psychological Stress and the Regulation of Pro-Inflammatory Cytokines: A Glucocorticoid-Resistance Model. Health Psychology 2002;21 531-541.

[128] Capelozzi MA, Leick-Maldonado EA, Parra ER, Martins MA, Tibério IF, Capelozzi VL: Morphological and Functional Determinants of Fluoxetine (Prozac)-Induced Pulmonary Disease in an Experimental Model. Respiratory Physiology and Neurobiology 2007;156 171-178.

[129] Leick EA, Reis FG, Honorio-Neves FA, Almeida-Reis R, Prado CM, Martins MA, Tibério IF. Effects of Repeated Stress on Distal Airway Inflammation, Remodeling and Mechanics in an Animal Model of Chronic Airway Inflammation. Neuroimmunomodulation 2012;19(1) 1-9.

[130] Marques RH, Reis FG, Starling CM, Cabido C, de Almeida-Reis R, Dohlnikoff M, Prado CM, Leick EA, Martins MA, Tibério IF. Inducible Nitric Oxide Synthase Inhibition Attenuates Physical Stress-Induced Lung Hyper-Responsiveness and Oxidative Stress in Animals with Lung Inflammation. Neuroimmunomodulation 2012;19(3) 158-170.

[131] Höglund CO, Axen J, Kemi C, Jernelov S, Grunewald J, Muller-Suur C, Smith Y, Gronneberg R, Eklund A, Stierna P, Lekander M: Changes in Immune Regulation in Response to Examination Stress in Atopic and Healthy Individuals. Clinical Experimental Allergy 2006;36 982-992.

[132] Karlsson MR, Kahu H, Hanson LA, Telemo E, Dahlgren UI. Tolerance and Bystander Suppression, with Involvement of CD25-Positive Cells, is Induced in Rats Receiving Serum from Ovalbumin-Fed Donors. Immunology 2000;100 326-333.

[133] Smith KM, Eaton AD, Finlayson LM, Garside P. Oral Tolerance. American Journal and Respiratoory Critical Care Medicine 2000;162 175-178.

[134] Weiner HL. Oral Tolerance: Immune Mechanisms and Treatment of Autoimmune Diseases. Immunology Today 1997;18 335-343.

[135] Faria AM,Weiner HL. Oral Tolerance. Immunology Review 2005;206 232-259.

[136] Garside P, Millington O, Smith KM. The Anatomy of Mucosal Immune Responses. Annual NY Academic Science 2004;1029 9-15.

[137] Dubois B, Goubier A, Joubert G, Kaiserlian D. Oral Tolerance and Regulation of Mucosal Immunity. Cellular Molecular Life Science 2005;62 1322-1332.

[138] Abramson MJ, Puy RM, Weiner JM. Allergen Immunotherapy for Asthma. American Journal and Respiratory and Critical Care Medicine 1999;160 1750-1757.

[139] Calamita Z, Saconato H, Pela AB, Atallah AN. Efficacy of Sublingual Immunotherapy in Asthma: Systematic Review of Randomized-Clinical Trials using the Cochrane Collaboration Method. Allergy 2006;61 1162-1172.

[140] Sopo SM, Macchiaiolo M, Zorzi G, Tripodi S. Sublingual Immunotherapy in Asthma and Rhinoconjunctivitis; Systematic Review of Paediatric Literature. Archives of Diseaes in Childhood 2004;89 620-624.

[141] Russo M, Jancar S, Siqueira ALP, Mengel J, Gomes E, Ficker SM, Faria AMC. Prevention of Lung Eosinophilic Inflammation by Oral Tolerance. Immunology Letters 1998;61 15-23.

[142] Russo M, Nahori MA, Lefort J, Gomes E, Keller AC, Rodriguez D, Ribeiro OG, Adriouch S, Gallois V, de Faria AM, Vargaftig BB. Suppression of Asthma-Like Responses in Different Mouse Strains by Oral Tolerance. American Journal and Respiratory and Cellular Molecular Biology 2001;24 518-526.

[143] Ruiz-Schütz VC, Drewiacki T, Nakashima AS, Arantes-Costa FM, Prado CM, Kasahara DI, EA, Martins M de A, Tibério IF. Oral Tolerance Attenuates Airway Inflammation and Remodeling in a Model of Chronic Pulmonary Allergic Inflammation. Respiratory Physiology and Neurobiology 2009;165 13-21.

[144] Chung Y, Cho J, Chang YS, Cho SH, Kang CY. Preventive and Therapeutic Effects of Oral Tolerance in a Murine Model of Asthma. Immunobiology 2002;206 408-423.

[145] Hasegawa M, Fujimoto M, Takehara K, Sato S. Pathogenesis of Systemic Sclerosis: Altered B Cell Function is the Key Linking Systemic Autoimmunity and Tissue Fibrosis. Journal of Dermatological Science 2005;39 1-7.

[146] Ludwig MS, Dallaire MJ. Structural Composition of Lung Parenchymal Strip and Mechanical Behavior during Sinusoidal Oscillation. Journal of Applied Physiology 1994;77 2029-2035.

[147] Wynn TA. Cellular and Molecular Mechanisms of Fibrosis. Journal of Pathology 2008;214 199-210.

[148] Mehrotra AK, Henderson WR Jr. The Role of Leukotrienes in Airway Remodeling. Currents in Molecular Medicine 2009:9(3) 383-391.

[149] Bergeron C, Al-Ramli W, Hamid Q. Remodeling in Asthma. Proceedings of American Thoracic Society 2009;6(3) 301-305.

[150] Swindle EJ, Collins JE, Davies DE. Breakdown in Epithelial Barrier Function in Patients with Asthma: Identification of Novel Therapeutic Approaches. Journal of Allergy Clinical Immunology 2009;124 23-34.

[151] Hackett TL. Epithelial-Mesenchymal Transition in the Pathophysiology of Airway Remodelling in Asthma. Currents Opinion Allergy Clinical Immunology 2012;12(1) 53-59.

[152] Nawijn MC, Hackett TL, Postma DS, et al. E-Cadherin: Gatekeeper of Airway Mucosa and Allergic Sensitization. Trends Immunology 2011;32 248-255.

[153] Holgate ST, Roberts G, Arshad HS, Howarth PH, Davies DE. The Role of the Airway Epithelium and its Interaction with Environmental Factors in Asthma Pathogenesis. *Proceedings of the American Thoracic Society* 2009;6(8) 655-659.

[154] Gerthoffer WT. Migration of Airway Smooth Muscle Cells. Proceedings of the American Thoracic Society 2008;5(1) 97-105.

[155] Zhang HY, Phan SH. Inhibition of Myofibroblast Apoptosis by Transforming Growth Factor Beta-1. American Journal of Respiratory and Cellular and Molecular Biology 1999;21 658-665.

[156] Benayoun L, Druilhe A, Dombret MC, Aubier M, and Pretolani M. Airway Structural Alterations Selectively Associated with Severe Asthma. American Journal of Respiratory and Critical Care Medicine 2003;167(10) 1360-1368.

[157] Woodruff PG, Dolganov GM, Ferrando RE et al. Hyperplasia of Smooth Muscle in Mild to Moderate Asthma without Changes in Cell Size or Gene Expression. American Journal of Respiratory and Critical Care Medicine 2004;169(9) 1001-1006.

[158] Trian T, Benard G, Begueret H et al. Bronchial Smooth Muscle Remodeling Involves Calcium-Dependent Enhanced Mitochondrial Biogenesis in Asthma. Journal of Experimental Medicine, 2007;204(13) 3173-3181.

[159] Jeffery P. Structural Alterations and Inflammation of Bronchi in Asthma. Int Journal Clin Pract Suppl 1998;96 5-14.

[160] Richter A, Puddicombe SM, Lordan JL, Bucchieri F, Wilson SJ, Djukanovic R. The Contribution of Interleukin IL-4 and IL-13 to the Epithelial-Mesenchymal Trophic Unit in Asthma. Am J Respir Cell Mol Biol 2001;25 385-391.

[161] James A. Remodelling of Airway Smooth Muscle in Asthma: What Sort do You Have? Clinical Experimental Allergy 2005;35(6) 703-707. Review.

[162] Laitinen LA, Laitinen A, Altraja A, et al. Bronchial Biopsy Findings in Intermittent or "Early" Asthma. Journal of Allergy and Clinical Immunology 1996;98(5 part 2) S33-S40.

[163] Roberts CR. Remodelling of the Extracellular Matrix in Asthma: Proteoglycan Synthesis and Degradation. Canadian Respiratory Journal 1998;5(1) 48-50.

[164] Bousquet J, Chanez P, Lacoste JY et al. Asthma: A Disease Remodeling the Airways. Allergy 1992;47(1) 3-11.

[165] Parameswaran K, Radford K, Zuo J, Janssen LJ, O' Byrne PM, Cox PG. Extracellular Matrix Regulates Human Airway Smooth Muscle Cell Migration. European Respiratory Journal 2004;24(4) 545-551.

[166] Fedorov IA, Wilson SJ, Davies DE, Holgate ST. Epithelial Stress and Structural Remodelling in Childhood Asthma. Thorax 2005;60 389-394.

[167] Girodet PO, Ozier A, Bara I, Tunon De Lara JM, Marthan R, Berger P. Airway Remodeling in Asthma: New Mechanisms and Potential for Pharmacological Intervention. Pharmacology Therapeutics 2011;130(3) 325-337

Adcock IM, Peters M, Gelder C, Shirasaki H, Brown CR, Barnes PJ. Increased Tachykinin Receptor Gene Expression in Asthmatic Lung and its Modulation by Steroids. *Journal of Molecular Endocrinology* 1993;11(1) 1-7.

Chapter 14

Regulation of Differentiated Phenotypes of Vascular Smooth Muscle Cells

Sho Shinohara, Satoko Shinohara, Takanori Kihara and Jun Miyake

Additional information is available at the end of the chapter

http://dx.doi.org/10.5772/48573

1. Introduction

Smooth muscle cells (SMCs) are found in many organs, including the blood vessels, trachea, stomach, small intestine, and uterus. SMC-like cells are found in some other organs, for example, hepatic stellate cells in the liver and mesangial cells in the kidney. These SMCs and SMC-like cells play an important role in the formation and function of the cardiovascular, digestive, respiratory, and urinary systems. Vascular SMCs, which generally exist in the tunica media, constitute a large portion of cells in blood vessels. A main function of vascular SMCs involves maintaining vessel structure by involving vessel contractile and relaxation activities to control blood pressure.

Vascular SMCs of each region are developed from different origins [1]. Vascular SMCs of large arteries near the heart originate from the neural crest cells of ectodermal origin, whereas other vascular SMCs are believed to differentiate from mesodermally derived mesenchymal cells. Among the mesodermally derived vascular SMCs, coronary SMCs are reported to come from the proepicardial organ [2]; and vascular SMCs of the root of the pulmonary artery and the lung artery stem from the second heart field [3]. Undifferentiated cells differentiate into progenitor cells or immature cells and ultimately differentiate into vascular SMCs with contractile ability.

Vascular SMCs show different phenotypes according to external conditions, such as developmental stage, angiogenesis state, and disease. Vascular SMCs existing within the tunica media are normally called contractile SMCs. On the other hand, vascular SMCs that are found in disease, the fetal period, and angiogenesis are called proliferative SMCs (Fig. 1). Proliferative SMCs have less contractive ability than contractile SMCs because of the lack of sufficient myofibrils inside the cells. Proliferative SMCs have the ability to proliferate and migrate, and they actively synthesize proteins and secrete extracellular matrices (ECMs) like collagen and elastin.

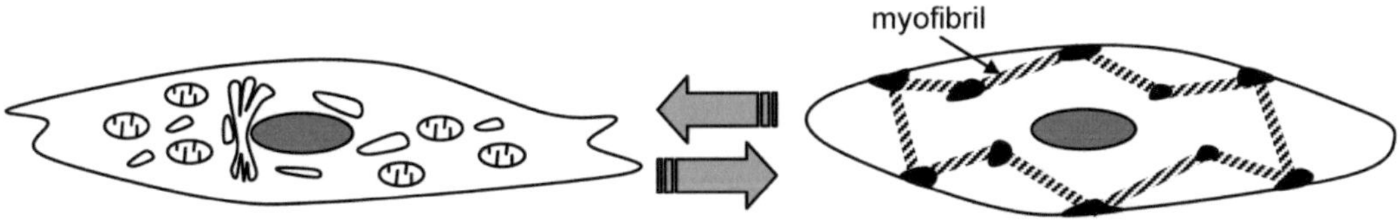

Figure 1. Phenotypes of vascular SMCs. There are 2 phenotypes of vascular SMCs, immature proliferative SMC and differentiated contractile SMC. Vascular SMCs transform their phenotypes in response to the surrounding environment. Proliferative immature SMCs have abilities to proliferate, migrate, and synthesize proteins well. On the other hand, contractile fully differentiated SMCs adhere each other and have contractile ability.

Because contractile SMCs change their phenotype into proliferative SMCs in response to the surrounding environment or growth factors and proliferative SMCs turn into contractile SMCs according to the surrounding environment, vascular SMCs are considered a unique cell type [4]. Proliferative SMCs and immature SMCs seen during the developmental period are considered identical. Therefore, the transformation from contractile SMCs to proliferative SMCs is considered the dedifferentiation process, whereas the transformation from proliferative SMCs to contractile SMCs is considered the differentiation process. These phenotype alterations of vascular SMCs are important for the regulation of angiogenesis, blood vessel remodeling, and homeostasis. In this chapter, we review the observation of regulatory mechanisms of the differentiated phenotypes of vascular SMCs.

2. Regulation of the vascular SMC phenotype *in vitro* by ECM

When contractile SMCs are collected from a body and cultured *in vitro,* they immediately transform into proliferative SMCs and then begin to proliferate under *in vitro* conditions. These transformed cells show the same characteristics as the proliferative SMCs *in vivo,* such as the inability to contract and secrete ECMs [5]. This transformation process decreases the expression of various actin-associated molecules that are seen in contractile SMCs and simultaneously increases the expression of proliferation-related proteins. On the other hand, it is difficult to retransform undifferentiated proliferative SMCs into contractile SMCs *in vitro.*

Many researchers have attempted to achieve the retransformation of proliferative SMCs into contractile SMCs. Koyama et al. reported that proliferation of the undifferentiated proliferative SMCs can be inhibited by culturing on type I collagen gel [6]. Pauly et al. reported that culturing proliferative SMCs on Matrigel extracted from basal lamina-like matrix, a product from mouse Engelbreth-Holm-Swarm tumor, enables the inhibition of proliferation and induction of differentiation [7]. These studies suggest that regulation of the vascular SMC retransformation has the potential to be achieved through control of their ECM conditions.

Hayashi et al. reported that during a primary culture of chick gizzard SMCs or rat aortic vascular SMCs, the *in vivo* contractile state can be maintained by seeding cells on laminin-coated dishes and adding insulin or insulin-like growth factor-1 (IGF-1) to the serum-free medium [8, 9]. However, once proliferative SMCs are transformed *in vitro* and then induced by the addition of a serum, platelet-derived growth factor-BB (PDGF-BB), or lysophosphatidic acid (LPA) to the maintaining medium, they do not redifferentiate into the contractile state despite being cultured under the previously mentioned condition [8, 10]. These studies suggest that SMC dedifferentiation is regulated by the extracellular environment and that extracellular signaling is an important factor in this differentiation and dedifferentiation process [8, 11].

Hirose et al. successfully induced redifferentiation of normal human aorta proliferative SMCs that were once dedifferentiated *in vitro* into a contractile state by culturing them on type IV collagen gel [12]. According to this report, SMCs take an elongated spindle-like structure and constructed network when cultured on type IV collagen gel (Fig. 2). At the same time, the expression levels of molecular markers of contractile SMCs, smooth muscle myosin heavy chain (SM-MHC) and smooth muscle α-actin (SM-α-actin), were increased, whereas comparable levels in proliferative SMCs were negligible or undetectable. Furthermore, elongated SMCs on type IV collagen gel could contract in response to stimulation by endothelin-I, a vessel contracting factor. Most important is that these phenomena were also observed under serum-added conditions. Primary SMC-like rat hepatic stellate and human kidney mesangial cells also showed elongated and network structures on type IV collagen gel [13]. These studies showed that it is possible to induce redifferentiation of proliferative SMCs into contractile SMCs *in vitro* and that the redifferentiation can be regulated by extracellular environments, especially by type IV collagen gel.

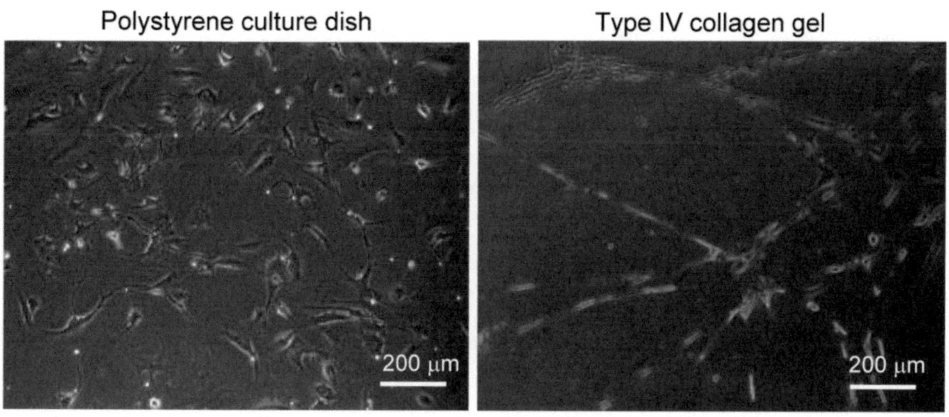

Figure 2. Morphology of human aortic vascular SMCs on different substrates. Proliferative normal human vascular SMCs cultured on polystyrene culture dish or type IV collagen gel. The cells spread flatly on culture dish. On the other hand, once proliferated SMCs on type IV collagen gel elongate and form mesh-like multicellular network by formation of cell-to-cell junction. This morphology is a characteristic of contractile phenotype of SMCs.

SMCs produce and deposit basal lamina components in their extracellular surroundings *in vivo*. They are covered by basal lamina and adhere to each other via the surrounding basal lamina. Major components of the basal lamina include type IV collagen, laminin, and proteoglycans like perlecan and nidogen. Type IV collagen is expected to work as a skeletal protein that consists of a micro meshwork at the basal lamina [14]. Therefore, the above-mentioned studies obviously indicate that the components of the basal lamina, especially type IV collagen, play an important role in maintaining the contractile state of SMCs *in vivo*.

Hirose et al. reported that when human proliferative SMCs were cultured on dishes coated with nongel type IV collagen, the cells retained their proliferative phenotype [12]. Hayashi et al. examined the detailed behavior of human proliferative SMCs on type IV collagen aggregates with a continuous change in the physicochemical properties [13]. They made a unique cell culture substrate, a hat-like-shaped gel on a cover glass using a type IV collagen solution. The central region of the hat-like-shaped gel has a domed gel structure surrounded by a broad brim-like region that consisted of a nongel form of type IV collagen aggregates. The proliferative SMCs in the domed gel region retained their initial round cell shape at the initial stage of culture (6 h) and eventually formed a multicellular meshwork at a later stage (24 h), as is seen with redifferentiated SMCs. However, the cells at the brim region started to adhere, spread, and proliferate soon after seeding. These results suggest that the physicochemical state of type IV collagen determines the vascular SMC phenotypes and that the gel form of type IV collagen, in particular, is essential to the induction of the redifferentiation of proliferative SMCs. Reports of inhibited proliferation of SMCs on type I collagen gel [6] and Matrigel [7] described earlier also indirectly imply the importance of the gel's physicochemical properties.

What factor of the gel form of type IV collagen supports redifferentiation of proliferative SMCs? As described previously, the proliferative SMCs cultured on dishes coated with type IV collagen aggregates remained in the proliferative state [12, 13]. It is assumed that the mechanical property of gel exercises an effect on SMC state. Some mechanical receptors that actually sense various mechanical stresses, such as shear stress, are found [15, 16]. Cells may also have made an essential morphological change as a result of transition to the physically steady state. It was revealed that the mechanical properties of ECMs have significant effects on cell proliferation or differentiation [17]. By changing the stiffness of a culture substrate, for example, the differentiation of mesenchymal stem cells into many kinds of cells can be controlled [18]. This finding implies that the ECM is not merely a functional molecule but works as an important factor for cell phenotype as a physical substrate. Thus, regulation of the dedifferentiation and redifferentiation of the vascular SMC phenotype by ECM is assumed to be a result of the ECM's physicochemical properties.

3. Regulation of gene expression of vascular SMC

Studies to clarify the regulatory mechanism of vascular SMC gene expressions have been performed by many researchers. Contractile SMCs express unique marker proteins, such as

SM α-actin, SM-MHC, SM22α (also kown as transgelin), high-molecular weight caldesmon (*h*-caldesmon), and calponin [19]. On the other hand, increased protein expressions, such as low-molecular weight caldesmon (*l*-caldesmon), c-fos, Egr-1, epiregulin, and SMemb MHC, are seen in proliferative SMCs [19, 20].

A promoter analysis of these proteins has revealed the associated transcription factors and their binding sites that regulate the protein expressions unique to contractile SMCs. The CArG box (CC(A/T)$_6$GG), one of these sites [19, 21], exists in the promoter region of proteins like SM22α, SM-MHC, SM α-actin, calponin, and caldesmon. It has been clarified that the expressions of contractile SMC-specific proteins are induced when the serum response factor (SRF), a ubiquitously expressed transcription factor, binds to the CArG box [19, 21]. Other than that, the E-box, a GATA-binding site, and an A/T-rich element are reported to regulate the gene expressions specific to contractile SMCs [22-24].

SRF was thought to be the main regulator of the SMC differentiation and dedifferentiation process because the CArG boxes exist in the promoter regions of most proteins expressed in contractile SMCs. However, the CArG boxes are found in the promoter region of proteins like c-fos or Egr-1, which are actively expressed by proliferative SMCs, and these proteins were also found to be regulated by the CArG box and SRF [25, 26]. These bipolar regulations of CArG box and SRF for the vascular SMC phenotypes have been given further explanations by the participation of transcriptional cofactors for SRF. In other words, SRF cofactors activate the gene expression specific to contractile SMCs either positively or negatively [27, 28].

The myocardin-related transcription factor (MRTF) family is attracting attention as the most sensible candidate for SRF cofactors that regulate vascular SMC differentiation the most [29]. The MRTF family consists of 3 SRF coactivators: myocardin, MKL1 (also called MAL, BSAC, or MRTF-A), and MKL2 (also called MAL16 or MRTF-B) [29-33]. Cysteine-rich proteins, CRP1 (also called CSRP1) and CRP2 (also called CSRP2 or SmLIM), were also reported to be SRF cofactors that promote contractile SMC-specific gene expression [34]. CRP1 and CRP2 associate with SRF and GATA proteins, forming SRF-GATA-CRP1/2 complexes that strongly activate SMC-specific gene targets [34]. Moreover, it is reported that SRF-Nkx3.2-GATA6 complex increases the SMC gene expression in chick gizzard SMCs [23]. As just described, several SRF cofactors have been reported to strongly activate the SMC-specific gene expression. These cofactors are assumed to play some roles in vascular SMC development and differentiation.

In addition, some cofactors, such as Elk-1, were reported to activate the expressions of specific proliferative SMC genes [35]. Elk-1 is a downstream protein of extracellular signal-related kinase (ERK) of mitogen-activated protein kinase (MAPK), whereas ERK directly activates it via phosphorylation [36, 37]. As Elk-1 binds to the Ets site on the genome as soon as it associates with SRF binding to the CArG box, it regulates the gene expression through promoters that have Ets site near the CArG box. Elk-1-induced gene expression activates several proteins, including c-fos [25]. Factors, such as serum or LPA, that transform contractile SMCs into proliferative SMCs are thought to activate Elk-1 through MAPK and promote the gene expressions of proliferation-associated proteins, such as c-fos.

Thus, it is widely accepted that the gene expressions involved in the SMC phenotype regulation are controlled by many cofactors through the transcriptional factor SRF, but it remains unclear how each factor functions *in vivo*.

4. Gene regulation by MRTF family

The MRTF family interacts with SRF and potently enhances the expression of SRF-dependent SMC genes. Myocardin is specifically expressed in the cardiac and circulation organs, whereas MKL1 and MKL2 expressions are widely distributed over various organs [29]. Myocardin-deficient mice died in the embryo stage, and vascular SMC differentiation was not observed [38]. *MKL1* null mice were born normal and bore children but exhibited failure to nurse their offspring because the mammary myoepithelial cells were undifferentiated [39, 40]. *MKL2* null mice had cardiovascular system defects, and the coronary SMCs that originated from the neural crest were undifferentiated [41, 42]. These results suggest that the MRTF family is widely involved in regulation of the SMC phenotype. Among the members of the MRTF family, MKL1 and MKL2, but not myocardin, are directly activated via the Rho-actin pathway [43, 44]. Myocardin and MKL1 strongly activate CArG box-dependent SMC gene transcription [29], whereas MKL2 is less effective in activating the SMC gene.

The MRTF family has many conserved domains (Fig. 3). The MRTF family binds to the MADS domain of SRF by the basic rich 1 (B1) domain, and the glutamine-rich (Q) domain supports this binding [45]. A powerful transcription activation domain (TAD) exists on the c-terminus region and functions with heterologous promoters [46]. Although the MRTF family and SRF bind singularly, the MRTF family forms a homo/heterodimer via the conserved leucine zipper (LZ) domain [43, 47] and preferentially binds SRF as a dimer, which then forms a dimer on the CArG box [48].

Myocardin is reported to regulate histone acetylation by binding p300 histone acetyltransferase and deacetylation by binding to class II histone deacetyltransferase [49]. The p300 histone acetyltransferase and the class II histone deacetyltransferase interact with the TAD and Q domains of myocardin, respectively. The N-terminus region of MKL1 directly binds to SPT16 and SSRP1, which are components of the facilitating chromatin transcription (FACT) complex [50]. The FACT complex functions as a histone chaperone and allows RNA polymerase II to traverse the nucleosomes by removing a H2A/H2B dimer [51]. Altering the repressive nature of the chromatin is necessary for the cell to implement all of the nuclear activities of the chromatin. Therefore, expression of the nucleosomal SMC-related gene is assumed to be activated by the MRTF family (Fig. 4). In this manner, the MRTF family positively and negatively regulates the nucleosomal dynamics of the SMC-specific gene.

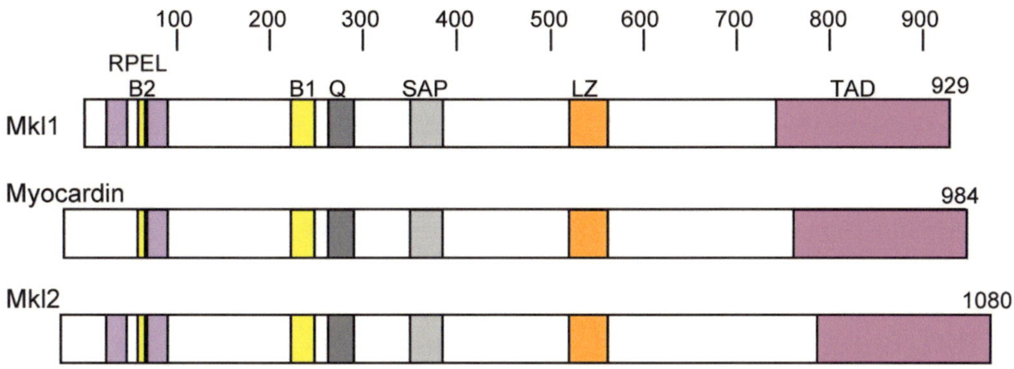

Figure 3. Structure of MRTF family. RPEL, RPEL motif; B1 and B2, basic region; Q, glutamine-rich domain; LZ, leucine zipper domain; TAD, transcription activation domain. The numbers on the right side indicate the number of amino acids in each protein.

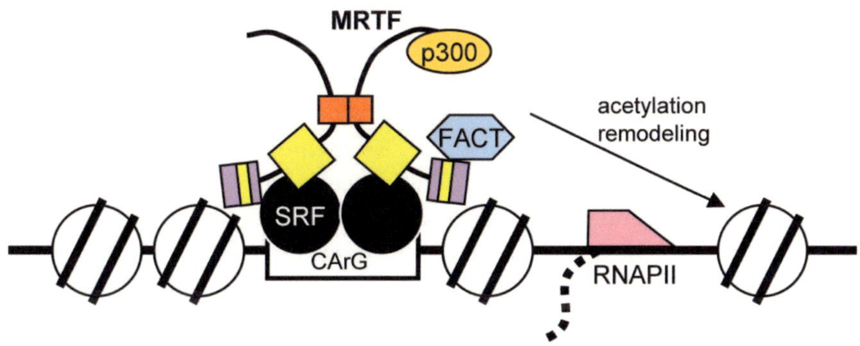

Figure 4. The model for nucleosomal gene activation by MRTF family. DNA is shown schematically as solid lines. The nucleosomal characteristic of the chromosomal site is indicated by closed circles. RNA is represented as a dotted line. MRTF associates with SRF and activates transcription of the nucleosomal genes via recruiting the FACT complex into the coding region. The FACT complex remodels the chromatin structure and facilitates the progression of RNA polymerase II (RNAPII). Furthermore, MRTF interact with p300 and loosen the nucleosomal structure by acetylating the histone.

The SMC gene activation function of the MRTF family can be regulated by other proteins. Elk-1, one of the TCF families, competitively blocks the binding of MRTF to the MADS domain of SRF [27, 45]. By SMC stimulation of PDGF-BB or serum, the C-terminus of Elk-1 gets phosphorylated by ERK, and phosphorylated Elk-1 then moves into the nucleus. In the nucleus, Elk-1 competitively inhibits the binding of myocardin and SRF by binding to the MADS domain of SRF; as a result, it inhibits the myocardin-activated gene expression [27]. It is assumed that PDGF-BB stimulation simultaneously recruits histone deacetyl transferase (HDAC) to the CArG box of the SMC-specific region, the acetyl group in histone H4 gets deacetylated by HDAC, and the promoters finally reach a stable "silencing state" [28].

Phosphoinositide-3-kinase (PI3K) and AKT signaling from insulin/IGF-1 is essential for maintaining primary culture of the chick gizzard contractile SMC phenotype [9]. Inhibiting the PI3K-AKT signal induces dedifferentiation of the contractile SMCs into the proliferative phenotype [9]. Especially in once-dedifferentiated proliferative SMCs, insulin receptor substrate 1 (IRS-1) gets phosphorylated by insulin/IGF-1 signaling, although IRS-1 phosphorylates Grb-2/SOS but not SHP-2 [11]. The different downstream molecules are then activated between the contractile and proliferative SMC states. Signaling from insulin/IGF-1 through PI3K-AKT promotes nuclear exports of Foxo4, which binds to myocardin in the nucleus and inhibits myocardin-activated transcription [52]. Therefore, PI3K-AKT signaling from insulin/IGF-1 enables myocardin function as an SRF cofactor in the nucleus to maintain the contractile state of SMCs.

5. CRP2 contributes to SMC differentiation

CRP family proteins consist of 2 LIM domains and 2 glycine-rich regions (Fig. 5). The LIM domain is a double zinc finger-like structure that mediates protein-protein interactions. The CRP family proteins CRP1, CRP2, and CRP3/MLP share high sequence homology [53]; however, their gene expression patterns differ. CRP1 is expressed in organs such as the arteries, stomach, and intestines, all of which contain abundant SMCs [54]. CRP2 is mainly expressed in vascular SMCs and is also found in the cardiac muscle in the developmental period [55]. CRP3 expression was confirmed in the striated heart and skeletal muscles [56]. As evidenced by their expression patterns, the CRP members are reported to be related to muscle cell differentiation [34, 55, 57]. CRP2, in particular, plays a role in the vascular SMC differentiation and dedifferentiation process. CRP2 expression is known to decrease when vascular SMCs dedifferentiate and proliferate in response to injury [58]. On the other hand, CRP2-deficient mice develop normally, and the expressions of the SMC-related proteins SM α-actin, SM22α, and calponin neither increase nor decrease [59]. In the CRP2-deficient mice, however, the effect of intimal regeneration or hypertrophy increases, which occurs when blood vessels gets injury. When vascular SMCs from wild-type and CRP2-deficient mice were stimulated by PDGF-BB *in vitro*, there were no differences in proliferation, but the migration ability was reported to be increased in CRP2-deficient mice.

Figure 5. Structure of CRP. Gly, glycine rich region. CRP consists of two LIM domains and two glycine rich regions. The LIM domain is a double zinc-finger like structure.

CRP2 localizes in the cell nucleus and cytoplasm, where it associates with the actin cytoskeleton [34, 60]. In chick embryo proepicardial cells, which are progenitor cells of SMCs in the coronary artery, endogeneous CRP2 localizes to the nucleus, whereas CRP2 translocates to the cytoskeleton as these cells fully differentiate into SMCs [34]. It is believed that CRP2 plays different roles in these different locations. In the nucleus, CRP2 associates with GATA proteins and SRF (CRP2-GATA-SRF) and acts as a transcriptional regulator of SRF-dependent SMC genes [34, 61]. On the other hand, in the cytoplasm, CRP2 directly associates with actin filaments, α-actinin, and zyxin *in vitro* [60, 62]. CRP1 also directly associates with actin filaments *in vitro* and *in vivo* and stabilizes actin filament formation *in vitro* [63, 64]. The distinct role of CRP2 in the cytoplasm is unclear, but CRP2 may be involved in the assembly and maintenance of the actin cytoskeleton in vascular SMCs.

We recently focused on the dynamics of CRP2 localization with respect to actin stress fiber formation during vascular SMC differentiation [65]. The vascular SMC differentiation process is a characteristic of the epithelial-to-mesenchymal transformation (EMT) [66]. The CRP2 localization dynamics during SMC differentiation is regulated by actin stress fiber formation accompanied by the EMT. In particular, nuclear CRP2 distribution is determined by the actin polymerization state [65]. These CRP2 localization dynamics can be interpreted from a simple *in silico* CRP2 localization kinetic model regulated by actin dynamics [65]. Reorganization of the actin cytoskeleton is able to affect vascular SMC differentiation progress through SRF activation and CRP2 translocation. The effects of cytoplasmic CRP2 for F-actin become more important for vascular SMC differentiation. We now speculate that actin-bound CRP2 plays direct and indirect roles in the stabilization of SMC differentiation.

6. Perspective

Phenotype alterations and differentiation of vascular SMC are important for angiogenesis, blood vessel remodeling, and homeostasis. These processes are regulated by extracellular signals. In particular, maintenance of the contractile SMC phenotype is highly supported by the basal lamina physicochemical properties, which are probably sensed by the actin cytoskeleton. On the other hand, vascular SMC differentiation and SMC-related gene expression are highly regulated by actin dynamics. Nuclear accumulation of MKL1 and MKL2 is controlled by the amounts of G-actin pool, and stimulation of F-actin formation activates contractile SMC-related gene expression by interacting with SRF and nuclear importing MKL1 and MKL2. However, the gene activation function of myocardin, the most important transcription factor of vascular SMC differentiation, is blocked competitively by Elk-1, which is activated by the extracellular signaling of serum and PDGF-BB. CRP2 localization is regulated by actin stress fiber formation, and nuclear and cytoplasmic CRP2 play a role in SMC differentiation. Therefore, the actin cytoskeleton is a key factor for vascular SMC differentiation and maintaining the contractile SMC phenotype. However, the details of the regulatory mechanism and process of SMC differentiation, as well as maintenance of the SMC phenotype, remain unclear. Future studies will address the integrated interrelationship among factors including ECM, extracellular signaling, actin dynamics, and SRF cofactors in the process of SMC differentiation and phenotype maintenance.

Author details

Sho Shinohara
Graduate School of Frontier Biosciences, Osaka University, Japan

Satoko Shinohara, Takanori Kihara and Jun Miyake
Graduate School of Engineering Science, Osaka University, Japan

Acknowledgement

This work was supported by grants from the ministry of Education, Culture, Sports, Science, and Technology of Japan (24106507 and 24700454 to T.K.).

7. References

[1] Hirschi KK, Majesky MW (2004) Smooth muscle stem cells. Anat Rec A Discov Mol Cell Evol Biol 276: 22-33.

[2] Mikawa T, Gourdie RG (1996) Pericardial mesoderm generates a population of coronary smooth muscle cells migrating into the heart along with ingrowth of the epicardial organ. Dev Biol 174: 221-232.

[3] Waldo KL, Hutson MR, Ward CC, Zdanowicz M, Stadt HA, Kumiski D, Abu-Issa R, Kirby ML (2005) Secondary heart field contributes myocardium and smooth muscle to the arterial pole of the developing heart. Dev Biol 281: 78-90.

[4] Ross R (1993) The pathogenesis of atherosclerosis: a perspective for the 1990s. Nature 362: 801-809.

[5] Chamley-Campbell J, Campbell GR, Ross R (1979) The smooth muscle cell in culture. Physiol Rev 59: 1-61.

[6] Koyama H, Raines EW, Bornfeldt KE, Roberts JM, Ross R (1996) Fibrillar collagen inhibits arterial smooth muscle proliferation through regulation of Cdk2 inhibitors. Cell 87: 1069-1078.

[7] Pauly RR, Passaniti A, Crow M, Kinsella JL, Papadopoulos N, Monticone R, Lakatta EG, Martin GR (1992) Experimental models that mimic the differentiation and dedifferentiation of vascular cells. Circulation 86: III68-73.

[8] Hayashi K, Saga H, Chimori Y, Kimura K, Yamanaka Y, Sobue K (1998) Differentiated phenotype of smooth muscle cells depends on signaling pathways through insulin-like growth factors and phosphatidylinositol 3-kinase. J Biol Chem 273: 28860-28867.

[9] Hayashi K, Takahashi M, Kimura K, Nishida W, Saga H, Sobue K (1999) Changes in the balance of phosphoinositide 3-kinase/protein kinase B (Akt) and the mitogen-activated protein kinases (ERK/p38MAPK) determine a phenotype of visceral and vascular smooth muscle cells. J Cell Biol 145: 727-740.

[10] Hayashi K, Takahashi M, Nishida W, Yoshida K, Ohkawa Y, Kitabatake A, Aoki J, Arai H, Sobue K (2001) Phenotypic modulation of vascular smooth muscle cells induced by unsaturated lysophosphatidic acids. Circ Res 89: 251-258.

[11] Hayashi K, Shibata K, Morita T, Iwasaki K, Watanabe M, Sobue K (2004) Insulin receptor substrate-1/SHP-2 interaction, a phenotype-dependent switching machinery of insulin-like growth factor-I signaling in vascular smooth muscle cells. J Biol Chem 279: 40807-40818.

[12] Hirose M, Kosugi H, Nakazato K, Hayashi T (1999) Restoration to a quiescent and contractile phenotype from a proliferative phenotype of myofibroblast-like human aortic smooth muscle cells by culture on type IV collagen gels. J Biochem 125: 991-1000.

[13] Hayashi T, Hirose M, Yamano H, Takeda Y, Kosugi H, Kihara T, Imamura Y, Mizuno K, Nakazato K, Yoshikawa K, Kajimura D, Takahashi S, Adachi E (2003) Regulation of phenotypes of human aorta endothelial cells and smooth muscle cells in culture by type IV collagen aggregates. In: Okazaki I, Ninomiya Y, Friedman SL, Tanikawa K, editors. Extracellular Matrix and the Liver. Academic Press. pp. 39-54.
[14] Adachi E, Takeda Y, Nakazato K, Muraoka M, Iwata M, Sasaki T, Imamura Y, Hopkinson I, Hayashi T (1997) Isolated collagen IV retains the potential to form an 18-nm sided polygonal meshwork of the lamina densa. J Electron Microsc (Tokyo) 46: 233-241.
[15] Sachs F (1988) Mechanical transduction in biological systems. Crit Rev Biomed Eng 16: 141-169.
[16] Hayakawa K, Tatsumi H, Sokabe M (2008) Actin stress fibers transmit and focus force to activate mechanosensitive channels. J Cell Sci 121: 496-503.
[17] Even-Ram S, Artym V, Yamada KM (2006) Matrix control of stem cell fate. Cell 126: 645-647.
[18] Engler AJ, Sen S, Sweeney HL, Discher DE (2006) Matrix elasticity directs stem cell lineage specification. Cell 126: 677-689.
[19] Sobue K, Hayashi K, Nishida W (1999) Expressional regulation of smooth muscle cell-specific genes in association with phenotypic modulation. Mol Cell Biochem 190: 105-118.
[20] Takahashi M, Hayashi K, Yoshida K, Ohkawa Y, Komurasaki T, Kitabatake A, Ogawa A, Nishida W, Yano M, Monden M, Sobue K (2003) Epiregulin as a major autocrine/paracrine factor released from ERK- and p38MAPK-activated vascular smooth muscle cells. Circulation 108: 2524-2529.
[21] Miano JM (2003) Serum response factor: toggling between disparate programs of gene expression. J Mol Cell Cardiol 35: 577-593.
[22] Katoh Y, Molkentin JD, Dave V, Olson EN, Periasamy M (1998) MEF2B is a component of a smooth muscle-specific complex that binds an A/T-rich element important for smooth muscle myosin heavy chain gene expression. J Biol Chem 273: 1511-1518.
[23] Nishida W, Nakamura M, Mori S, Takahashi M, Ohkawa Y, Tadokoro S, Yoshida K, Hiwada K, Hayashi K, Sobue K (2002) A triad of serum response factor and the GATA and NK families governs the transcription of smooth and cardiac muscle genes. J Biol Chem 277: 7308-7317.
[24] Owens GK, Kumar MS, Wamhoff BR (2004) Molecular regulation of vascular smooth muscle cell differentiation in development and disease. Physiol Rev 84: 767-801.
[25] Treisman R (1992) The serum response element. Trends Biochem Sci 17: 423-426.
[26] Arsenian S, Weinhold B, Oelgeschlager M, Ruther U, Nordheim A (1998) Serum response factor is essential for mesoderm formation during mouse embryogenesis. Embo J 17: 6289-6299.
[27] Wang Z, Wang DZ, Hockemeyer D, McAnally J, Nordheim A, Olson EN (2004) Myocardin and ternary complex factors compete for SRF to control smooth muscle gene expression. Nature 428: 185-189.

[28] Yoshida T, Gan Q, Shang Y, Owens GK (2007) Platelet-derived growth factor-BB represses smooth muscle cell marker genes via changes in binding of MKL factors and histone deacetylases to their promoters. Am J Physiol Cell Physiol 292: C886-895.

[29] Wang DZ, Li S, Hockemeyer D, Sutherland L, Wang Z, Schratt G, Richardson JA, Nordheim A, Olson EN (2002) Potentiation of serum response factor activity by a family of myocardin-related transcription factors. Proc Natl Acad Sci U S A 99: 14855-14860.

[30] Chen J, Kitchen CM, Streb JW, Miano JM (2002) Myocardin: a component of a molecular switch for smooth muscle differentiation. J Mol Cell Cardiol 34: 1345-1356.

[31] Du KL, Ip HS, Li J, Chen M, Dandre F, Yu W, Lu MM, Owens GK, Parmacek MS (2003) Myocardin is a critical serum response factor cofactor in the transcriptional program regulating smooth muscle cell differentiation. Mol Cell Biol 23: 2425-2437.

[32] Wang Z, Wang DZ, Pipes GC, Olson EN (2003) Myocardin is a master regulator of smooth muscle gene expression. Proc Natl Acad Sci U S A 100: 7129-7134.

[33] Yoshida T, Sinha S, Dandre F, Wamhoff BR, Hoofnagle MH, Kremer BE, Wang DZ, Olson EN, Owens GK (2003) Myocardin is a key regulator of CArG-dependent transcription of multiple smooth muscle marker genes. Circ Res 92: 856-864.

[34] Chang DF, Belaguli NS, Iyer D, Roberts WB, Wu SP, Dong XR, Marx JG, Moore MS, Beckerle MC, Majesky MW, Schwartz RJ (2003) Cysteine-rich LIM-only proteins CRP1 and CRP2 are potent smooth muscle differentiation cofactors. Dev Cell 4: 107-118.

[35] Treisman R (1994) Ternary complex factors: growth factor regulated transcriptional activators. Curr Opin Genet Dev 4: 96-101.

[36] Janknecht R, Ernst WH, Pingoud V, Nordheim A (1993) Activation of ternary complex factor Elk-1 by MAP kinases. Embo J 12: 5097-5104.

[37] Marais R, Wynne J, Treisman R (1993) The SRF accessory protein Elk-1 contains a growth factor-regulated transcriptional activation domain. Cell 73: 381-393.

[38] Li S, Wang DZ, Wang Z, Richardson JA, Olson EN (2003) The serum response factor coactivator myocardin is required for vascular smooth muscle development. Proc Natl Acad Sci U S A 100: 9366-9370.

[39] Li S, Chang S, Qi X, Richardson JA, Olson EN (2006) Requirement of a myocardin-related transcription factor for development of mammary myoepithelial cells. Mol Cell Biol 26: 5797-5808.

[40] Sun Y, Boyd K, Xu W, Ma J, Jackson CW, Fu A, Shillingford JM, Robinson GW, Hennighausen L, Hitzler JK, Ma Z, Morris SW (2006) Acute myeloid leukemia-associated Mkl1 (Mrtf-a) is a key regulator of mammary gland function. Mol Cell Biol 26: 5809-5826.

[41] Li J, Zhu X, Chen M, Cheng L, Zhou D, Lu MM, Du K, Epstein JA, Parmacek MS (2005) Myocardin-related transcription factor B is required in cardiac neural crest for smooth muscle differentiation and cardiovascular development. Proc Natl Acad Sci U S A 102: 8916-8921.

[42] Oh J, Richardson JA, Olson EN (2005) Requirement of myocardin-related transcription factor-B for remodeling of branchial arch arteries and smooth muscle differentiation. Proc Natl Acad Sci U S A 102: 15122-15127.

[43] Miralles F, Posern G, Zaromytidou AI, Treisman R (2003) Actin dynamics control SRF activity by regulation of its coactivator MAL. Cell 113: 329-342.
[44] Kuwahara K, Barrientos T, Pipes GC, Li S, Olson EN (2005) Muscle-specific signaling mechanism that links actin dynamics to serum response factor. Mol Cell Biol 25: 3173-3181.
[45] Zaromytidou AI, Miralles F, Treisman R (2006) MAL and ternary complex factor use different mechanisms to contact a common surface on the serum response factor DNA-binding domain. Mol Cell Biol 26: 4134-4148.
[46] Wang D, Chang PS, Wang Z, Sutherland L, Richardson JA, Small E, Krieg PA, Olson EN (2001) Activation of cardiac gene expression by myocardin, a transcriptional cofactor for serum response factor. Cell 105: 851-862.
[47] Du KL, Chen M, Li J, Lepore JJ, Mericko P, Parmacek MS (2004) Megakaryoblastic leukemia factor-1 transduces cytoskeletal signals and induces smooth muscle cell differentiation from undifferentiated embryonic stem cells. J Biol Chem 279: 17578-17586.
[48] Pellegrini L, Tan S, Richmond TJ (1995) Structure of serum response factor core bound to DNA. Nature 376: 490-498.
[49] Cao D, Wang Z, Zhang CL, Oh J, Xing W, Li S, Richardson JA, Wang DZ, Olson EN (2005) Modulation of smooth muscle gene expression by association of histone acetyltransferases and deacetylases with myocardin. Mol Cell Biol 25: 364-376.
[50] Kihara T, Kano F, Murata M (2008) Modulation of SRF-dependent gene expression by association of SPT16 with MKL1. Exp Cell Res 314: 629-637.
[51] Reinberg D, Sims RJ, 3rd (2006) de FACTo nucleosome dynamics. J Biol Chem 281: 23297-23301.
[52] Liu ZP, Wang Z, Yanagisawa H, Olson EN (2005) Phenotypic modulation of smooth muscle cells through interaction of Foxo4 and myocardin. Dev Cell 9: 261-270.
[53] Weiskirchen R, Gunther K (2003) The CRP/MLP/TLP family of LIM domain proteins: acting by connecting. Bioessays 25: 152-162.
[54] Henderson JR, Macalma T, Brown D, Richardson JA, Olson EN, Beckerle MC (1999) The LIM protein, CRP1, is a smooth muscle marker. Dev Dyn 214: 229-238.
[55] Jain MK, Kashiki S, Hsieh CM, Layne MD, Yet SF, Sibinga NE, Chin MT, Feinberg MW, Woo I, Maas RL, Haber E, Lee ME (1998) Embryonic expression suggests an important role for CRP2/SmLIM in the developing cardiovascular system. Circ Res 83: 980-985.
[56] Arber S, Hunter JJ, Ross J, Jr., Hongo M, Sansig G, Borg J, Perriard JC, Chien KR, Caroni P (1997) MLP-deficient mice exhibit a disruption of cardiac cytoarchitectural organization, dilated cardiomyopathy, and heart failure. Cell 88: 393-403.
[57] Arber S, Halder G, Caroni P (1994) Muscle LIM protein, a novel essential regulator of myogenesis, promotes myogenic differentiation. Cell 79: 221-231.
[58] Jain MK, Fujita KP, Hsieh CM, Endege WO, Sibinga NE, Yet SF, Kashiki S, Lee WS, Perrella MA, Haber E, Lee ME (1996) Molecular cloning and characterization of SmLIM, a developmentally regulated LIM protein preferentially expressed in aortic smooth muscle cells. J Biol Chem 271: 10194-10199.

[59] Wei J, Gorman TE, Liu X, Ith B, Tseng A, Chen Z, Simon DI, Layne MD, Yet SF (2005) Increased neointima formation in cysteine-rich protein 2-deficient mice in response to vascular injury. Circ Res 97: 1323-1331.

[60] Grubinger M, Gimona M (2004) CRP2 is an autonomous actin-binding protein. FEBS Lett 557: 88-92.

[61] Chang DF, Belaguli NS, Chang J, Schwartz RJ (2007) LIM-only protein, CRP2, switched on smooth muscle gene activity in adult cardiac myocytes. Proc Natl Acad Sci U S A 104: 157-162.

[62] Louis HA, Pino JD, Schmeichel KL, Pomies P, Beckerle MC (1997) Comparison of three members of the cysteine-rich protein family reveals functional conservation and divergent patterns of gene expression. J Biol Chem 272: 27484-27491.

[63] Tran TC, Singleton C, Fraley TS, Greenwood JA (2005) Cysteine-rich protein 1 (CRP1) regulates actin filament bundling. BMC Cell Biol 6: 45.

[64] Jang HS, Greenwood JA (2009) Glycine-rich region regulates cysteine-rich protein 1 binding to actin cytoskeleton. Biochem Biophys Res Commun 380: 484-488.

[65] Kihara T, Shinohara S, Fujikawa R, Sugimoto Y, Murata M, Miyake J (2011) Regulation of cysteine-rich protein 2 localization by the development of actin fibers during smooth muscle cell differentiation. Biochem Biophys Res Commun 411: 96-101.

[66] Landerholm TE, Dong XR, Lu J, Belaguli NS, Schwartz RJ, Majesky MW (1999) A role for serum response factor in coronary smooth muscle differentiation from proepicardial cells. Development 126: 2053-2062.

Chapter 15

Structure and Function of Smooth Muscle with Special Reference to Mast Cells

Angel Vodenicharov

Additional information is available at the end of the chapter

http://dx.doi.org/10.5772/48566

1. Introduction

The structure of the smooth muscle tissue is quite different from that of other muscle tissue subtypes. The primary smooth muscle structural and functional unit – the smooth muscle myocyte (*Myocytus nonstriatus)* has unique structure, arrangement and innervation. The spindle shape of the mononucleated smooth muscle cell permits a close contact among cells in the splanchnic and vascular walls. Regardless of its location in the body, the communication between tightly packed adjacent spindle-shaped mononucleated smooth muscle cells occurs via a specific junction, referred to as *Macula communicans* (nexus, gap junction). The space between the different macular connexons is about 2 nm, which allows low-molecular compounds to pass from one cell to another. This type of junction is analogous to the plasmodesma (pl. plasmodesmata), encountered in cells of plants. The basement membrane covering each smooth muscle cell, is absent at gap junction sites.

The contractility, proper of smooth muscle cells, is influenced by the autonomic nervous system, hormones and local metabolites, which alter the contractility in a way that adapts to the new functional requirements. Smooth muscle cell contraction could be modulated via surface receptors, activating internal second messenger systems. The expression of a variety of receptors accounts for the response of smooth muscle cells in different areas to a number of hormones [1].

It is known that for the major part of smooth musculature (except for the iris, vas deferens and large blood vessels) one autonomic nerve fibre innervates a group of 15-20 smooth muscle cells, as the stimulus to one cell is transmitted via nexus contacts to the other [2] assuming a simultaneous contraction of the pack of cells. This pathway of stimulus transmission naturally raises the question whether the excitation potential decreases towards more distant cells and whether the smooth muscle contraction and relaxation is mediated by biologically active substances, released by adjacent cells. In this connection, the proof of [3] that *in vitro* equine cecal smooth muscle contractions could occur without the nervous system participation is important to understand the biology of smooth muscle cells. Later, other researchers ([4] added new important data about the involvement of an unique, in their opinion, adhesion molecule of mast cells – the cell adhesion molecule-1 (CADM-1), which mediates the functional communication between mast cells and nerves, as well as between smooth muscle and mast cells. They outlined that in bronchial asthma, mast cells infiltrated the smooth muscles of airways and interacted directly with smooth muscle cells, presuming a role of mastocytes in the pathogenesis of airway obstruction.

The interaction between mast cells and vascular smooth muscle cells that are largely involved in vascular wall motility is of special interest for the cardiovascular system function. The presence of mast cells in blood vessels' wall and especially in the tunica media (the muscle layer), supposes a participation in smooth muscle cell activity modulation via release of vasoactive mediators. A similar effect could be hypothesized when mast cells are located close to the blood vessels from the microcirculatory bed and to arterioles in particular, as it is recognized that arterial blood pressure is largely influenced by the smooth muscle tone in the wall of arterioles.

Besides the classic and more recent data about biologically active substances released by mast cells, and implicated in smooth muscle cell activity, this chapter presents concisely some original information about the localization of mast cells in muscle layers of blood vessels and visceral organs. On the basis of this information, the role of known mast cell mediators modulating smooth muscle cells' activity in studied organs and tissues is discussed.

2. Brief characteristics of the mast cell and its role for smooth muscle

The mast cell (*Mastocytus*) or tissue basophilic granulocyte (*Granulocytus basophilus textus*) was observed for the first time about 150 years ago and since then, is described as a connective tissue cell. Mastocytes are usually present in loose connective tissue of organs, communicating with the environment – digestive and respiratory organs, skin.

It is acknowledged that the structure and the function of mast cells are similar to those of blood basophils, but the two cell types have different precursors [5]. The main biological function of mast cells is the release of inflammatory mediators and cytokines [6]. They are outlined as cells with paracrine secretion (messenger-producing cells), whose products disseminate in the extracellular fluid and act on adjacent target cells [7]. Furthermore, mast cells are rich in histamine, heparin and proteases. *Histamine* is a vasoconstrictor increasing the permeability of small venules and stimulator of small-airway smooth muscle contraction. *Heparin* acts as an anticoagulant and is thought to stimulate angiogenesis [8]. Heparin proteoglycan, released by activated mast cells, inhibits the proliferation of smooth muscle cells in arterial tunica media and uterine myometrium [9, 10].

The accumulated evidence on biologically active substances, synthesized, stored and released by mast cells over the last 2-3 decades not only added to the information about their heterogeneity and biology, but also gives reason to re-evaluate their participation in important processes as the homeostasis, immune response, allergy, neurotransmission, vasomotor activity and motility of smooth muscle tissue. With regard to the blood flow to organs and blood circulation, the research on vasoactive substance released by mast cells and playing a key role in the vascular motorics, is particularly important. This role is primarily related to blood flow regulation in various functional states – both physiological and pathological. The tentative role of mast cells in the modulation of splanchnic wall smooth muscle tissue, accomplished by specific mediators, is comparable.

Despite the abundant literature data related to the morphology, localization, histochemical behaviour, species-related features and involvement in systemic homeostasis, immune response, allergy, anaphylaxis etc. of tissue basophil granulocytes (mastocytes), the information about their presence in smooth muscle tissue is still scarce.

Among the nearly 70 biologically active substances found in mast cells so far, it could be affirmed that mast cell-derived ligands, which act as mediators of smooth muscle tissue motility, belongs to four groups, namely:

Biogenic amines (incl. catecholamines): histamine, dopamine, serotonin
Polypeptides (including neuropeptides): vasoactive intestinal polypeptide (VIP), endothelin
Proteoglycans: heparin
Free radicals: nitric oxide

The content and expression of mentioned biologically active substances outline mastocytes as cells, involved in the functioning of the smooth muscles of internal organs and blood vessels.

In our studies, histochemical and immunohistochemical methods were used on paraffin sections, cryostat, semi-thin and ultrathin sections.

3. Considering selected substances

3.1. Biogenic amines

There are literature reports about mast cells containing biogenic amines, established by the method of [11] via alcian blue/safranin staining with relatively low pH of the staining solution (1.42). Such cells were identified in bovine trachea [12, 13], as well as in several porcine organs [14]. In porcine renal blood vessels, renal pelvis and ureter, the specific staining for biogenic amines and glycosaminoglycans allowed to determine the localization and counts of alcian-positive mast cells in the tunica media, on its boundary with the adventitia and less frequently on its boundary with the intima Fig. 1. It should be noted that alcian-positive mast cells in renal pelvis and ureter were preferentially located in the muscle layer compared to the other layers of the wall, with statistically significant differences [15, 16].

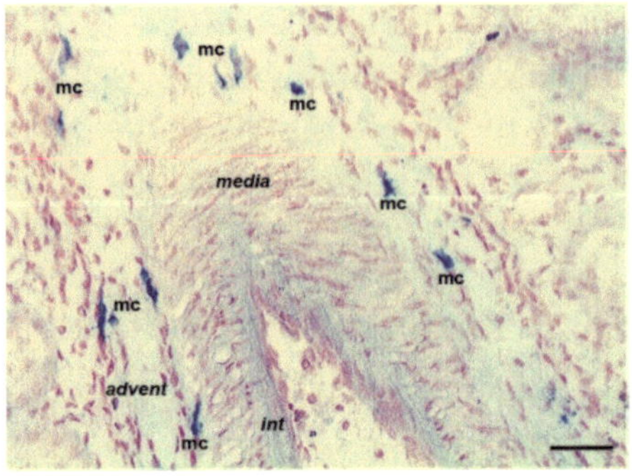

Figure 1. Longitudinal section from a part of the wall of the porcine kidney interlobar artery. **int** – intima; **media** – middle shell; **advent** - adventitia. The A+ mast cells (**mc**) are located mainly between the media and adventitia. Alcian blue – safranin. Bar = 50 μm.

The observed findings were confirmed by electron microscopy as well Fig. 2. No direct contact, neither specialized contact differentiations between mast cell and smooth muscle cell plasmalemmas, were noted in any case.

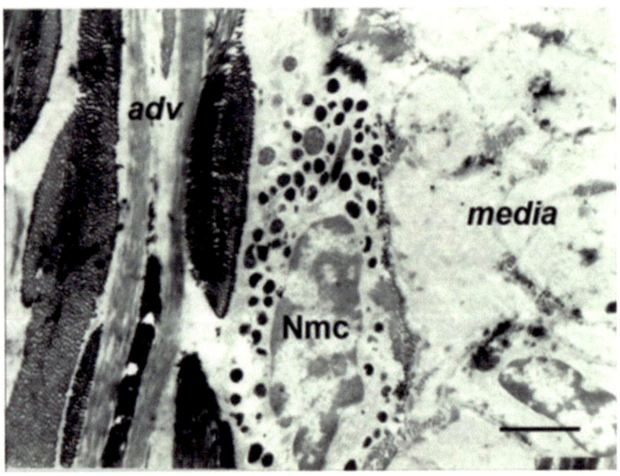

Figure 2. Mast cell (**Nmc**), located between middle shell (**media**) and adventitia (**adv**) of porcine renal interlobar artery. TEM picture. Bar = 5μm.

3.1.1. Histamine

Histamine is a biogenic (vasoactive) amine, mediator of inflammation, gastric hydrochloric acid secretion and smooth muscle contraction [17]. Histamine is detected in mast cells, nerve and neuroendocrine cells, lungs, kidneys, cerebrovascular endothelial cells, peripheral nervous system [18]. It elicits a contractile response in smooth muscles and lowers blood pressure [19].

Histamine is released by mast cells in response to allergic reactions or tissue damage. The close vicinity of mast cells to blood vessels, together with the strong vascular effect of histamine suggests that it could influence blood flow, including that of the brain [20]. Histamine acts upon visceral (smooth) musculature by contracting it [8].

Histamine has a marked cardiovascular effect as well. It provokes dilation of terminal arterioles and the other vessels from the microcirculatory vascular bed, increases the permeability of capillaries during oedema formation and causes contraction of smooth muscle cells of large arteries and veins. The relative predominance of these effects is species-dependent. For instance, the histamine-induced contraction of arterioles is strong in rodents, less pronounced in cats, while they are dilated in dogs, non-human primates, and humans [19]. It was initially thought that vascular effects of histamine were mediated only by H_1 receptors. Later data confirmed the presence of H_2 receptors. Histamine-induced changes in the permeability of small vessels were obviously mediated by H_1 receptors, whereas the role of H_2 receptors in small arteriole contractions is still uncertain. The exact ratio of H_1- and H_2- receptor involvements in vascular responses to histamine in the different animal species is variable [21].

In general, the effect of histamine on a specific regional vasculature could be best described as a result of its multiple effects on smooth muscle and the lining endothelium. H_1 and H_2 receptors on vascular smooth muscle mediate direct constriction and relaxation, respectively, while endothelial H_1 receptors promote vasorelaxation via release of endothelial-derived relaxing factor (EDRF: i.e. nitric oxide) and/or prostacyclin. It is supposed that the transmembrane signalling mechanisms are involved in the different effects of histamine on vascular smooth muscles [22, 23].

The role of histamine as a chemical mediator of renal autoregulation in some animal species is long acknowledged. H_1 receptors mediate the autoregulation of both renal blood flow and glomerular filtration rate in dogs, whereas in rabbits, both effects are present – H_1-mediated contraction and H_2-mediated relaxation of the renal artery [24-26].

Immunohistochemically, histamine-positive mast cells were found in all layers of renal blood vessels in pigs [16]. They were most numerous in the tunica media, as well as at the boundary between media and adventitia (Fig. 3). In the tunica media of arcuate arteries, histamine-positive mastocytes were relatively few, while in arcuate veins were observed only as single findings.

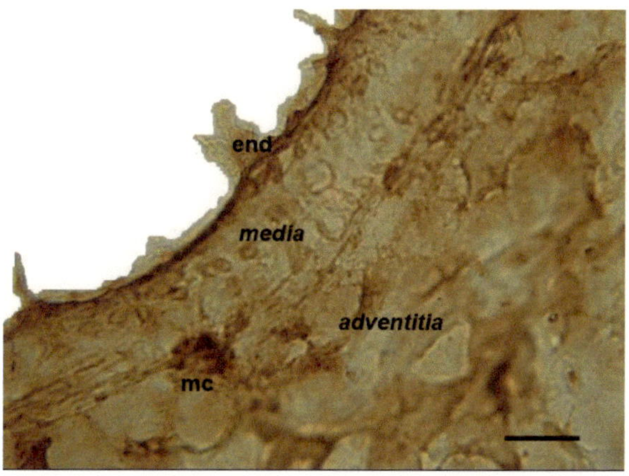

Figure 3. Histamine positive mast cell (**mc**) between media and adventitia of porcine interlobar artery. Bar = 30 μm

Histamine-positive mast cells were also detected in the middle layer of the renal pelvis and the ureter of the pig [15, 16], (Fig. 4).

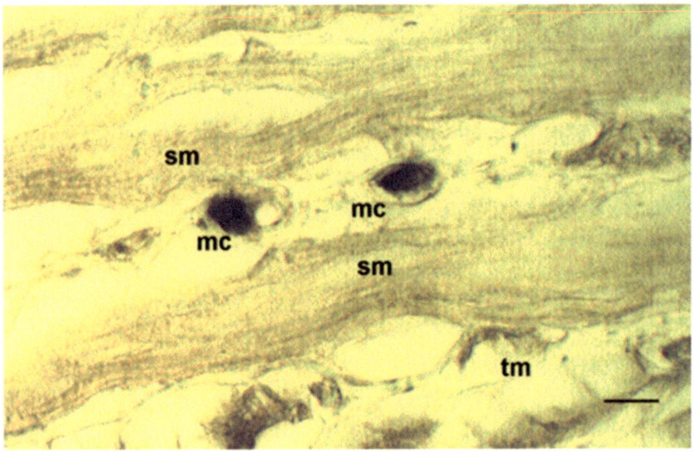

Figure 4. Two histamine positive mast cells **(mc)** in the circular smoothlayer (**sm**) of porcine ureter. **tm** – mucosal sheet. Bar = 25 μm

3.1.2. Dopamine

Dopamine belongs to the group of catecholamines, which are direct-acting sympathomimetic amines. Adrenergic nerves are not required for their effects because they activate the receptors of effector cells. This is mainly valid for already synthesized and exogenous catecholamines [27]. Before, dopamine was believed to be important only as an immediate precursor of norepinephrine (noradrenalin, levarterenol, arterenol), but later some of its important physiological functions in mammals were revealed and thus, it was considered more thoroughly in some clinical states in men [28]. Apart from its significance for nervous system, dopamine is tightly related in cardiovascular activity. Cardiovascular effects of dopamine depend on the activation of different types of catecholamine receptors. Its pressor effect is inhibited by an α blocker (such as phenoxybenzamine). Cardiac stimulating effects however, could be inhibited by a β blocker (for example, propranolol) [29].

Dopaminergic receptors of vascular beds could be considered a fifth adrenergic receptor subtype. Although the physiological significance of dopamine receptors is unknown, this type is important for clinical pharmacology as it is involved in vasodilator responses in renal, coronary and brain circulatory beds. The activation of these receptors by dopamine is highly selective, while the agonistic activity of other catecholamines is minor [27].

By opinion of [29] dopamine induced a reduction of vascular resistance and increases the blood flow to kidneys and mesenteries blood vessels, together with myocardial stimulation. This partial effect of dopamine could be advantageous in the treatment of shock compared to conventional catecholamines, because norepinephrine and epinephrine induce a marked contraction of renal and mesenterial arteries secondary to α-receptor effects.

The selective vasodilation of renal and visceral beds by dopamine suggested its use in clinical cardiac dysfunctions. It is successfully used in treatment of shock as it dilates renal arteries through activation of dopamine-1 (DA_1) receptors and enhances cardiac activity via activation of cardiac β-adrenergic receptors [20, 27]. Dopamine receptors on vascular smooth muscles are classified by [30, 31] as DA_1.

With regard to the presence of dopamine in mast cells, data available so far describe it as mastocyte amine mediator only in ruminants [32].

Using a histochemical reaction for the detection of tyrosine hydroxylase (TH) – a primary enzyme in dopamine synthesis pathway in porcine kidney, it was found out that TH-positive mast cells were predominantly localized at two sites: the renal sinus and in glomeruli of the superficial, middle and juxtamedullary cortical zones [16]. TH-positive mast cells in the renal sinus were usually seen as single cells, less frequently as clusters of several cells near the large blood vessels (Fig. 5).

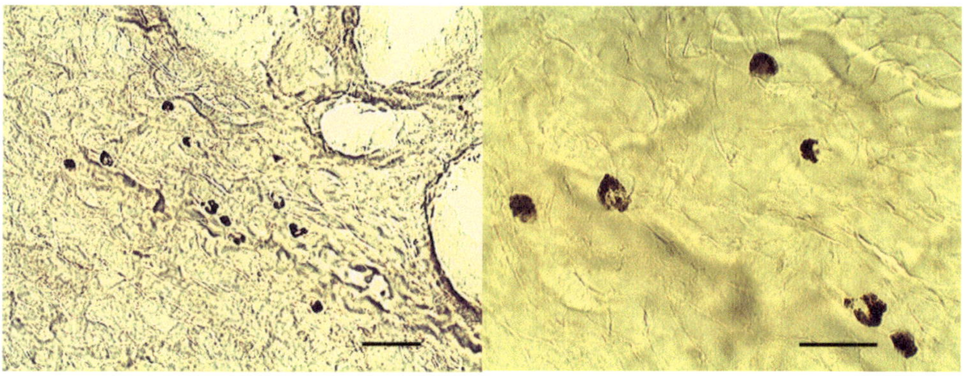

A B

Figure 5. A. TH-positive mast cells in the renal sinus, gathered near the blood vessels. Bar = 100 μm
B. Some of TH-positive mast cells (upper half of 5A at a higher magnification). Bar = 40 μm

TH-positive mast cells were not detected in the cortex. In our view, the absence of TH-positive mast cells around and within cortical vessels is probably one of the reasons for the application of dopamine in spasms of the interlobular arteries to relieve anuria due to a variety of causes. Nevertheless, as outlined by [33], the application of dopamine in humans and some animal species with renal failure did not provoke the anticipated effect.

3.2. Polypeptides

3.2.1. Vasoactive intestinal polypeptide (VIP)

The vasoactive intestinal polypeptide is a multifunctional peptide built of 29 amino acids, first isolated from porcine duodenum [34] and initially considered a potential vasodilator.

The extensive research on VIP in late 1980's revealed that this peptide is a physiological regulator of essential body functions, namely brain metabolism and blood flow, gastrointestinal motility and secretion, neuroendocrine secretion, immune response, sexual activity and reproduction. Along its importance in carcinogenesis, VIP is also related to diseases such as bronchial asthma, urinary bladder fibrosis and AIDS [35].

VIP was first described in mastocytes of rats and mice by [36], and its relationship to histamine was shown. Later, VIP was described in basophils as well [37]. As mast cells are abundant in the connective tissue of different organs, the contained VIP could influence the local or regional blood flow in both normal and pathological states [36].

It is acknowledged that VIP, some peptides and histamine secreted by mast cells are actively involved in the motility of smooth muscle (including vascular) cells and modulate the motility of smooth muscles in organs and the vascular wall through specific receptors [38–44].

Mast cells, immunopositive to VIP were observed in large blood vessels, including arcuate arteries and veins of pigs [16]. Relatively high mast cells counts were established within the media of these vessels, with highest density in the renal artery. Less VIP-positive cells were observed on the boundary between the media and the adventitia. It should be emphasized that in arcuate arteries, VIP-positive mast cells were present only in the middle shell, with relatively regular distribution along the circumference of the blood vessel (Fig. 6).

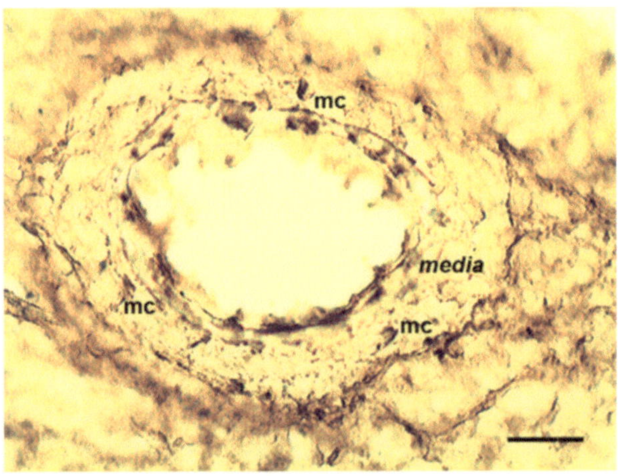

Figure 6. VIP positive mast cells (**mc**) in the middle shell (**media**) of porcine arcuate artery. Bar = 80 μm

3.2.2. Endothelin

Endothelin (ET) is a brain-vascular peptide [45], with three isopeptides – ET-1, ET-2 and ET-3 [46]. Its various biological actions include, apart from the regulation of vasoconstriction and neurotransmission, the regulation of cytokine-regulated cell growth [47]. Out of the three ET isopeptides, ET-1 has marked effect mainly on smooth muscle cells of the vascular wall.

ET-1 is a 21-amino acid peptide with a strong vasoconstrictor activity, first isolated from the supernatant of cultured endothelial cells. There is evidence that vascular smooth muscle cells are also capable to produce endothelin [48, 49]. Apart being a potent vasoconstrictor, ET-1 also exhibits hypertrophic, mitogenic and anti-apoptotic effects on vascular smooth muscle cells [50-52]. ET-1 induces strong and prolonged contractile responses in vascular smooth muscle cells in different systems, with a special effect on the renal vascular bed [48, 53-56]. Initially described as a vasoconstrictor, ET-1 is now acknowledged to participate in the pathogenesis of a number of disorders, i.e. vascular, inflammatory, fibrotic diseases via its multifunctional effects on mast cells, but under certain conditions [57].

ET-1 is widely spread in various tissues and organs, including the gastrointestinal tract. Its three isopeptides were detected mainly in mast cells and less frequently, in macrophages of *Lamina propria* of the stomach, small intestine and colon in Wistar rats. Apart the confirmed synthesis and secretion of ET by mast cells, its role as a new cytokine factor in these cells was suggested [46].

ET was reported to be present in the basilar artery of the rat [58, 59] and post mortem, in human cerebrovascular nerves of the middle cerebral artery [60]. Immunoreactive ET-1 was also detected in endothelial cells of the intima, vascular smooth muscle cells and macrophages of the media and neointima, and in perivascular nerves (axons) varicosities at the boundary between media and adventitia of the middle cerebral artery in a patient with multiple system atrophy with autonomic deficiency [61]. ET-1 positivity of uterine smooth muscle cells and mast cells was reported in a post partum mouse, but its functions remained unclear [62]. More recent data [3] from *in vitro* investigations showed the contractile effect of ET-1 on longitudinal smooth muscle of the equine cecum, mediated by ET_A and ET_B receptors. It was therefore concluded that the spontaneous contraction of equine caecal smooth musculature most probably originated in smooth muscle cells, and not in enteric nervous system. Having investigated the vasoconstrictor effect of ET-1 on resistant renal blood vessels in a rabbit through *in vitro* microperfusion of afferent and efferent glomerular arterioles, [63] established a dose-dependent decrease in their lumen.

Endothelin-immunopositive mast cells were present in the wall of both extrarenal and intrarenal blood vessels of the pig [16]. The largest amount of mast cells was observed in peripheral layers of tunica media, as well as on the boundary between media and adventitia. Endothelin-positive mastocytes were also observed in deeper layers of the media, in the connective tissue among smooth muscle cells. The detected mast cells were of different shape and with well visualized immunopositive granules. ET-positive mast cells in the wall of the renal vein were more rarely observed. Their localisation was similar to that in the renal artery. ET-positive mast cells were detected occasionally in the wall of intrarenal blood vessels, but only as single findings in arcuate and interlobular arteries and veins.

The research of [64] on mast cells in the wall of canine sublobular hepatic veins by transmission electron microscopy has shown that ET-1 was present in both the cytoplasmic matrix and cytoplasmic granules. According to the author, the coexistence of ET-1 and histamine in mast cell granules was closely related to the strong vasoconstrictor effect on venous sphincters of canine liver.

3.3. Proteoglycans

3.3.1. Heparin

The strong correlation between the density of mast cells in tissues and the efficacy of tissue extracts to prevent blood coagulation, it was supposed long ago that mast cells contained the potent anticoagulant heparin. According to [65] heparin is the only large glycosaminoglycan and its amount in the cell is about 20 pg.

Activated mast cells, releasing heparin proteoglycans, inhibit the proliferation of smooth muscle cells in the tunica media of human arterial wall, while histamine stimulates it [10]. It was demonstrated by [9] the inhibitory effect of heparin on human myometrium proliferation, suggesting that it could induce the differentiation of uterine smooth muscle cells and to influence tissue remodelling and reconstruction in different physiological and pathophysiological events.

Glycosaminoglycan-positive, including heparin-positive mast cells were observed by us [16] in the walls of extra and intrarenal blood vessels in pigs (Fig. 7). To determine the amount of heparin-containing mast cells, they were initially stained with 0.02 w/v aqueous berberine neutral sulfate solution and then, with 0.1% toluidine blue in McIlvane's buffer, pH 3 and observed by light microscopy. Calculating the ratio of berberine-positive (heparin) to toluidine positive (glycosaminoglycans) with well expressed γ-metachromasia mast cells, it was found out that 42% of mast cells in the renal vein media were berberine-positive, i.e. contained heparin in their granules.

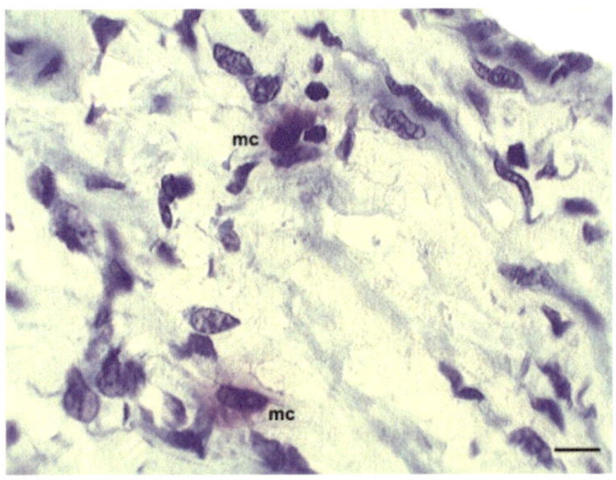

Figure 7. Mast cells (**mc**) in the media of porcine renal vein with well expressed γ-metachromasia. Bar = 30 μm

Similar studies with mast cells localized in the internal anal sphincter part, adjacent to the paranal sinus wall in the dog, showed that 100% of the mast cells were berberine (heparin) positive [66]. Positive reaction was detected both in mast cells in the connective tissue stroma among smooth muscle cell clusters, as well as within packs, in close vicinity to smooth muscle cells (Fig. 8). These data, compared to previously cited repots allowed assuming that heparin-containing mast cells regulated the growth of smooth muscle cells not only in the vascular wall, but also in the external anal sphincter in dogs [16, 66].

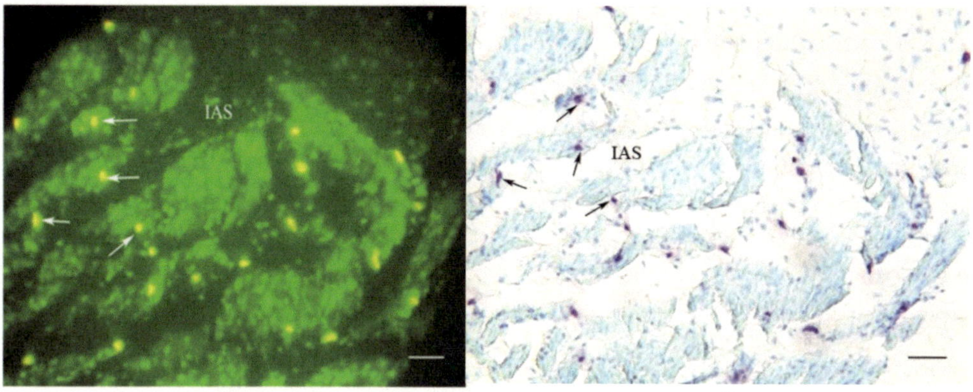

Figure 8. A. Heparin contained- berberin positive mast cells (arrows). IAS- internal anal sphincter. GA- apocrine glands. Male 4 years dog. Bar = 40 μm. (Courtesy by Dr I. Stefanov).
B. Toluidine blue staining of the same area - 8A. Mast cells with γ-ma metachromasia – glycosaminoglycans (arrows). Bar = 40 μm. (Courtesy by Dr I. Stefanov).

3.4. Radicals

3.4.1. Nitric oxide

Nitric oxide (NO) is a molecular free-radical gas, an important mediator with a variety of functions [67]. NO is biosynthesized from the guanidine nitrogen (N^G) of L-arginine through conversion of the intermediate compound N^{ω}-hydroxy L-arginine by nitric oxide synthase (NOS) enzyme family. As this family of enzymes incorporates molecular oxygen, NOS are classified as dioxygenases with similar features with cytochrome P-450 reductase. Cofactors needed for NO synthesis include flavin adenine dinucleotide (FAD), flavin mononucleotide, nicotinamide adenine dinucleotide phosphate (NADPH), haemoglobin and tetrahydrobiopterin.

Three NOS isoforms are known– NOS 1 (neuronal, nNOS), NOS 2 (inducible, iNOS) and NOS 3 (endothelial, eNOS). NOS 1 and NOS 3 (nNOS and eNOS) are also described as constitutive NOS (cNOS), the other original form is iNOS. NOS-coding genes are localised in different chromosomes. The structure of enzymes is common, i.e. oxygenase, reductase domain and calmodulin-binding site, with 51 to 57% homology in amino acid sequences. The C-terminal reductase domain possesses binding sites for FMN-, FAD- and NADPH, and is linked to the oxygenase domain by a calmodulin-binding site [68, 69].

Apart regulating the mast cell phenotype and function [70], NO is also produced by mast cells. There are numerous data confirming the production of this important signalling molecule by mast cells [71], but also data opposing to this statement [72].

There is increasing evidence in support of the fact that the release of mediators by mast cells is regulated by NO, most probably through posttranslational modification of proteins. The three isoforms of NOS are expressed in mast cells, although other cells also produce NO that could regulate mast cells' function. Probably, each of NOS isoforms in mast cells coordinates a specific outcome depending on the mast cell phenotype, subcellular distribution of the enzyme and the presence of various cofactors [72].

By [68] it was demonstrated that NO generated via eNOS is the main vasodilator responsible for hypotension in anaphylactic shock. The authors believe that soluble guanylate cyclase is the main vasorelaxing mediator of NO. These data support the surprising hypothesis that not iNOS but eNOS was the principal factor of vascular dilation in a state of anaphylaxis. Thus, eNOS and/or PI3K and Akt are defined as novel possible targets in anaphylaxis treatment.

Our studies on renal blood vessels in pigs showed that there were no mast cells, positive for NADPH-d and NOS [16]. Mastocytes positive to these ligands were detected in the internal anal sphincter of the dog [66]. In sexually mature animals, NADPH-d-positive mast cells were statistically significantly more numerous per 0.1 mm^2 than NOS-positive cells – 3.27±0.78, vs. 1.65±0.62 ($p< 0.001$).

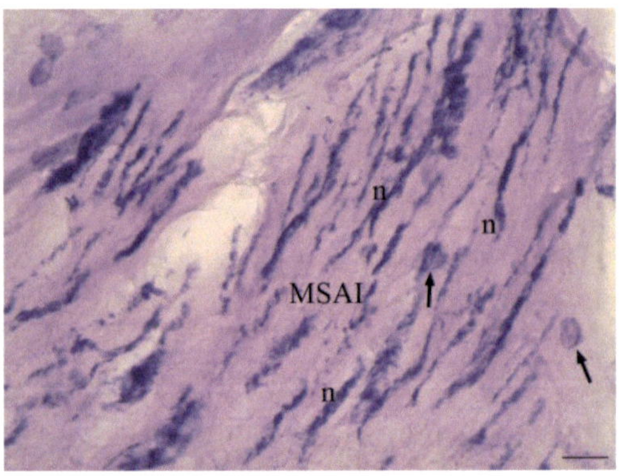

Figure 9. NADPH-d positive mast cells (arrows) and nerves (n) in dog's internal anal sphincter (MSAI). Bar = 15μm. (Courtesy by Dr I. Stefanov)

4. Some morphofunctional aspects of the interaction between smooth muscle cells and mast cells

The presence of mast cells containing biogenic amines and other mediators, the histochemical evidence for aforementioned ligands and more recent literature data suggesting that the nervous system is perhaps not the only mediator of smooth muscle contractility [3], allowed to hypothesize that mast cells were actively involved in smooth muscle cells motility. In support of this assumption, a detailed study of mast cells in the ureteral wall was conducted [15]. The results convincingly demonstrated that their role was substantial and incontestable. The highest mast cell density per mm^2 in the muscular layer of the ureter (16.7±3.3) compared to mucosa and adventitia (7.3±1.8 and 11.1±2.7, respectively) was probably related to the marked activity of closely situated smooth muscle cells. This assumption is based upon the opinion of [13] about bovine tracheal mast cells. Ureteral smooth musculature activity is manifested, in the first place, by periodic contractions sending urine into the urinary bladder (due to the horizontal position of the animal) and second, by blood flow regulation through vasoconstriction control [73] and/or modulation of smooth muscle activity [38, 41, 43]. This primary activity of mast cells in the porcine ureter is different to that in other organs such as lungs, alimentary organs and the skin, where mast cells are involved in immune response to external antigens, easily invading the body in large amounts [6]. This could be attributed to the specific retroperitoneal localisation of the ureter [74], which is relatively unexposed to such antigens.

It is well acknowledged that autonomic nerve fibres did not connect each smooth muscle cell in a large part of organs and blood vessels [2, 75]. Therefore, signals could be transmitted by nexus ion channel contact between mast cells and muscle cells – most probably via substances like Ca^{2+}, Mn^{2+} [76], K^{+} [77], Cl^{-} [78, 79], enzymes [73], biologically active substances as the vasoactive intestinal polypeptide – VIP [36, 37], histamine [13, 73, 80], and substance P [81], all of them responsible for signal transduction.

The detailed elucidation of mast cells effects on smooth muscle cells requires special attention on the tentative role of mastocytes in myometrium recovery [82], emergence of smooth muscle contraction [3], and the specificity of autonomic innervation of smooth muscle tissue [1, 75].

5. Conclusion

The presented data convincingly show that mast cells are normally resident cells in the smooth muscle layer of the blood vessels and the visceral organs. Established biologic active substances in observed mast cells with emphasized importance for smooth muscle motility give a reason to believe that mast cells via these substances take part in modulation of smooth muscle cells activity. This presumption could be useful for further investigations, which apparently will allow more detailed elucidation on mast cells function in smooth muscle tissue, as whole.

Author details

Angel Vodenicharov
Department of Veterinary Anatomy, Histology and Embryology, Faculty of Veterinary Medicine, Trakia University, Stara Zagora, Bulgaria

6. References

[1] Young B, Lowe J. S, Stevens A, Heath J. W (2006). Basic tissue types. In: Wheather's Functional Histology, A Text and Colour Atlas, 5th ed, Churchill Livingstone Elsevier, pp. 78-79.
[2] Chouchkov Ch (1995) Smooth muscle tissue. In: Koichev K, editor. Koichev Anatomy of the Man. Sofia: Meditzina and Fizkultura Publishing House, part I, pp.103-104.

[3] Chidambaram R, Eades S, Moore R, Hosgood G, Venugopal Ch (2005) Characterization of the in vitro responses of equine cecal longitudinal smooth muscle to endothelin-1. American Journal of Veterinary Research 66: 1202-1208.

[4] Ito A, Hagiyama M, Oonuma J (2008) Nerve-mast cell and smooth muscle-mast cell interaction mediated by cell adhesion molecule-1, CADM1. Journal of Smooth Muscle Research. 44: 83-93.

[5] Shiohara M, Koike K (2005) Regulation of mast cell development. Chemistry Immunology Allegry. 87: 1-21.

[6] Hill P.B, Martin, R. J (1998) A review of mast cell biology. Veterinary Dermatology. 9: 145-166.

[7] Junqueira L. C, Carneiro J, Kelly R. O (1992) Basic Histology. 7th ed., East Nowalk, Connecticut: Appletion & Lange. pp. 82, 111-113.

[8] Eurell J. C, Van Sickle D. C (1998) Connective and Supportive Tissues. Resident Cells of Connective Tissue. Mast cells. In:. Dellmann H. D, Aurell J. C, editors. Textbook of Veterinary Histology, Vth ed., Baltimore, USA: Lippincott Williams & Wilkins. p.35.

[9] Horiuchi, A, Nikaido T, Ya-Li Z, Ito K, Orii A, Fujii S (1999). Heparin inhibits proliferation of myometrial and leiomyomal smooth muscle cells through the induction of α-smooth muscle actin, calponin h1 and p27. Molecular Human Reproduction. 5: 139-145.

[10] Wang Y, Kovanen P.T (1999). Heparin proteoglycans released from rat serosal mast cells inhibit proloferation of rat aortic smooth muscle cells in culture. Circulation Research. 84: 74-83.

[11] Csaba G (1990) Alcian blue – Safranin method for mast cells. In: Bancroft J. D, Steven A. editors. Theory and Practice of Histological Techniques. New York: Livingstone, p. 639.

[12] Hunt T, Campbell A, Robinson C, Holgate S (1991) Structural and secretory characteristics of bovine lung and skin mast cells: Evidence for the existence of heterogeneity. Clinical and Experimental Allergy. 21: 173-182.

[13] Harris W, Marshall J, Yamashiro S, Shaikh N (1999) Mast cells in bovine trachea: Staining characteristics, dispersion techniques and response to secretagogues. Canadian Journal of Veterinary Research, 63: 5 –12.

[14] Xu R, Carr M, Bland A, Hall A (1993) Histochemistry and morphology of porcine mast cells. Histochemical Journal. 25: 516–522.

[15] Vodenicharov A, Leiser R, Gulubova M, Vlaykova (2005) Morphological and immunocytochemical investigations on mast cells in porcine ureter. Anatomia Histologia Embriologia. 34: 343–349.

[16] Vodenicharov A (2008) Morphological investigations on the role of mast cells mediators, other vasoactive substances and the glomerular arterioles in the renal hemodynamics of domestic swine. Doctor Vet. Med. Science Dissertation, Faculty of Veterinary Medicine, Trakia University, Stara Zagora, pp. 77-197.

[17] Schwartz L. B (1994) Mast cells: function and contents. Current opinion in Immunology. 6: 91-97.

[18] Panula P, Airaksinen M. S, Pirvola U, Kotilainen E (1990) A histamine-containing neuronal system in human brain. Neuroscience. 28: 585-610.

[19] Adams, H. R (2009) Histamine, Serotonin and their Antagonists. In: Riviere J. E, Papich M. G, editors. Veterinary Pharmacology and Therapeutics, 9th ed. Ames, Iowa: Wiley-Blackwell, pp.411-427.

[20] Purves D, Augustine G, Fitzpatrick D, Katz L, La Mantia A-S, Namara J. D, Williams S. M (2001) Neuroscience, 2nd ed., I. Neural Signaling, 6. Neurotransmitters. Sunderland (MA): Sinaer Associates, Inc., pp. 275-296.

[21] Hirschowitz B. I (1979) H-2 histamine receptors. Annual Review of Pharmacology. 19: 203-244.

[22] Krstić M. K, Stepanović R. M,. Krstić S. K, Katušić Z. C (1989). Endothelium-dependent relaxation of the rat renal artery caused by activation of histamin H_1-receptors. Pharmacology. 38: 113-120.

[23] [23]Levi R, Rubin L. E, Gross S. S (1991) Histamine in Cardiovascular Function and Dysfunction: Recent Developments. Effects of Histamine on the Vasculature. In: Uvnäs B, editor.Histamine and Histamine Antagonists, Vol. 97. Berlin: Springer – Verlag. pp. 354-367.

[24] Banks, R. O, Inscho E. W, Jacobson E. D (1984) Histamine H_1 receptor antagonists inhibit autoregulation of renal blood flow in the dog. Circulation Research, 54: 248-252.

[25] Robinson C. P, Maxson S, (1982) Differences in histamine H_1 and H_2 receptor responses in several rabbit arteries. Res Commun Chemistry, Pathology and Pharmacology. 36: 355-366.

[26] Tayo F. M, Bevan J. A (1986). Pharmacological characterization of histamine receptors in the rabbit renal artery. European Journal of Pharmacology, 121: 129-133.

[27] Adams H. R (2009) Adrenergic agonists and antagonists. In: Riviere J. E, Papich M. G, editors. Veterinary Pharmacology and Therapeutics, 9th ed. Ames, Iowa: Wiley-Blackwell, pp.125-155.

[28] Caccavelli L, Cussac D, Pellegrini I, Audinot V, Jaquet P, Enjalbert A (1992) D_2 dopaminergic receptors: normal and abnormal transduction mechanisms. Horm Research. 38: 78-83.

[29] Adams H. R, Parker J. L (1979). Pharmacologic management of circulatory shock: cardiovascular drugs and corticosteroids. Journal of American Veterinary Medical Association. 175: 86-92.

[30] Goldberg L. I, Rajfer S. I (1985) Dopamine receptors: Application in clinical cardiology. *Circulation*, 72: 245-248.

[31] Murphy, M. B, Murray C, Shorten G. D (2001) Fenoldopam-A selective peripheral dopamine receptor agonist for the treatment of severe hypertension. N. Eng. J Med 345:1548-1557.

[32] Falck B, Nystedt T, Rosengren E, Stenflo J (1964) Dopamine and mast cells in ruminants. Acta Pharmacologica (Copenhagen). 21: 51-58.

[33] Sigrist N. E (2007) Use of dopamine in acute renal failure. J Vet Emerg Crit Care. 17: 117-126.

[34] Said, S. I, Mutt V (1970) Polypeptide with broad biological activity: isolation from small intestine. Science, 169: 1217-1218.

[35] Said, S. I (1988). This Week's Citation Classic. Current Contents, 20, 16 May, 16 (11416).

[36] Cutz, E., Chan W, Track N. S, Goth A, Said S (1978). Release of vasoactive intestinal polypeptide in mast cells by histamine liberators. Nature 275: 661-662.

[37] Goetzl, E. J, Sreedharan S. P, Turck C. W (1988). Structurally distinctive vasoactive intestinal polypeptides from rat basophilic leukemia cells. Journal of Biology and Chemistry 263: 9083-9086.

[38] Ishikawa, S., Sperelakis N., 1987. A novel class (H_3) of histamine receptors on perivascular nerve terminals. Nature, 327: 158-160.

[39] Mori, T, Kawashima T, Beppu Y., Takagi K (1994). Histamine release induced by pituitary adenylate cyclase activating polypeptide from rat peritoneal mast cells. Arzneimittelforschung: 44: 1044-1046.

[40] Seebeck, J, M.Kruse, A. Schmidt-Choudhury, J. Schmidtmayer, W. E. Schmidt, 1998. Pituitary adenylate cyclase-activating polypeptide induces multiple signaling pathways in rat peritoneal mast cells. European Journal of Pharmacology, 352: 343-350.

[41] Champion, H, Bivalacqua T, Lambert D, Abassi R, Kadoitz P (1999) Analysis of vasoconstrictor responses to histamine in the hindlimb vascular bed of the rabbit. American Journal of Physiology, Regul. Intergr. Comp. Physiol. 277: R1179-R1187.

[42] Mirabella, N, Squillacioti C, Collitti M, Germano G, Pelagali A, Paino G (2002) Pituitary adenilate cyclase activating peptide (PACAP) immunoreactivity and mRNA expression in the duck gastrointestinal tract. Cell and Tissue Research 308: 347-359.

[43] Varty, L, Hey J (2002) Histamine H-3 receptor activation inhibits neurogenic sympathetic vasoconstriction in porcine nasal mucosa. European Journal of Pharmacology 452: 339-345.

[44] Squillacioti, C, Mirabella N, Colitti M, Esposito V, Paino G (2003) Expression and distribution of Pacap m-RNA in the duck gastrointestinal tract. Acta Veterinaria Brno 72 (Suppl. 7), S51.

[45] Yamada H, Kurokawa K (1994) Histochemical analysis of endothelin and its role in the central nervous system. In: H. Takahashi, editor. Central nervous system and blood pressure control. Yubunsha Publishing; Tokyo, pp. 105-114.

[46] Liu Y, Yamada H, Ochi J (1998) Immunocytochemical studies on endothelin in mast cells and macrophages in the gastrointestinal tract. Histochemistry and Cell Biology, 109: 301-307.

[47] Yamada H, Ochi J (1995). Histocytochemical and functional aspects on the "brain-vascular peptides". Kaibogaku Zasshi, 70: 422-435.

[48] Yanagisawa M, Kurihara H, Kimura S, Tomobe Y, Kobayashi M, Mitsui Y, Yazaki Y, Goto K, Masaki T (1988): A novel potent vasoconstrictor peptide produced by vascular endothelial cells. Nature 332: 411-415.

[49] Miller, R.C, Pelton J.T, Huggins J.P (1993) Endothelins: From receptors to medicine. TIPS 14: 54 – 60.

[50] Komuro I, Kurihara H, Sugiyama T, Yoshizumi M Takaku, F, Yazaki Y (1988) Endothelin stimulates c-fos and c-myc expression and proliferation of vascular smooth muscle cells. FEBS Letters, 238: 249-252.

[51] Tasaka, K, Kitazumi K (1994) The control of endothelin-1 secretion. Genetics and Pharmacology, 25: 1056-1069.

[52] Sharifi, A, Schiffrin E (1999) Apoptosis in aorta of deoxycorticosterone acetate-salt hypertensive rats: effect of endothelin receptor antagonism. Journal of Hypertension, 15: 1441-1448.
[53] Schulz E, Ruschitzka F, Lueders S, Heydebluth R, Schrader J, Muller G.A (1995): Effects of endothelin on hemodynamics, prostaglandins, blood coagulation and renal function. Kidney International, 47: 795-801.
[54] Goto, K, Hama H, Kasuya Y (1996): Molecular pharmacology and pathophysiological significance of endothelin. Japanese Journal of Pharmacology 72: 261-290.
[55] Navar L. G, Incsho E. W,. Majid D. S. A, Imig J. D, Harison-Bernard L. M, Mitchel K. D (1996) Paracrine regulation of the renal microcirculation. Physiological Reviews 3: 425-536.
[56] Kishi, F, Minami K, Okishima N, Murakami M, Mori S, Yano M, Niwa Y, Nakaya Y, H. Kido (1998) Novel 31-amino acid length endothelins cause contraction of vascular smooth muscle. Biochemical and Biophysical Research Communications 248: 387-390.
[57] Matsushima H, Yamada N, Matsue H. Shimada S (2004) The effects of endothelin-1 on degranulation, cytokine, and growth factor production by skin-derived mast cells. European Journal of Immunology 34: 1910-1919.
[58] Loesch A, Milner P, Burnstock G (1998) Endothelin in perivascular nerves. An electron-immunocytochemical study of rat basilar artery. Neuroreport 9: 3903-3906.
[59] Milner P, Loesch A, Burnstock G (2000) Neural endothelin in hypertension: immunoreactivity in ganglia and nerves to cerebral arteries of the spontaneously hypertensive rat. Journal of Vascular Research 37: 39-49.
[60] Loesch A, Burnstock G (2002) Endothelin in human cerebrovascular nerves. Clinical Science, 103: (Suppl. 48), 404S-407S.
[61] Mickey I, Kilford L, Kingbury A, Loesch A (2002) Endothelin in middle cerebral artery: A case of multiple system atrophy. The Histochemical Journal, 34: 469-477.
[62] Uchide T, Uchide T, Adur J, Yoshioka K, Sasaki T, Temma K, Saida K (2001) Endothelin-1 in smooth muscle cells and mast cells of mouse uterus after parturition. Journal of Molecular Endocrinology 27: 165-173.
[63] Ozawa Y, Hasegawa T, Tsuchiya K, Yoshizumi M, Tamaki T (2003) Effect of endothelin-1 (1-31) on the renal resistance vessels. The Journal of Medical Investigation, 50: 87-94.
[64] Yamamoto K (2000) Electron microscopy of mast cells in the venous wall of canine liver. Journal of Veterinary Medicine Science 62: 1183 – 1188.
[65] Bloom W, Fawsett D (1975) A Textbook of Histology., Philadelphia: W. B. Saunders Company. pp. 185-187.
[66] Stefanov I (2011). Morphofunctional aspects of dog's paranal sinus (Sinus paranalis). PhD Dissertation, Faculty of veterinary Medicine, Trakia University, Stara Zagora, pp. 104-152.
[67] Bredt D. S (2003) Nitric oxide signaling specifity – the heart of the problem. Journal of Cell Science 116: 9-15.
[68] Cauwels A, Janssen B, Buys E, Sips P, Brouckaert P (2006). Anaphylactic shock depends on PI3K and eNOS-derived NO. The Journal of Clinical Investigation, 116: 2244-2251.

[69] Adams H. R (2009) Introduction to Neurohumoral transmission and the Autonomic Nervous System. In: Riviere J. E, Papich M. G, editors. Veterinary Pharmacology and Therapeutics, 9th ed, Wiley-Blackwell, pp.101-123.

[70] Koranteng R. D, Dearman R. J, Kimber I, Coleman J. V (2000) Phenotyp variation in mast cell responsivenes to the inhibitory action of nitric oxide. Inflammation Research 49: 240-246.

[71] Swindle, E. J, Metcalfe D. D, Coleman J. W (2004) Rodent and human mast cells produce functionally significant intracellular reactive oxygen species but not nitric oxide. Journal of Biology and Chemisry 279: 48751-48759.

[72] Sekar Y, Moon T. C, Muñoz S, Befus A. D (2005) Role of nitric oxide in mast cells: controversies, current knowledge, and future applications. Immunologic Research 33: 223-240.

[73] Barret K, Pearce F (1991) Mast cells heterogeneity. In: Barret K, Pearce F, editors. Immunopharmacology of mast Cells and Basophils. London: Academic Press, pp. 29-38.

[74] Vollmerhaus, B., 1999: Harnorgane. In: Nickel R, Schummer A, Seiferle E, Lehrbuch der Anatomie der Hausteire, Bd. II, Einweide, 8th ed. Berlin: Parey Buchervald, pp. 319-320.

[75] Fawcett D, Jensh R. P (2002) Mast cells, Smooth muscle. In: Bloom & Fawcett's Concise Histology, 2nd ed., London: Arnold, pp. 70-71, 113-115.

[76] Fasolato C, Hoth M, Matthews G, Penner R (1993) Ca^{2+} and Mn^{2+} influx through receptor – mediated activation of nonspecific cation channels in mast cells. Proceeding of the National Academy of Sciences of USA, 90: 3068-3072.

[77] Quin Y, McCloskey M (1993) Activation of mast cell K^+ channels through multiple G protein-linked receptors. Proceeding of the National Academy of Sciences of USA, 90: 7844-7888.

[78] Romanin C, Reinsprecht M, Pecht I, Schndler H (1991) Immunologically activated chloride channels involved in degranulation of rat mucosal mast cells. EMBO Journal, 10: 3603-3608.

[79] Dietrich J, Lindau M (1994) Chloride channels in mast cells: block by DIDS and role in exocytosis. Journal of General Physiology, 104: 1099-1111.

[80] Matsumoto Y, Inoue Y, Shimada T, Aikima T (2001) Brain mast cells act as an immune gate to the hypothalamic-pituitary-adrenal axis in dog. Journal of Experimental Medicine, 194: 71-78.

[81] Toyoda M, Makino T, Kagoura M, Horohachi M (2000) Immunological study of substance P in human skin mast cells. Archiv of Dermatology Research, 292: 418-421.

[82] Uchide, T, Uchide T, Adur J, Yoshioka K, Sasaki T, Temma K, Saida K (2001) Endothelin-1 in smooth muscle cells and mast cells of mouse uterus after parturition. Journal of Molecular Endocrinology, 27: 165-173.

Chapter 16

Role of Prokineticin in Epicardial Progenitor Cell Differentiation to Regenerate Heart

Canan G. Nebigil

Additional information is available at the end of the chapter

http://dx.doi.org/10.5772/48234

1. Introduction

Cardiovascular diseases are one of the most common health-care problems throughout the world and carry a high rate of mortality (Zannad, et al., 2009). New strategies are urgently needed to replace cardiomyocytes and increase circulatory support for the treatment of cardiovascular diseases.

Over the last decade, stem/progenitor-cell therapy has emerged as an innovative approach to provide cardiac repair and regeneration (Zimmermann, et al., 2006). Several stem- and progenitor-cell types from autologous and allogeneic donors have been analyzed to find the most appropriate candidate. Although embryonic stem (ES) cells can differentiate into most cardiac cell types (Mummery, et al., 2002) , their clinical use is severely limited due to ethical concerns and immunogenic and teratogenic side effects (Blum and Benvenisty, 2008). Adult bone marrow-derived stem cells avoid the ethical and clinical issues associated with ES cells (Bianco, et al., 2001). However, animal studies have demonstrated a variable degree of cardiomyogenesis, and improvement in heart function by bone marrow-derived stem cells (Murry, et al., 2004). Thus, the utility of adult bone marrow-derived stem cells is hampered by their limited population size and restricted potential for cardiovascular differentiation (Assmus, et al., 2010).

Recently, therapies based on cardiac progenitor cells (CPC) have emerged as promising potential cardiac therapeutics (Gonzales and Pedrazzini, 2009). For cardiovascular therapy, pluripotent cardiac progenitor cells (CPCs) resident in the epicardium offer distinct advantages over other adult stem-cell types (Wessels and Perez-Pomares, 2004). They are autologous, tissue-specific and pre-committed (Dube, et al., 2012) to a cardiac fate, and display a greater propensity to differentiate towards cardiovascular lineages (Cai, et al., 2008), (Smart and Riley, 2012). Epicardial derived cardiac progenitor cells (EPDCs) exist in the heart of several species, including mice (Limana, et al., 2007) and humans (van Tuyn, et al., 2007). Due to cardiogenic and angiogenic abilities, epicardial CPCs represent an ideal candidate for cardiac regeneration. However, we do not know the mechanisms underlying epicardial CPC self renewal, proliferation and differentiation, which are prerequisites for cardiac regenerative therapy. An optimal paradigm of cardiovascular therapy may therefore consist of identifying the most effective factors that trigger the restoration of epicardial CPCs for healing heart injuries, with an emphasis on small molecule-based therapy over cell-based therapy.

It is therefore imperative to obtain a better understanding of the biology and regenerative potential of endogenous epicardial CPCs. The race is still on to find the "best" factor or drugs to reprogram endogenous epicardial CPCs to reconstitute the myocardium and improve function after myocardial damage.

2. Epicardium as a source of multipotent progenitor cells

Epicardium derived from proepicardium has an essential modulating role in the differentiation of the compact ventricular layer of the myocardium and the development of cardiac vessels during embryogenesis (Zhou, et al., 2008). Deletions of selected genes expressed in the epicardium (i.e. VCAM-1, α4-integrin) resulted in severe defects in the developing heart and its vasculature. The zebrafish epicardium promotes cardiac regeneration through epithelial to mesenchymal transition (EMT) and subsequent migration into the myocardium to form neovasculorization (Lepilina, et al., 2006). Signalling from the myocardium to the epicardium (i.e. Tβ4, FOG-2) (Smart, et al., 2007; Tevosian, et al., 2000) also leads undeveloped ventricle with vascularisation defects.

The epicardium through EMT generates a population of Epicardial Derived Progenitor Cells (EPDCs) that invade the underlying myocardium, and differentiate into various cardiac lineages (Smart and Riley, 2012; Zhou, et al., 2008). Williams Tumour (WT1) gene has been shown to regulate epicardial EMT through beta-catenin (Zamora, et al., 2007) and retinoic acid signaling pathways (von Gise, et al., 2011). EPDCs can either form endothelial cells, in response to a combination of myocardial vascular endothelial growth factor and basic-fibroblast growth factor signalling (van Wijk, et al., 2009), or differentiate into smooth muscle cells, upon exposure to platelet-derived growth factor (Kang, et al., 2008), transforming growth factor beta and bone morphogenetic protein-2 (Sanchez and Barnett, 2012).

However, Tβ4 (Smart, et al., 2007) and PKR1 (Urayama, et al., 2008) signaling appear to be a necessary and sufficient signaling factor for adult EPDC differentiation into the endothelial and smooth muscle cells to induce neovascularization. Thymosin beta-4 can activate adult epicardial cells (Bock-Marquette, et al., 2009) acting through reactivation of embryonic signalling pathways (Smart, et al., 2007).

In a regenerative context, the adult epicardial progenitor cell population also mediates cardiac repair after injury. Tβ4 can activate adult epicardial cells (Bock-Marquette, et al., 2009; Smart, et al., 2007) to promote revascularization of the injured mammalian heart by forming endothelial and vascular smooth muscle cells. Tβ4 treatment before myocardial infarction alters the responsiveness and fate of activated epicardial cells (WT1+ progenitor cells), to differentiate into cardiomyocytes (Smart, et al., 2011). However Tβ4 treatment after myocardial infarction induces epicardial expansion and coronary capillary density without affecting migration or alteration of $WT1^+$ progenitor cell fate into cardiomyocytes (Zhou, et al., 2012). Tβ4 treatment of mice after MI activates cardiac progenitor cell fate to induce cardiomyocyte linage (Bock-Marquette, et al., 2009). However, the cardiac progenitor subpopulation remains to be characterized. Further, a sub-population of adult epicardial cells retains the potential to give rise to cardiac precursors or endothelial cells (Limana, et al., 2007). The regenerative potential of EPDCs has been tested in the injured myocardium. The injection of human EPDCs was reported to enhance cardiac repair (Winter, et al., 2007). When the cardiomyocyte progenitors were co-transplanted with EPDCs into infarcted myocardial tissues, they improved functional repair as compare to single cell type supplementation (Zhou, et al., 2011). The effect was shown to be caused by paracrine effects from both cell types. Nevertheless, signals and cellular contributions from the EPDCs are indispensable for the establishment of normal coronary vasculature and myocardial architecture (Smart and Riley, 2012; Winter, et al., 2009).

3. GPCRs and cardiovascular system

Many hormones and neurotransmitters use GPCRs to exert their cardiovascular effects (Marinissen and Gutkind, 2001; Tang and Insel, 2004). Relatively little information is available regarding the role of GPCRs in the functional activities of cardiac stem/progenitor cells, both in normal and disease conditions. The well-studied cardiac role of GPCRs via Gαq signalling (Gutkind and Offermanns, 2009) is to promote cardiac hypertrophy (Wettschureck, et al., 2001) or protect cardiomyocytes against hypoxic insult (Nebigil, et al., 2003). Gα12 signaling can interact with the cytoplasmic domain of cadherins (Kaplan, et al., 2001), resulting in the release of the transcriptional activator β-catenin. Gα13 signaling is involved in vessel formation (Offermanns, et al., 1997). Gαs signaling regulates heart rate and contractility in response to catecholamine stimulation, but excessive Gαs signaling in heart eventually induces myocardial hypertrophy, fibrosis and necrosis (Gaudin, et al., 1995). Given the important roles of GPCRs in cardiac regulation, a key question is how many different GPCRs exist in the heart and what is their physiologic significance? Since forty percent of these GPCRs represent viable drug targets (Schlyer and Horuk, 2006) and also many of GPCR is involved in regulating cardiovascular system, unraveling of novel GPCR in cardiac progenitor/stem cells is very important to develop novel therapies for limit cardiovascular disease.

3.1. Prokineticins and cognate receptors:

Prokineticins are structurally homologues of amphibian or reptilian peptide toxins (Kaser, et al., 2003). They were first identified in the gastrointestinal tract as potent agents mediating muscle contraction (Hoogerwerf, 2006; Li, et al., 2001), and have been isolated from bovine milk (Masuda, et al., 2002). They comprise two classes: Prokineticin-1 (PK1), originally called endocrine gland-derived vascular endothelial growth factor (EG-VEGF)

(LeCouter and Ferrara, 2002) based on the functional similarity to VEGF and prokineticin-2 (PK2, also called Bv8). PK1 and PK2 are approximately 50% homologous and contain carboxyl-terminal cysteine-rich domains that form five disulfide bridges (Bullock, et al., 2004). N terminal hexapeptide (AVITGA) and cysteine residues in the carboxy-terminal domain are crucial for their biological activities . Prokineticins and their receptor are widely distributed in mammalian tissues (Soga, et al., 2002). Prokineticins induce cell excitability such as gut spasmogen (Wade, et al., 2009), pain sensitization (Negri, et al., 2006), circadian rhythm (Li, et al., 2006), and sleep (Hu, et al., 2007)). They also induce cell motility such as angiogenesis (LeCouter and Ferrara, 2002), neurogenesis (Ng, et al., 2005), hemotopoiesis (LeCouter, et al., 2004), neovasculogenesis (Urayama, et al., 2008). Prokineticins regulate complex behaviors such as feeding (Negri, et al., 2004), drinking (Negri, et al., 2004), anxiolity (Li, et al., 2009). Moreover, prokineticins are potent survival/mitogenic factors for various cells including endothelial cells , neuronal cells (Kisliouk, et al., 2005; Ngan, et al., 2007a), lymphocytes, hematopoietic stem cells (LeCouter, et al., 2004), and cardiomyocytes (Nebigil, 2009). Table 1 summarize the involvement of prokineticin in the diseases.

Prokineticins bind to two cognate 7-transmembrane G-protein-coupled receptors. PKR1 and PKR2 share about 85% amino acid identity and encoded within distinct chromosomes in both mouse and human (Masuda, et al., 2002). Prokineticin-2 is the most potent agonist for both receptors (Masuda, et al., 2002). PKR2 is the dominant receptor in the adult brain, particularly in the hypothalamus, the olfactory ventricular regions, and the limbic system. However, PKR1 is widely distributed in the periphery. These receptors couple to Gαq, Gαi and Gαs to mediate intracellular calcium mobilization, activation of MAPK, Akt kinases and cAMP accumulation, respectively (Ngan and Tam, 2008). Although prokineticin signaling has been implicated as a survival/mitogenic factor for various cells including endothelial cells (Guilini, et al., 2010), neuronal cells (Ngan, et al., 2007b), enteric neural crest cells (Ngan, et al., 2007a), granulocytic (Giannini, et al., 2009)and monocytic lineage (Dorsch, et al., 2005) , lymphocytes and hematopoietic stem cells (LeCouter, et al., 2004), until recently, little was known about the underlying molecular and cellular events to regulate cardiovascular function.

3.1.1. A novel role for prokineticin in regulating cardiovascular system

PK2/PKR1 signaling pathway seems an important cardiovascular regulatory pathway, because of the following aspects: Prokineticins are potent angiogenic factors (LeCouter and Ferrara, 2003), which have beneficial effects on cardiac repair by inducing angiogenesis to improve coronary circulation or regenerating the cardiomyocytes (Bellomo, et al., 2000). They exert their biological effects via activating GPCRs that couple to diverse G proteins. Mutations in the gene encoding prokineticin-2 cause Kallmann syndrome (hypogonadotropic hypogonadism) in human (Abreu, et al., 2008; Canto, et al., 2009; Cole, et al., 2008), with congestive heart failure and dilated cardiomyopathy. Prokineticins induce differentiation of murine and human bone marrow cells into the monocyte/macrophage lineage and activate monocyte proliferation, differentiation and macrophage migration (Denison, et al.,

2008; Dorsch, et al., 2005; Giannini, et al., 2009). In human end-stage failing heart samples, reduced PKR1 and prokineticin-2 transcripts and protein levels implicate a more important role for PK2/PKR1 signaling in heart (Urayama, et al., 2007). Therefore, we reasoned that PK2/PKR1 signaling should contribute to heart repair by inducing angiogenesis or repairing cardiomyocytes.

3.1.2. Role of PKR1 signaling in cardiovascular system

In cultured capillary endothelial cells derived from heart, PK2 via PKR1 induces proliferation, migration and vessel-like formation, activating Gα11/MAPK and Akt kinases (Guilini, et al., 2010). In cardiomyocytes, activation of overexpression of PKR1 protects cardiomyocytes against hypoxic insult, activating the PI3/Akt pathway (Urayama, et al., 2007).

Transient PKR1 gene transfer after coronary ligation in the mouse model of myocardial infarction reduces mortality and preserves heart function by promoting cardiac angiogenesis and cardiomyocyte survival. This result suggests that PKR1 may represent a novel therapeutic target to limit myocardial injury following ischemic events (Urayama, et al., 2007).

Transgenic mice overexpressing PKR1 specifically in the heart under the control of cardiac α-myosin heavy chain (α-MHC) promoter displayed no spontaneous abnormalities of cardiomyocytes, but showed increased neovascularisation (Urayama, et al., 2008). Thus, these data suggest that PKR1 is involved in post-natal de novo vascularization, rather than vasculogenesis during embryogenesis.

Genetic inactivation of PKR1 in mice (PKR1-knockout mice) exhibit dilated cardiomyopathy and reduced angiogenesis in heart (Boulberdaa, et al., 2011). The heart pathology in PKR1 knockout mice is due to increased apoptosis in cardiomyocytes and reduced epicardial progenitor cell numbers. These data was consistent with an endogenous role of PKR1 signalling in stimulating epicardial progenitor cell proliferation and differentiation. All together these findings show that PKR1 signalling is involved in regulating cardiomyocyte survival signalling, and progenitor cell proliferation and differentiation.

3.1.3. Role of PKR2 signaling in cardiovascular system

Since PKR1 and PKR2 are 85% identical and are both expressed in cardiovascular tissues, PKR2 may also contribute to cardiomyocyte growth and vascularization. Transgenic mice overexpressing PKR2 specifically in the heart under the control of cardiac (α-MHC) promoter exhibit eccentric hypertrophy in an autocrine regulation and impaired endothelial integrity in a paracrine regulation without inducing angiogenesis (Urayama, et al., 2009). These transgenic PKR2 mice may provide a new genetic model for heart diseases. We found that in the endothelial cells PKR2 couples to Gα12 signaling pathway and downregulates ZO-1, thereby inducing endothelial cell fenestration (Urayama, et al., 2009).

3.1.4. Prokineticin signaling in cardiac stem/progenitor cell activation

Prokineticin-2 has been shown to modulate mobilization of bone morrow-derived cells and also promote angiogenesis. Systemic exposure to prokineticins promoted the survival of hematopoietic cells and enhanced progenitor mobilization (LeCouter, et al., 2004). Recently, we found that prokineticin-2 induces significant outgrowth from mouse epicardial explants and quiescent EPDCs, restoring epicardial pluripotency and triggering differentiation of endothelial and vascular smooth muscle cells (Urayama, et al., 2008). Co-culturing EPDCs with cardiomyocytes overexpressing PKR1 increased prokineticin-2 levels as a paracrine factor, thereby promoting EPDC differentiation, mimicking our PKR1-transgenic mice model (Urayama, et al., 2008). These prokineticin-2 effects were abolished in EPDC derived from PKR1-null mutant hearts, demonstrating PKR1 involvement. Prokineticin/PKR1 signaling can reprogram adult EPDCs to induce neovascularization. These studies provided novel insight for possible therapeutic strategies aiming at restoring pluripotency of adult EPDCs to promote neovasculogenesis, by induction of cardiomyocyte- PKR1 signaling. Whether epicardial-PKR1 signaling contributes cardiomyocyte function and metabolism, and it determines lineage choice decision in EPDCs remained to be investigated.

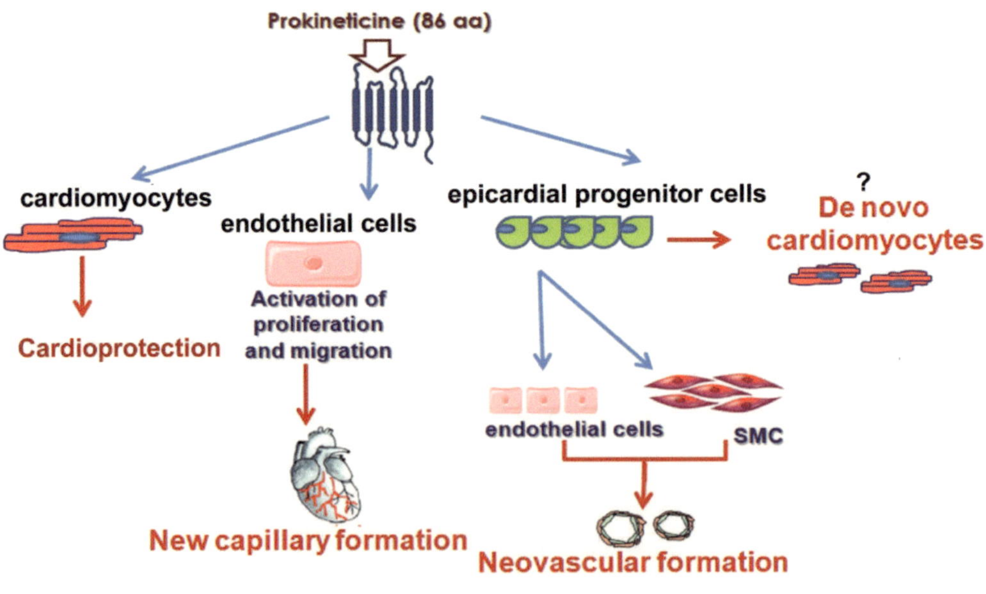

Figure 1. Role of prokineticin PKR1 signaling in cardiac regeneration.

PKR1 signaling protects cardiomyocyte against hypoxia-mediated apoptosis, activates endothelial cells for angiogenesis, activates EPDC differentiation into vasculogenic cell type to induce neovascular formation, activates EPDC differentiation into new cardiomyocytes.

DOMAIN	ROLE/EXPRESSION in human organs	REFERENCE
Reproduction		
Menstrual cycle	Progesterone induces elevation of prokineticin-1 expression during the secretory phase indicating a role of prokineticins and their receptors in endometrial vascular function	(Battersby, et al., 2004)
	Prokineticin-1 is derived from granulosa lutein cells and its synthesis is elevated during the mid- to late luteal phase	(Fraser, et al., 2005)
	Alteration of prokineticin-1 can induces several biochimical abnormalities characterizing eutopic endometrium in endometriosis	(Tiberi, et al., 2009)
Placentation and pregnancy	Prokineticin-1 and PKR1 expression is elevated in human decidua during early pregnancy. Prokineticin-1 via PKR1 regulates expression of host implatation-related gene.	(Evans, et al., 2008)
	Dysregulation of Prokineticin signaling in fallopian tube could affect fallopian tube smooth muscle cells contractility and embryo-tubal transport providing a potential cause for ectopic pregnancy	(Shaw, et al., 2010)
	Prokineticin-1 and its receptor gene polymorphism and haplotype were associated with idiopathic recurrent pregnancy loss. These three gene contribute to recurent pregnancy loss in the Taiwanese Han population	(Su, et al.)
Kallman syndrome	Insufficient prokineticin signaling leads to abnormal development of the olfactory system and reproductive axis in man	(Dode, et al., 2006)
	Mutation in prokineticin-2 and PKR2 genes underlie both Kallman sydrome and idiopathic hypogonadotropic hypogonadism	(Cole, et al., 2008)
Behaviour	Prokineticin-2 may play a role in the pathophysiology of mood disorders in the Japanese population	(Kishi, et al., 2009)
	Prokineticin-2 may play a role in the pathophysiology of methamphetamine dependance in the Japanese population	(Kishi, et al., 2010)
Cancer	Prokineticins and their receptors are expressed in human prostate and their levels increased with prostate malignancy	(Pasquali, et al., 2006)
	Prokineticin-1 favors neuroblastoma progression	(Ngan, et al., 2007b)

DOMAIN	ROLE/EXPRESSION in human organs	REFERENCE
	Prokineticin-1 derived from islet and/or pancreatic stellate cells act through its receptor on endothelial cells to increase angiogenesis in pancreatic disease	
	Prokineticin-2 play a role in pathophysiological in human tumors and inflammatory disorders	(Zhong, et al., 2009)
	Prokineticin-1 is significantly increased in papillary thyroid cancer and its expression in papillary thyroid cancer is related to BRAF oncogen	(Pasquali, et al., 2011)
Vascular	Prokineticin-2 is involved in immune and inflammatory response at abdominal aortic aneurysms site	(Choke, et al., 2009)
Inflammation	Prokineticin-1 was found in the controls in the patients with temporomandibular joint disorders	(Herr, et al.)
Cardiology	Prokineticin-2 and PKR1 were reduced in human end stage failure heart sample	(Urayama, et al., 2007)

Table 1. Involvement of prokineticins in human diseases

4. Conclusion

All together these data showed that PK2 via PKR1 signaling has important roles on heart physiology and pathophysiology. PKR1 is involved in postnatal cardiac vascularization by activating epicardial progenitor cells. These studies also raise numerous questions for further investigation. Do EPDCs differentiate into functional (beating) cardiomyocytes in vitro or in vivo? Do EPDCs differentiate into cardiac lineages in vivo in the damaged adult? Does the activity or potential of EPDCs decline with age? The identification of factors which stimulate endogenous cardiac progenitor cells to induce neovascularization and cardiomyocyte replacement is an evolving paradigm towards therapeutic intervention in cardiac diseases. The race is to facilitate drug discovery for targets acting on cardiomyocytes or EPDCs to invoke new coronary vessels and cardiac tissues as a significant step toward cardioprotection and cardiovascular regeneration.

Author details

Canan G. Nebigil
University of Strasbourg/CNRS , UMR7242, France

Acknowledgement

I acknowledge all the members of my laboratory and Dr. Laurent Désaubry for their fruitful discussion during preparation of this article.

5. References

Abreu, A.P., Trarbach, E.B., de Castro, M., Frade Costa, E.M., Versiani, B., Matias Baptista, M.T., Garmes, H.M., Mendonca, B.B. and Latronico, A.C., (2008). 'Loss-of-function mutations in the genes encoding prokineticin-2 or prokineticin receptor-2 cause autosomal recessive Kallmann syndrome'. *J Clin Endocrinol Metab*, 93 (10):4113-4118.

Assmus, B., Rolf, A., Erbs, S., Elsasser, A., Haberbosch, W., Hambrecht, R., Tillmanns, H., Yu, J., Corti, R., Mathey, D.G., Hamm, C.W., Suselbeck, T., Tonn, T., Dimmeler, S., Dill, T., Zeiher, A.M. and Schachinger, V., (2010). 'Clinical outcome 2 years after intracoronary administration of bone marrow-derived progenitor cells in acute myocardial infarction'. *Circ Heart Fail*, 3 (1):89-96.

Battersby, S., Critchley, H.O., Morgan, K., Millar, R.P. and Jabbour, H.N., (2004). 'Expression and regulation of the prokineticins (endocrine gland-derived vascular endothelial growth factor and Bv8) and their receptors in the human endometrium across the menstrual cycle'. *J Clin Endocrinol Metab*, 89 (5):2463-2469.

Bellomo, D., Headrick, J.P., Silins, G.U., Paterson, C.A., Thomas, P.S., Gartside, M., Mould, A., Cahill, M.M., Tonks, I.D., Grimmond, S.M., Townson, S., Wells, C., Little, M., Cummings, M.C., Hayward, N.K. and Kay, G.F., (2000). 'Mice lacking the vascular endothelial growth factor-B gene (Vegfb) have smaller hearts, dysfunctional coronary vasculature, and impaired recovery from cardiac ischemia'. *Circ Res*, 86 (2):E29-35.

Bianco, P., Riminucci, M., Gronthos, S. and Robey, P.G., (2001). 'Bone marrow stromal stem cells: nature, biology, and potential applications'. *Stem Cells*, 19 (3):180-192.

Blum, B. and Benvenisty, N., (2008). 'The tumorigenicity of human embryonic stem cells'. *Adv Cancer Res*, 100:133-158.

Bock-Marquette, I., Shrivastava, S., Pipes, G.C., Thatcher, J.E., Blystone, A., Shelton, J.M., Galindo, C.L., Melegh, B., Srivastava, D., Olson, E.N. and DiMaio, J.M., (2009). 'Thymosin beta4 mediated PKC activation is essential to initiate the embryonic coronary developmental program and epicardial progenitor cell activation in adult mice in vivo'. *J Mol Cell Cardiol*, 46 (5):728-738.

Boulberdaa, M., Turkeri, G., Urayama, K., Dormishian, M., Szatkowski, C., Zimmer, L., Messaddeq, N., Laugel, V., Dolle, P. and Nebigil, C.G., (2011). 'Genetic inactivation of prokineticin receptor-1 leads to heart and kidney disorders'. *Arterioscler Thromb Vasc Biol*, 31 (4):842-850.

Bullock, C.M., Li, J.D. and Zhou, Q.Y., (2004). 'Structural determinants required for the bioactivities of prokineticins and identification of prokineticin receptor antagonists'. *Mol Pharmacol*, 65 (3):582-588.

Cai, C.L., Martin, J.C., Sun, Y., Cui, L., Wang, L., Ouyang, K., Yang, L., Bu, L., Liang, X., Zhang, X., Stallcup, W.B., Denton, C.P., McCulloch, A., Chen, J. and Evans, S.M., (2008). 'A myocardial lineage derives from Tbx18 epicardial cells'. *Nature*, 454 (7200):104-108.

Canto, P., Munguia, P., Soderlund, D., Castro, J.J. and Mendez, J.P., (2009). 'Genetic analysis in patients with Kallmann syndrome: coexistence of mutations in prokineticin receptor 2 and KAL1'. *J Androl*, 30 (1):41-45.

Choke, E., Cockerill, G.W., Laing, K., Dawson, J., Wilson, W.R., Loftus, I.M. and Thompson, M.M., (2009). 'Whole genome-expression profiling reveals a role for immune and inflammatory response in abdominal aortic aneurysm rupture'. *Eur J Vasc Endovasc Surg*, 37 (3):305-310.

Cole, L.W., Sidis, Y., Zhang, C., Quinton, R., Plummer, L., Pignatelli, D., Hughes, V.A., Dwyer, A.A., Raivio, T., Hayes, F.J., Seminara, S.B., Huot, C., Alos, N., Speiser, P., Takeshita, A., Van Vliet, G., Pearce, S., Crowley, W.F., Jr., Zhou, Q.Y. and Pitteloud, N., (2008). 'Mutations in prokineticin 2 and prokineticin receptor 2 genes in human gonadotrophin-releasing hormone deficiency: molecular genetics and clinical spectrum'. *J Clin Endocrinol Metab*, 93 (9):3551-3559.

Denison, F.C., Battersby, S., King, A.E., Szuber, M. and Jabbour, H.N., (2008). 'Prokineticin-1: a novel mediator of the inflammatory response in third-trimester human placenta'. *Endocrinology*, 149 (7):3470-3477.

Dode, C., Teixeira, L., Levilliers, J., Fouveaut, C., Bouchard, P., Kottler, M.L., Lespinasse, J., Lienhardt-Roussie, A., Mathieu, M., Moerman, A., Morgan, G., Murat, A., Toublanc, J.E., Wolczynski, S., Delpech, M., Petit, C., Young, J. and Hardelin, J.P., (2006). 'Kallmann syndrome: mutations in the genes encoding prokineticin-2 and prokineticin receptor-2'. *PLoS Genet*, 2 (10):e175.

Dorsch, M., Qiu, Y., Soler, D., Frank, N., Duong, T., Goodearl, A., O'Neil, S., Lora, J. and Fraser, C.C., (2005). 'PK1/EG-VEGF induces monocyte differentiation and activation'. *J Leukoc Biol*, 78 (2):426-434.

Dube, K.N., Bollini, S., Smart, N. and Riley, P.R., (2012). 'Thymosin beta4 protein therapy for cardiac repair'. *Curr Pharm Des*, 18 (6):799-806.

Evans, J., Catalano, R.D., Morgan, K., Critchley, H.O., Millar, R.P. and Jabbour, H.N., (2008). 'Prokineticin 1 signaling and gene regulation in early human pregnancy'. *Endocrinology*, 149 (6):2877-2887.

Fraser, H.M., Bell, J., Wilson, H., Taylor, P.D., Morgan, K., Anderson, R.A. and Duncan, W.C., (2005). 'Localization and quantification of cyclic changes in the expression of endocrine gland vascular endothelial growth factor in the human corpus luteum'. *J Clin Endocrinol Metab*, 90 (1):427-434.

Gaudin, C., Ishikawa, Y., Wight, D.C., Mahdavi, V., Nadal-Ginard, B., Wagner, T.E., Vatner, D.E. and Homcy, C.J., (1995). 'Overexpression of Gs alpha protein in the hearts of transgenic mice'. *J Clin Invest*, 95 (4):1676-1683.

Giannini, E., Lattanzi, R., Nicotra, A., Campese, A.F., Grazioli, P., Screpanti, I., Balboni, G., Salvadori, S., Sacerdote, P. and Negri, L., (2009). 'The chemokine Bv8/prokineticin 2 is up-regulated in inflammatory granulocytes and modulates inflammatory pain'. *Proc Natl Acad Sci U S A*, 106 (34):14646-14651.

Gonzales, C. and Pedrazzini, T., (2009). 'Progenitor cell therapy for heart disease'. *Exp Cell Res*, 315 (18):3077-3085.

Guilini, C., Urayama, K., Turkeri, G., Dedeoglu, D.B., Kurose, H., Messaddeq, N. and Nebigil, C.G., (2010). 'Divergent roles of prokineticin receptors in the endothelial cells: angiogenesis and fenestration'. *Am J Physiol Heart Circ Physiol*, 298 (3):H844-852.

Gutkind, J.S. and Offermanns, S., (2009). 'A new G(q)-initiated MAPK signaling pathway in the heart'. *Dev Cell*, 16 (2):163-164.

Herr, M.M., Fries, K.M., Upton, L.G. and Edsberg, L.E., 'Potential biomarkers of temporomandibular joint disorders'. *J Oral Maxillofac Surg*, 69 (1):41-47.

Hoogerwerf, W.A., (2006). 'Prokineticin 1 inhibits spontaneous giant contractions in the murine proximal colon through nitric oxide release'. *Neurogastroenterol Motil*, 18 (6):455-463.

Hu, W.P., Li, J.D., Zhang, C., Boehmer, L., Siegel, J.M. and Zhou, Q.Y., (2007). 'Altered circadian and homeostatic sleep regulation in prokineticin 2-deficient mice'. *Sleep*, 30 (3):247-256.

Kang, J., Gu, Y., Li, P., Johnson, B.L., Sucov, H.M. and Thomas, P.S., (2008). 'PDGF-A as an epicardial mitogen during heart development'. *Dev Dyn*, 237 (3):692-701.

Kaplan, D.D., Meigs, T.E. and Casey, P.J., (2001). 'Distinct regions of the cadherin cytoplasmic domain are essential for functional interaction with Galpha 12 and beta-catenin'. *J Biol Chem*, 276 (47):44037-44043.

Kaser, A., Winklmayr, M., Lepperdinger, G. and Kreil, G., (2003). 'The AVIT protein family. Secreted cysteine-rich vertebrate proteins with diverse functions'. *EMBO Rep*, 4 (5):469-473.

Kishi, T., Kitajima, T., Tsunoka, T., Okumura, T., Ikeda, M., Okochi, T., Kinoshita, Y., Kawashima, K., Yamanouchi, Y., Ozaki, N. and Iwata, N., (2009). 'Possible association of prokineticin 2 receptor gene (PROKR2) with mood disorders in the Japanese population'. *Neuromolecular Med*, 11 (2):114-122.

Kishi, T., Kitajima, T., Tsunoka, T., Okumura, T., Kawashima, K., Okochi, T., Yamanouchi, Y., Kinoshita, Y., Ujike, H., Inada, T., Yamada, M., Uchimura, N., Sora, I., Iyo, M., Ozaki, N. and Iwata, N., (2010). 'Lack of association between prokineticin 2 gene and Japanese methamphetamine dependence'. *Curr Neuropharmacol*, 133-136.

Kisliouk, T., Podlovni, H., Spanel-Borowski, K., Ovadia, O., Zhou, Q.Y. and Meidan, R., (2005). 'Prokineticins (endocrine gland-derived vascular endothelial growth factor and BV8) in the bovine ovary: expression and role as mitogens and survival factors for corpus luteum-derived endothelial cells'. *Endocrinology*, 146 (9):3950-3958.

LeCouter, J. and Ferrara, N., (2002). 'EG-VEGF and the concept of tissue-specific angiogenic growth factors'. *Semin Cell Dev Biol*, 13 (1):3-8.

LeCouter, J. and Ferrara, N., (2003). 'EG-VEGF and Bv8. a novel family of tissue-selective mediators of angiogenesis, endothelial phenotype, and function'. *Trends Cardiovasc Med*, 13 (7):276-282.

LeCouter, J., Zlot, C., Tejada, M., Peale, F. and Ferrara, N., (2004). 'Bv8 and endocrine gland-derived vascular endothelial growth factor stimulate hematopoiesis and hematopoietic cell mobilization'. *Proc Natl Acad Sci U S A*, 101 (48):16813-16818.

Lepilina, A., Coon, A.N., Kikuchi, K., Holdway, J.E., Roberts, R.W., Burns, C.G. and Poss, K.D., (2006). 'A dynamic epicardial injury response supports progenitor cell activity during zebrafish heart regeneration'. *Cell*, 127 (3):607-619.

Li, J.D., Hu, W.P., Boehmer, L., Cheng, M.Y., Lee, A.G., Jilek, A., Siegel, J.M. and Zhou, Q.Y., (2006). 'Attenuated circadian rhythms in mice lacking the prokineticin 2 gene'. *J Neurosci*, 26 (45):11615-11623.

Li, J.D., Hu, W.P. and Zhou, Q.Y., (2009). 'Disruption of the circadian output molecule prokineticin 2 results in anxiolytic and antidepressant-like effects in mice'. *Neuropsychopharmacology*, 34 (2):367-373.

Li, M., Bullock, C.M., Knauer, D.J., Ehlert, F.J. and Zhou, Q.Y., (2001). 'Identification of two prokineticin cDNAs: recombinant proteins potently contract gastrointestinal smooth muscle'. *Mol Pharmacol*, 59 (4):692-698.

Limana, F., Zacheo, A., Mocini, D., Mangoni, A., Borsellino, G., Diamantini, A., De Mori, R., Battistini, L., Vigna, E., Santini, M., Loiaconi, V., Pompilio, G., Germani, A. and Capogrossi, M.C., (2007). 'Identification of myocardial and vascular precursor cells in human and mouse epicardium'. *Circ Res*, 101 (12):1255-1265.

Marinissen, M.J. and Gutkind, J.S., (2001). 'G-protein-coupled receptors and signaling networks: emerging paradigms'. *Trends Pharmacol Sci*, 22 (7):368-376.

Masuda, Y., Takatsu, Y., Terao, Y., Kumano, S., Ishibashi, Y., Suenaga, M., Abe, M., Fukusumi, S., Watanabe, T., Shintani, Y., Yamada, T., Hinuma, S., Inatomi, N., Ohtaki, T., Onda, H. and Fujino, M., (2002). 'Isolation and identification of EG-VEGF/prokineticins as cognate ligands for two orphan G-protein-coupled receptors'. *Biochem Biophys Res Commun*, 293 (1):396-402.

Mummery, C., Ward, D., van den Brink, C.E., Bird, S.D., Doevendans, P.A., Opthof, T., Brutel de la Riviere, A., Tertoolen, L., van der Heyden, M. and Pera, M., (2002). 'Cardiomyocyte differentiation of mouse and human embryonic stem cells'. *J Anat*, 200 (Pt 3):233-242.

Murry, C.E., Soonpaa, M.H., Reinecke, H., Nakajima, H., Nakajima, H.O., Rubart, M., Pasumarthi, K.B., Virag, J.I., Bartelmez, S.H., Poppa, V., Bradford, G., Dowell, J.D., Williams, D.A. and Field, L.J., (2004). 'Haematopoietic stem cells do not transdifferentiate into cardiac myocytes in myocardial infarcts'. *Nature*, 428 (6983):664-668.

Nebigil, C.G., (2009). 'Prokineticin receptors in cardiovascular function: foe or friend?' *Trends Cardiovasc Med*, 19 (2):55-60.

Nebigil, C.G., Etienne, N., Messaddeq, N. and Maroteaux, L., (2003). 'Serotonin is a novel survival factor of cardiomyocytes: mitochondria as a target of 5-HT2B receptor signaling'. *FASEB J*, 17 (10):1373-1375.

Negri, L., Lattanzi, R., Giannini, E., Colucci, M., Margheriti, F., Melchiorri, P., Vellani, V., Tian, H., De Felice, M. and Porreca, F., (2006). 'Impaired nociception and inflammatory pain sensation in mice lacking the prokineticin receptor PKR1: focus on interaction between PKR1 and the capsaicin receptor TRPV1 in pain behavior'. *J Neurosci*, 26 (25):6716-6727.

Negri, L., Lattanzi, R., Giannini, E., De Felice, M., Colucci, A. and Melchiorri, P., (2004). 'Bv8, the amphibian homologue of the mammalian prokineticins, modulates ingestive behaviour in rats'. *Br J Pharmacol*, 142 (1):181-191.

Ng, K.L., Li, J.D., Cheng, M.Y., Leslie, F.M., Lee, A.G. and Zhou, Q.Y., (2005). 'Dependence of olfactory bulb neurogenesis on prokineticin 2 signaling'. *Science*, 308 (5730):1923-1927.

Ngan, E.S., Lee, K.Y., Sit, F.Y., Poon, H.C., Chan, J.K., Sham, M.H., Lui, V.C. and Tam, P.K., (2007a). 'Prokineticin-1 modulates proliferation and differentiation of enteric neural crest cells'. *Biochim Biophys Acta*, 1773 (4):536-545.

Ngan, E.S., Sit, F.Y., Lee, K., Miao, X., Yuan, Z., Wang, W., Nicholls, J.M., Wong, K.K., Garcia-Barcelo, M., Lui, V.C. and Tam, P.K., (2007b). 'Implications of endocrine gland-derived vascular endothelial growth factor/prokineticin-1 signaling in human neuroblastoma progression'. *Clin Cancer Res*, 13 (3):868-875.

Ngan, E.S. and Tam, P.K., (2008). 'Prokineticin-signaling pathway'. *Int J Biochem Cell Biol*, 40 (9):1679-1684.

Offermanns, S., Mancino, V., Revel, J.P. and Simon, M.I., (1997). 'Vascular system defects and impaired cell chemokinesis as a result of Galpha13 deficiency'. *Science*, 275 (5299):533-536.

Pasquali, D., Rossi, V., Staibano, S., De Rosa, G., Chieffi, P., Prezioso, D., Mirone, V., Mascolo, M., Tramontano, D., Bellastella, A. and Sinisi, A.A., (2006). 'The endocrine-gland-derived vascular endothelial growth factor (EG-VEGF)/prokineticin 1 and 2 and receptor expression in human prostate: Up-regulation of EG-VEGF/prokineticin 1 with malignancy'. *Endocrinology*, 147 (9):4245-4251.

Pasquali, D., Santoro, A., Bufo, P., Conzo, G., Deery, W.J., Renzullo, A., Accardo, G., Sacco, V., Bellastella, A. and Pannone, G., (2011). 'Upregulation of endocrine gland-derived vascular endothelial growth factor in papillary thyroid cancers displaying infiltrative patterns, lymph node metastases, and BRAF mutation'. *Thyroid*, 21 (4):391-399.

Sanchez, N.S. and Barnett, J.V. (2012). 'TGFbeta and BMP-2 regulate epicardial cell invasion via TGFbetaR3 activation of the Par6/Smurf1/RhoA pathway'. *Cell Signal*, 24 (2):539-548.

Schlyer, S. and Horuk, R., (2006). 'I want a new drug: G-protein-coupled receptors in drug development'. *Drug Discov Today*, 11 (11-12):481-493.

Shaw, J.L., Denison, F.C., Evans, J., Durno, K., Williams, A.R., Entrican, G., Critchley, H.O., Jabbour, H.N. and Horne, A.W., (2010). 'Evidence of prokineticin dysregulation in fallopian tube from women with ectopic pregnancy'. *Fertil Steril*, 94 (5):1601-1608 e1601.

Smart, N., Bollini, S., Dube, K.N., Vieira, J.M., Zhou, B., Davidson, S., Yellon, D., Riegler, J., Price, A.N., Lythgoe, M.F., Pu, W.T. and Riley, P.R., (2011). 'De novo cardiomyocytes from within the activated adult heart after injury'. *Nature*, 474 (7353):640-644.

Smart, N. and Riley, P.R., (2012). 'The epicardium as a candidate for heart regeneration'. *Future Cardiol*, 8 (1):53-69.

Smart, N., Risebro, C.A., Melville, A.A., Moses, K., Schwartz, R.J., Chien, K.R. and Riley, P.R., (2007). 'Thymosin beta4 induces adult epicardial progenitor mobilization and neovascularization'. *Nature*, 445 (7124):177-182.

Soga, T., Matsumoto, S., Oda, T., Saito, T., Hiyama, H., Takasaki, J., Kamohara, M., Ohishi, T., Matsushime, H. and Furuichi, K., (2002). 'Molecular cloning and characterization of prokineticin receptors'. *Biochim Biophys Acta*, 1579 (2-3):173-179.

Su, M.T., Lin, S.H., Lee, I.W., Chen, Y.C., Hsu, C.C., Pan, H.A. and Kuo, P.L., 'Polymorphisms of endocrine gland-derived vascular endothelial growth factor gene and its receptor genes are associated with recurrent pregnancy loss'. *Hum Reprod*, 25 (11):2923-2930.

Tang, C.M. and Insel, P.A., (2004). 'GPCR expression in the heart; "new" receptors in myocytes and fibroblasts'. *Trends Cardiovasc Med*, 14 (3):94-99.

Tevosian, S.G., Deconinck, A.E., Tanaka, M., Schinke, M., Litovsky, S.H., Izumo, S., Fujiwara, Y. and Orkin, S.H., (2000). 'FOG-2, a cofactor for GATA transcription factors, is essential for heart morphogenesis and development of coronary vessels from epicardium'. *Cell*, 101 (7):729-739.

Tiberi, F., Tropea, A., Apa, R., Romani, F., Lanzone, A. and Marana, R., (2009). 'Prokineticin 1 mRNA expression in the endometrium of healthy women and in the eutopic endometrium of women with endometriosis'. *Fertil Steril*, 93 (7):2145-2149.

Urayama, K., Dedeoglu, D.B., Guilini, C., Frantz, S., Ertl, G., Messaddeq, N. and Nebigil, C.G., (2009). 'Transgenic myocardial overexpression of prokineticin receptor-2 (GPR73b) induces hypertrophy and capillary vessel leakage'. *Cardiovasc Res*, 81 (1):28-37.

Urayama, K., Guilini, C., Messaddeq, N., Hu, K., Steenman, M., Kurose, H., Ert, G. and Nebigil, C.G., (2007). 'The prokineticin receptor-1 (GPR73) promotes cardiomyocyte survival and angiogenesis'. *FASEB J*, 21 (11):2980-2993.

Urayama, K., Guilini, C., Turkeri, G., Takir, S., Kurose, H., Messaddeq, N., Dierich, A. and Nebigil, C.G., (2008). 'Prokineticin receptor-1 induces neovascularization and epicardial-derived progenitor cell differentiation'. *Arterioscler Thromb Vasc Biol*, 28 (5):841-849.

van Tuyn, J., Atsma, D.E., Winter, E.M., van der Velde-van Dijke, I., Pijnappels, D.A., Bax, N.A., Knaan-Shanzer, S., Gittenberger-de Groot, A.C., Poelmann, R.E., van der Laarse, A., van der Wall, E.E., Schalij, M.J. and de Vries, A.A., (2007). 'Epicardial cells of human adults can undergo an epithelial-to-mesenchymal transition and obtain characteristics of smooth muscle cells in vitro'. *Stem Cells*, 25 (2):271-278.

van Wijk, B., van den Berg, G., Abu-Issa, R., Barnett, P., van der Velden, S., Schmidt, M., Ruijter, J.M., Kirby, M.L., Moorman, A.F. and van den Hoff, M.J., (2009). 'Epicardium and myocardium separate from a common precursor pool by crosstalk between bone morphogenetic protein- and fibroblast growth factor-signaling pathways'. *Circ Res*, 105 (5):431-441.

von Gise, A., Zhou, B., Honor, L.B., Ma, Q., Petryk, A. and Pu, W.T., (2011). 'WT1 regulates epicardial epithelial to mesenchymal transition through beta-catenin and retinoic acid signaling pathways'. *Dev Biol*, 356 (2):421-431.

Wade, P.R., Palmer, J.M., Mabus, J., Saunders, P.R., Prouty, S., Chevalier, K., Gareau, M.G., McKenney, S. and Hornby, P.J., (2009). 'Prokineticin-1 evokes secretory and contractile activity in rat small intestine'. *Neurogastroenterol Motil*, 22 (5):e152-161.

Wessels, A. and Perez-Pomares, J.M., (2004). 'The epicardium and epicardially derived cells (EPDCs) as cardiac stem cells'. *Anat Rec A Discov Mol Cell Evol Biol*, 276 (1):43-57.

Wettschureck, N., Rutten, H., Zywietz, A., Gehring, D., Wilkie, T.M., Chen, J., Chien, K.R. and Offermanns, S., (2001). 'Absence of pressure overload induced myocardial hypertrophy after conditional inactivation of Galphaq/Galpha11 in cardiomyocytes'. *Nat Med*, 7 (11):1236-1240.

Winter, E.M., Grauss, R.W., Hogers, B., van Tuyn, J., van der Geest, R., Lie-Venema, H., Steijn, R.V., Maas, S., DeRuiter, M.C., deVries, A.A., Steendijk, P., Doevendans, P.A., van der Laarse, A., Poelmann, R.E., Schalij, M.J., Atsma, D.E. and Gittenberger-de Groot, A.C., (2007). 'Preservation of left ventricular function and attenuation of remodeling after transplantation of human epicardium-derived cells into the infarcted mouse heart'. *Circulation*, 116 (8):917-927.

Winter, E.M., van Oorschot, A.A., Hogers, B., van der Graaf, L.M., Doevendans, P.A., Poelmann, R.E., Atsma, D.E., Gittenberger-de Groot, A.C. and Goumans, M.J., (2009). 'A new direction for cardiac regeneration therapy: application of synergistically acting epicardium-derived cells and cardiomyocyte progenitor cells'. *Circ Heart Fail*, 2 (6):643-653.

Zamora, M., Manner, J. and Ruiz-Lozano, P., (2007). 'Epicardium-derived progenitor cells require beta-catenin for coronary artery formation'. *Proc Natl Acad Sci U S A*, 104 (46):18109-18114.

Zannad, F., Agrinier, N. and Alla, F., (2009). 'Heart failure burden and therapy'. *Europace*, 11 Suppl 5:v1-9.

Zhong, C., Qu, X., Tan, M., Meng, Y.G. and Ferrara, N., (2009). 'Characterization and regulation of bv8 in human blood cells'. *Clin Cancer Res*, 15 (8):2675-2684.

Zhou, B., Honor, L.B., He, H., Ma, Q., Oh, J.H., Butterfield, C., Lin, R.Z., Melero-Martin, J.M., Dolmatova, E., Duffy, H.S., Gise, A., Zhou, P., Hu, Y.W., Wang, G., Zhang, B., Wang, L., Hall, J.L., Moses, M.A., McGowan, F.X. and Pu, W.T., (2011). 'Adult mouse epicardium modulates myocardial injury by secreting paracrine factors'. *J Clin Invest*, 121 (5):1894-1904.

Zhou, B., Honor, L.B., Ma, Q., Oh, J.H., Lin, R.Z., Melero-Martin, J.M., von Gise, A., Zhou, P., Hu, T., He, L., Wu, K.H., Zhang, H., Zhang, Y. and Pu, W.T., (2012). 'Thymosin beta 4 treatment after myocardial infarction does not reprogram epicardial cells into cardiomyocytes'. *J Mol Cell Cardiol*, 52 (1):43-47.

Zhou, B., Ma, Q., Rajagopal, S., Wu, S.M., Domian, I., Rivera-Feliciano, J., Jiang, D., von Gise, A., Ikeda, S., Chien, K.R. and Pu, W.T., (2008). 'Epicardial progenitors contribute to the cardiomyocyte lineage in the developing heart'. *Nature*, 454 (7200):109-113.

Zimmermann, W.H., Didie, M., Doker, S., Melnychenko, I., Naito, H., Rogge, C., Tiburcy, M. and Eschenhagen, T., (2006). 'Heart muscle engineering: an update on cardiac muscle replacement therapy'. *Cardiovasc Res*, 71 (3):419-429.

Chapter 17

Hypoxic Pulmonary Vascular Smooth Muscle Cell Proliferation

Shiro Mizuno, Hirohisa Toga and Takeshi Ishizaki

Additional information is available at the end of the chapter

http://dx.doi.org/10.5772/45979

1. Introduction

Pulmonary arterial smooth muscle cell (PASMC) proliferation in response to hypoxia is thought to be a key component of the vascular remodeling that occurs in chronic hypoxic pulmonary hypertension. Many pulmonary disorders, including chronic obstructive pulmonary disease, are associated with chronic hypoxia, and when pulmonary hypertension and right heart failure develop due to pulmonary vascular remodeling, patient survival is impaired. Hypoxia is one of the factors known to cause secondary pulmonary hypertension and pulmonary vascular remodeling [1]. According to a WHO statement in 1996, there were approximately 140 million people living at altitudes above 2500m and there are several areas of permanent habitation at altitudes in excess of 4000 m. After several weeks of exposure to high altitude, lowlanders develop pulmonary hypertension, which is not completely reversed by supplemental oxygen [2], suggesting development of vascular remodeling of the lung [3]. Secondary pulmonary hypertension is characterized by proliferation of vascular smooth muscle cells and pulmonary arterial fibroblasts in small pulmonary vessels [4-6]. These results suggest that hypoxic enhancement of PASMC proliferation contributes to the progression of hypoxia-associated small pulmonary arterial remodeling and secondary pulmonary hypertension. In animals, hypoxia has been shown to cause pulmonary vascular wall thickening by inducing PASMC proliferation [7-9]. Most commonly, pulmonary vascular remodeling is studied in rats or mice exposed to 10% oxygen hypoxia for 2 to 8 weeks [9,10]. In the animals exposed to chronic hypoxia, muscular arteries increase the thickness and distal extension and migration of PASMC into normally non-muscular arteries can be observed (Figure 1).

In addition, many *in vitro* studies have also addressed the proliferation of vascular smooth muscle cells upon exposure to hypoxia [11,12]. Increased levels of growth factors derived from the accumulation of hypoxia-inducible factor 1α (HIF-1α) are thought to regulate

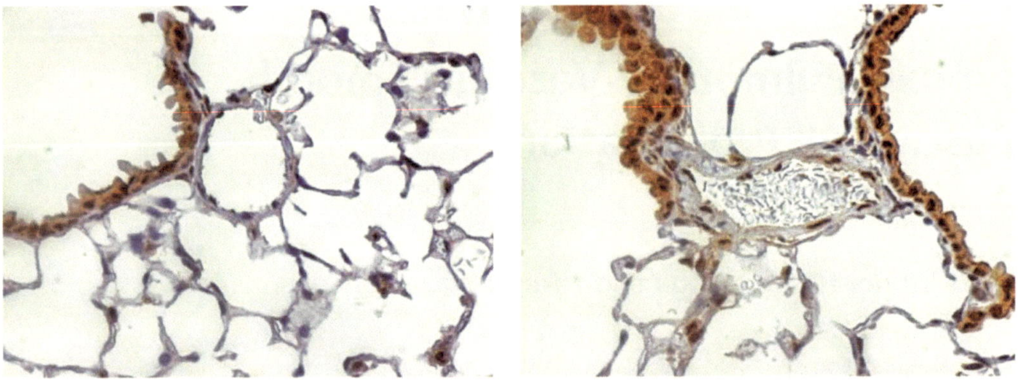

Figure 1. Proliferating cell nuclear antigen (PCNA) staining of small pulmonary arteries from mice exposed to chronic hypoxia (10% oxygen) for 8 weeks. PCNA positive cells were found in the smooth muscle layer of small pulmonary arteries (right picture) compared with normoxic control (left picture).

PASMC proliferation under hypoxic conditions, since a partial HIF-1α deficiency decreases muscularization of pulmonary arterioles in animals exposed to chronic hypoxia [7]. However, it is unclear whether hypoxia is a direct mitogen or indirect mitogen induced by the mediators from endothelial cells or fibroblasts, because some investigators have shown that hypoxia is not a direct stimulus of PASMC proliferation [13,14]. This discrepancy may be explained by the severity of hypoxia. The investigators who have found PASMC proliferation by hypoxic exposure usually used the moderate hypoxia (1 – 5% oxygen) [1,11,12,15,16].

Although HIF-1α regulates various transcriptional genes for angiogenic factors, severe hypoxia and iron depletion induce cell growth arrest. In contrast to severe hypoxia, moderate hypoxia can also enhances the proliferation of airway-smooth muscle cells, lung fibroblasts and mesangial cells [17,18]. Proliferation of PASMCs, which causes pulmonary vascular remodeling, requires re-entry of the cells into the cell cycle. We confirmed that the cultured PASMC cell cycle progresses more quickly in hypoxia and that severe hypoxia or iron depletion using an iron chelator, which mimics anoxia, caused inhibition of cell cycle progress compared to the normoxic conditions (Figure 2).

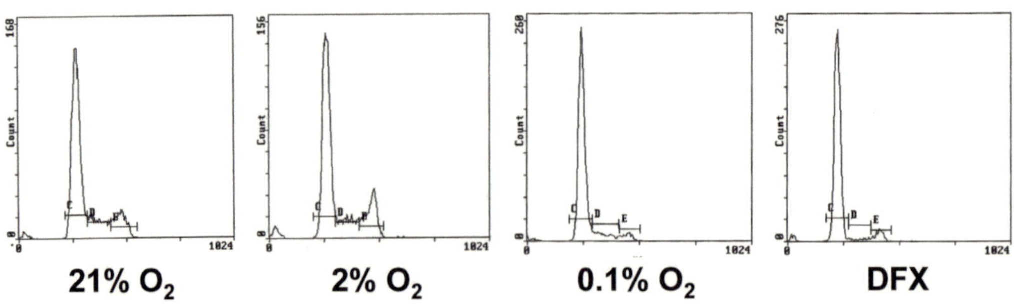

Figure 2. Cell cycle analysis of human pulmonary arterial smooth muscle cells (HPASMC) cultured in various concentration of oxygen (0.1%, 2%, 21%) or 100 μM of iron chelator, desferroxamine (DFX) using flow cytometric analyses with propidium iodide staining. S+M phases are increased in moderate hypoxia (2% oxygen) compared to the normoxic condition. Severe hypoxia (0.1% oxygen) and iron chelator decreased the S+M phases in cultured HPASMCs.

2. Cell cycle regulation of hypoxic PASMC proliferation

Under normal physiological conditions, the majority of the pulmonary vascular cells are in a quiescent state. The most important molecular event necessary for progression of the cell cycle is phosphorylation of the retinoblastoma protein by cyclin-dependent kinase (CDK)-cyclin complexes. Cell cycle progression requires the coordinated interaction of CDK and its regulatory subunits, the cyclins, to drive cells through G1 into S phase to ultimately result in cell division. Cyclin–CDK complexes activate transcription factors important in cell cycle progression. The cyclin–CDK complexes include cyclin D–CDK4/ CDK6 and cyclin E–CDK2 which inactivate retinoblastoma, an antitumor and antiproliferative protein which limits E2F-mediated gene transcription [19]. CDK inhibitors are proteins that bind cyclin–CDK complexes, inhibit hyperphosphorylation of retinoblastoma, cause G1 arrest, and suppress cell proliferation [20]. CDK activity can be inhibited by CDK inhibitors, which arrest the cell cycle at each corresponding phase and inhibit cell proliferation. Two families of CDK inhibitors have been shown to regulate vascular smooth muscle cell proliferation, p21 and p27. The loss of CDK inhibitors has been implicated in tumor development [21,22], and is closely related with the state of the tumor suppressor p53 [23,24].

2.1. Role of tumor suppressor p53 and CDK inhibitor p21

The endogenous CDK inhibitor p21 plays an important role in PASMC proliferation via induction of the tumor suppressor p53 [23,25], and has been identified as a key regulator of the cell cycle in cells exposed to hypoxia and oxidative stress [26-28]. In tumors expressing wild-type p53, apoptosis occurs in hypoxic regions, whereas tumors expressing mutant p53 exhibit lower levels of apoptosis in hypoxic regions [29]. p53-/- mouse embryo fibroblasts are more resistant to hypoxia-induced apoptosis, and have selective growth advantages compared to wild-type p53 cells [30]. In addition, hypoxic p53 accumulation has been linked to the hypoxia-inducible factor-1α (HIF-1α), which is known as a central transcriptional factor operating during hypoxia toward angiogenesis [31,32]. We recently reported that hypoxic p53 accumulation has been linked to the hypoxia inducible factor-1α (HIF-1α) [9]. These results support the view that the p53 protein opposes cell proliferation under hypoxia, and p53 plays a critical role as a modulator of hypoxia-induced small pulmonary arterial remodeling. Decreased expression of p21 and increased expression of HIF-1α via suppression of p53 protein may mitigate hypoxic pulmonary arterial remodeling and PASMC proliferation. Recently, several groups identified microRNAs (miRNAs) regulated by p53 [33,34]. miRNAs are non-coding RNA molecules which modulate gene expression by binding to complementary sequences in the coding - or the 3`-untranslated region of target mRNAs. miRNAs can regulate cell proliferation, differentiation and apoptosis [35,36]. The miR34a has been shown to be the most significant miRNA induced by p53, which is closely related to induction of apoptosis and cell cycle arrest in cancer cells [37]. We have shown that this miRNA is also associated with HIF-1α expression both in animal and human lung tissues [9,38,39].

p21 has been shown to regulate cell cycle progression through both p53-dependent and -independent pathways [24,25,40]. However, it is known that nitric oxide (NO) donors suppress proliferation of cultured PASMC via the expression of p53 and p21 [23]. NO is synthesized from L-arginine via nitric oxide synthase (NOS), and endothelial NOS plays an important regulatory role in hypertrophic and hyperplastic growth of PASMC *in vivo* and *in vitro*. Several studies have suggested that NO derived from endothelial NOS has a protective effect toward arterial smooth muscle cell proliferation [41].[42]. It is possible that the anti-proliferative effect of NO derived from pulmonary arterial endothelial cells is depend on the status of p53 in PASMC.

In our recent data from p53 knockout mice, chronic hypoxia increased p21 expression and induced medial wall thickening of small pulmonary arteries in wild type mice, and the deletion of the p53 gene prevented the hypoxic induction of p21 expression [9]. These results indicate that under hypoxic conditions, induction of the p53-p21 signaling pathway serves as a negative feed-back to prevent excessive vascular cell proliferation and vascular remodeling. Using cultured PASMCs, we confirmed that the anti-proliferative NO pathway was intact in the hypoxic condition and the protein expression of p21 was associated with HPASMC proliferation (Figure 3).

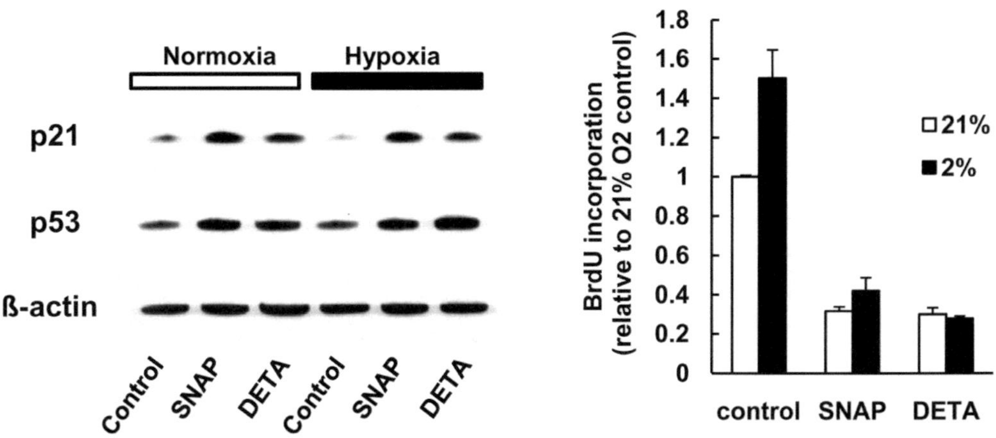

Figure 3. Western blot analysis of p21 and p53 (left photographs) and BrdU incorporations in cultured PASMC exposed hypoxia (2% oxygen) and NO donors (SNAP and DETANO) (right graph). Moderate hypoxia decreased p21 and the NO donors increased the both p21 and p53 protein expressions. The NO donors suppressed DNA amplification in cultured PASMCs during hypoxia.

2.2. CDK inhibitor p27

The suppressive effect of hypoxia on p27 expression has been demonstrated in mice with pulmonary hypertension induced by hypoxia [8]. However, the expression of p27, which blocks the cell cycle at the G0/1 phase, is regulated by several mechanisms including transcription, protein degradation and translation [43-45]. We reported that hypoxia-induced down-regulation of p27 was not mediated by hypoxia *per se*, but rather by mitogenic factors including PDGF and hypoxia enhanced p27 protein degradation. The moderate hypoxia enhanced the proliferation of serum-stimulated PASMC in accordance with promoted p27 protein degradation, probably via the induction of growth factors [12]. The prostacyclin analogue suppressed PASMC proliferation under both hypoxic and normoxic conditions by blocking p27 mRNA degradation through an increase in intracellular cAMP. Hypoxia may activate HIF-1α regulated growth factors and cell growth signaling such as mitogen activated protein kinase (MAPK). We also found that hypoxic exposure and p53 regulate MAPK activation in cultured PASMCs (Figure 4). To clarify the effect of p53 on the activation of MAPK in hypoxic PASMC proliferation, we performed gene silencing of p53 to the cultured PASMC. The gene silencing of p53 suppressed MAPK activation, which indicates hypoxic activation of MAPK and p53 are also associated with the degradation of p27 and hypoxic PASMC proliferation.

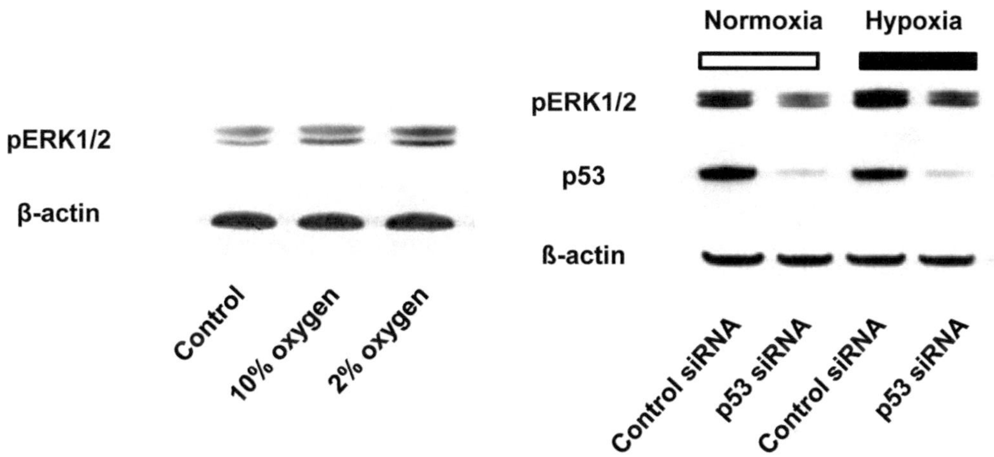

Figure 4. Left figure shows protein expressions of phosphorylated ERK1/2 in cultured PASMC exposed hypoxia (2 - 10% oxygen). Right figure shows effect of p53 gene silencing on ERK1/2 phosphorylation in cultured PASMC exposed to hypoxia (2% oxygen). Hypoxic exposure induced MAPK activation, and the p53 protein expression are suppressing the MAPK activation.

Previous findings have suggested that p27 mRNA stability is controlled by interactions between MAPK [46] and Rho-dependent translation[47]. Further, cAMP induces cell relaxation through Rho GTPase activation [48,49], which might be an important target of hypoxic pulmonary vascular remodeling [50,51]. These reports imply that the Rho and MAPK interaction contributes to p27 mRNA stability during exposure to agents that elevate cAMP and hypoxia. These interactions may be also possible in smooth muscle cells. The Rho inhibitor Y-27632 inhibited human aortic smooth muscle proliferation in response to platelet derived growth factor and markedly suppressed neointima formation associated with decreased expression of p27 in rat carotid artery [52]. Regarding the effect of Rho on the expression of p27 in PASMC, we simply measure the p27 protein expression using Rho inhibitor in cultured PASMC and found that the p27 protein expression was increased by Y-27632 (Figure 5).

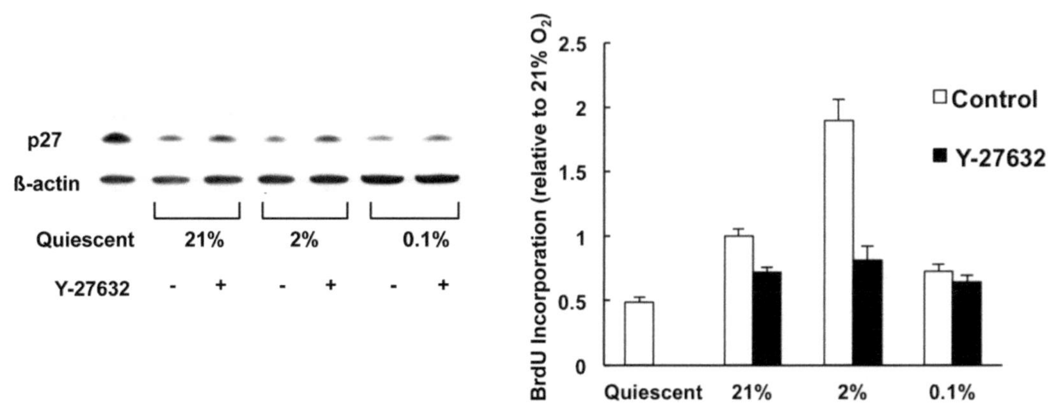

Figure 5. p27 protein expression in cultured PASMC exposed to hypoxia with 10 μM of Rho inhibitor Y-27632 (left figure) and BrdU incorporations in cultured PASMC exposed to hypoxia with or without Y-27632 (right graph). The p27 protein expression was increased by the treatment of Rho inhibitor both in the normoxic and hypoxic conditions with association of the cell proliferation.

3. Conclusion

It is well accepted that hypoxia is a cause of pulmonary vascular remodeling and PASMC proliferation. In the aggregate, research conducted by us and others suggests that decreased oxygen levels affect PASMC proliferation. Decreased expression of p27 and signal transduction via p53 and p21 play critical roles in the fine-tuning of hypoxic PASMC proliferation. *In vitro* studies have demonstrated that hypoxia has direct mitogenic effect on cultured PASMC by increased production of growth factors and decreased expression of CDK inhibitors. The cell cycle regulation by p53 may be closely related with the severity of hypoxia and HIF-1α status (Figure 6).

However, the hypoxia induced remodeling of the pulmonary circulation including PASMC proliferation is a highly complex process, which may have numerous interactions between the vascular cells, especially between endothelial cells and lung fibroblasts. Because of that, it is difficult to explain the *in vivo* pulmonary vascular remodeling using single cell culture experiments. For example, hypoxic endothelial cells and adventitial fibroblasts around PASMC may also be able to release mitogenic factors for PASMC proliferation, and damage of vascular endothelial cells by hypoxia can lead the decrease of anti-proliferative mediator production resulting PASMC proliferation. In addition, hypoxic proliferation of lung fibroblasts can secrete matrix proteins, which may play an important role for the proliferation of PASMC [53]. Further studies are necessary to characterize the role of cell cycle regulations on the hypoxic vascular cell proliferation, and to clarify the interactions between PASMC, pulmonary vascular endothelial cells and fibroblasts. We believe that a better understanding of the genetic and cellular mechanisms of hypoxic pulmonary remodeling will lead to improved modes of therapy for hypoxia-associated changes in lung tissue structure and aid in the remodeling of pulmonary hypertension.

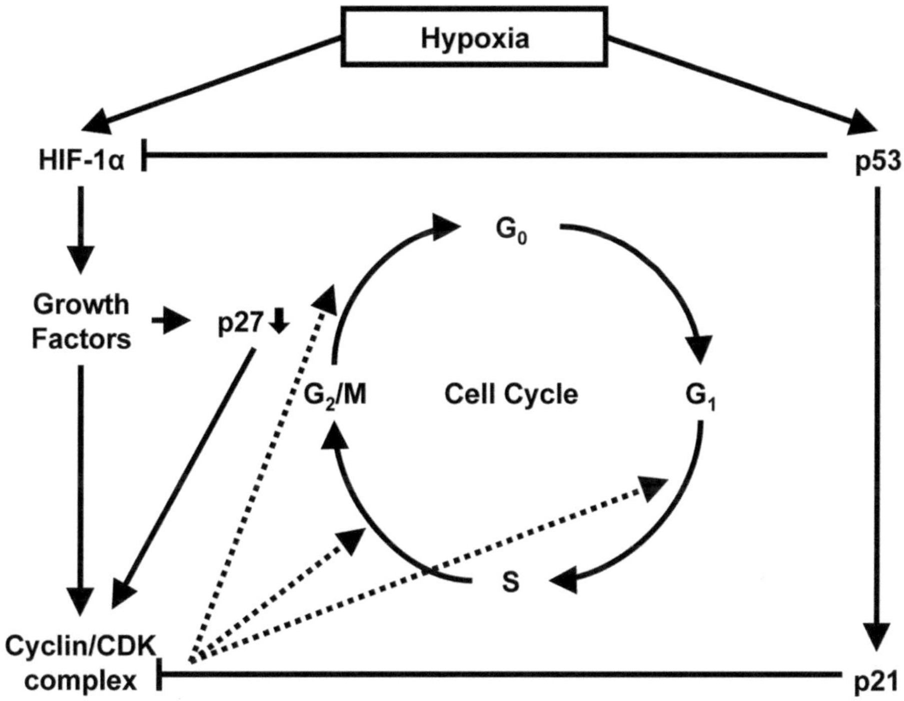

Figure 6. Schematic depicting molecular interactions in the hypoxic PASMC proliferation: Hypoxic exposure increases HIF-1α and p53 protein expression. Up-regulated p53 induces p21 expression and the p21 inhibit G1/0 transition via inhibition of cyclin/CDK complex. In contrast, HIF-1α activation causes up-regulation of growth factors that reduce the expression of p27. Reduced expression of p27 and increased HIF-1α transactivation induces cyclin/CDK complex, which causes progress of cell cycle of pulmonary smooth muscle cell.

Author details

Shiro Mizuno[1,*], Hirohisa Toga[1] and Takeshi Ishizaki[2]
[1]*Division of Respiratory Disease, Kanazawa Medical University, Ishikawa, Japan*
[2]*Department of Fundamental Nursing, University of Fukui, Fukui, Japan*

Acknowledgement

We wish to thank Prof. Norbert F Voelkel, Virginia Commonwealth University, Richmond, VA, USA and Dr. Herman J. Bogaard, VU University Medical Center, Amsterdam, the Netherlands, for their critical reading of this manuscript.

4. References

[1] A. Cogo, G. Napolitano, M.C. Michoud, D.R. Barbon, M. Ward, J.G. Martin (2003) Effects of hypoxia on rat airway smooth muscle cell proliferation. J. Appl. Physiol. 94: 1403–1409.

[2] B.M. Groves, J.T. Reeves, J.R. Sutton, P.D. Wagner, A. Cymerman, M.K. Malconian, P.B. Rock, P.M. Young, C.S. Houston (1987) Operation Everest II: elevated high-altitude pulmonary resistance unresponsive to oxygen. J. Appl. Physiol. 63: 521–530.

[3] R. Hainsworth, M.J. Drinkhill (2007) Cardiovascular adjustments for life at high altitude. Respir Physiol Neurobiol. 158: 204–211.

[4] W. Biernacki, D.C. Flenley, A.L. Muir, W. MacNee (1988) Pulmonary hypertension and right ventricular function in patients with COPD. Chest. 94: 1169–1175.

[5] W. MacNee (1994) Pathophysiology of cor pulmonale in chronic obstructive pulmonary disease Part One. Am. J. Respir. Crit. Care Med. 150: 833–852.

[6] M. Semmens, L. Reid (1974) Pulmonary arterial muscularity and right ventricular hypertrophy in chronic bronchitis and emphysema. Br J Dis Chest. 68: 253–263.

[7] A.Y. Yu, L.A. Shimoda, N.V. Iyer, D.L. Huso, X. Sun, R. McWilliams, T. Beaty, J.S. Sham, C.M. Wiener, J.T. Sylvester, G.L. Semenza (1999) Impaired physiological responses to chronic hypoxia in mice partially deficient for hypoxia-inducible factor 1alpha. J. Clin. Invest. 103: 691–696.

[8] L. Yu, D.A. Quinn, H.G. Garg, C.A. Hales (2005) Cyclin-dependent kinase inhibitor p27Kip1, but not p21WAF1/Cip1, is required for inhibition of hypoxia-induced pulmonary hypertension and remodeling by heparin in mice. Circ. Res. 97: 937–945.

[9] S. Mizuno, H.J. Bogaard, D. Kraskauskas, A. Alhussaini, J. Gomez-Arroyo, N.F. Voelkel, T. Ishizaki (2011) p53 Gene deficiency promotes hypoxia-induced pulmonary hypertension and vascular remodeling in mice. Am. J. Physiol. Lung Cell Mol. Physiol. 300: L753–61.

[10] H.J. Bogaard, R. Natarajan, S. Mizuno, A. Abbate, P.J. Chang, V.Q. Chau, N.N. Hoke, D. Kraskauskas, M. Kasper, F.N. Salloum, N.F. Voelkel (2010) Adrenergic receptor

* Corresponding Author

blockade reverses right heart remodeling and dysfunction in pulmonary hypertensive rats. Am. J. Respir. Crit. Care Med. 182: 652–660.

[11] A.L. Cooper, D. Beasley (1999) Hypoxia stimulates proliferation and interleukin-1alpha production in human vascular smooth muscle cells. Am. J. Physiol. 277: H1326–37.

[12] M. Kadowaki, S. Mizuno, Y. Demura, S. Ameshima, I. Miyamori, T. Ishizaki (2007) Effect of hypoxia and Beraprost sodium on human pulmonary arterial smooth muscle cell proliferation: the role of p27kip1. Respir. Res. 8: 77.

[13] S. Eddahibi, V. Fabre, C. Boni, M.P. Martres, B. Raffestin, M. Hamon, S. Adnot (1999) Induction of serotonin transporter by hypoxia in pulmonary vascular smooth muscle cells Relationship with the mitogenic action of serotonin. Circ. Res. 84: 329–336.

[14] L. Stiebellehner, M.G. Frid, J.T. Reeves, R.B. Low, M. Gnanasekharan, K.R. Stenmark (2003) Bovine distal pulmonary arterial media is composed of a uniform population of well-differentiated smooth muscle cells with low proliferative capabilities. Am. J. Physiol. Lung Cell Mol. Physiol. 285: L819–28.

[15] M. Tamm, M. Bihl, O. Eickelberg, P. Stulz, A.P. Perruchoud, M. Roth (1998) Hypoxia-induced interleukin-6 and interleukin-8 production is mediated by platelet-activating factor and platelet-derived growth factor in primary human lung cells. Am. J. Respir. Cell Mol. Biol. 19: 653–661.

[16] D.B. Frank, A. Abtahi, D.J. Yamaguchi, S. Manning, Y. Shyr, A. Pozzi, H.S. Baldwin, J.E. Johnson, M.P. de Caestecker (2005) Bone morphogenetic protein 4 promotes pulmonary vascular remodeling in hypoxic pulmonary hypertension. Circ. Res. 97: 496–504.

[17] S. Krick, J. Hänze, B. Eul, R. Savai, U. Seay, F. Grimminger, J. Lohmeyer, W. Klepetko, W. Seeger, F. Rose (2005) Hypoxia-driven proliferation of human pulmonary artery fibroblasts: cross-talk between HIF-1alpha and an autocrine angiotensin system. FASEB J. 19: 857–859.

[18] A. Sahai, C. Mei, T.A. Pattison, R.L. Tannen (1997) Chronic hypoxia induces proliferation of cultured mesangial cells: role of calcium and protein kinase C. Am. J. Physiol. 273: F954–60.

[19] S.E. Shackney, T.V. Shankey (1999) Cell cycle models for molecular biology and molecular oncology: exploring new dimensions. Cytometry. 35: 97–116.

[20] C.J. Sherr, J.M. Roberts (2004) Living with or without cyclins and cyclin-dependent kinases. Genes Dev. 18: 2699–2711.

[21] R.V. Lloyd, L.A. Erickson, L. Jin, E. Kulig, X. Qian, J.C. Cheville, B.W. Scheithauer (1999) p27kip1: a multifunctional cyclin-dependent kinase inhibitor with prognostic significance in human cancers. Am. J. Pathol. 154: 313–323.

[22] A.L. Gartel, A.L. Tyner (1998) The growth-regulatory role of p21 (WAF1/CIP1). Prog. Mol. Subcell. Biol. 20: 43–71.

[23] S. Mizuno, M. Kadowaki, Y. Demura, S. Ameshima, I. Miyamori, T. Ishizaki. (2004) p42/44 Mitogen-activated protein kinase regulated by p53 and nitric oxide in human pulmonary arterial smooth muscle cells. Am. J. Respir. Cell Mol. Biol. 31: 184–192.

[24] S. Mizuno, H.J. Bogaard, N.F. Voelkel, Y. Umeda, M. Kadowaki, S. Ameshima, I. Miyamori, T. Ishizaki (2009) Hypoxia regulates human lung fibroblast proliferation via p53-dependent and -independent pathways. Respir. Res. 10: 17.

[25] M.A. O'Reilly, R.J. Staversky, R.H. Watkins, C.K. Reed, K.L. de Mesy Jensen, J.N. Finkelstein, P.C. Keng (2001) The cyclin-dependent kinase inhibitor p21 protects the lung from oxidative stress. Am. J. Respir. Cell Mol. Biol. 24: 703–710.

[26] S. Adachi, H. Ito, M. Tamamori-Adachi, Y. Ono, T. Nozato, S. Abe, Ikeda Ma, F. Marumo, M. Hiroe (2001) Cyclin A/cdk2 activation is involved in hypoxia-induced apoptosis in cardiomyocytes. Circ. Res. 88: 408–414.

[27] S.A. McGrath-Morrow, J. Stahl (2001) Growth arrest in A549 cells during hyperoxic stress is associated with decreased cyclin B1 and increased p21(Waf1/Cip1/Sdi1) levels. Biochim. Biophys. Acta. 1538: 90–97.

[28] S. Roy, S. Khanna, A.A. Bickerstaff, S.V. Subramanian, M. Atalay, M. Bierl, S. Pendyala, D. Levy, N. Sharma, M. Venojarvi, A. Strauch, C.G. Orosz, C.K. Sen (2003) Oxygen sensing by primary cardiac fibroblasts: a key role of p21(Waf1/Cip1/Sdi1). Circ. Res. 92: 264–271.

[29] T.G. Graeber, C. Osmanian, T. Jacks, D.E. Housman, C.J. Koch, S.W. Lowe, A.J. Giaccia (1996) Hypoxia-mediated selection of cells with diminished apoptotic potential in solid tumours. Nature. 379: 88–91.

[30] J. Yu, Z. Wang, K.W. Kinzler, B. Vogelstein, L. Zhang (2003) PUMA mediates the apoptotic response to p53 in colorectal cancer cells. Proc. Natl. Acad. Sci. U.S.A. 100: 1931–1936.

[31] W.G. An, M. Kanekal, M.C. Simon, E. Maltepe, M.V. Blagosklonny, L.M. Neckers (1998) Stabilization of wild-type p53 by hypoxia-inducible factor 1alpha. Nature. 392: 405–408.

[32] M. Sano, T. Minamino, H. Toko, H. Miyauchi, M. Orimo, Y. Qin, H. Akazawa, K. Tateno, Y. Kayama, M. Harada, I. Shimizu, T. Asahara, H. Hamada, S. Tomita, J.D. Molkentin, Y. Zou, I. Komuro (2007) p53-induced inhibition of Hif-1 causes cardiac dysfunction during pressure overload. Nature. 446: 444–448.

[33] R. Brosh, R. Shalgi, A. Liran, G. Landan, K. Korotayev, G.H. Nguyen, E. Enerly, H. Johnsen, Y. Buganim, H. Solomon, I. Goldstein, S. Madar, N. Goldfinger, A.-L. Børresen-Dale, D. Ginsberg, C.C. Harris, Y. Pilpel, M. Oren, V. Rotter (2008) p53-Repressed miRNAs are involved with E2F in a feed-forward loop promoting proliferation. Mol. Syst. Biol. 4: 229.

[34] H. Hermeking (2007) p53 enters the microRNA world. Cancer Cell. 12: 414–418.

[35] W.P. Kloosterman, R.H.A. Plasterk (2006) The diverse functions of microRNAs in animal development and disease. Dev. Cell. 11: 441–450.

[36] N. Bushati, S.M. Cohen (2007) microRNA functions. Annu. Rev. Cell Dev. Biol. 23: 175–205.

[37] V. Tarasov, P. Jung, B. Verdoodt, D. Lodygin, A. Epanchintsev, A. Menssen, G. Meister, H. Hermeking (2007) Differential regulation of microRNAs by p53 revealed by massively parallel sequencing: miR-34a is a p53 target that induces apoptosis and G1-arrest. Cell Cycle. 6: 1586–1593.

[38] S. Mizuno, M. Yasuo, H.J. Bogaard, D. Kraskauskas, R. Natarajan, N.F. Voelkel (2011) Inhibition of histone deacetylase causes emphysema. Am. J. Physiol. Lung Cell Mol. Physiol. 300: L402–13.
[39] S. Mizuno, H.J. Bogaard, J. Gomez-Arroyo, A. Alhussaini, D. Kraskauskas, C.D. Cool, N.F. Voelkel (2012) MicroRNA-199a-5p is associated with hypoxia inducible factor-1α expression in the lung from COPD patients. Chest. in press.
[40] M.B. Datto, Y. Li, J.F. Panus, D.J. Howe, Y. Xiong, X.F. Wang (1995) Transforming growth factor beta induces the cyclin-dependent kinase inhibitor p21 through a p53-independent mechanism. Proc. Natl. Acad. Sci. U.S.A. 92: 5545–5549.
[41] Y. Mitani, K. Maruyama, M. Sakurai (1997) Prolonged administration of L-arginine ameliorates chronic pulmonary hypertension and pulmonary vascular remodeling in rats. Circulation. 96: 689–697.
[42] M. Ananthakrishnan, F.E. Barr, M.L. Summar, H.A. Smith, M. Kaplowitz, G. Cunningham, J. Magarik, Y. Zhang, C.D. Fike (2009) L-Citrulline ameliorates chronic hypoxia-induced pulmonary hypertension in newborn piglets. Am. J. Physiol. Lung Cell Mol. Physiol. 297: L506–11.
[43] I. Eto (2006) Nutritional and chemopreventive anti-cancer agents up-regulate expression of p27Kip1, a cyclin-dependent kinase inhibitor, in mouse JB6 epidermal and human MCF7, MDA-MB-321 and AU565 breast cancer cells. Cancer Cell Int. 6: 20.
[44] M. Loda, B. Cukor, S.W. Tam, P. Lavin, M. Fiorentino, G.F. Draetta, J.M. Jessup, M. Pagano (1997) Increased proteasome-dependent degradation of the cyclin-dependent kinase inhibitor p27 in aggressive colorectal carcinomas. Nat. Med. 3: 231–234.
[45] J. Philipp-Staheli, K.-H. Kim, D. Liggitt, K.E. Gurley, G. Longton, C.J. Kemp (2004) Distinct roles for p53, p27Kip1, and p21Cip1 during tumor development. Oncogene. 23: 905–913.
[46] K. Sakakibara, K. Kubota, B. Worku, E.J. Ryer, J.P. Miller, A. Koff, K.C. Kent, B. Liu (2005) PDGF-BB regulates p27 expression through ERK-dependent RNA turn-over in vascular smooth muscle cells. J. Biol. Chem. 280: 25470–25477.
[47] A. Vidal, S.S. Millard, J.P. Miller, A. Koff (2002) Rho activity can alter the translation of p27 mRNA and is important for RasV12-induced transformation in a manner dependent on p27 status. J. Biol. Chem. 277: 16433–16440.
[48] J.M. Dong, T. Leung, E. Manser, L. Lim (1998) cAMP-induced morphological changes are counteracted by the activated RhoA small GTPase and the Rho kinase ROKalpha. J. Biol. Chem. 273: 22554–22562.
[49] U. Laufs, D. Marra, K. Node, J.K. Liao (1999) 3-Hydroxy-3-methylglutaryl-CoA reductase inhibitors attenuate vascular smooth muscle proliferation by preventing rho GTPase-induced down-regulation of p27(Kip1). J. Biol. Chem. 274: 21926–21931.
[50] J.-M. Hyvelin, K. Howell, A. Nichol, C.M. Costello, R.J. Preston, P. McLoughlin (2005) Inhibition of Rho-kinase attenuates hypoxia-induced angiogenesis in the pulmonary circulation. Circ. Res. 97: 185–191.
[51] M.J. Connolly, P.I. Aaronson (2011) Key role of the RhoA/Rho kinase system in pulmonary hypertension. Pulmonary Pharmacology & Therapeutics. 24: 1–14.

[52] N. Sawada, H. Itoh, K. Ueyama, J. Yamashita, K. Doi, T.H. Chun, M. Inoue, K. Masatsugu, T. Saito, Y. Fukunaga, S. Sakaguchi, H. Arai, N. Ohno, M. Komeda, K. Nakao (2000) Inhibition of rho-associated kinase results in suppression of neointimal formation of balloon-injured arteries. Circulation. 101: 2030–2033.

[53] K.R. Stenmark, R.P. Mecham (1997) Cellular and molecular mechanisms of pulmonary vascular remodeling. Annu. Rev. Physiol. 59: 89–144.